TURNPIKE PROPERTIES IN THE CALCULUS OF VARIATIONS AND OPTIMAL CONTROL

Nonconvex Optimization and Its Applications

VOLUME 80

TURNPIKE PROPERTIES IN THE CALCULUS OF VARIATIONS AND OPTIMAL CONTROL

By

ALEXANDER J. ZASLAVSKI
The Technion—Israel Institute of Technology, Haifa, Israel

 Springer

Library of Congress Cataloging-in-Publication Data

Zaslavski, Alexander J.
   Turnpike properties in the calculus of variations and optimal control / by Alexander J.
      Zaslavski.
      p. cm. — (Nonconvex optimization and its applications ; v. 80)
   Includes bibliographical references and index.

   1. Calculus of variations. 2. Mathematical optimization. I. Title. II. Series

   QA316.Z37 2005
   515´.64—dc22

                                                                          2005050039

AMS Subject Classifications: 49-02

                                        e-ISBN-10: 0-387-28154-1
ISBN-13: 978-1-4419-3924-1   e-ISBN-13: 978-0-387-28154-4

© 2010 Springer Science+Business Media, Inc.

Printed in the United States of America.

springeronline.com

Contents

Preface

This monograph is devoted to recent progress in the turnpike theory. Turnpike properties are well known in mathematical economics. The term was first coined by Samuelson who showed that an efficient expanding economy would for most of the time be in the vicinity of a balanced equilibrium path (also called a von Neumann path) [78, 79]. These properties were studied by many authors for optimal trajectories of a Neumann–Gale model determined by a superlinear set-valued mapping. In the monograph we discuss a number of results concerning turnpike properties in the calculus of variations and optimal control which were obtained by the author in the last ten years. These results show that the turnpike properties are a general phenomenon which holds for various classes of variational problems and optimal control problems.

Turnpike properties are studied for optimal control problems on finite time intervals $[T_1, T_2]$ of the real line. Solutions of such problems (trajectories) always depend on the time interval $[T_1, T_2]$, an optimality criterion which is usually determined by a cost function, and on data which is some initial conditions. In the turnpike theory we are interested in the structure of solutions of optimal problems. We study the behavior of solutions when an optimality criterion is fixed while T_1, T_2 and the data vary. To have turnpike properties means, roughly speaking, that the solutions of a problem are determined mainly by the optimality criterion (a cost function), and are essentially independent of the choice of time interval and data, except in regions close to the endpoints of the time interval. If a point t does not belong to these regions, then the value of a solution at t is closed to a trajectory ("turnpike") which is defined on the infinite time interval and depends only on the optimality criterion. This phenomenon has the following interpretation. If one wishes to reach a point A from a point B by a car in an optimal way, then one should enter onto a turnpike, spend most of one's time on it and then leave the turnpike to reach the required point.

The turnpike phenomenon was discovered by Samuelson in a specific situation. In further numerous studies turnpike properties were established under strong assumptions on an optimality criterion (a cost function). The usual assumptions were that a cost function is time independent and is convex as a function of all its variables. Under these assumptions the "turnpike" is a stationary trajectory (a singleton). The simple form of the "turnpike" with a convex cost function allowed one to discover the turnpike property in this case. Since convexity plays an important role in mathematical economics, turnpike theory has many applications in this area of research. It should be mentioned that there are several interesting results concerning turnpike properties without convexity assumptions. In these results convexity was replaced by other assumptions. The verification of these assumptions was rather difficult and they hold for a narrow class of problems. Thus the turnpike phenomenon was considered by experts as an interesting property of some very particular problems arising in mathematical economics for which a "turnpike" was usually a singleton or a half-ray. This situation has changed in the last ten years. In this monograph we discuss results which were obtained during this period and allow us today to think about turnpike properties as a general phenomenon which holds for various classes of variational problems and optimal control problems. To establish these properties we do not need convexity of a cost function and its time independence.

It was my great pleasure to receive on October 2000 the following letter from Paul A. Samuelson, the discoverer of the turnpike phenomenon.

Dear Professor Zaslavski:

I note with interest your long paper "The Turnpike Property ...Functions" in *Nonlinear Analysis* 42 (2000), 1465-98.

It may be of interest to report that this property and name originated just over half a century ago when, as a Guggenheim Fellow on a 1948-49 sabbatical leave from MIT, I conjectured it in a memo written at the RAND Corporation in Santa Monica, California. In *The Collected Scientific Papers of Paul A. Samuelson*, MIT Press, 1966, 1972, 1977, 1986, it is reproduced. R. Dorfman, P.A. Samuelson, R.M. Solow, *Linear Programming and Economic Analysis*, McGraw-Hill, 1958 gives a pre-Roy Radner exposition. I believe that somewhere Lionel McKenzie has given a nice survey of the relevant mathematical-economics literature.

With admiration,
Paul A. Samuelson

Our studies are based on the following ideas. A "turnpike" is not necessarily a singleton or a half-ray. It can be an absolutely continuous time-dependent function (trajectory) or a compact subset of R^n. To establish a turnpike property we consider a space of cost functions equipped with a natural complete metric and show that a turnpike property holds for most elements of this space in the sense of Baire categories. We obtain a turnpike theorem in the following way. We consider an optimality criterion (a cost function f) and show that for a problem with this criterion there exists an optimal trajectory, say X_f, on an infinite time interval. Then we perturb our cost function by some nonnegative small perturbation which is zero only on X_f. We show that for our new cost function $\bar{f}$ the trajectory X_f is a turnpike, and that optimal solutions of the problem with a cost function g which is closed to $\bar{f}$, are also most of the time close to X_f.

ALEXANDER J. ZASLAVSKI

June 2005

Introduction

Let us consider the following problem of the calculus of variations:

$$\int_0^T f(v(t), v'(t))dt \to \min, \tag{P_0}$$

$v : [0, T] \to R^n$ is an absolutely continuous function

such that $v(0) = y, \ v(T) = z$.

Here T is a positive number, y and z are elements of the n-dimensional Euclidean space R^n and an integrand $f : R^n \times R^n \to R^1$ is a continuous function.

We are interested in the structure of solutions of the problem (P_0) when y, z and T vary and T is sufficiently large.

Assume that the function f is strictly convex and differentiable and satisfies the following growth condition:

$$f(y, z)/(|y| + |z|) \to \infty \text{ as } |y| + |z| \to \infty.$$

Here we denote by $|\cdot|$ the Euclidean norm in R^n and by $<\cdot, \cdot>$ the scalar product in R^n. In order to analyse the structure of minimizers of the problem (P_0) we consider the auxiliary minimization problem:

$$f(y, 0) \to \min, \ y \in R^n. \tag{P_1}$$

It follows from the growth condition and the strict convexity of f that the problem (P_1) has a unique solution which will be denoted by $\bar{y}$. Clearly,

$$\partial f/\partial y(\bar{y}, 0) = 0.$$

Define an integrand $L : R^n \times R^n \to R^1$ by

$$L(y, z) = f(y, z) - f(\bar{y}, 0) - <\nabla f(\bar{y}, 0), (y, z) - (\bar{y}, 0) >$$

$$= f(y, z) - f(\bar{y}, 0) - <(\partial f/\partial z)(\bar{y}, 0), z > .$$

Clearly L is also differentiable and srictly convex and satisfies the same growth condition as f:

$$L(y,z)/(|y| + |z|) \to \infty \text{ as } |y| + |z| \to \infty.$$

Since f and L are strictly convex we obtain that

$$L(y,z) \geq 0 \text{ for all } (y,z) \in R^n \times R^n$$

and

$$L(y,z) = 0 \text{ if and only if } y = \bar{y}, \ z = 0.$$

Consider the following auxiliary problem of the calculus of variations:

$$\int_0^T L(v(t), v'(t))dt \to \min, \qquad\qquad (P_2)$$

$$v : [0,T] \to R^n \text{ is an absolutely continuous function}$$

$$\text{such that } v(0) = y, \ v(T) = z,$$

where $T > 0$ and $y, z \in R^n$. It is easy to see that for any absolutely continuous function $x : [0,T] \to R^n$ with $T > 0$,

$$\int_0^T L(x(t), x'(t))dt$$

$$= \int_0^T [f(x(t), x'(t)) - f(\bar{y},0) - \, < (\partial f / \partial z)(\bar{y},0), x'(t) >] dt$$

$$= \int_0^T f(x(t), x'(t))dt + Tf(\bar{y},0) - \, < (\partial f / \partial z)(\bar{y}), x(T) - x(0) > .$$

These equations imply that the problems (P_0) and (P_2) are equivalent: a function $x : [0,T] \to R^n$ is a solution of the problem (P_0) if and only if it is a solution of the problem (P_2).

The integrand $L : R^n \times R^n \to R^1$ has the following property:

(C) If $\{(y_i, z_i)\}_{i=1}^{\infty} \subset R^n \times R^n$ satisfies $\lim_{i \to \infty} L(y_i, z_i) = 0$, then $\lim_{i \to \infty} y_i = \bar{y}$ and $\lim_{i \to \infty} z_i = 0$.

Indeed, assume that

$$\{(y_i, z_i)\}_{i=1}^{\infty} \subset R^n \times R^n \text{ and } \lim_{i \to \infty} L(y_i, z_i) = 0.$$

By the growth condition the sequence $\{(y_i, z_i)\}_{i=1}^{\infty}$ is bounded. Let (y,z) be a limit point of the sequence $\{(y_i, z_i)\}_{i=1}^{\infty}$. Then,

$$L(y,z) = \lim_{i \to \infty} L(y_i, z_i) = 0$$

$$\text{and } (y, z) = (\bar{y}, 0).$$

This implies that $(\bar{y}, 0) = \lim_{i \to \infty} (y_i, z_i)$.

Let $y, z \in R^n$, $T > 2$ and a function $\bar{x} : [0, T] \to R^n$ be an optimal solution of the problem (P_0). Then $\bar{x}$ is also an optimal solution of the problem (P_2). We will show that

$$\int_0^T L(\bar{x}(t), \bar{x}'(t)) dt \leq 2c_0(|y|, |z|)$$

where $c_0(|y|, |z|)$ is a constant which depends only on $|y|$ and $|z|$.

Define a function $x : [0, T] \to R^n$ by

$$x(t) = y + t(\bar{y} - y), \ t \in [0, 1], \ x(t) = \bar{y}, \ t \in [1, T - 1],$$

$$x(t) = \bar{y} + (t - (T - 1))(z - \bar{y}), \ t \in [T - 1, T].$$

It follows from the definition of $\bar{x}$ and x that

$$\int_0^T L(\bar{x}(t), \bar{x}'(t)) dt \leq \int_0^T L(x(t), x'(t)) dt$$

$$= \int_0^1 L(x(t), \bar{y} - y) dt + \int_1^{T-1} L(\bar{y}, 0) dt + \int_{T-1}^T L(x(t), z - \bar{y}) dt$$

$$= \int_0^1 L(x(t), \bar{y} - y) dt + \int_{T-1}^T L(x(t), z - \bar{y}) dt.$$

It is not difficult to see that the integrals

$$\int_0^1 L(x(t), \bar{y} - y) dt \text{ and } \int_{T-1}^T L(x(t), z - \bar{y}) dt$$

do not exceed a constant $c_0(|y|, |z|)$ which depends only on $|y|, |z|$. Thus

$$\int_0^T L(\bar{x}(t), \bar{x}'(t)) dt \leq 2c_0(|y|, |z|).$$

It is very important that in this inequality the constant $c_0(|y|, |z|)$ does not depend on T.

We denote by $\text{mes}(E)$ the Lebesgue measure of a Lebesgue mesurable set $E \subset R^1$.

Now let ϵ be a positive number. By the property (C) there is $\delta > 0$ such that if $(y, z) \in R^n \times R^n$ and $L(y, z) \leq \delta$, then $|y - \bar{y}| + |z| \leq \epsilon$. Then by the choice of δ and the inequality $\int_0^T L(\bar{x}(t), \bar{x}'(t)) dt \leq 2c_0(|y|, |z|)$,

$$\text{mes}\{t \in [0, T] : |(\bar{x}(t), \bar{x}'(t)) - (\bar{y}, 0)| > \epsilon\}$$

$$\leq \operatorname{mes}\{t \in [0, T] : L(\bar{x}(t), \bar{x}'(t)) > \delta\}$$

$$\leq \delta^{-1} \int_0^T L(\bar{x}(t), \bar{x}'(t)) dt \leq \delta^{-1} 2 c_0(|y|, |z|)$$

and

$$\operatorname{mes}\{t \in [0, T] : |\bar{x}(t) - \bar{y}| > \epsilon\} \leq \delta^{-1} 2 c_0(|y|, |z|).$$

Therefore the optimal solution $\bar{x}$ spends most of the time in an ϵ-neighbor- hood of the point $\bar{y}$. The Lebesgue measure of the set of all points t, for which $\bar{x}(t)$ does not belong to this ϵ-neighborhood, does not exceed the constant $2\delta^{-1} c_0(|y|, |z|)$ which depends only on $|y|, |z|$ and ϵ and does not depend on T. Following the tradition, the point $\bar{y}$ is called the turnpike. Moreover we can show that the set

$$\{t \in [0, T] : |\bar{x}(t) - \bar{y}| > \epsilon\}$$

is contained in the union of two intervals $[0, \tau_1] \cup [T - \tau_2, T]$, where $0 < \tau_1, \tau_2 \leq 2\delta^{-1} c_0(|y|, |z|)$.

Under the assumptions posed on f, the structure of optimal solutions of the problem (P_0) is rather simple and the turnpike $\bar{y}$ is calculated easily. On the other hand the proof is strongly based on the convexity of f and its time independence. The approach used in the proof cannot be employed to extend the turnpike result for essentially larger classes of variational problems. For such extensions we need other approaches and ideas. The question of what happens if the integrand f is nonconvex and nonautonomous seems very interesting. What kind of turnpike and what kind of convergence to the turnpike do we have for general nonconvex nonautonomous integrands? The following example helps to understand the problem.

Let

$$f(t, x, u) = (x - \cos(t))^2 + (u + \sin(t))^2, \quad (t, x, u) \in R^1 \times R^1 \times R^1$$

and consider the family of the variational problems

$$\int_{T_1}^{T_2} [(v(t) - \cos(t))^2 + (v'(t) + \sin(t))^2] dt \to \min, \qquad (P_3)$$

$$v : [T_1, T_2] \to R^1 \text{ is an absolutely continuous function}$$

$$\text{such that } v(T_1) = y, \ v(T_2) = z,$$

where $y, z, T_1, T_2 \in R^1$ and $T_2 > T_1$. The integrand f depends on t, for each $t \in R^1$ the function $f(t, \cdot, \cdot) : R^2 \to R^1$ is convex, and for each $x, u \in R^1 \setminus \{0\}$ the functon $f(\cdot, x, u) : R^1 \to R^1$ is nonconvex. Thus the function $f : R^1 \times R^1 \times R^1 \to R^1$ is also nonconvex and depends on t.

Assume that $y, z, T_1, T_2 \in R^1$, $T_2 > T_1 + 2$ and $\hat{v} : [T_1, T_2] \to R^1$ is an optimal solution of the problem (P_3). Note that the problem (P_3) has a solution since f is continuous and $f(t, x, \cdot) : R^1 \to R^1$ is convex and grows superlinearly at infinity for each $(t, x) \in [0, \infty) \times R^1$.

Define $v : [T_1, T_2] \to R^1$ by

$$v(t) = y + (\cos(1) - y)(t - T_1), \ t \in [T_1, T_1 + 1],$$

$$v(t) = \cos(t), \ t \in [T_1 + 1, T_2 - 1],$$

$$v(t) = \cos(T_2 - 1) + (t - T_2 + 1)(z - \cos(T_2)), \ t \in [T_2 - 1, T_2].$$

It is easy to see that

$$\int_{T_1+1}^{T_2-1} f(t, v(t), v'(t)) dt = 0$$

and

$$\int_{T_1}^{T_2} f(t, \hat{v}(t), \hat{v}'(t)) dt \le \int_{T_1}^{T_2} f(t, v(t), v'(t)) dt$$

$$= \int_{T_1}^{T_1+1} f(t, v(t), v'(t)) dt + \int_{T_2-1}^{T_2} f(t, v(t), v'(t)) dt$$

$$\le 2 \sup\{|f(t, x, u)| : \ t, x, u \in R^1, \ |x|, |u| \le |y| + |z| + 1\}.$$

Thus

$$\int_{T_1}^{T_2} f(t, \hat{v}(t), \hat{v}'(t)) dt \le c_1(|y|, |z|),$$

where

$$c_1(|y|, |z|) = 2 \sup\{|f(t, x, u)| : \ t, x, u \in R^1, \ |x|, |u| \le |y| + |z| + 1\}.$$

For any $\epsilon \in (0, 1)$ we have

$$\text{mes}\{t \in [T_1, T_2] : \ |\hat{v}(t) - \cos(t)| > \epsilon\}$$

$$\le \epsilon^{-2} \int_{T_1}^{T_2} f(t, \hat{v}(t), \hat{v}'(t)) dt \le \epsilon^{-2} c_1(|y|, |z|).$$

Since the constant $c_1(|y|, |z|)$ does not depend on T_2 and T_1 we conclude that if $T_2 - T_1$ is sufficiently large, then the optimal solution $\hat{v}(t)$ is equal to $\cos(t)$ up to ϵ for most $t \in [T_1, T_2]$. Again, as in the case of convex time independent problems we can show that

$$\{t \in [T_1, T_2] : \ |x(t) - \cos(t)| > \epsilon\} \subset [T_1, T_1 + \tau] \cup [T_2 - \tau, T_2]$$

where $\tau > 0$ is a constant which depends only on ϵ, $|y|$ and $|z|$.

This example shows that there exist nonconvex time dependent integrands which have the turnpike property with the same type of convergence as in the case of convex autonomous variational problems. The difference is that the turnpike is not a singleton but an absolutely continuous time dependent function defined on the infinite interval $[0, \infty)$. This leads us to the following definition of the turnpike property for general integrands.

Let us consider the following variational problem:

$$\int_{T_1}^{T_2} f(t, v(t), v'(t))dt \rightarrow \min, \qquad (P)$$

$v : [T_1, T_2] \rightarrow R^n$ is an absolutely continuous function

such that $v(T_1) = y$, $v(T_2) = z$.

Here $T_1 < T_2$ are real numbers, y and z are elements of the n-dimensional Euclidean space R^n and an integrand $f : [0, \infty) \times R^n \times R^n \rightarrow R^1$ is a continuous function.

We say that the integrand f has the *turnpike property* if there exists a locally absolutely continuous function $X_f : [0, \infty) \rightarrow R^n$ (called the "turnpike") which depends only on f and satisfies the following condition:

For each bounded set $K \subset R^n$ and each $\epsilon > 0$ there exists a constant $T(K, \epsilon) > 0$ such that for each $T_1 \geq 0$, each $T_2 \geq T_1 + 2T(K, \epsilon)$, each $y, z \in K$ and each optimal solution $v : [T_1, T_2] \rightarrow R^n$ of variational problem (P), the inequality $|v(t) - X_f(t)| \leq \epsilon$ holds for all $t \in [T_1 + T(K, \epsilon), T_2 - T(K, \epsilon)]$.

The turnpike property is very important for applications. Suppose that the integrand f has the turnpike property, K and ϵ are given, and we know a finite number of "approximate" solutions of the problem (P). Then we know the turnpike X_f, or at least its approximation, and the constant $T(K, \epsilon)$ which is an estimate for the time period required to reach the turnpike. This information can be useful if we need to find an "approximate" solution of the problem (P) with a new time interval $[T_1, T_2]$ and the new values $y, z \in K$ at the end points T_1 and T_2. Namely instead of solving this new problem on the "large" interval $[T_1, T_2]$ we can find an "approximate" solution of problem (P) on the "small" interval $[T_1, T_1 + T(K, \epsilon)]$ with the values $y, X_f(T_1 + T(K, \epsilon))$ at the end points and an "approximate" solution of problem (P) on the "small" interval $[T_2 - T(K, \epsilon), T_2]$ with the values $X_f(T_2 - T(K, \epsilon)), z$ at the end points. Then the concatenation of the first solution, the function $X_f : [T_1 + T(K, \epsilon), T_2 - T(K, \epsilon)]$ and the second solution is an "approximate" solution of problem (P) on the interval $[T_1, T_2]$ with the values y, z at the end points.

We begin our monograph with a discussion of the problem (P). In Chapter 1 we introduce a space $\mathcal{M}$ of continuous integrands $f : [0, \infty) \times R^n \times R^n \to R^1$. This space is equipped with a natural complete metric. We show that for any initial condition $x_0 \in R^n$ there exists a locally absolutely continuous function $x : [0, \infty) \to R^n$ with $x(0) = x_0$ such that for each $T_1 \geq 0$ and $T_2 > T_1$ the function $x : [T_1, T_2] \to R^n$ is a solution of problem (P) with $y = x(T_1)$ and $z = x(T_1)$. We also establish that for every bounded set $E \subset R^n$ the $C([T_1, T_2])$ norms of approximate solutions $x : [T_1, T_2] \to R^n$ for the problem (P) with $y, z \in E$ are bounded by some constant which does not depend on T_1 and T_2.

In Chapter 2 we establish the turnpike property stated above for a generic integrand $f \in \mathcal{M}$. We establish the existence of a set $\mathcal{F} \subset \mathcal{M}$ which is a countable intersection of open everywhere dense sets in $\mathcal{M}$ such that for each $f \in \mathcal{F}$ the turnpike property holds. Moreover we show that the turnpike property holds for approximate solutions of variational problems with a generic integrand f and that the turnpike phenomenon is stable under small pertubations of a generic integrand f.

In Chapters 3-5 we study turnpike properties for autonomous problems (P) with integrands $f : R^n \times R^n \to R^1$ which do not depend on t. Since the turnpike theorems of Chapter 2 are of generic nature and the subset of $\mathcal{M}$ which consists of all time independent integrands are nowhere dense, the results of Chapter 2 can not be applied for this subset. Moreover, we cannot expect to obtain the turnpike property stated above for the general autonomous case. Indeed, if an integrand f does not depend on t and has a turnpike, then this turnpike should also be time independent. It means that the turnpike is a stationary trajectory (a singleton). But it is not true when a time independent integrand f is not a convex function.

Consider the following example. Let

$$f(x_1, x_2, u_1, u_2) = (x_1^2 + x_2^2 - 1)^2 + (u_1 + x_2)^2 + (u_2 - x_1)^2,$$

$$(x_1, x_2, u_1, u_2) \in R^2 \times R^2$$

and consider the family of the variational problems

$$\int_0^T f(v_1(t), v_2(t), v_1'(t), v_2'(t))dt \to \min, \qquad (P_4)$$

$(v_1, v_2) : [0, T] \to R^2$ is an absolutely continuous function

such that $(v_1, v_2)(0) = y$, $(v_1, v_2)(T) = z$,

where $y = (y_1, y_2)$, $z = (z_1, z_2) \in R^2$ and $T > 0$. The integrand f does not depend on t. Since f is continuous and for each $x = (x_1, x_2) \in R^2$ the function $f(x, \cdot) : R^2 \to R^1$ is convex and grows superlinearly at infinity, the problem (P_4) has a solution for each $T > 0$ and each $y, z \in R^2$. Clearly, if $T > 0$, $y = (\cos(0), \sin(0))$ and $z = (\cos(T), \sin(T))$, then the function

$$\widehat{x}_1(t) = \cos(t), \ \widehat{x}_2(t) = \sin(t), \ t \in [0, T]$$

is a solution of the problem (P_4). Thus, if the integrand f has a turnpike property, then the turnpike is not a singleton.

Let $T > 2$, $y, z \in R^2$ and let $\bar{v} = (\bar{v}_1, \bar{v}_2) : [0, T] \to R^2$ be a solution of the problem (P_4). Define a function $v = (v_1, v_2) : [0, T] \to R^n$ by

$$v(t) = y + t((\cos(1), \sin(1)) - y), \ t \in [0, 1],$$

$$v(t) = (\cos(t), \sin(t)), \ t \in [1, T - 1],$$

$$v(t) = (\cos(T - 1), \sin(T - 1)) + (t - T + 1)(z - (\cos(T - 1), \sin(T - 1))),$$

$$t \in [T - 1, T].$$

Then

$$\int_1^{T-1} f(v(t), v'(t)) dt = 0$$

and

$$\int_0^T (\bar{v}_1(t)^2 + \bar{v}_2(t)^2 - 1)^2 dt \leq \int_0^T f(\bar{v}(t), \bar{v}'(t)) dt$$

$$\leq \int_0^T f(v(t), v'(t)) dt$$

$$= \int_0^1 f(v(t), v'(t)) dt + \int_{T-1}^T f(v(t), v'(t)) dt$$

$$\leq \sup\{f(x_1, x_2, u_1, u_2) : \ x_1, x_2, u_1, u_2 \in R^1$$

$$\text{and } |x_i|, |u_i| \leq 2|y| + 2|z| + 2, \ i = 1, 2\}.$$

Thus

$$\int_0^T (\bar{v}_1(t)^2 + \bar{v}_2(t)^2 - 1)^2 dt \leq c_2(|y|, |z|)$$

with

$$c_2(|y|, |z|) = \sup\{f(x_1, x_2, u_1, u_2) : \ x_1, x_2, u_1, u_2 \in R^1$$

$$\text{and } |x_i|, |u_i| \leq 2|y| + 2|z| + 2\}.$$

Here $c_2(|y|, |z|)$ depends only on $|y|, |z|$ and does not depend on T. For any $\epsilon \in (0, 1)$ we have

$$\text{mes}\{t \in [0, T] : ||(\bar{v}_1(t), \bar{v}_2(t))| - 1| > \epsilon\}$$

$$\leq \mathrm{mes}\{t \in [0,T] : |\bar{v}_1(t)^2 + \bar{v}_2(t)^2 - 1| > \epsilon^2\}$$

$$\leq \epsilon^{-4} \int_0^T (\bar{v}_1(t)^2 + \bar{v}_2^2 - 1)^2 dt$$

$$\leq \epsilon^{-4} c_2(|y|, |z|).$$

It means that for most $t \in [0,T]$, $\bar{v}(t)$ belongs to the ϵ-neighborhood of the set $\{x \in R^2 : |x| = 1\}$. Thus we can say that the integrand f has a weakened version of the turnpike property and the set $\{|x| = 1\}$ can be considered as the turnpike for f.

For a general autonomous nonconvex problem (P) we also have a version of the turnpike property in which a turnpike is a compact subset of R^n. This subset depends only on the integrand f.

Consider the following autonomous variational problem:

$$\int_0^T f(z(t), z'(t))dt \rightarrow \min, \; z(0 = x, \; z(T) = y, \qquad (P_a)$$

$z : [0,T] \rightarrow R^n$ is an absolutely continuous function

where $T > 0$, $x, y \in R^n$ and $f : R^{2n} \rightarrow R^1$ is an integrand.

We say that a time independent integrand $f = f(x, u) \in C(R^{2n})$ has the *turnpike property* if there exists a compact set $H(f) \subset R^n$ such that for each bounded set $K \subset R^n$ and each $\epsilon > 0$ there exist numbers $L_1 > L_2 > 0$ such that for each $T \geq 2L_1$, each $x, y \in K$ and an optimal solution $v : [0,T] \rightarrow R^n$ for the variational problem (P_a), the relation

$$\mathrm{dist}(H(f), \{v(t) : \; t \in [\tau, \tau + L_2]\}) \leq \epsilon$$

holds for each $\tau \in [L_1, T - L_1]$. (Here $\mathrm{dist}(\cdot, \cdot)$ is the Hausdorff metric).

We also consider a weak version of this turnpike property for a time independent integrand $f(x, u)$. In this weak version, for an optimal solution of the problem (P_a) with $x, y \in R^n$ and large enough T, the relation

$$\mathrm{dist}(H(f), \{v(t) : \; t \in [\tau, \tau + L_2]\}) \leq \epsilon$$

with L_2, which depends on ϵ and $|x|, |y|$ and a compact set $H(f) \subset R^n$ depending only on the integrand f, holds for each $\tau \in [0,T] \setminus E$ where $E \subset [0,T]$ is a measurable subset such that the Lebesgue measure of E does not exceed a constant which depends on ϵ and on $|x|, |y|$.

These two turnpike properties for autonomous problems (P_a) are considered in Chapters 3-5.

In Chapter 3 we consider the space $\mathcal{A}$ of all time independent integrands $f \in \mathcal{M}$. We establish the existence of a set $\mathcal{F} \subset \mathcal{A}$ which is a countable intersection of open everywhere dense sets in $\mathcal{A}$ such that for each $f \in \mathcal{F}$ the weakened version of the turnpike property holds.

The turnpike property for time independent integrands is established in Chapter 5 for a generic element of a subset $\mathcal{N}$ of the space $\mathcal{A}$. The space $\mathcal{N}$ is a subset of all integrands $f \in \mathcal{A}$ which satisfy some differentiability assumptions.

In the other chapters of the monograph we establish a number of turnpike results (generic and individual) for various classes of optimal control problems. We study optimal control of linear periodic systems with convex integrands (Chapter 6) and optimal solutions of linear systems with convex nonperiodic integrands (Chapter 7). In Chapter 8 we establish turnpike theorems for discrete-time control systems in Banach spaces and in complete metric spaces. Infinite-dimensional continuous-time optimal control problems in a Hilbert space are studied in Chapter 9. A turnpike theorem for a class of differential inclusions arising in economic dynamics is proved in Chapter 10 and structure of optimal trajectories of convex processes is studied in Chapter 11. In Chapter 12 we establish a turnpike property for a dynamic discrete-time zero-sum game.

Chapter 1

INFINITE HORIZON
VARIATIONAL PROBLEMS

In this chapter we study existence and uniform boundedness of extremals of variational problems with integrands which belong to a complete metric space of functions. We establish that for every bounded set $E \subset R^n$ the $C([0,T])$ norms of approximate solutions $x : [0,T] \to R^n$ for the minimization problem on an interval $[0,T]$ with $x(0), x(T) \in E$ are bounded by some constant which does not depend on T. Given an $x_0 \in R^n$ we study the infinite horizon problem of minimizing the expression $\int_0^T f(t, x(t), x'(t))dt$ as T grows to infinity, where $x : [0, \infty) \to R^n$ satisfies the initial condition $x(0) = x_0$. We analyse the existence and the properties of approximate solutions for every prescribed initial value x_0.

1.1. Preliminaries

Variational and optimal control problems defined on infinite intervals are of interest in many areas of mathematics and its applications [10, 11, 16, 32, 62, 63, 88, 89, 95]. These problems arise in engineering [1, 3], in models of economic growth [14, 26, 27, 28, 29, 45, 46, 49-52, 60, 61, 67, 68, 72, 74, 80, 86, 94], in dynamic games theory [15, 17], in infinite discrete models of solid-state physics related to dislocations in one-dimensional crystals [6, 85] and in the theory of thermodynamical equilibrium of materials [20, 44, 53-55, 90-92, 95].

We consider the infinite horizon problem of minimizing the expression

$$\int_0^T f(t, x(t), x'(t))dt$$

as T grows to infinity where a function $x : [0, \infty) \to K$ is locally absolutely continuous (a.c.) and satisfies the initial condition $x(0) = x_0$, $K \subset R^n$ is a closed convex set and f belongs to a complete metric space of functions to be described below.

We say that an a.c. function $x : [0, \infty) \to K$ is (f)-overtaking optimal if

$$\limsup_{T \to \infty} \int_0^T [f(t, x(t), x'(t)) - f(t, y(t), y'(t))]dt \leq 0$$

for any a.c. function $y : [0, \infty) \to K$ satisfying $y(0) = x(0)$.

This notion, known as the overtaking optimality criterion, was introduced in the economics literature by Atsumi [4], Gale [33] and von Weizsacker [81] and has been used in control theory [3, 13, 14, 16, 39, 40]. In general, overtaking optimal solutions may fail to exist. Most studies that are concerned with their existence assume convex integrands [13, 40, 72].

Another type of optimality criterion for infinite horizon problems was introduced by Aubry and Le Daeron [6] in their study of the discrete Frenkel–Kontorova model related to dislocations in one-dimensional crystals. More recently this optimality criterion was used in [44, 65, 66, 85]. A similar notion was introduced in Halkin [34] for his proof of the maximum principle.

Let I be either $[0, \infty)$ or $(-\infty, \infty)$. We say that an a.c. function $x : I \to K$ is an (f)-minimal solution if

$$\int_{T_1}^{T_2} f(t, x(t), x'(t))dt \leq \int_{T_1}^{T_2} f(t, y(t), y'(t))dt \leq 0$$

for each $T_1 \in I$, $T_2 > T_1$ and each a.c. function $y : [T_1, T_2] \to K$ which satisfies $y(T_i) = x(T_i)$, $i = 1, 2$.

It is easy to see that every (f)-overtaking optimal function is an (f)-minimal solution.

In this chapter we consider a functional space of integrands $\mathcal{M}$ described in Section 1.1. We show that for each $f \in \mathcal{M}$ and each $z \in R^n$ there exists a bounded (f)-minimal solution $Z : [0, \infty) \to R^n$ satisfying $Z(0) = z$ such that any other a.c. function $Y : [0, \infty) \to R^n$ is not "better" than Z. We also establish that given $f \in \mathcal{M}$ and a bounded set $E \subset R^n$ the $C([0, T])$ norms of approximate solutions $x : [0, T] \to R^n$ for the minimization problem on an interval $[0, T]$ with $x(0), x(T) \in E$ are bounded by some constant which depends only on f and E.

1.2. Main results

Let $a > 0$ be a constant and $\psi : [0, \infty) \to [0, \infty)$ be an increasing function such that $\psi(t) \to \infty$ as $t \to \infty$.

Let $K \subset R^n$ be a closed convex set. Denote by $|\cdot|$ the Euclidean norm in R^n and denote by $\mathcal{M}$ the set of continuous functions $f : [0, \infty) \times K \times R^n \to R^1$ which satisfy the following assumptions:

A(i) for each $(t, x) \in [0, \infty) \times K$ the function $f(t, x, \cdot) : R^n \to R^1$ is convex;

A(ii) the function f is bounded on $[0, \infty) \times E$ for any bounded set $E \subset K \times R^n$;

A(iii) for each $(t, x, u) \in [0, \infty) \times K \times R^n$,

$$f(t, x, u) \geq \max\{\psi(|x|), \psi(|u|)|u|\} - a;$$

A(iv) for each $M, \epsilon > 0$ there exist $\Gamma, \delta > 0$ such that

$$|f(t, x_1, u_1) - f(t, x_2, u_2)| \leq \epsilon \max\{f(t, x_1, u_1), f(t, x_2, u_2)\}$$

for each $t \in [0, \infty)$, each $u_1, u_2 \in R^n$ and each $x_1, x_2 \in K$ which satisfy

$$|x_i| \leq M, \ |u_i| \geq \Gamma, \ i = 1, 2, \quad \max\{|x_1 - x_2|, |u_1 - u_2|\} \leq \delta;$$

A(v) for each $M, \epsilon > 0$ there exist $\delta > 0$ such that

$$|f(t, x_1, u_1) - f(t, x_2, u_2)| \leq \epsilon$$

for each $t \in [0, \infty)$, each $u_1, u_2 \in R^n$ and each $x_1, x_2 \in K$ which satisfy

$$|x_i|, |u_i| \leq M, \ i = 1, 2, \quad \max\{|x_1 - x_2|, |u_1 - u_2|\} \leq \delta.$$

When $K = R^n$ it is an elementary exercise to show that an integrand $f = f(t, x, u) \in C^1([0, \infty) \times R^n \times R^n)$ belongs to $\mathcal{M}$ if f satisfies Assumptions A(i), A(iii),

$$\sup\{|f(t, 0, 0)| : t \in [0, \infty)\} < \infty$$

and there exists an increasing function $\psi_0 : [0, \infty) \to [0, \infty)$ such that

$$\sup\{|\partial f/\partial x(t, x, u)|, \ |\partial f/\partial u(t, x, u)|\} \leq \psi_0(|x|)(1 + \psi(|u|)|u|)$$

for each $t \in [0, \infty)$, $x, u \in R^n$.

Therefore the space $\mathcal{M}$ contains many functions.

Example 1. It is not difficult to see that if $\psi(t) = t$ for all $t \geq 0$, $n = 1$, $K = R^1$, if functions $h_1, h_2, h_3 \in C^1(R^1)$ satisfy

$$h_1(t) \geq 0, \ t \in [0, \infty), \ \sup\{h_1(t) : t \in [0, \infty)\} < \infty,$$

$$h_2(x) \geq |x| + 1, \; x \in R^1$$

and if the function $h_3 : R^1 \to R^1$ is convex and

$$u^2 + 1 \leq h_3(u) \leq c_0(u^2 + 1), \; |h_3'(u)| \leq c_0(u^2 + 1)$$

for all $u \in R^1$, where c_0 is a positive constant, then the function

$$f(t, x, u) = h_1(t) + h_2(x)h_3(u), \; (t, x, u) \in [0, \infty) \times R^1 \times R^1$$

belongs to $\mathcal{M}$.

In Chapters 1-5 we consider variational problems with integrands belonging to the space $\mathcal{M}$ or to its subspaces. The Assumption A(i) and the inequality $f(t, x, u) \geq \psi(|u|)|u| - a$ in the Assumption A(iii) guarantee the existence of minimizers of the variational problems. These assumptions are common in the literature. We need the inequality $f(t, x, u) \geq \psi(|x|) - a$ in A(iii) in order to show that for every bounded set $E \subset R^n$ the $C([0, T])$ norms of approximate solutions $x : [0, T] \to R^n$ for the variational problems on intervals $[0, T]$ with $x(0), x(T) \in E$ are bounded by some constant which does not depend on T. We need the Assumptions A(ii) and A(v) in order to obtain certain properties of approximate solutions for variational problems on intervals $[T_1, T_2]$ which depend on $T_2 - T_1$ and do not depend of T_1 and T_2. Note that if a function f is Frechet differentiable, then the Assumption A(v) means that the growth of the partial derivatives of f does not exceed the growth of f. We use it in order to establish the continuity of the function U^f which is defined below.

We equip the set $\mathcal{M}$ with the uniformity which is determined by the following base:

$$E(N, \epsilon, \lambda) = \{(f, g) \in \mathcal{M} \times \mathcal{M} : |f(t, x, u) - g(t, x, u)| \leq \epsilon \qquad (2.1)$$

for each $t \in [0, \infty)$, each $u \in R^n$ each $x \in K$ satisfying $|x|, |u| \leq N\}$

$$\cap \{(f, g) \in \mathcal{M} \times \mathcal{M} : (|f(t, x, u)| + 1)(|g(t, x, u)| + 1)^{-1} \in [\lambda^{-1}, \lambda]$$

for each $t \in [0, \infty)$, each $u \in R^n$ and each $x \in K$ satisfying $|x| \leq N\}$

where $N > 0$, $\epsilon > 0$, $\lambda > 1$ [37].

Clearly, the uniform space $\mathcal{M}$ is Hausdorff and has a countable base. Therefore $\mathcal{M}$ is metrizable. We will prove in Secton 1.3 that the uniform space $\mathcal{M}$ is complete.

Put

$$I^f(T_1, T_2, x) = \int_{T_1}^{T_2} f(t, x(t), x'(t)) dt \qquad (2.2)$$

where $f \in \mathcal{M}$, $0 \le T_1 < T_2 < \infty$ and $x : [T_1, T_2] \to K$ is an a.c. function.

For $f \in \mathcal{M}$, $a, b \in K$ and numbers T_1, T_2 satisfying $0 \le T_1 < T_2$, put

$$U^f(T_1, T_2, a, b) = \inf\{I^f(T_1, T_2, x) : \; x : [T_1, T_2] \to K \quad (2.3)$$

is an a.c. function satisfying $x(T_1) = a$, $x(T_2) = b\}$,

$$\sigma^f(T_1, T_2, a) = \inf\{U^f(T_1, T_2, a, b) : \; b \in K\}. \quad (2.4)$$

It is easy to see that $-\infty < U^f(T_1, T_2, a, b) < \infty$ for each $f \in \mathcal{M}$, each $a, b \in K$ and each pair of numbers T_1, T_2 satisfying $0 \le T_1 < T_2$.

Let $f \in \mathcal{M}$. We say that an a.c. function $x : [0, \infty) \to K$ is an (f)-good function if for any a.c. function $y : [0, \infty) \to K$,

$$\inf\{I^f(0, T, y) - I^f(0, T, x) : \; T \in (0, \infty)\} > -\infty. \quad (2.5)$$

In this chapter we study the set of (f)-good functions and prove the following results.

THEOREM 1.2.1 *For each $h \in \mathcal{M}$ and each $z \in K$ there exists an (h)-good function $Z^h : [0, \infty) \to K$ satisfying $Z^h(0) = z$ such that:*

1. For each $f \in \mathcal{M}$, each $z \in K$ and each a.c. function $y : [0, \infty) \to K$ one of the following properties holds:

(i) $I^f(0, T, y) - I^f(0, T, Z^f) \to \infty$ as $T \to \infty$;

(ii) $\sup\{|I^f(0, T, y) - I^f(0, T, Z^f)| : \; T \in (0, \infty)\} < \infty$,

$$\sup\{|y(t)| : \; t \in [0, \infty)\} < \infty.$$

2. For each $f \in \mathcal{M}$ and each number $M > \inf\{|u| : \; u \in K\}$ there exist a neighborhood U of f in $\mathcal{M}$ and a number $Q > 0$ such that

$$\sup\{|Z^g(t)| : \; t \in [0, \infty)\} \le Q$$

for each $g \in U$ and each $z \in K$ satisfying $|z| \le M$.

3. For each $f \in \mathcal{M}$ and each number $M > \inf\{|u| : \; u \in K\}$ there exist a neighborhood U of f in $\mathcal{M}$ and a number $Q > 0$ such that for each $g \in U$, each $z \in K$ satisfying $|z| \le M$, each $T_1 \ge 0$, $T_2 > T_1$ and each a.c. function $y : [T_1, T_2] \to K$ satisfying $|y(T_1)| \le M$ the following relation holds:

$$I^g(T_1, T_2, Z^g) \le I^g(T_1, T_2, y) + Q.$$

4. If $K = R^n$, then for each $f \in \mathcal{M}$ and each $z \in R^n$ the function $Z^f : [0, \infty) \to R^n$ is an (f)-minimal solution.

COROLLARY 1.2.1 *Let $f \in \mathcal{M}$, $z \in K$ and let $y : [0, \infty) \to K$ be an a.c. function. Then y is an (f)-good function if and only if condition (ii) of Assertion 1 of Theorem 1.2.1 holds.*

THEOREM 1.2.2 *For each $f \in \mathcal{M}$ there exist a neighborhood U of f in $\mathcal{M}$ and a number $M > 0$ such that for each $g \in U$ and each (g)-good function $x : [0, \infty) \to K$,*

$$\limsup_{t \to \infty} |x(t)| < M.$$

Our next result shows that for every bounded set $E \subset K$ the $C([0, T])$ norms of approximate solutions $x : [0, T] \to K$ for the minimization problem on an interval $[0, T]$ with $x(0), x(T) \in E$ are bounded by some constant which does not depend on T.

THEOREM 1.2.3 *Let $f \in \mathcal{M}$ and M_1, M_2, c be positive numbers. Then there exist a neighborhood U of f in $\mathcal{M}$ and a number $S > 0$ such that for each $g \in U$, each $T_1 \in [0, \infty)$ and each $T_2 \in [T_1 + c, \infty)$ the following properties hold:*

(i) if $x, y \in K$ satisfy $|x|, |y| \leq M_1$ and if an a.c. function $v : [T_1, T_2] \to K$ satisfies

$$v(T_1) = x, \ v(T_2) = y, \ I^g(T_1, T_2, v) \leq U^g(T_1, T_2, x, y) + M_2,$$

then

$$|v(t)| \leq S, \ t \in [T_1, T_2]; \tag{2.6}$$

(ii) if $x \in K$ satisfies $|x| \leq M_1$ and if an a.c. function $v : [T_1, T_2] \to K$ satisfies

$$v(T_1) = x, \ I^g(T_1, T_2, v) \leq \sigma^g(T_1, T_2, x) + M_2,$$

then the inequality (2.6) is valid.

Theorems 1.2.1-1.2.3 have been proved in [98].
In the sequel we use the following notation:

$$B(x, r) = \{ y \in R^n : |y - x| \leq r \}, \ x \in R^n, \ r > 0, \tag{2.7}$$

$$B(r) = B(0, r), \ r > 0.$$

Chapter 1 is organized as follows. In Section 1.3 we study the space $\mathcal{M}$ and the dependence of the functionals U^f and I^f of f. In Section 1.4 we associate with any $f \in \mathcal{M}$ a related discrete-time control system and study its approximate solutions. Theorems 1.2.1-1.2.3 are proved in Section 1.5.

1.3. Auxiliary results

In this section we study the space $\mathcal{M}$ and continuity properties of the functionals I^f and U^f. The next proposition follows from Assumption A(iv).

PROPOSITION 1.3.1 *Let $f \in \mathcal{M}$. Then for each pair of positive numbers M and ϵ there exist $\Gamma, \delta > 0$ such that the following property holds:*
If $t \in [0, \infty)$ and if $u_1, u_2 \in R^n$ and $x_1, x_2 \in K$ satisfy

$$|x_i| \leq M, \ |u_i| \geq \Gamma, \ i = 1, 2, \ |u_1 - u_2|, \ |x_1 - x_2| \leq \delta, \qquad (3.1)$$

then

$$|f(t, x_1, u_1) - f(t, x_2, u_2)| \leq \epsilon \min\{f(t, x_1, u_1), f(t, x_2, u_2)\}.$$

Proof. Let $M, \epsilon > 0$. Choose

$$\epsilon_0 \in (0, 8^{-1} \inf\{1, \epsilon\}). \qquad (3.2)$$

It follows from Assumption A(iv) that there exist $\Gamma, \delta > 0$ such that the following property holds:
If $t \in [0, \infty)$ and if $u_1, u_2 \in R^n$ and $x_1, x_2 \in K$ satisfy (3.1), then

$$|f(t, x_1, u_1) - f(t, x_2, u_2)| \leq \epsilon_0 \sup\{f(t, x_1, u_1), f(t, x_2, u_2)\}. \qquad (3.3)$$

Assume that $t \in [0, \infty)$, $u_1, u_2 \in R^n$ and $x_1, x_2 \in K$ satisfy (3.1). By the definition of Γ, δ, (3.2) and (3.3),

$$\min\{f(t, x_1, u_1), \ f(t, x_2, u_2)\} \geq (1 - \epsilon_0) \max\{f(t, x_1, u_1), f(t, x_2, u_2)\}$$

$$\geq (1 - \epsilon_0)\epsilon_0^{-1}|f(t, x_1, u_1) - f(t, x_2, u_2)| \geq \epsilon^{-1}|f(t, x_1, u_1) - f(t, x_2, u_2)|.$$

Proposition 1.3.1 is proved.

PROPOSITION 1.3.2 *The uniform space $\mathcal{M}$ is complete.*

Proof. Assume that $\{f_i\}_{i=1}^{\infty} \subset \mathcal{M}$ is a Cauchy sequence. Clearly, for each $(t, x, u) \in [0, \infty) \times K \times R^n$ the sequence $\{f_i(t, x, u)\}_{i=1}^{\infty}$ is a Cauchy sequence. Then there exists a function $f : [0, \infty) \times K \times R^n \to R^1$ such that

$$f(t, x, u) = \lim_{i \to \infty} f_i(t, x, u) \qquad (3.4)$$

for each $(t, x, u) \in [0, \infty) \times K \times R^n$.

In order to prove the proposition it is sufficient to show that f satisfies Assumption A(iv).

Let M, ϵ be positive numbers. Choose a number $\lambda > 1$ for which

$$\lambda^2 - 1 < 8^{-1}\epsilon. \tag{3.5}$$

Since $\{f_i\}_{i=1}^\infty$ is a Cauchy sequence there exists an integer $j \geq 1$ such that

$$(f_i, f_j) \in E(M, \epsilon, \lambda) \text{ for any integer } i \geq j. \tag{3.6}$$

By (3.5) and the properties of ψ there exists a number Γ_0 such that

$$\Gamma_0 > 1, \ \psi(\Gamma_0) \geq 2a, \ \lambda^2(1 + 2\psi(\Gamma_0)^{-1})^2 - 1 < 8^{-1}\epsilon. \tag{3.7}$$

Choose $\epsilon_1 > 0$ such that

$$8\epsilon_1[\lambda(1 + 2\psi(\Gamma_0)^{-1})]^2 < \epsilon. \tag{3.8}$$

By Proposition 1.3.1 there exist numbers $\Gamma, \delta > 0$ such that

$$\Gamma > \Gamma_0$$

and that for each $t \in [0, \infty)$, each $u_1, u_2 \in R^n$ and each $x_1, x_2 \in K$ which satisfy (3.1) the inequality

$$|f_j(t, x_1, u_1) - f_j(t, x_2, u_2)| \leq \epsilon_1 \min\{f_j(t, x_1, u_1), f_j(t, x_2, u_2)\} \tag{3.9}$$

is true.

Assume that $t \in [0, \infty)$, $u_1, u_2 \in R^n$, $x_1, x_2 \in K$ satisfy (3.1). Then the inequality (3.9) follows from the definition of Γ, δ. (2.1), (3.4), (3.6) and (3.1) imply that

$$(|f(t, x_i, u_i)| + 1)(|f_j(t, x_i, u_i)| + 1)^{-1} \in [\lambda^{-1}, \lambda], \ i = 1, 2. \tag{3.10}$$

It follows from Assumption A(iii), (3.1), (3.7) and (3.9) that

$$\min\{f(t, x_i, u_i), f_j(t, x_i, u_i)\} \geq 2^{-1}\psi(\Gamma_0), \ i = 1, 2. \tag{3.11}$$

By (3.11) and (3.10),

$$f(t, x_i, u_i)f_j(t, x_i, u_i)^{-1}$$

$$\in [(\lambda(1 + 2\psi(\Gamma_0)^{-1}))^{-1}, \lambda(1 + 2\psi(\Gamma_0)^{-1})], \ i = 1, 2. \tag{3.12}$$

We may assume without loss of generality that

$$f(t, x_1, u_1) \geq f(t, x_2, u_2). \tag{3.13}$$

It follows from (3.12), (3.9), (3.8) and (3.7) that

$$f(t, x_1, u_1) - f(t, x_2, u_2) \leq \lambda(1 + 2\psi(\Gamma_0)^{-1})f_j(t, x_1, u_1)$$

$$-(\lambda(1 + 2\psi(\Gamma_0)^{-1}))^{-1}f_j(t, x_2, u_2)$$
$$= \lambda(1 + 2\psi(\Gamma_0)^{-1})[f_j(t, x_1, u_1) - f_j(t, x_2, u_2)]$$
$$+f_j(t, x_2, u_2)[\lambda(1 + 2\psi(\Gamma_0)^{-1}) - (\lambda(1 + 2\psi(\Gamma_0)^{-1}))^{-1}]$$
$$\leq \lambda(1 + 2\psi(\Gamma_0)^{-1})\epsilon_1 f_j(t, x_2, u_2) + f_j(t, x_2, u_2)[\lambda(1 + 2\psi(\Gamma_0)^{-1})$$
$$-(\lambda(1 + 2\psi(\Gamma_0)^{-1}))^{-1}] \leq \epsilon_1[\lambda(1 + 2\psi(\Gamma_0)^{-1})]^2 f(t, x_2, u_2)$$
$$+f(t, x_2, u_2)[\lambda^2(1 + 2\psi(\Gamma_0)^{-1})^2 - 1] \leq \epsilon f(t, x_2, u_2).$$

Therefore the function f satisfies Assumption A(iv). This completes the proof of the proposition.

The next auxiliary result will be used in order to establish the continuous dependence of the functional $U^f(T_1, T_2, y, z)$ of T_1, T_2, y, z and the continuous dependence of the functional $I^f(T_1, T_2, x)$ of f.

PROPOSITION 1.3.3 *Let $M_1 > 0$ and let $0 < \tau_0 < \tau_1$. Then there exists a number $M_2 > 0$ such that the following property holds:*
If $f \in \mathcal{M}$, numbers T_1, T_2 satisfy

$$0 \leq T_1, \; T_2 \in [T_1 + \tau_0, T_1 + \tau_1] \tag{3.14}$$

and if an a.c. function $x : [T_1, T_2] \to K$ satisfies

$$I^f(T_1, T_2, x) \leq M_1, \tag{3.15}$$

then

$$|x(t)| \leq M_2, \; t \in [T_1, T_2]. \tag{3.16}$$

Proof. By Assumption A(iii) and the properties of the function ψ there exists a number $c_0 > 0$ such that

$$f(t, x, u) \geq |u| \tag{3.17}$$

for each $f \in \mathcal{M}$ and each $(t, x, u) \in [0, \infty) \times K \times R^n$ satisfying $|u| \geq c_0$, and

$$f(t, x, u) \geq 2M_1(\min\{1, \tau_0\})^{-1} \tag{3.18}$$

for each $f \in \mathcal{M}$ and each $(t, x, u) \in [0, \infty) \times K \times R^n$ satisfying $|x| \geq c_0$. Fix a number

$$M_2 > 1 + M_1 + a\tau_1 + c_0(1 + \tau_1) \tag{3.19}$$

(recall a in Assumption A(iii)).

Let $f \in \mathcal{M}$, T_1, T_2 be numbers satisfying (3.14) and let $x : [T_1, T_2] \to K$ be an a.c. function satisfying (3.15). We will show that (3.16) holds.

Assume the contrary. Then there exists $t_0 \in [T_1, T_2]$ such that

$$|x(t_0)| > M_2. \tag{3.20}$$

By the definition of c_0, (3.18), (3.14) and (3.15) there exists $t_1 \in [T_1, T_2]$ satisfying

$$|x(t_1)| \leq c_0. \tag{3.21}$$

Set

$$E = [\inf\{t_0, t_1\}, \sup\{t_0, t_1\}], \ E_1 = \{t \in E : |x'(t)| \geq c_0\}, \ E_2 = E \setminus E_1. \tag{3.22}$$

By the definition of c_0, Assumption A(iii), (3.15), (3.22), (3.14) and (3.17),

$$|x(t_1) - x(t_0)| \leq \int_{E_1} |x'(t)| dt + \int_{E_2} |x'(t)| dt$$

$$\leq \tau_1 c_0 + \int_{E_1} |x'(t)| dt \leq \tau_1 c_0 + \int_{E_1} f(t, x(t), x'(t)) dt$$

$$\leq \tau_1 c_0 + I^f(T_1, T_2, x) + a\tau_1 \leq \tau_1(c_0 + a) + M_1.$$

It follows from this inequality, (3.20) and (3.21) that

$$M_2 - c_0 \leq \tau_1(c_0 + a) + M_1.$$

This is contradictory to (3.19). The obtained contradiction proves the proposition.

The following propositon establishes an important property which will be used in Chapter 2.

PROPOSITION 1.3.4 *Let $M_1, \epsilon > 0$ and let $0 < \tau_0 < \tau_1$. Then there exists a positive number δ such that for each $f \in \mathcal{M}$ and each pair of numbers T_1, T_2 satisfying (3.14) the following property holds:*

If an a.c. function $x : [T_1, T_2] \to K$ satisfies (3.15) and if $t_1, t_2 \in [T_1, T_2]$ satisfies $|t_1 - t_2| \leq \delta$, then $|x(t_1) - x(t_2)| \leq \epsilon$.

Proof. By Assumption A(iii) and the properties of the function ψ there exists a number $c_0 > 0$ such that for each $f \in \mathcal{M}$ and each $(t, x, u) \in [0, \infty) \times K \times R^n$ satisfying $|u| \geq c_0$ the inequality

$$f(t, x, u) \geq 4\epsilon^{-1}(M_1 + 2 + a\tau_1)|u| \tag{3.23}$$

is true. Choose

$$\delta \in (0, 8^{-1}(c_0 + 1)^{-1}\epsilon). \tag{3.24}$$

Assume that $f \in \mathcal{M}$, numbers T_1, T_2 satisfy (3.14), an a. c. function $x : [T_1, T_2] \to K$ satisfies (3.15) and

$$t_1, t_2 \in [T_1, T_2], \ 0 < |t_1 - t_2| \le \delta. \tag{3.25}$$

Set

$$E = [\min\{t_1, t_2\}, \ \max\{t_1, t_2\}],$$

$$E_1 = \{t \in E : |x'(t)| \ge c_0\}, \ E_2 = E \setminus E_1.$$

By Assumption A(iii), the choice of c_0, (3.14), (3.25) and (3.23),

$$|x(t_2) - x(t_1)| \le \int_{E_1} |x'(t)| dt + \int_{E_2} |x'(t)| dt \le \delta c_0 + \int_{E_1} |x'(t)| dt$$

$$\le \delta c_0 + [4(M_1 + 2 + a\tau_1)]^{-1} \epsilon \int_{E_1} f(t, x(t), x'(t)) dt$$

$$\le \delta c_0 + [4(M_1 + 2 + a\tau_1)]^{-1} \epsilon (I^f(T_1, T_2, x) + a\tau_1).$$

Combined with (3.15), (3.14) and (3.24) this inequality implies that

$$|x(t_2) - x(t_1)| \le \delta c_0 + 4^{-1}\epsilon \le \epsilon.$$

This completes the proof of the proposition.

We have the following result (see [9]).

PROPOSITION 1.3.5 *Assume that $f \in \mathcal{M}$, $M_1 > 0$, $0 \le T_1 < T_2$, $x_i : [T_1, T_2] \to K$, $i = 1, 2, \ldots$ is a sequence of a.c. functions such that*

$$I^f(T_1, T_2, x_i) \le M_1, \ i = 1, 2, \ldots.$$

Then there exists a subsequence $\{x_{i_k}\}_{k=1}^{\infty}$ and an a.c. function $x : [T_1, T_2] \to K$ such that

$$I^f(T_1, T_2, x) \le M_1, \ x_{i_k} \to x(t) \ \text{as } k \to \infty \ \text{uniformly in } [T_1, T_2] \ \text{and}$$

$$x_{i_k}' \to x' \ \text{as } k \to \infty \ \text{weakly in } L^1(R^n; (T_1, T_2)).$$

COROLLARY 1.3.1 *For each $f \in \mathcal{M}$, each pair of numbers T_1, T_2 satisfying $0 \le T_1 < T_2$ and each $z_1, z_2 \in K$ there exists an a.c. function $x : [T_1, T_2] \to K$ such that $x(T_i) = z_i$, $i = 1, 2$, $I^f(T_1, T_2, x) = U^f(T_1, T_2, z_1, z_2)$.*

COROLLARY 1.3.2 *For each $f \in \mathcal{M}$, each T_1, T_2 satisfying $0 \le T_1 < T_2$ and each $z \in K$ there exists an a.c. function $x : [T_1, T_2] \to K$ such that $x(T_1) = z$, $I^f(T_1, T_2, x) = \sigma^f(T_1, T_2, z)$.*

It is an elementary exercise to prove the following result.

PROPOSITION 1.3.6 *Let* $f \in \mathcal{M}$, $0 < c_1 < c_2 < \infty$ *and let* $c_3 > 0$. *Then there exists a neighborhood* U *of* f *in* $\mathcal{M}$ *such that the set*

$$\{U^g(T_1, T_2, z_1, z_2) : g \in U, T_1 \in [0, \infty), T_2 \in [T_1 + c_1, T_1 + c_2],$$

$$z_1, z_2 \in K \cap B(c_3), i = 1, 2\}$$

is bounded.

The next auxiliary result establishes the continuity of the functional $(T_1, T_2, y, z) \to U^f(T_1, T_2, y, z)$.

PROPOSITION 1.3.7 *Assume that* $K = R^n$, $f \in \mathcal{M}$, $0 < c_1 < c_2 < \infty$ *and* $M, \epsilon > 0$. *Then there exists* $\delta > 0$ *such that the following property holds:*

If $T_1, T_2 \geq 0$ *satisfy*

$$T_2 \in [T_1 + c_1, T_1 + c_2] \tag{3.26}$$

and if $y_1, y_2, z_1, z_2 \in R^n$ *satisfy*

$$|y_i|, |z_i| \leq M, \ i = 1, 2, \quad \sup\{|y_1 - y_2|, |z_1 - z_2|\} \leq \delta, \tag{3.27}$$

then

$$|U^f(T_1, T_2, y_1, z_1) - U^f(T_1, T_2, y_2, z_2)| \leq \epsilon. \tag{3.28}$$

Proof. By Proposition 1.3.6 there exists a number

$$M_0 > \sup\{|U^f(T_1, T_2, y, z)| : T_1 \in [0, \infty), T_2 \in [T_1 + c_1, T_1 + c_2], \tag{3.29}$$

$$y, z \in B(M)\}.$$

It follows from Proposition 1.3.3 that there exists a number $M_1 > 0$ such that the following property holds:

If a pair of numbers $T_1, T_2 \geq 0$ satisfies (3.26) and an a.c. function $x : [T_1, T_2] \to R^n$ satisfies

$$I^f(T_1, T_2, x) \leq 4M_0 + 1,$$

then

$$|x(t)| \leq M_1, \ t \in [T_1, T_2]. \tag{3.30}$$

Choose a number $\delta_1 > 0$ such that

$$4\delta_1(2c_2 + 2a + 4ac_2 + 1 + M_0) < \epsilon \tag{3.31}$$

(see Assumption A(iii)). By Proposition 1.3.1 there exist

$$\Gamma_0 > 2 \text{ and } \delta_2 \in (0, 8^{-1}) \tag{3.32}$$

such that

$$|f(t, x_1, u_1) - f(t, x_2, u_2)| \leq \delta_1 \inf\{f(t, x_1, u_1), \ f(t, x_2, u_2)\} \tag{3.33}$$

for each $t \in [0, \infty)$ and each $u_1, u_2, x_1, x_2 \in R^n$ which satisfy

$$|x_i| \leq M_1 + 1, \ |u_i| \geq \Gamma_0 - 1, \ i = 1, 2, |u_1 - u_2|, |x_1 - x_2| \leq \delta_2. \tag{3.34}$$

By Assumption A(iv) there exists

$$\delta_3 \in (0, 4^{-1} \inf\{\delta_1, \delta_2\}) \tag{3.35}$$

such that

$$|f(t, x_1, u_1) - f(t, x_2, u_2)| \leq \delta_1 \tag{3.36}$$

for each $t \in [0, \infty)$, each $u_1, u_2, x_1, x_2 \in R^n$ which satisfy

$$|x_i|, |u_i| \leq \Gamma_0 + M_1 + 4, \ i = 1, 2, \ \sup\{|x_1 - x_2|, |u_1 - u_2|\} \leq \delta_3. \tag{3.37}$$

Choose a number $\delta > 0$ for which

$$8(c_1^{-1} + 1)\delta < \delta_3. \tag{3.38}$$

Assume that numbers $T_1, T_2 \geq 0$ satisfy (3.26) and $y_1, y_2, z_1, z_2 \in R^n$ satisfy (3.27). By Corollary 1.3.1 there exists an a.c. function $x_1 : [T_1, T_2] \to R^n$ such that

$$x_1(T_1) = y_1, \ x_1(T_2) = z_1, \ I^f(T_1, T_2, x_1) = U^f(T_1, T_2, y_1, z_1). \tag{3.39}$$

Put

$$x_2(t) = x_1(t) + y_2 - y_1 + (t - T_1)(T_2 - T_1)^{-1}(z_2 - z_1 - y_2 + y_1), \ t \in [T_1, T_2]. \tag{3.40}$$

Clearly

$$x_2(T_1) = y_2, \ x_2(T_2) = z_2. \tag{3.41}$$

It follows from (3.26), (3.27), (3.39), (3.29) and the definition of M_1 that

$$|x_1(t)| \leq M_1, \ t \in [T_1, T_2]. \tag{3.42}$$

(3.40), (3.27) and (3.26) imply that

$$|x_1(t) - x_2(t)| \leq 3\delta, \ |x_1'(t) - x_2'(t)| \leq 2c_1^{-1}\delta, \ t \in [T_1, T_2]. \tag{3.43}$$

Set

$$E_1 = \{t \in [T_1, T_2] : |x_1'(t)| \geq \Gamma_0\}, \ E_2 = [T_1, T_2] \setminus E_1. \tag{3.44}$$

We have

$$|I^f(T_1, T_2, x_2) - I^f(T_1, T_2, x_1)| \leq \sigma_1 + \sigma_2 \tag{3.45}$$

where

$$\sigma_j = \int_{E_j} |f(t, x_1(t), x_1'(t)) - f(t, x_2(t), x_2'(t))| dt, \ j = 1, 2. \tag{3.46}$$

We will estimate σ_1, σ_2 separately. By (3.42), (3.43), (3.44), (3.38), (3.35), (3.32) and the definition of δ_2 for each $t \in E_1$,

$$|f(t, x_1(t), x_1'(t)) - f(t, x_2(t), x_2'(t))|$$

$$\leq \delta_1 f(t, x_1(t), x_1'(t))$$

and

$$\sigma_1 \leq \delta_1 \int_{E_1} f(t, x_1(t), x_1'(t)) dt.$$

It follows from this inequality, (3.39), (3.27), (3.29), (3.26) and Assumption A(iii) that

$$\sigma_1 \leq \delta_1 (I^f(T_1, T_2, x_1) + a(T_2 - T_1)) \leq \delta_1(M_0 + ac_2). \tag{3.47}$$

By the definition of δ_3, (3.42), (3.43), (3.38) and (3.44),

$$|f(t, x_1(t), x_1'(t)) - f(t, x_2(t), x_2'(t))| \leq \delta_1$$

for each $t \in E_2$ and

$$\sigma_2 \leq \delta_1 c_2. \tag{3.48}$$

Combining (3.45), (3.47), (3.48) and (3.31) we obtain that

$$|I^f(T_1, T_2, x_2) - I^f(T_1, T_2, x_1)| \leq \delta_1(M_0 + ac_2 + c_2) \leq \epsilon.$$

Together with (3.39) and (3.41) this implies that

$$U^f(T_1, T_2, y_2, z_2) \leq U^f(T_1, T_2, y_1, z_1) + \epsilon.$$

This completes the proof of the proposition.

The next proposition is an important tool which will be used in Chapters 1-5. It establishes that the integral functional $I^f(T_1, T_2, x)$ depends continuously on f.

PROPOSITION 1.3.8 *Let $f \in \mathcal{M}$, $0 < c_1 < c_2 < \infty$, $D, \epsilon > 0$. Then there exists a neighborhood V of f in $\mathcal{M}$ such that for each $g \in V$, each* *pair of numbers* $T_1, T_2 \geq 0$ *satisfying* $T_2 - T_1 \in [c_1, c_2]$ *the following property holds:*

If an a. c. function $x : [T_1, T_2] \to K$ *satisfies*

$$\inf\{I^f(T_1, T_2, x), I^g(T_1, T_2, x)\} \leq D, \tag{3.49}$$

then

$$|I^f(T_1, T_2, x) - I^g(T_1, T_2, x)| \leq \epsilon.$$

Proof. It follows from Proposition 1.3.3 there exists a number $S > 0$ such that

$$|x(t)| \leq S, \; t \in [T_1, T_2] \tag{3.50}$$

for each $g \in \mathcal{M}$, each $T_1, T_2 \geq 0$ satisfying $T_2 - T_1 \in [c_1, c_2]$ and each a.c. function $x : [T_1, T_2] \to K$ which satisfies $I^g(T_1, T_2, x) \leq D + 1$.

Choose $\delta \in (0, 1)$, $N > S$ and $\Gamma > 1$ such that

$$\delta(c_2 + 1) \leq 4^{-1}\epsilon, \; \psi(N)N > 4a, \; (\Gamma - 1)(c_2 + D + ac_2 + 1) \leq 4^{-1}\epsilon \tag{3.51}$$

and put

$$V = \{g \in \mathcal{M} : \; (f, g) \in E(N, \delta, \Gamma)\}$$

(see (2.1)). Assume that $g \in V$,

$$T_1, T_2 \geq 0, \; T_2 - T_1 \in [c_1, c_2] \tag{3.52}$$

and $x : [T_1, T_2] \to K$ is an a.c. function satisfying (3.49). By the choice of S the inequality (3.50) is true. Put

$$E_1 = \{t \in [T_1, T_2] : \; |x'(t)| \leq N\}, \; E_2 = [T_1, T_2] \setminus E_1.$$

It follows from (3.50) and the choice of V and N that

$$|f(t, x(t), x'(t)) - g(t, x(t), x'(t))| \leq \delta, \; t \in E_1. \tag{3.53}$$

Define

$$h(t) = \inf\{f(t, x(t), x'(t)), \; g(t, x(t), x'(t))\}, \; t \in [T_1, T_2]. \tag{3.54}$$

It follows from (3.50), (3.51), Assumption A(iii) and the definition of V, N that for $t \in E_2$

$$(f(t, x(t), x'(t)) + 1)(g(t, x(t), x'(t)) + 1)^{-1} \in [\Gamma^{-1}, \Gamma], \tag{3.55}$$

$$|f(t, x(t), x'(t)) - g(t, x(t), x'(t))| \leq (\Gamma - 1)(h(t) + 1).$$

By (3.53), (3.52), (3.55), (3.49), (3.54), Assumption A(iii) and (3.51),

$$|I^f(T_1, T_2, x) - I^g(T_1, T_2, x)| \leq \int_{E_1} |f(t, x(t), x'(t)) - g(t, x(t), x'(t))| dt$$

$$+ \int_{E_2} |f(t, x(t), x'(t)) - g(t, x(t), x'(t))| dt \leq \delta c_2 + (\Gamma - 1) \int_{E_2} (h(t) + 1) dt$$

$$\leq \delta c_2 + (\Gamma - 1) c_2 + (\Gamma - 1)(D + a c_2) \leq \epsilon.$$

The proposition is proved.

The next result establishes that the functional $U^f(T_1, T_2, y, z)$ depends continuously on f. It is also an important tool which will be used in Chapters 1-5.

PROPOSITION 1.3.9 *Let $f \in \mathcal{M}$, $0 < c_1 < c_2 < \infty$, $c_3, \epsilon > 0$. Then there exists a neighborhood V of f in $\mathcal{M}$ such that*

$$|U^f(T_1, T_2, y, z) - U^g(T_1, T_2, y, z)| \leq \epsilon$$

for each $g \in V$, each $T_1, T_2 \geq 0$ satisfying $T_2 - T_1 \in [c_1, c_2]$ and each $y, z \in K \cap B(c_3)$.

Proof. By Proposition 1.3.6 there exist a neighborhood V_1 of f in $\mathcal{M}$ and a positive number D_0 such that

$$|U^g(T_1, T_2, z_1, z_2)| + 1 < D_0$$

for each $g \in V_1$, each $T_1 \in [0, \infty)$, $T_2 \in [T_1 + c_1, T_1 + c_2]$ and each $z_1, z_2 \in K \cap B(c_3)$, $i = 1, 2$. It follows from Proposition 1.3.8 that there is a neighborhood V of f in $\mathcal{M}$ such that $V \subset V_1$ and that

$$|I^f(T_1, T_2, x) - I^g(T_1, T_2, x)| \leq \inf\{1, \epsilon\}$$

for each $g \in V$, each $T_1, T_2 \geq 0$ satisfying $T_2 - T_1 \in [c_1, c_2]$ and each a.c. function $x : [T_1, T_2] \to K$ which satisfy

$$\min\{I^f(T_1, T_2, x), \ I^g(T_1, T_2, x)\} \leq D_0 + 2.$$

The validity of the proposition now follows from the equality

$$U^g(T_1, T_2, y, z) = \inf\{I^g(T_1, T_2, x) :$$

$$x : [T_1, T_2] \to K \text{ is an a.c. function}$$

$$\text{satisfying } x(T_1) = y, \ x(T_2) = z, \ I^g(T_1, T_2, x) \leq D_0 + 1\}$$

which holds for $g \in V$, $T_1 \geq 0$, $T_2 \in [T_1 + c_1, T_1 + c_2]$ and $y, z \in K$ satisfying $|y|, |z| \leq c_3$.

1.4.   Discrete-time control systems

In this section we associate with $f \in \mathcal{M}$ a related discrete-time control system. We establish a boundedness of approximate solutions of this system (see Proposition 1.4.2). This result plays a crucial role in the proof of Theorem 1.2.3.

Let $f \in \mathcal{M}$, $\bar{z} \in K$ and let $0 < c_1 < c_2 < \infty$. It follows from Proposition 1.3.6 that there exist a positive number M_0 and a neighborhood U_0 of f in $\mathcal{M}$ such that

$$|U^g(T_1, T_2, y, z)| \leq M_0 \text{ for each } g \in U_0,$$

$$\text{each } T_1 \in [0, \infty), \; T_2 \in [T_1 + c_1, T_1 + c_2] \tag{4.1}$$

$$\text{and each } y, z \in K \cap B(2|\bar{z}| + 1).$$

Proposition 1.3.3 implies that there is a number $M_1 > 0$ such that

$$2M_0 + 2 \leq U^g(T_1, T_2, y, z) \text{ for each } g \in \mathcal{M},$$

$$\text{each } T_1 \in [0, \infty), \; T_2 \in [T_1 + c_1, T_1 + c_2], \tag{4.2}$$

$$\text{and each } y, z \in K \text{ satisfying } |y| + |z| \geq M_1.$$

PROPOSITION 1.4.1 *Let a number $M_1 > 0$ satisfy (4.2) and let $M_2 > 0$. Then there are an integer $N > 2$ and a neighborhood U of f in $\mathcal{M}$ such that for each $g \in U$, each $\Delta \in [0, \infty)$, each $T \in [c_1, c_2]$ and each pair of integers q_1, q_2 satisfying $0 \leq q_1 < q_2$, $q_2 - q_1 \geq N$ the following assertions hold:*

1. If $\{z_i\}_{i=q_1}^{q_2} \subset K$ satisfies

$$\{i \in \{q_1, \ldots, q_2\} : \; |z_i| \leq M_1\} = \{q_1, q_2\}$$

and if $y_i = z_i, i = q_1, q_2, \; y_i = \bar{z}, i = q_1 + 1, \ldots, q_2 - 1$, then

$$\sum_{i=q_1}^{q_2-1} [U^g(\Delta + iT, \Delta + (i+1)T, z_i, z_{i+1})$$

$$- U^g(\Delta + iT, \Delta + (i+1)T, y_i, y_{i+1})] \geq M_2; \tag{4.3}$$

2. If $\{z_i\}_{i=q_1}^{q_2} \subset K$ satisfies

$$\{i \in \{q_1, \ldots, q_2\} : |z_i| \leq M_1\} = \{q_1\}$$

and if $y_{q_1} = z_{q_1}, \; y_i = \bar{z}, \; i = q_1 + 1, \ldots, q_2$, then the inequality (4.3) is valid.

Proof. It follows from Proposition 1.3.6 that there exist a positive number M_3 and a neighborhood U of f in $\mathcal{M}$ such that

$$U \subset U_0,$$

$$|U^g(T_1, T_2, y, z)| \le M_3 \text{ for each } g \in U, \text{ each } T_1 \in [0, \infty),$$

$$T_2 \in [T_1 + c_1, T_1 + c_2] \text{ and each } y, z \in K \cap B(2|\bar{z}| + 1 + 2M_1).$$

Fix an integer $N \ge M_2 + 4M_3 + 4$. The validity of the proposition now follows from the definition of U, M_3, N, (4.1) and (4.2).

PROPOSITION 1.4.2 *Assume that a positive number M_1 satisfies (4.2) and $M_3 > 0$. Then there exist a neighborhood V of f in $\mathcal{M}$ and a number $M_4 > M_1$ such that for each $g \in V$, each $\Delta \in [0, \infty)$, each $T \in [c_1, c_2]$ and each pair of integers q_1, q_2 satisfying $0 \le q_1 < q_2$ the following assertions hold:*
1. If a sequence $\{z_i\}_{i=q_1}^{q_2} \subset K$ satisfies

$$|z_{q_1}|, |z_{q_2}| \le M_1, \ \max\{|z_i| : \ i = q_1, \dots, q_2\} > M_4, \qquad (4.4)$$

then there is a sequence $\{y_i\}_{i=q_1}^{q_2} \subset K$ such that $y_{q_j} = z_{q_j}$, $j = 1, 2$ and

$$\sum_{i=q_1}^{q_2-1} [U^g(\Delta + iT, \Delta + (i+1)T, z_i, z_{i+1})$$

$$- U^g(\Delta + iT, \Delta + (i+1)T, y_i, y_{i+1})] \ge M_3. \qquad (4.5)$$

2. If a sequence $\{z_i\}_{i=q_1}^{q_2} \subset K$ satisfies

$$|z_{q_1}| \le M_1, \ \max\{|z_i| : \ i = q_1, \dots, q_2\} > M_4, \qquad (4.6)$$

then there is a sequence $\{y_i\}_{i=q_1}^{q_2} \subset K$ such that $y_{q_1} = z_{q_1}$ and the inequality (4.5) is true.

Proof. There exist a neighborhood $U \subset U_0$ of f in $\mathcal{M}$ and an integer $N > 2$ such that Proposition 1.4.1 holds with $M_2 = 4(M_3 + 1)$. It follows from Proposition 1.3.6 that there exist a positive number r_1 and a neighborhood V of f in $\mathcal{M}$ such that

$$V \subset U, \ |U^g(T_1, T_2, y, z)| + 1 < r_1 \text{ for each } g \in V, \text{ each } T_1 \in [0, \infty), \tag{4.7}$$

$$T_2 \in [T_1 + c_1, T_1 + c_2] \text{ and each } y, z \in K \cap B(|\bar{z}| + 1 + M_1).$$

By Proposition 1.3.3 there exists $M_4 > M_1$ such that

$$\inf\{U^g(T_1, T_2, y, z) : \ g \in \mathcal{M}, \ T_1 \in [0, \infty), \ T_2 \in [T_1 + c_1, T_1 + c_2], \tag{4.8}$$

$$y, z \in K, \ |y| + |z| \geq M_4\} > 3r_1 N + 4 + 4M_3 + 3ac_2 N$$

(recall a in Assumption A(iii)).

Let $g \in V$, $\Delta \in [0, \infty)$, $T \in [c_1, c_2]$, $0 \leq q_1 < q_2$, $\{z_i\}_{i=q_1}^{q_2} \subset K$. We prove Assertion 1. Assume that (4.4) holds. Then there is $j \in \{q_1, \ldots, q_2\}$ such that $|z_j| > M_4$. Set

$$i_1 = \max\{i \in \{q_1, \ldots, j\} : |z_i| \leq M_1\},$$

$$i_2 = \min\{i \in \{j, \ldots, q_2\} : |z_i| \leq M_1\}.$$

If $i_2 - i_1 \geq N$, then by the definition of V, U, N and Proposition 1.4.1 there exists a sequence $\{y_i\}_{i=q_1}^{q_2} \subset K$ which satisfies (4.5) and $y_{q_i} = z_{q_i}$, $i = 1, 2$.

Now assume that $i_2 - i_1 < N$. Put

$$y_i = z_i, \ i \in \{q_1, \ldots, i_1\} \cup \{i_2, \ldots, q_2\}, \ y_i = \bar{z}, \ i = i_1 + 1, \ldots, i_2 - 1. \quad (4.9)$$

It follows from (4.9), (4.7), Assumption A(iii) and the definition of i_1, i_2, j that

$$\sum_{i=q_1}^{q_2 - 1} [U^g(\Delta + iT, \Delta + (i+1)T, z_i, z_{i+1}) - U^g(\Delta + iT, \Delta + (i+1)T, y_i, y_{i+1})]$$

$$\quad (4.10)$$

$$= \sum_{i=i_1}^{i_2 - 1} [U^g(\Delta + iT, \Delta + (i+1)T, z_i, z_{i+1}) - U^g(\Delta + iT, \Delta + (i+1)T, y_i, y_{i+1})]$$

$$\geq U^g(\Delta + (j-1)T, \Delta + jT, z_{j-1}, z_j) - a(i_2 - i_1 - 1)c_2 - (i_2 - i_1)r_1.$$

By this relation and the definition of j, M_4 (see (4.8))

$$\sum_{i=q_1}^{q_2 - 1} [U^g(\Delta + iT, \Delta + (i+1)T, z_i, z_{i+1}) \quad (4.11)$$

$$-U^g(\Delta + iT, \Delta + (i+1)T, y_i, y_{i+1})] \geq 4M_3 + 4.$$

This completes the proof of Assertion 1.

We prove Assertion 2. Assume that (4.6) holds. Then there is $j \in \{q_1, \ldots, q_2\}$ such that $|z_j| > M_4$. Set $i_1 = \sup\{i \in \{q_1, \ldots, j\} : |z_i| \leq M_1\}$.

There are two cases: 1) $|z_i| > M_1$, $i = j, \ldots, q_2$; 2) $\inf\{|z_i| : i = j, \ldots, q_2\} \leq M_1$. Consider the first case. We set

$$y_i = z_i, \ i = q_1, \ldots, i_1, \ y_i = \bar{z}, \ i = i_1 + 1, \ldots, q_2.$$

If $q_2 - i_1 \geq N$, then (4.5) follows from the definition of V, U, N and Proposition 1.4.1. If $q_2 - i_1 < N$, then (4.5) follows from the definition of $\{y_i\}_{i=q_1}^{q_2}$, i_1, j, M_4, (4.7) (see (4.10), (4.11) with $i_2 = q_2$).

Consider the second case. Set $i_2 = \inf\{i \in \{j, \ldots, q_2\} : |z_i| \leq M_1\}$. If $i_2 - i_1 \geq N$, then by the definition of V, U, N and Proposition 1.4.1 there exists a sequence $\{y_i\}_{i=q_1}^{q_2} \subset K$ which satisfies (4.5) and $y_{q_i} = z_{q_i}$, $i = 1, 2$. If $i_2 - i_1 < N$ we define a sequence $\{y_i\}_{i=q_1}^{q_2} \subset K$ by (4.9). Then (4.10) and (4.11) follows from (4.9), the definition of i_1, i_2, j, M_4, (4.7). Assertion 2 is proved. This completes the proof of the proposition.

1.5.   Proofs of Theorems 1.1-1.3

Construction of a neighborhood U. Let $f \in \mathcal{M}$, $\bar{z} \in K$, $M > 2|\bar{z}|$. It follows from Proposition 1.3.6 that there exist a positive number M_0 and a neighborhood U_0 of f in $\mathcal{M}$ such

$$|U^g(T_1, T_2, y, z)| \leq M_0 \text{ for each } g \in U_0, \text{ each } T_1 \in [0, \infty), \quad (5.1)$$

$$T_2 \in [T_1 + 4^{-1}, T_1 + 4] \text{ and each } y, z \in K \cap B(2|\bar{z}| + 1).$$

It follows from Proposition 1.3.3 that there exists a number $M_1 > M$ for which

$$2M_0 + 1 < \inf\{U^g(T_1, T_2, y, z) : g \in \mathcal{M},$$

$$T_1 \in [0, \infty), \ T_2 \in [T_1 + 4^{-1}, T_1 + 4],$$

$$y, z \in K, \ |y| + |z| \geq M_1\}. \quad (5.2)$$

By (5.1), (5.2) there exists a neighborhood U_1 of f in $\mathcal{M}$ and a number M_2 such that

$$U_1 \subset U_0, \ M_2 > M_1 \text{ and Proposition 1.4.2 holds with } M_3 = 1, \quad (5.3)$$

$$c_1 = 4^{-1}, \ c_2 = 4, \ V = U_1, \ M_4 = M_2.$$

Proposition 1.3.6 implies that there exist a positive number Q_0 and a neighborhood U_2 of f in $\mathcal{M}$ such that

$$U_2 \subset U_1, \ |U^g(T_1, T_2, y, z)| + 1 \leq Q_0 \text{ for each } g \in U_2, \text{ each } T_1 \in [0, \infty),$$
$$\quad (5.4)$$

$$T_2 \in [T_1 + 4^{-1}, T_1 + 4] \text{ and each } y, z \in K \cap B(M_2 + 1).$$

By Proposition 1.3.3 there exists a number

$$Q_1 > Q_0 + M_2 + 1 \quad (5.5)$$

such that the following property holds:

If $g \in \mathcal{M}$, T_1, T_2 satisfy

$$0 \leq T_1 < T_2, \ T_2 - T_1 \in [4^{-1}, 4]$$

and if an a.c. function $x : [T_1, T_2] \rightarrow K$ satisfies $I^g(T_1, T_2, x) \leq 2Q_0 + 2$, then

$$|x(t)| \leq Q_1, \ t \in [T_1, T_2]. \tag{5.6}$$

It follows from Proposition 1.3.6 that there exist a positive number Q_2 and a neighborhood U of f in $\mathcal{M}$ such that

$$U \subset U_2, \ Q_2 > Q_1,$$

$$|U^g(T_1, T_2, y, z)| + 1 < Q_2 \text{ for each } g \in U, \tag{5.7}$$

each $T_1 \in [0, \infty)$, $T_2 \in [T_1 + 4^{-1}, T_1 + 4]$ and each $y, z \in K \cap B(2Q_1 + 4)$.

We may assume without loss of generality that there exists a positive number Q_3 such that

$$|g(t, y, u)| + 1 < Q_3 \text{ for each } g \in U, \text{ each } t \in [0, \infty) \tag{5.8}$$

and each $y \in K \cap B(2M_2 + 2)$, $u \in B(2M_2 + 2)$.

Construction of a function $Z^g : [0, \infty) \rightarrow K$. Let $g \in U$, $z \in K$, $|z| \leq M$. By Corollary 1.3.2 for any integer $q \geq 1$ there exists an a. c. function $Z_q^g : [0, q] \rightarrow K$ such that

$$Z_q^g(0) = z, \ I^g(0, q, Z_q^g) = \sigma^g(0, q, z). \tag{5.9}$$

It follows from Proposition 1.4.2 and the definition of Z_q^g, U_1, M_2 that

$$|Z_q^g(i)| \leq M_2, \ i = 0, \ldots, q, \ q = 1, 2, \ldots. \tag{5.10}$$

There exists a subsequence $\{Z_{q_j}^g\}_{j=1}^{\infty}$ such that for any integer $i \geq 0$ there exists

$$z_i^g = \lim_{j \to \infty} Z_{q_j}^g(i). \tag{5.11}$$

By Corollary 1.3.1 there exists an a.c. function $Z^g : [0, \infty) \rightarrow K$ such that for each integer $i \geq 0$,

$$Z^g(i) = z_i^g, \ I^g(i, i+1, Z^g) = U^g(i, i+1, z_i^g, z_{i+1}^g). \tag{5.12}$$

It follows from (5.9), (5.10) and (5.4) that

$$I^g(i, i+1, Z_q^g) < Q_0, \ i = 0, \ldots, q-1, \ q = 1, 2, \ldots. \tag{5.13}$$

(5.10), (5.11), (5.12) and (5.4) imply that

$$I^g(i, i+1, Z^g) < Q_0, \ i = 0, 1, \dots. \tag{5.14}$$

By (5.13), (5.14) and the definition of Q_1 (see (5.5), (5.6))

$$|Z_q^g(t)| \le Q_1, \ t \in [0, q], \ q = 1, 2, \dots, \quad |Z^g(t)| \le Q_1, \ t \in [0, \infty). \tag{5.15}$$

Therefore for each $g \in U$ and each $z \in K$ satisfying $|z| \le M$ we define a.c. functions $Z_q^g : [0, q] \to K$, $q = 1, 2, \dots$ and $Z^g : [0, \infty) \to K$ satisfying (5.9)-(5.15).

The next auxiliary result shows that the sequence $\{z_i^q\}_{i=0}^{\infty}$ is (g)-good for each $g \in U$.

LEMMA 1.5.1 *Let $g \in U$, $z \in K$, $|z| \le M$ and let a pair of integers q_1, q_2 satisfy $0 \le q_1 < q_2$. Then if a sequence $\{y_i\}_{i=q_1}^{q_2} \subset K$ satisfies $|y_{q_1}| \le M_1$, then*

$$\sum_{i=q_1}^{q_2-1} [U^g(i, i+1, z_i^g, z_{i+1}^g) - U^g(i, i+1, y_i, y_{i+1})] \le 4 + 4Q_2. \tag{5.16}$$

Proof. Assume that a sequence $\{y_i\}_{i=q_1}^{q_2} \subset K$ satisfies $|y_{q_1}| \le M_1$. We will show that (5.16) holds.

Let us assume the converse. Then

$$\sum_{i=q_1}^{q_2-1} [U^g(i, i+1, z_i^g, z_{i+1}^g) - U^g(i, i+1, y_i, y_{i+1})] > 4 + 4Q_2. \tag{5.17}$$

By Corollaries 1.3.1 and 1.3.2 we may assume without loss of generality that if a sequence $\{\bar{y}_i\}_{i=q_1}^{q_2} \subset K$ satisfies $\bar{y}_{q_1} = y_{q_1}$, then

$$\sum_{i=q_1}^{q_2-1} [U^g(i, i+1, y_i, y_{i+1}) - U^g(i, i+1, \bar{y}_i, \bar{y}_{i+1})] \le 0.$$

(5.3) and (5.5) imply that

$$|y_i| \le M_2 < Q_1, \ i = q_1, \dots, q_2. \tag{5.18}$$

By Proposition 1.3.5, (5.9), (5.11) and (5.13) for any integer $i \ge 0$,

$$U^g(i, i+1, z_i^g, z_{i+1}^g) \le \liminf_{j \to \infty} U^g(i, i+1, Z_{q_j}^g(i), Z_{q_j}^g(i+1)).$$

Therefore there exists an integer $q > q_2 + 1$ such that

$$\sum_{i=q_1}^{q_2} [U^g(i, i+1, z_i^g, z_{i+1}^g) - U^g(i, i+1, Z_q^g(i), Z_q^g(i+1))] \le 1. \tag{5.19}$$

We define a sequence $\{h_i\}_{i=0}^{q} \subset K$ as follows:

$$h_i = Z_q^g(i), \ i \in \{0, \ldots, q_1\} \cup \{q_2 + 1, \ldots, q\}, \ h_i = y_i, \ i = q_1 + 1, \ldots, q_2. \tag{5.20}$$

It follows from (5.20), (5.9), Corollary 1.3.1, (5.19) and (5.17) that

$$0 \geq \sum_{i=0}^{q-1} [U^g(i, i+1, Z_q^g(i), Z_q^g(i+1)) - U^g(i, i+1, h_i, h_{i+1})]$$

$$= \sum_{i=q_1}^{q_2} [U^g(i, i+1, Z_q^g(i), Z_q^g(i+1)) - U^g(i, i+1, h_i, h_{i+1})]$$

$$= \sum_{i=q_1}^{q_2} [U^g(i, i+1, Z_q^g(i), Z_q^g(i+1)) - U^g(i, i+1, z_i^g, z_{i+1}^g)]$$

$$+ \sum_{i=q_1}^{q_2} U^g(i, i+1, z_i^g, z_{i+1}^g) - \sum_{i=q_1}^{q_2-1} U^g(i, i+1, y_i, y_{i+1})$$

$$+ U^g(q_1, q_1 + 1, y_{q_1}, y_{q_1+1})$$

$$- U^g(q_1, q_1 + 1, h_{q_1}, h_{q_1+1}) - U^g(q_2, q_2 + 1, h_{q_2}, h_{q_2+1}) \geq 3 + 4Q_2$$

$$+ U^g(q_2, q_2 + 1, z_{q_2}^g, z_{q_2+1}^g) + U^g(q_1, q_1 + 1, y_{q_1}, y_{q_1+1})$$

$$- U^g(q_1, q_1 + 1, h_{q_1}, h_{q_1+1}) - U^g(q_2, q_2 + 1, h_{q_2}, h_{q_2+1}).$$

Combined with (5.20), (5.18), (5.10), (5.11), (5.5) and (5.7) this relation implies that

$$0 \geq 3 + 4Q + U^g(q_2, q_2 + 1, z_{q_2}^g, z_{q_2+1}^g) + U^g(q_1, q_1 + 1, y_{q_1}, y_{q_1+1})$$

$$- U^g(q_1, q_1 + 1, Z_q^g(q_1), y_{q_1+1})$$

$$- U^g(q_2, q_2 + 1, y_{q_2}, Z_q^g(q_2 + 1)) \geq 3 + 4Q_2 - 4Q_2.$$

The contradiction we have reached proves the lemma.

Lemma 1.5.1 implies the following result.

LEMMA 1.5.2 *Let* $g \in U$, $z \in K$, $|z| \leq M$, *an integer* $q \geq 0$ *and let* $T \in (q, \infty)$. *Then*

$$I^g(q, T, Z^g) \leq I^g(q, T, x) + 4 + 4Q_2 + Q_0 + 2a \tag{5.21}$$

for each a.c. function $x : [q, T] \to K$ *such that* $|x(q)| \leq M_1$ *(recall* a *in Assumption A (iii))*.

Proof. Assume that an a.c. function $x : [q, T] \to K$ satisfies $|x(q)| \le M_1$. There exists an integer $q_1 \ge q$ such that $q_1 < T \le q_1 + 1$. It follows from Lemma 1.5.1 and (5.12) that

$$I^g(q, q_1, Z^g) \le I^g(q, q_1, x) + 4 + 4Q_2. \tag{5.22}$$

By Assumption A(iii) and (5.14)

$$I^g(q_1, T, x) \ge -a, \quad I^g(q_1, T, Z^g) \le Q_0 + a. \tag{5.23}$$

(5.22) and (5.23) imply (5.21). The lemma is proved.

The next lemma, which follows from Lemma 1.5.2, establishes that the function Z^g is (g)-good for each $g \in U$ and each $z \in K$ satisfying $|z| \le M$.

LEMMA 1.5.3 *Let* $g \in U$, $z \in K$, $|z| \le M$ *and* $0 \le T_1 < T_2$. *Then*

$$I^g(T_1, T_2, Z^g) \le I^g(T_1, T_2, x) + 4 + 4Q_2 + Q_0 + Q_3 + 3a \tag{5.24}$$

for each a.c. function $x : [T_1, T_2] \to K$ *such that* $|x(T_1)| \le M_1$.

Proof. Assume that an a.c. function $x : [T_1, T_2] \to K$ satisfies $|x(T_1)| \le M_1$.

There exists an integer $q \ge 0$ such that $q \le T_1 < q + 1$. Set

$$x_1(t) = x(T_1),\ t \in [q, T_1],\ x_1(t) = x(t),\ t \in [T_1, T_2]. \tag{5.25}$$

By Lemma 1.5.2,

$$I^g(q, T_2, Z^g) \le I^g(q, T_2, x_1) + 4 + 4Q_2 + Q_0 + 2a. \tag{5.26}$$

By Assumption A(iii) and (5.26),

$$I^g(T_1, T_2, Z^g) = I^g(q, T_2, Z^g) - I^g(q, T_1, Z^g) \le I^g(q, T_2, Z^g) + a \tag{5.27}$$

$$\le I^g(q, T_2, x_1) + 4 + 4Q_2 + Q_0 + 3a.$$

It follows from (5.25) and (5.8) that $|I^g(q, T_1, x_1)| \le Q_3$. (5.24) now follows from this relation and (5.27), (5.25). The lemma is proved.

The following auxiliary result shows that a sequence $\{y_i\}_{i=0}^{\infty} \subset K$ is not good if $\limsup_{i \to \infty} |y_i| > M_2$. It plays an important role in the proof of Theorem 1.2.2.

LEMMA 1.5.4 *Let* $g \in U$, $z \in K$, $|z| \le M$, $\{y_i\}_{i=0}^{\infty} \subset K$,

$$\limsup_{i \to \infty} |y_i| > M_2. \tag{5.28}$$

Then

$$\sum_{i=0}^{N-1} [U^g(i, i+1, y_i, y_{i+1}) - U^g(i, i+1, z_i^g, z_{i+1}^g)] \to \infty \text{ as } N \to \infty. \quad (5.29)$$

Proof. There are two cases:

$$\text{a) } \liminf_{i \to \infty} |y_i| > 2^{-1} M_1; \quad \text{b) } \liminf_{i \to \infty} |y_i| \leq 2^{-1} M_1.$$

Consider the case a). Set $h_i = \bar{z}$ for $i = 0, 1, \ldots$. It follows from (5.1), (5.2) that

$$U^g(i, i+1, y_i, y_{i+1}) - U^g(i, i+1, h_i, h_{i+1}) \geq M_0 + 1$$

for all large i. (5.29) now follows from this relation and Lemma 1.5.1.

Consider the case b). By (5.28) there exists a subsequence $\{y_{i_k}\}_{k=1}^{\infty}$ such that

$$0 < i_1, \ |y_{i_k}| < M_1, \ \sup\{|y_j| : \ j = i_k, \ldots, i_{k+1}\} > M_2, \ k = 1, 2, \ldots. \quad (5.30)$$

It follows from (5.3), (5.30) and Proposition 1.4.2 that for any integer $k \geq 1$ there exists a sequence $\{h_j\}_{j=i_k}^{i_{k+1}} \subset K$ such that $h_j = y_j$, $j \in \{i_k, i_{k+1}\}$,

$$\sum_{j=i_k}^{i_{k+1}-1} [U^g(j, j+1, y_j, y_{j+1}) - U^g(j, j+1, h_j, h_{j+1})] \geq 1. \quad (5.31)$$

Fix an integer $q \geq 4$. By (5.30), Lemma 1.5.1 and (5.31) for an integer $N > i_q$

$$\sum_{j=i_q}^{N-1} [U^g(j, j+1, z_j^g, z_{j+1}^g) - U^g(j, j+1, y_j, y_{j+1})] \leq 4 + 4Q_2,$$

$$\sum_{j=i_1}^{i_q-1} [U^g(j, j+1, z_j^g, z_{j+1}^g) - U^g(j, j+1, h_j, h_{j+1})] \leq 4 + 4Q_2,$$

$$\sum_{j=0}^{N-1} [U^g(j, j+1, z_j^g, z_{j+1}^g) - U^g(j, j+1, y_j, y_{j+1})]$$

$$= \sum_{j=0}^{i_1-1} [U^g(j, j+1, z_j^g, z_{j+1}^g) - U^g(j, j+1, y_j, y_{j+1})]$$

$$+ \sum_{j=i_1}^{i_q-1} [U^g(j, j+1, z_j^g, z_{j+1}^g) - U^g(j, j+1, h_j, h_{j+1})]$$

$$+ \sum_{j=i_1}^{i_q-1} [U^g(j, j+1, h_j, h_{j+1}) - U^g(j, j+1, y_j, y_{j+1})]$$

$$+ \sum_{j=i_q}^{N-1} [U^g(j, j+1, z_j^g, z_{j+1}^g) - U^g(j, j+1, y_j, y_{j+1})]$$

$$\leq \sum_{j=0}^{i_1-1} [U^g(j, j+1, z_j^g, z_{j+1}^g) - U^g(j, j+1, y_j, y_{j+1})] + 2(4 + 4Q_2) - (q-1).$$

This completes the proof of the lemma.

The next lemma implies Theorem 1.2.2. Its proof is based on Lemma 1.5.4.

LEMMA 1.5.5 *Assume that* $g \in U$, $z \in K$, $|z| \leq M$ *and* $y : [0, \infty) \to K$ *is an a.c. function which satisfies*

$$\limsup_{t \to \infty} |y(t)| > Q_1. \tag{5.32}$$

Then

$$I^g(0, T, y) - I^g(0, T, Z^g) \to \infty \ as \ T \to \infty. \tag{5.33}$$

Proof. There are two cases:

a) $\limsup_{i \to \infty} |y(i)| > M_2$; b) $\limsup_{i \to \infty} |y(i)| \leq M_2$

where i is an integer. Consider the case a). It follows from Lemma 1.5.4, (5.12) that

$$I^g(0, q, y) - I^g(0, q, Z^g) \to \infty \ as \ an \ integer \ q \to \infty. \tag{5.34}$$

Let $T > 0$. There exists an integer $q(T) \geq 0$ such that

$$q(T) < T \leq q(T) + 1. \tag{5.35}$$

By Assumption A(iii) and (5.14),

$$I^g(q(T), T, y) \geq -a, \ I^g(q(T), T, Z^g) = I^g(q(T), q(T) + 1, Z^g) \tag{5.36}$$

$$-I^g(T, q(T) + 1, Z^g) \leq Q_0 + a.$$

Together with (5.34) these relations imply that

$$I^g(0, T, y) - I^g(0, T, Z^g) \geq I^g(0, q(T), y)$$

$$-I^g(0, q(T), Z^g) - Q_0 - 2a \to \infty \text{ as } T \to \infty.$$

Consider the case b). There exists an integer $i_0 \geq 2$ such that

$$|y(i)| \leq M_2 + 2^{-1} \text{ for all integers } i \geq i_0. \tag{5.37}$$

By (5.37), (5.32), (5.4) and the definition of Q_1 (see (5.5)),

$$\sum_{i=0}^{N} [I^g(i, i+1, y) - U^g(i, i+1, y(i), y(i+1))] \to \infty \text{ as } N \to \infty. \tag{5.38}$$

Define a sequence $\{d_i\}_{i=i_0}^{\infty} \subset K$ as

$$d_{i_0} = z, \; d_i = y(i) \text{ for all integers } i > i_0.$$

By Lemma 1.5.1 and the definition of $\{d_i\}_{i=i_0}^{\infty}$ for any integer $N \geq i_0 + 1$

$$\sum_{i=i_0+1}^{N} [U^g(i, i+1, y(i), y(i+1)) - U^g(i, i+1, z_i^g, z_{i+1}^g)]$$

$$= \sum_{i=i_0}^{N} [U^g(i, i+1, d_i, d_{i+1}) - U^g(i, i+1, z_i^g, z_{i+1}^g)]$$

$$+ U^g(i_0, i_0 + 1, z_{i_0}^g, z_{i_0+1}^g) - U^g(i_0, i_0 + 1, z, y(i_0 + 1))$$

$$\geq -4 - 4Q_2 + U^g(i_0, i_0 + 1, z_{i_0}^g, z_{i_0+1}^g) - U^g(i_0, i_0 + 1, z, y(i_0 + 1)).$$

Together with (5.28), (5.12) this implies that

$$\sum_{i=0}^{N} [I^g(i, i+1, y) - I^g(i, i+1, Z^g)] \to +\infty \text{ as } N \to \infty. \tag{5.39}$$

Let $T > 0$. There exists an integer $q(T) \geq 0$ satisfying (5.35). Clearly (5.36) holds. (5.33) now follows from (5.36) and (5.39). The lemma is proved.

The following lemma implies Assertion 1 of Theorem 1.2.1.

LEMMA 1.5.6 *Let* $g \in U$, $z \in K$, $|z| \leq M$ *and let* $y : [0, \infty) \to K$ *be an a.c. function. Then one of the relations below holds:*
(i) $I^g(0, T, y) - I^g(0, T, Z^g) \to \infty$ *as* $T \to \infty$;
(ii) $\sup\{|I^g(0, T, y) - I^g(0, T, Z^g)| : T \in (0, \infty)\} < \infty.$

Proof. By Lemma 1.5.5 we may assume that $\limsup_{t \to \infty} |y(t)| \leq Q_1$. There exists an integer $i_0 > 0$ such that

$$|y(t)| \leq Q_1 + 2^{-1}, \; t \in [i_0, \infty). \tag{5.40}$$

Fix an integer $i > i_0$. By Corollary 1.3.1 there exists an a.c. function $\bar{y} : [i-1, \infty) \to K$ such that

$$\bar{y}(i-1) = z, \; \bar{y}(t) = y(t), \; t \in [i, \infty), \; I^g(i-1, i, \bar{y}) = U^g(i-1, i, z, y(i)). \tag{5.41}$$

(5.7), (5.41), (5.40), (5.5) imply that $|U^g(i-1, i, z, y(i))| \leq Q_2$. It follows from this relation, (5.41), Lemma 1.5.2 and Assumption A(iii) that for each $T > i$,

$$I^g(i, T, y) - I^g(i, T, Z^g) = I^g(i-1, T, \bar{y}) - I^g(i-1, T, Z^g) \tag{5.42}$$

$$-I^g(i-1, i, \bar{y}) + I^g(i-1, i, Z^g) \geq -4 - 4Q_2 - Q_0 - 2a$$

$$-I^g(i-1, i, \bar{y}) + I^g(i-1, i, Z^g) \geq -4 - 5Q_2 - Q_0 - 3a.$$

(5.42) holds for each integer $i > i_0$ and each $T > i$.

Let $S > i_0 + 1$, $T > S + 1$. There exists an integer $i > i_0 + 1$ such that $i - 1 \leq S < i$. Clearly (5.42) holds. By Assumption A(iii) and (5.14),

$$I^g(S, i, y) \geq -a, \; I^g(S, i, Z^g) = I^g(i-1, i, Z^g) - I^g(i-1, S, Z^g) \leq Q_0 + a.$$

Together with (5.42) this implies that

$$I^g(S, T, y) - I^g(S, T, Z^g)$$

$$= I^g(i, T, y) - I^g(i, T, Z^g) + I^g(S, i, y) - I^g(S, i, Z^g) \tag{5.43}$$

$$\geq -4 - 5Q_2 - 2Q_0 - 5a.$$

We established (5.43) for each $S > i_0 + 1$ and each $T > S + 1$.

Assume that (ii) does not hold. It follows from (5.14), Assumption A(iii) and (5.43) which holds for each $S > i_0 + 1$, $T > S + 1$ that

$$\inf\{I^g(0, T, y) - I^g(0, T, Z^g) : T \in (0, \infty)\} > -\infty.$$

Therefore $\sup\{I^g(0, T, y) - I^g(0, T, Z^g) : T \in (0, \infty)\} = \infty$. By Assumption A(iii) and (5.14) $\sup\{I^g(0, i, y) - I^g(0, i, Z^g) : i = 1, 2, \ldots\} = \infty$. Together with (5.43) which holds for each $S > i_0 + 1$, $T > S + 1$ this implies (i). The lemma is proved.

The next auxiliary result implies Assertion 4 of Theorem 1.2.1.

LEMMA 1.5.7 *Let $K = R^n$, $g \in U$ and let $z \in K$ satisfy $|z| \leq M$. Then the function Z^g is a (g)-minimal solution.*

Proof. Let us assume the converse. Then there exist $T_1 \geq 0$ and $T_2 > T_1$ such that

$$I^g(T_1, T_2, Z^g) > U^g(T_1, T_2, Z^g(T_1), Z^g(T_2)).$$

Choose a number

$$\epsilon \in (0, 8^{-1}[I^g(T_1, T_2, Z^g) - U^g(T_1, T_2, Z^g(T_1), Z^g(T_2))] \qquad (5.44)$$

and an integer $q_0 > T_2 + 5$. By Corollary 1.3.1 there exists an a.c. function $y : [T_1, T_2] \to K$ such that

$$y(T_i) = Z^g(T_i), \; i = 1, 2, \quad I^g(T_1, T_2, y) = U^g(T_1, T_2, Z^g(T_1), Z^g(T_2)). \qquad (5.45)$$

It follows from (5.10), (5.11), (5.12) and Proposition 1.3.7 that there exists an integer $k > 2q_0 + 4$ for which

$$|U^g(i, i + 1, Z^g(i), Z^g(i + 1)) - U^g(i, i + 1, Z_k^g(i), Z_k^g(i + 1))| \le \qquad (5.46)$$

$$(2q_0 + 1)^{-1}\epsilon, \; i = 0, \ldots, 2q_0 + 1,$$

$$|U^g(q_0, q_0 + 1, Z^g(q_0), Z_k^g(q_0 + 1)) \qquad (5.47)$$

$$-U^g(q_0, q_0 + 1, Z_k^g(q_0), Z_k^g(q_0 + 1))| \le (2q_0 + 1)^{-1}\epsilon.$$

By Corollary 1.3.1 and (5.45) there exists an a.c. function $x : [0, k] \to K$ such that

$$x(t) = Z^g(t), \; t \in [0, T_1] \cup [T_2, q_0], \; x(t) = y(t), \; t \in [T_1, T_2], \qquad (5.48)$$

$$x(t) = Z_k^g(t), \; t \in [q_0 + 1, k],$$

$$I^g(q_0, q_0 + 1, x) = U^g(q_0, q_0 + 1, x(q_0), x(q_0 + 1)).$$

It follows from (5.48), (5.9) that

$$I^g(0, k, x) \ge I^g(0, k, Z_k^g). \qquad (5.49)$$

By (5.48), (5.9), (5.12), (5.46), (5.47) and (5.44),

$$I^g(0, k, x) - I^g(0, k, Z_k^g) = I^g(0, q_0 + 1, x) - I^g(0, q_0 + 1, Z_k^g)$$

$$= (I^g(0, q_0, x) - I^g(0, q_0, Z^g)) + (I^g(0, q_0, Z^g) - I^g(0, q_0, Z_k^g))$$

$$+I^g(q_0, q_0 + 1, x) - I^g(q_0, q_0 + 1, Z_k^g) \le I^g(T_1, T_2, y) - I^g(T_1, T_2, Z^g)$$

$$+ \sum_{i=0}^{q_0-1} [U^g(i, i + 1, Z^g(i), Z^g(i + 1)) - U^g(i, i + 1, Z_k^g(i), Z_k^g(i + 1))]$$

$$+U^g(q_0, q_0 + 1, Z^g(q_0), Z_k^g(q_0 + 1)) - U^g(q_0, q_0 + 1, Z_k^g(q_0), Z_k^g(q_0 + 1))$$

$$\le I^g(T_1, T_2, y) - I^g(T_1, T_2, Z^g) + \epsilon.$$

It follows from this relation, (5.44), (5.45) that

$$I^g(0, k, x) - I^g(0, k, Z_k^g) \le I^g(T_1, T_2, y) - I^g(T_1, T_2, Z^g) + \epsilon < -\epsilon.$$

This is contradictory to (5.49). The obtained contradiction proves the lemma.

Proof of Theorem 1.2.1. At the begining of Section 1.5 for each $f \in \mathcal{M}$ and each $M > 2|\bar{z}|$ we constructed a neighborhood U of f in $\mathcal{M}$ and for each $g \in U$ and each $z \in K$ satisfying $|z| \leq M$ we defined a.c. functions $Z^g : [0, \infty) \to K$, $Z_q^g : [0, q] \to K$, $q = 1, 2, \ldots$ satisfying (5.9)-(5.15). Clearly an a.c. function $Z^f : [0, \infty) \to K$ was defined for every $f \in \mathcal{M}$ and every $z \in K$. By Lemmas 1.5.5, 1.5.6 for each $f \in \mathcal{M}$ and each $z \in K$ the function Z^f is (f)-good and Assertion 1 of Theorem 1.2.1 holds.

Assertion 2 of Theorem 1.2.1 follows from (5.15) which holds for every $g \in U$ (U is a neighborhood of f in $\mathcal{M}$) and each $z \in K$ satisfying $|z| \leq M$.

Assertion 3 of Theorem 1.2.1 follows from Lemma 1.5.3. Lemma 1.5.7 implies Assertion 4 of Theorem 1.2.1. Theorem 1.2.1 is proved.

Theorem 1.2.2 follows from Lemma 1.5.5.

Proof of Theorem 1.2.3 Fix $\bar{z} \in K$. It follows from Proposition 1.3.6 that there exists a positive number M_0 and a neighborhood U_0 of f in $\mathcal{M}$ such that

$$|U^g(T_1, T_2, y, z)| \leq M_0 \text{ for each } g \in U_0,$$

$$\text{each } T_1 \in [0, \infty), \ T_2 \in [T_1 + c, T_1 + 2c + 2], \tag{5.50}$$

$$\text{and each } y, z \in K \cap B(2|\bar{z}| + 1).$$

By Proposition 1.3.3 we may assume without loss of generality that

$$\inf\{U^g(T_1, T_2, y, z) : \ g \in \mathcal{M}, \ T_1 \in [0, \infty), \ T_2 \in [T_1 + c, T_1 + 2c + 2], \tag{5.51}$$

$$y, z \in K, \ |y| + |z| \geq M_1\} > 2M_0 + 1.$$

There exist a positive number S_1 and a neighborhood U_1 of f in $\mathcal{M}$ such that

$$U_1 \subset U_0, \ S_1 > M_1 \text{ and Proposition 1.4.2 holds with} \tag{5.52}$$

$$M_3 = M_2 + 2, \ M_4 = S_1, \ V = U_1, \ c_1 = c, \ c_2 = 2c + 2.$$

It follows from Proposition 1.3.6 that there exist a positive number M_3 and a neighborhood U of f in $\mathcal{M}$ such that

$$U \subset U_1, \ |U^g(T_1, T_2, y, z)| + 1 < M_3 \text{ for each } g \in U, \text{ each } T_1 \in [0, \infty), \tag{5.53}$$

$$T_2 \in [T_1 + c, T_1 + 2c + 2] \text{ and } y, z \in K \cap B(S_1).$$

It follows from Proposition 1.3.3 that there exists $S > S_1 + 1$ such that the following property holds:

If $g \in \mathcal{M}$, $T_1 \in [0, \infty)$, $T_2 \in [T_1 + c, T_1 + 2c + 2]$ and if an a.c. function $v : [T_1, T_2] \to K$ satisfies $I^g(T_1, T_2, v) \le 2M_3 + 2M_2 + 2$, then

$$|v(t)| \le S, \ t \in [T_1, T_2].$$

Assume that $g \in U$, $T_1 \in [0, \infty)$, $T_2 \ge c + T_1$. We will show that property (i) holds.

Let $x, y \in K$, $|x|, |y| \le M_1$ and let $v : [T_1, T_2] \to K$ be an a.c. function which satisfies

$$v(T_1) = x, \ v(T_2) = y, \ I^g(T_1, T_2, v) \le U^g(T_1, T_2, x, y) + M_2. \quad (5.54)$$

There is a natural number p such that $pc \le T_2 - T_1 < (p+1)c$. Set $T = p^{-1}(T_2 - T_1)$. Clearly $T \in [c, 2c]$. By (5.54) and Corollary 1.3.1,

$$\sum_{i=0}^{p-1} [U^g(T_1 + iT, T_1 + (i+1)T, v(T_1 + iT), v(T_1 + (i+1)T))$$

$$- U^g(T_1 + iT, T_1 + (i+1)T, y_i, y_{i+1})] \le M_2$$

for each sequence $\{y_i\}_{i=0}^p \subset K$ satisfying $y_0 = v(T_1)$, $y_p = v(T_2)$. It follows from this, (5.52), (5.54) and Proposition 1.4.2 that

$$|v(T_1 + iT)| \le S_1, \ i = 0, \dots, p.$$

By this relation and (5.54), (5.53) for $i = 0, \dots, p-1$,

$$I^g(T_1 + iT, T_1 + (i+1)T, v) \le$$

$$U^g(T_1 + iT, T_1 + (i+1)T, v(T_1 + iT), v(T_1 + (i+1)T)) + M_2 < M_3 + M_2.$$

It follows from this relation and the definition of S that

$$|v(t)| \le S, \ t \in [T_1, T_2].$$

Therefore property (i) holds. Analogously to this we can show that property (ii) holds. The theorem is proved.

Chapter 2

EXTREMALS
OF NONAUTONOMOUS PROBLEMS

In this chapter we show that the turnpike property is a general phenomenon which holds for a large class of nonautonomous variational problems with nonconvex integrands. We consider the complete metric space of integrands $\mathcal{M}$ introduced in Section 1.1 and establish the existence of a set $\mathcal{F} \subset \mathcal{M}$ which is a countable intersection of open everywhere dense sets in $\mathcal{M}$ such that for each $f \in \mathcal{F}$ and each $z \in R^n$ the following properties hold:

(i) there exists an (f)-overtaking optimal function $Z^f : [0, \infty) \to R^n$ satisfying $Z^f(0) = z$;

(ii) the integrand f has the turnpike property with the trajectory $\{Z^f(t) : t \in [0, \infty)\}$ being the turnpike.

Moreover we show that the turnpike property holds for approximate solutions of variational problems with a generic integrand f and that the turnpike phenomenon is stable under small perturbations of a generic integrand f.

2.1. Main results

Let $a > 0$ be a constant and let $\psi : [0, \infty) \to [0, \infty)$ be an increasing function such that

$$\psi(t) \to +\infty \text{ as } t \to \infty.$$

Denote by $|\cdot|$ the Euclidean norm in R^n. We consider the space of integrands $\mathcal{M}$ introduced in Section 1.1. This space consists of all continuous functions $f : [0, \infty) \times R^n \times R^n \to R^1$ which satisfy the following assumptions:

A(i) for each $(t, x) \in [0, \infty) \times R^n$ the function $f(t, x, \cdot) : R^n \to R^1$ is convex;

A(ii) the function f is bounded on $[0, \infty) \times E$ for any bounded set $E \subset R^n \times R^n$;

A(iii)

$$f(t, x, u) \geq \max\{\psi(|x|), \psi(|u|)|u|\} - a$$

for each $(t, x, u) \in [0, \infty) \times R^n \times R^n$;

A(iv) for each pair of positive numbers M, ϵ there exist $\Gamma, \delta > 0$ such that if $t \in [0, \infty)$ and if $u_1, u_2, x_1, x_2 \in R^n$ satisfy

$$|x_i| \leq M, \ |u_i| \geq \Gamma, \ i = 1, 2, \quad \max\{|x_1 - x_2|, |u_1 - u_2|\} \leq \delta,$$

then

$$|f(t, x_1, u_1) - f(t, x_2, u_2)| \leq \epsilon \max\{f(t, x_1, u_1), f(t, x_2, u_2)\};$$

A(v) for each pair of positive numbers M, ϵ there is a positive number δ such that if $t \in [0, \infty)$ and if $u_1, u_2, x_1, x_2 \in R^n$ satisfy

$$|x_i|, |u_i| \leq M, \ i = 1, 2, \quad \max\{|x_1 - x_2|, |u_1 - u_2|\} \leq \delta,$$

then

$$|f(t, x_1, u_1) - f(t, x_2, u_2)| \leq \epsilon.$$

We equip the set $\mathcal{M}$ with two topologies where one is weaker than the other. We refer to them as the weak and the strong topologies, respectively. For the set $\mathcal{M}$ we consider the uniformity determined by the following base:

$$E_s(\epsilon) = \{(f, g) \in \mathcal{M} \times \mathcal{M} : |f(t, x, u) - g(t, x, u)| \leq \epsilon$$

for each $t \in [0, \infty)$ and each $x, u \in R^n\},$

where $\epsilon > 0$. It is not difficult to see that the uniform space $\mathcal{M}$ with this uniformity is metrizable and complete. This uniformity generates in $\mathcal{M}$ the strong topology.

We also equip the set $\mathcal{M}$ with the uniformity which is determined by the following base:

$$E(N, \epsilon, \lambda) = \{(f, g) \in \mathcal{M} \times \mathcal{M} : |f(t, x, u) - g(t, x, u)| \leq \epsilon$$

for each $t \in [0, \infty)$ and each $x, u \in R^n$ satisfying $|x|, |u| \leq N$,

$$(|f(t, x, u)| + 1)(|g(t, x, u)| + 1)^{-1} \in [\lambda^{-1}, \lambda]$$

for each $t \in [0, \infty)$ and each $x, u \in R^n$ satisfying $|x| \leq N\},$

where $N > 0$, $\epsilon > 0$, $\lambda > 1$. This uniformity which was introduced in Section 1.2, generates in $\mathcal{M}$ the weak topology. By Proposition 1.3.2 the space $\mathcal{M}$ with this uniformity is complete.

We consider functionals of the form

$$I^f(T_1, T_2, x) = \int_{T_1}^{T_2} f(t, x(t), x'(t))dt \qquad (1.1)$$

where $f \in \mathcal{M}$, $0 \le T_1 < T_2 < +\infty$ and $x : [T_1, T_2] \to R^n$ is an a.c. function.

For each $f \in \mathcal{M}$, each pair of vectors $y, z \in R^n$, each $T_1 \ge 0$ and each $T_2 > T_1$ we set

$$U^f(T_1, T_2, y, z) = \inf\{I^f(T_1, T_2, x) : x : [T_1, T_2] \to R^n \qquad (1.2)$$

$$\text{is an a.c. function satisfying } x(T_1) = y, \ x(T_2) = z\},$$

$$\sigma^f(T_1, T_2, y) = \inf\{U^f(T_1, T_2, y, u) : u \in R^n\}. \qquad (1.3)$$

It is not difficult to see that $U^f(T_1, T_2, y, z)$ is finite for each $f \in \mathcal{M}$, each $y, z \in R^n$ and all numbers T_1, T_2 satisfying $0 \le T_1 < T_2$.

Recall the definition of an overtaking optimal function given in Section 1.1 and the definition of a good function introduced in Section 1.2.

Let $f \in \mathcal{M}$. An a.c. function $x : [0, \infty) \to R^n$ is called (f)-overtaking optimal if for any a.c. function $y : [0, \infty) \to R^n$ satisfying $y(0) = x(0)$,

$$\limsup_{T \to \infty} \int_0^T [f(t, x(t), x'(t)) - f(t, y(t), y'(t))]dt \le 0.$$

Let $f \in \mathcal{M}$. We say that an a.c. function $x : [0, \infty) \to R^n$ is an (f)-*good function* if for any a.c. function $y : [0, \infty) \to R^n$, the function

$$T \to I^f(0, T, y) - I^f(0, T, x), \ T \in (0, \infty)$$

is bounded from below.

In this chapter we establish the existence of a set $\mathcal{F} \subset \mathcal{M}$ which is a countable intersection of open (in the weak topology) everywhere dense (in the strong topology) subsets of $\mathcal{M}$ such that the following theorems are valid.

THEOREM 2.1.1 1. *For each $g \in \mathcal{F}$ and each pair of (g)-good functions $v_i : [0, \infty) \to R^n$, $i = 1, 2$,*

$$|v_2(t) - v_1(t)| \to 0 \ as \ t \to \infty.$$

2. *For each $g \in \mathcal{F}$ and each $y \in R^n$ there exists a (g)-overtaking optimal function $Y : [0, \infty) \to R^n$ satisfying $Y(0) = y$.*

3. Let $g \in \mathcal{F}$, $\epsilon > 0$ and $Y : [0, \infty) \to R^n$ be a (g)-overtaking optimal function. Then there exists a neighborhood $\mathcal{U}$ of g in $\mathcal{M}$ with the weak topology such that the following property holds:

If $h \in \mathcal{U}$ and if $v : [0, \infty) \to R^n$ is an (h)-good function, then

$$|v(t) - Y(t)| \leq \epsilon \text{ for all large } t.$$

THEOREM 2.1.2 *Let $g \in \mathcal{F}$, $M, \epsilon > 0$ and let $Y : [0, \infty) \to R^n$ be a (g)-overtaking optimal function. Then there exists a neighborhood $\mathcal{U}$ of g in $\mathcal{M}$ with the weak topology and a number $\tau > 0$ such that for each $h \in \mathcal{U}$ and each (h)-overtaking optimal function $v : [0, \infty) \to R^n$ satisfying $|v(0)| \leq M$,*

$$|v(t) - Y(t)| \leq \epsilon \text{ for all } t \in [\tau, \infty).$$

Theorems 2.1.1 and 2.1.2 establish the existence of (g)-overtaking optimal functions and describe the asymptotic behavior of (g)-good functions for $g \in \mathcal{F}$.

THEOREM 2.1.3 *Let $g \in \mathcal{F}$, $S_1, S_2, \epsilon > 0$ and let $Y : [0, \infty) \to R^n$ be a (g)-overtaking optimal function. Then there exists a neighborhood $\mathcal{U}$ of g in $\mathcal{M}$ with the weak topology, a number $L > 0$ and an integer $Q \geq 1$ such that if $h \in \mathcal{U}$, $T_1 \in [0, \infty)$, $T_2 \in [T_1 + LQ, \infty)$ and if an a.c. function $v : [T_1, T_2] \to R^n$ satisfies one of the following relations below:*

(a) $|v(T_i)| \leq S_1$, $i = 1, 2$, $I^h(T_1, T_2, v) \leq U^h(T_1, T_2, v(T_1), v(T_2)) + S_2$;

(b) $|v(T_1)| \leq S_1$, $I^h(T_1, T_2, v) \leq \sigma^h(T_1, T_2, v(T_1)) + S_2$,

then the following property holds:

There exist sequences of numbers $\{d_i\}_{i=1}^q$, $\{b_i\}_{i=1}^q \subset [T_1, T_2]$ such that

$$q \leq Q, \ b_i < d_i \leq b_i + L, \ i = 1, \ldots, q,$$

$$|v(t) - Y(t)| \leq \epsilon \text{ for each } t \in [T_1, T_2] \setminus \cup_{i=1}^q [b_i, d_i].$$

THEOREM 2.1.4 *Let $g \in \mathcal{F}$, $S, \epsilon > 0$ and let $Y : [0, \infty) \to R^n$ be a (g)-overtaking optimal function. Then there exist a neighborhood $\mathcal{U}$ of g in $\mathcal{M}$ with the weak topology and numbers $\delta, L > 0$ such that for each $h \in \mathcal{U}$, each pair of numbers $T_1 \in [0, \infty)$, $T_2 \in [T_1 + 2L, \infty)$ and each a.c. function $v : [T_1, T_2] \to R^n$ which satisfies one of the following relations below:*

(a) $|v(T_i)| \leq S$, $i = 1, 2$, $I^h(T_1, T_2, v) \leq U^h(T_1, T_2, v(T_1), v(T_2)) + \delta$;

(b) $|v(T_1)| \leq S$, $I^h(T_1, T_2, v) \leq \sigma^h(T_1, T_2, v(T_1)) + \delta$

the inequality $|v(t) - Y(t)| \leq \epsilon$ is valid for all $t \in [T_1 + L, T_2 - L]$.

Theorem 2.1.4 establishes the turnpike property for any $g \in \mathcal{F}$. The results of this chapter have been established in [103]. In the sequel we use the notation

$$B(x, r) = \{y \in R^n : |y - x| \leq r\}, \ x \in R^n, \ r > 0, \tag{1.4}$$

$$B(r) = B(0, r), \ r > 0.$$

Chapter 2 is organized as follows. In Section 2.2 for a given $f \in \mathcal{M}$ and a given neighborhood of f in $\mathcal{M}$ with the strong topology we construct an integrand f^* which belongs to this neighborhood and establishes the turnpike property for f^*. We also study the structure of approximate solutions of variational problems with integrands belonging to a small neighborhood of f^* in the weak topology. Theorems 2.1.1-2.1.4 are proved in Section 2.3. In Section 2.4 we discuss analogs of Theorems 2.1.1-2.1.4 for a class of periodic variational problems. In Section 2.5 we show that Theorems 2.1.1-2.1.4 also hold for certain subspaces of $\mathcal{M}$ which consist of smooth integrands. In Section 2.6 we consider an example of an integrand which has the turnpike property and an example of an integrand which does not have the turnpike property.

2.2. Preliminary lemmas

Fix $f \in \mathcal{M}$ and $z_* \in R^n$. Let $\epsilon > 0$, $M > |z_*|$ and let an a.c. function $Z_*^f : [0, \infty) \to R^n$ be as guaranteed by Theorem 1.2.1. We have that Z_*^f is an (f)-good function, $Z_*^f(0) = z_*$ and for each $T_1 \geq 0$, $T_2 > T_1$,

$$U^f(T_1, T_2, Z_*^f(T_1), Z_*^f(T_2)) = I^f(T_1, T_2, Z_*^f). \tag{2.1}$$

First we define functions f_r^M for $r > 0$ such that $f_r^M \to f$ as $r \to 0^+$ in the strong topology and such that each f_r^M has the turnpike property.

Fix a continuous bounded function $\phi^M : [0, \infty) \times R^n \to [0, \infty)$ which satisfies the following assumptions:

B(i) $\{(t, x) \in [0, \infty) \times R^n : \phi^M(t, x) = 0\} = \{(t, Z_*^f(t)) :$
$$t \in [0, \infty)\} \cup \{(t, x) \in [0, \infty) \times R^n : |x| \geq M + 2\};$$

B(ii) for any $\delta > 0$ there is a positive number γ such that if $t \in [0, \infty)$ and if $x_1, x_2 \in R^n$ satisfy $|x_1 - x_2| \leq \gamma$, then

$$|\phi^M(t, x_1) - \phi^M(t, x_2)| \leq \delta;$$

B(iii) for any positive number δ there is $\gamma > 0$ such that if $t \in [0, \infty)$ and if $x \in R^n$ satisfies

$$|x - Z_*^f(t)| \geq \delta \text{ and } |x| \leq M + 1,$$

then $\phi^M(t, x) \geq \gamma$.

REMARK 2.2.1 *Consider a continuous function* $\theta : R^1 \to [0, 1]$ *for which*

$$\theta(t) = 1, \ t \in (-\infty, M+1], \ \theta(t) = 0, \ t \in [M+2, \infty),$$

$$\theta(t) > 0, \ t \in (M+1, M+2).$$

Let q be a natural number.
   Define a bounded continuous function $\phi^M : [0, \infty) \times R^n \to R^1$ *by*

$$\phi^M(t, x) = |x - Z_*^f(t)|^q \theta(|x|), \quad t \in [0, \infty), \ x \in R^n. \qquad (2.2).$$

It is easy to verify that the function ϕ^M *satisfies assumption (B).*

   Define a function $f_\epsilon^M : [0, \infty) \times R^n \times R^n \to R^1$ by

$$f_\epsilon^M(t, x, u) = f(t, x, u) + \epsilon \phi^M(t, x), \ t \in [0, \infty), \ x, u \in R^n. \qquad (2.3)$$

It is easy to verify that $f_\epsilon^M \in \mathcal{M}$ and to prove the following result.

LEMMA 2.2.1 *Let $M > |z_*|$ and V be a neighborhood of f in $\mathcal{M}$ with the strong topology. Then there exists a number $r_0 > 0$ such that $f_r^M \in V$ for every number $r \in (0, r_0)$.*

   Fix a natural number p. It follows from Theorem 1.2.1 and Theorem 1.2.2 that there exist a number $M^f > 0$ and a neighborhood W^f of f in $\mathcal{M}$ with the weak topology such that

$$|z_*|, \ \sup\{|Z_*^f(t)| : t \in [0, \infty)\} < M^f \qquad (2.4)$$

and

$$\limsup_{t \to \infty} |x(t)| < M^f \qquad (2.5)$$

for each $g \in W^f$ and each (g)-good function $x : [0, \infty) \to R^n$. There exist a positive number $M_0(f, p)$ and an open neighborhood $W_0(f, p)$ of f in $\mathcal{M}$ with the weak topology such that

$$W_0(f, p) \subset W^f, \ M_0(f, p) > 2M^f + 2p + 2 \qquad (2.6)$$

and Theorem 1.2.3 holds with

$$M_1, M_2 = 2M^f + 2p + 2, c = 4^{-1}, \ S = M_0(f, p), \ \mathcal{U} = W_0(f, p). \qquad (2.7)$$

There exists a neighborhood $W(f, p)$ of f in $\mathcal{M}$ with the weak topology and a number $M(f, p)$ such that

$$W(f, p) \subset W_0(f, p), \ M(f, p) > 2M_0(f, p) + 2 \tag{2.8}$$

and Theorem 1.2.3 holds with

$$M_1, M_2 = 2M_0(f, p) + 2, \ c = 4^{-1}, \ S = M(f, p), \ \mathcal{U} = W(f, p). \tag{2.9}$$

It follows from Lemma 2.2.1 that there is a positive number $r(f, p)$ such that

$$f_r^{M(f,p)+1} \in W(f, p) \text{ for each } r \in (0, r(f, p)). \tag{2.10}$$

Fix $r \in (0, r(f, p))$ and set

$$f^* = f_r^{M(f,p)+1}. \tag{2.11}$$

We study the structure of approximate solutions of variational problems with integrands belonging to a small neighborhood of f^* in the strong topology. We show that the integrand f^* has the turnpike property and the function Z_*^f is its turnpike. The next lemma establishes that each approximate solution defined on an interval $[T_1, T_2]$ is close enough to the turnpike Z_*^f at a certain point of $[T_1, T_2]$ if the integrand is close enough to f^* and $T_2 - T_1$ is large enough.

LEMMA 2.2.2 *Let $\epsilon_0 \in (0, 1)$. Then there exist a neighborhood $\mathcal{U}$ of f^* in $\mathcal{M}$ with the weak topology and an integer $N \geq 8$ such that if $g \in \mathcal{U}$, $T \geq 0$ and if an a.c. function $v : [T, T + N] \to R^n$ satisfies*

$$\max\{|v(T)|, |v(T + N)|\} \leq 2M_0(f, p) + 2, \tag{2.12}$$

$$I^g(T, T + N, v) \leq U^g(T, T + N, v(T), v(T + N)) + 2M_0(f, p) + 2,$$

then there is an integer $i_0 \in [0, N - 6]$ such that

$$|v(t) - Z_*^f(t)| \leq \epsilon_0, \ t \in [i_0 + T, i_0 + T + 6]. \tag{2.13}$$

Proof. It follows from Proposition 1.3.6 that there exist a positive number S_0 and an open neighborhood $\mathcal{U}_0$ of f^* in $\mathcal{M}$ with the weak topology such that

$$\mathcal{U}_0 \subset W(f, p), \ |U^g(T_1, T_2, y_1, y_2)| + 1 < S_0 \text{ for each } g \in \mathcal{U}_0,$$

$$\text{each } T_1 \in [0, \infty), \ T_2 \in [T_1 + 4^{-1}, T_1 + 8] \tag{2.14}$$

$$\text{and each } y_1, y_2 \in B(M(f, p) + 1), \ i = 1, 2.$$

By Theorem 1.2.1 there is a positive number S_1 such that the inequality

$$I^f(T_1, T_2, Z_*^f) \leq I^f(T_1, T_2, v) + S_1$$

holds for each $T_1 \geq 0$, $T_2 > T_1$ and each a.c. function $v : [T_1, T_2] \to R^n$ which satisfies $|v(T_1)| \leq M(f, p) + 1$.

(2.4), Proposition 1.3.6 and Assertion 4 of Theorem 1.2.1 imply that there exists

$$S_2 > \sup\{|I^f(T_1, T_2, Z_*^f)| :$$

$$T_1 \in [0, \infty),\ T_2 \in [T_1 + 4^{-1}, T_1 + 8]\} + 2M(f, p). \tag{2.15}$$

By Proposition 1.3.4 there exists $\delta \in (0, 8^{-1})$ such that for each $g \in \mathcal{M}$, each $T_1, T_2 \in [0, \infty)$ satisfying $4^{-1} \leq T_2 - T_1 \leq 8$ and each a.c. function $v : [T_1, T_2] \to R^n$ satisfying

$$I^g(T_1, T_2, v) \leq 2S_0 + 2S_1 + 2S_2 + 2, \tag{2.16}$$

the following property holds:
  If $t_1, t_2 \in [T_1, T_2]$ and if $|t_1 - t_2| \leq \delta$, then

$$|v(t_1) - v(t_2)| \leq 16^{-1}\epsilon_0. \tag{2.17}$$

There exists a number $\epsilon_1 \in (0, 4^{-1}\epsilon_0)$ such that Assumption B(iii) holds with $M = M(f, p) + 1$, $\delta = 4^{-1}\epsilon_0$, $\gamma = \epsilon_1$. Fix a natural number $N > 48$ for which

$$4^{-1}(6^{-1}N - 8)\delta\epsilon_1 r > 2M(f, p) + 2S_0 + 6a + 4 + S_1 \tag{2.18}$$

(recall a in Assumption A(iii)). By Proposition 1.3.6 there exist a number $S_3 > 0$ and a neighborhood $\mathcal{U}_1$ of f^* in $\mathcal{M}$ with the weak topology such that

$$\mathcal{U}_1 \subset \mathcal{U}_0,$$

$$|U^g(T_1, T_2, y_1, y_2)| + 1 < S_3 \text{ for each } g \in \mathcal{U}_1 \text{ each } T_1 \in [0, \infty), \tag{2.19}$$

$T_2 \in [T_1 + 4^{-1}, T_1 + N + 4]$ and each $y_1, y_2 \in B(M(f, p) + 2)$, $i = 1, 2$.

By Propositions 1.3.8 and 1.3.9 there is an open neighborhood $\mathcal{U}$ of f^* in $\mathcal{M}$ with the weak topology such that $\mathcal{U} \subset \mathcal{U}_1$ and that for each $g \in \mathcal{U}$ and each $T_1 \in [0, \infty)$, $T_2 \in [T_1 + 4^{-1}, T_1 + N + 4]$ the following properties hold:
  a) if an a.c. function $v : [T_1, T_2] \to R^n$ satisfies

$$\min\{I^{f^*}(T_1, T_2, v),\ I^g(T_1, T_2, v)\} \leq S_3 + 2M(f, p) + 2,$$

then $|I^{f^*}(T_1, T_2, v) - I^g(T_1, T_2, v)| \leq 4^{-1};$

b)
$$|U^{f^*}(T_1, T_2, y_1, y_2) - U^g(T_1, T_2, y_1, y_2)| \leq 4^{-1}$$

for each $y_1, y_2 \in B(M(f,p) + 2)$, $i = 1, 2$.

Assume that $g \in \mathcal{U}$, $T \in [0, \infty)$ and $v : [T, T + N] \to R^n$ is an a.c. function which satisfies (2.12). We show that (2.13) holds with an integer $i_0 \in [0, N - 6]$.

Let us assume the converse. Then for any integer $i \in [0, N - 6]$,

$$\sup\{|v(t) - Z_*^f(t)| : t \in [i + T, i + T + 6]\} > \epsilon_0. \tag{2.20}$$

By the inequalities (2.12), Theorem 1.2.3 and the choice of $W(f,p)$, $M(f,p)$ (see (2.8) and (2.9)),

$$|v(t)| \leq M(f,p), \ t \in [T, T + N]. \tag{2.21}$$

(2.12), (2.19) and (2.21) imply that

$$I^g(T, T + N, v) \leq U^g(T, T + N, v(T), v(T + N)) + 2M_0(f,p) + 2$$

$$\leq 2M_0(f,p) + S_3 + 2. \tag{2.22}$$

It follows from property (b) and the inequality (2.12) that

$$|U^{f^*}(T, T + N, v(T), v(T + N)) - U^g(T, T + N, v(T), v(T + N))| \leq 4^{-1}. \tag{2.23}$$

Combined with the inequality (2.22) the property (a) implies that

$$|I^{f^*}(T, T + N, v) - I^g(T, T + N, v)| \leq 4^{-1}. \tag{2.24}$$

By (2.12), (2.23) and (2.24),

$$I^{f^*}(T, T+N, v) \leq U^{f^*}(T, T+N, v(T), v(T+N)) + 2M_0(f,p) + 3. \tag{2.25}$$

There exists an integer j_1 such that

$$j_1 - 2 < T \leq j_1 - 1. \tag{2.26}$$

Fix an integer $i \in [0, N - 6]$. By (2.20) there exists a number t_i such that

$$t_i \in [i + T, i + T + 6], \ |v(t_i) - Z_*^f(t_i)| > \epsilon_0. \tag{2.27}$$

The inequality (2.15) implies that

$$|I^f(T + i, T + i + 6, Z_*^f)| \leq S_2. \tag{2.28}$$

It follows from (2.12), (2.21) and (2.14) that

$$I^g(T + i, T + i + 6, v)$$

$$\leq U^g(T+i, T+i+6, v(T+i), v(T+i+6)) + 2M_0(f,p) + 2 \quad (2.29)$$
$$\leq 2M_0(f,p) + S_0 + 2.$$

By (2.28), (2.29), (2.15), (2.27) and the definition of δ (see (2.16), (2.17)), for each

$$t \in [i+T, i+T+6] \cap [t_i - \delta, t_i + \delta] \quad (2.30)$$

the inequalitites

$$|v(t_i) - v(t)| \leq 16^{-1}\epsilon_0, \ |Z_*^f(t_i) - Z_*^f(t)| \leq 16^{-1}\epsilon_0, \ |v(t) - Z_*^f(t)| \geq 3 \cdot 4^{-1}\epsilon_0$$

are true. By these inequalities, the definition of ϵ_1, (2.21) and assumption B(iii),

$$\phi^{M(f,p)+1}(t, v(t)) \geq \epsilon_1 \quad (2.31)$$

for each integer $i \in [0, N-6]$ and each number t sastisfying (2.30). (2.11), (2.3) and (2.21) imply that

$$I^{f^*}(T, T+N, v) = I^f(T, T+N, v) + r \int_T^{T+N} \phi^{M(f,p)+1}(t, v(t))dt.$$

By this equality and (2.31) which is true for each integer $i \in [0, N-6]$ and each t satisfying (2.30),

$$I^{f^*}(T, T+N, v) \geq I^f(T, T+N, v) + r(N6^{-1} - 2)\delta\epsilon_1. \quad (2.32)$$

By Corollary 1.3.1 there exists an a.c. function $w : [T, T+N] \to R^n$ for which

$$w(T) = v(T), \ w(T+N) = v(T+N), \ w(t) = Z_*^f(t), \ t \in [j_1, j_1 + N - 3],$$
$$(2.33)$$
$$I^{f^*}(T, j_1, w) = U^{f^*}(T, j_1, w(T), w(j_1)), \quad I^{f^*}(j_1 + N - 3, T+N, w)$$
$$= U^{f^*}(j_1 + N - 3, T+N, w(j_1 + N - 3), w(T+N)).$$

Combined with the inequality (2.25) the relations (2.33) imply that

$$I^{f^*}(T, T+N, v) \leq I^{f^*}(T, T+N, w) + 2M_0(f,p) + 3. \quad (2.34)$$

By (2.33), (2.21), (2.4) and (2.14),

$$I^{f^*}(T, T+N, w) \leq I^{f^*}(j_1, j_1 + N - 3, Z_*^f) + 2S_0. \quad (2.35)$$

It follows from Assumption A(iii), the definition of S_1 and the inequalities (2.26) and (2.21) that

$$I^f(T, T+N, v) \geq I^f(j_1, j_1 + N - 3, v) - 6a$$

$$\geq -6a + I^f(j_1, j_1 + N - 3, Z_*^f) - S_1. \tag{2.36}$$

Combined with (2.32) and (2.35) the inequality (2.36) implies that

$$I^{f^*}(T, T+N, v) \geq r(N6^{-1} - 2)\delta\epsilon_1 - 6a + I^f(j_1, j_1 + N - 3, Z_*^f) - S_1$$

$$\geq I^{f^*}(T, T+N, w) + r(N6^{-1} - 2)\delta\epsilon_1 - 6a - S_1 - 2S_0.$$

Together with (2.34) this inequality implies that

$$2M_0(f, p) + 3 \geq r(N6^{-1} - 2)\delta\epsilon_1 - 6a - S_1 - 2S_0.$$

This is contradictory to (2.18). The contradiction we have reached proves the lemma.

The following auxiliary result shows that an approximate solution of a variational problem defined on an interval $[T_1, T_2]$ is close to the turnpike Z_*^f at any point of $[T_1, T_2]$ if it is close enough to the turnpike at the points T_1 and T_2.

LEMMA 2.2.3 *For each $\epsilon \in (0, 1)$ there is $\delta \in (0, \epsilon)$ such that the following property holds:*

If $T_1 \in [0, \infty)$, $T_2 \in [T_1 + 1, \infty)$ and if an a.c. function $v : [T_1, T_2] \to R^n$ satisfies

$$|v(T_i) - Z_*^f(T_i)| \leq \delta, \ i = 1, 2,$$

$$I^{f^*}(T_1, T_2, v) \leq U^{f^*}(T_1, T_2, v(T_1), v(T_2)) + \delta, \tag{2.37}$$

then

$$|v(t) - Z_*^f(t)| \leq \epsilon, \ t \in [T_1, T_2]. \tag{2.38}$$

Proof. Let $\epsilon \in (0, 1)$. It follows from Proposition 1.3.6 that there exists a positive number S_0 such that

$$|U^{f^*}(T_1, T_2, y_1, y_2)| + 6 < S_0 \text{ for each } T_1 \in [0, \infty), \ T_2 \in [T_1 + 4^{-1}, T_1 + 10] \tag{2.39}$$

$$\text{and each } y_1, y_2 \in B(M(f, p) + 1), \ i = 1, 2.$$

It follows from (2.4), Proposition 1.3.6 and Assertion 4 of Theorem 1.2.1 that there is a positive number S_1 such that

$$|I^f(T_1, T_2, Z_*^f)| + 1 < S_1 \text{ for each } T_1 \in [0, \infty), \ T_2 \in [T_1 + 4^{-1}, T_1 + 10]\}. \tag{2.40}$$

By Proposition 1.3.4 there is $\delta_0 \in (0, 8^{-1})$ such that for each $g \in \mathcal{M}$, each pair of numbers $T_1, T_2 \in [0, \infty)$ satisfying $4^{-1} \leq T_2 - T_1 \leq 10$ and each a.c. function $v : [T_1, T_2] \to R^n$ which satisfies

$$I^g(T_1, T_2, v) \leq 2S_0 + 2S_1 + 2, \tag{2.41}$$

the following property holds:

$$|v(t_1) - v(t_2)| \le 16^{-1}\epsilon \qquad (2.42)$$

for each $t_1, t_2 \in [T_1, T_2]$ satisfying $|t_1 - t_2| \le \delta_0$.

There exists a number $\epsilon_1 \in (0, 4^{-1}\epsilon)$ such that Assumption B(iii) holds with $M = M(f, p) + 1$, $\delta = 4^{-1}\epsilon$, $\gamma = \epsilon_1$. By Proposition 1.3.7 there exists a number

$$\delta \in (0, \inf\{\delta_0, 8^{-1}\epsilon_1, 8^{-1}\epsilon_1\delta_0 r\}) \qquad (2.43)$$

such that if $T_1 \ge 0$, $T_2 \in [T_1 + 4^{-1}, T_1 + 10]$ and if

$$y_i, x_i \in B(M(f, p) + 2), \ i = 1, 2, \ \max\{|y_1 - y_2|, |x_1 - x_2|\} \le \delta, \qquad (2.44)$$

then

$$|U^f(T_1, T_2, y_1, x_1) - U^f(T_1, T_2, y_2, x_2)|, \qquad (2.45)$$
$$|U^{f^*}(T_1, T_2, y_1, x_1) - U^{f^*}(T_1, T_2, y_2, x_2)| \le 2^{-7}\epsilon_1\delta_0 r.$$

Assume that $T_1 \in [0, \infty)$, $T_2 \in [T_1 + 1, \infty)$ and an a.c. function $v : [T_1, T_2] \to R^n$ satisfies (2.37). We show that the inequality (2.38) is valid.

Let us assume the converse. Then there exists a number t_1 for which

$$t_1 \in [T_1, T_2], \ |v(t_1) - Z_*^f(t_1)| > \epsilon. \qquad (2.46)$$

(2.37), (2.7), Theorem 1.2.3 and (2.4) imply that

$$|v(t)| \le M_0(f, p), \ t \in [T_1, T_2]. \qquad (2.47)$$

It is not difficult to see that there exist $d_1, d_2 \in [T_1, T_2]$ such that

$$d_2 - d_1 = 1, \ t_1 \in [d_1, d_2]. \qquad (2.48)$$

It follows from the inequalities (2.37), (2.47) and (2.39) that

$$I^{f^*}(d_1, d_2, v) \le U^{f^*}(d_1, d_2, v(d_1), v(d_2)) + \delta \le S_0 + 1. \qquad (2.49)$$

The inequality (2.40) implies that

$$I^f(d_1, d_2, Z_*^f) < S_1. \qquad (2.50)$$

By the choice of δ_0 (see (2.41), (2.42)), (2.50) and (2.49)

$$|v(t) - v(t_1)| \le 16^{-1}\epsilon, \ |Z_*^f(t) - Z_*^f(t_1)| \le 16^{-1}\epsilon$$

for each

$$t \in [d_1, d_2] \cap [t_1 - \delta_0, t_1 + \delta_0].$$

Combined with (2.46) this fact implies that

$$|v(t) - Z_*^f(t)| \geq 3 \cdot 4^{-1}\epsilon, \ t \in [d_1, d_2] \cap [t_1 - \delta_0, t_1 + \delta_0].$$

It follows from this inequality, assumption B(iii), the definition of ϵ_1 and the inequality (2.47) that

$$\phi^{M(f,p)+1}(t, v(t)) \geq \epsilon_1 \text{ for each } t \in [d_1, d_2] \cap [t_1 - \delta_0, t_1 + \delta_0]. \quad (2.51)$$

We prove that the inequality

$$|U^g(T_1, T_2, v(T_1), v(T_2)) - U^g(T_1, T_2, Z_*^f(T_1), Z_*^f(T_2))| \leq \delta + 32^{-1}\epsilon_1\delta_0 r \quad (2.52)$$

is true with $g = f, f^*$. Corollary 1.3.1 implies that for $g = f, f^*$ there exist a.c. functions $v_i^g : [T_1, T_2] \to R^n$, $i = 1, 2$ such that

$$v_1^g(T_i) = Z_*^f(T_i), \ i = 1, 2, \ v_1^g(t) = v(t), \ t \in [T_1 + 2^{-1}, T_2 - 2^{-1}], \quad (2.53)$$

$$I^g(S, S+2^{-1}, v_1^g) = U^g(S, S+2^{-1}, v_1^g(S), v_1^g(S+2^{-1})), \ S = T_1, T_2 - 2^{-1},$$

$$v_2^g(T_i) = v(T_i), \ i = 1, 2, \ v_2^g(t) = Z_*^f(t), \ t \in [T_1 + 2^{-1}, T_2 - 2^{-1}],$$

$$I^g(S, S+2^{-1}, v_2^g) = U^g(S, S+2^{-1}, v_2^g(S), v_2^g(S+2^{-1})), \ S = T_1, T_2 - 2^{-1}.$$

By the definition of δ (see (2.43)) and the inequalities (2.53), (2.47), (2.4) and (2.37),

$$|U^g(S, S+2^{-1}, v_2^g(S), v_2^g(S+2^{-1})) - U^g(S, S+2^{-1}, Z_*^f(S), Z_*^f(S+2^{-1}))| \quad (2.54)$$

$$\leq 2^{-7}\epsilon_1\delta_0 r, \quad g = f, f^*, \ S = T_1, T_2 - 2^{-1},$$

$$|U^g(S, S+2^{-1}, v_1^g(S), v_1^g(S+2^{-1})) - U^g(S, S+2^{-1}, v(S), v(S+2^{-1}))| \quad (2.55)$$

$$\leq 2^{-7}\epsilon_1\delta_0 r, \quad g = f, f^*, \ S = T_1, T_2 - 2^{-1}.$$

It follows from (2.3), (2.53), the inequalities (2.54) and (2.55) and Assertion 4 of Theorem 1.2.1 that for $g = f, f^*$,

$$U^g(T_1, T_2, v(T_1), v(T_2)) - U^g(T_1, T_2, Z_*^f(T_1), Z_*^f(T_2)) \quad (2.56)$$

$$\leq I^g(T_1, T_2, v_2^g) - I^g(T_1, T_2, Z_*^f)$$

$$= I^g(T_1, T_1 + 2^{-1}, v_2^g) - I^g(T_1, T_1 + 2^{-1}, Z_*^f)$$

$$+ I^g(T_2 - 2^{-1}, T_2, v_2^g) - I^g(T_2 - 2^{-1}, T_2, Z_*^f)$$

$$= U^g(T_1, T_1 + 2^{-1}, v_2^g(T_1), v_2^g(T_1 + 2^{-1}))$$

$$- U^g(T_1, T_1 + 2^{-1}, Z_*^f(T_1), Z_*^f(T_1 + 2^{-1}))$$

$$+U^g(T_2 - 2^{-1}, T_2, v_2^g(T_2 - 2^{-1}), v_2^g(T_2))$$
$$-U^g(T_2 - 2^{-1}, T_2, Z_*^f(T_2 - 2^{-1}), Z_*^f(T_2))$$
$$\leq 2^{-6}\epsilon_1\delta_0 r.$$

(2.55), (2.37) and (2.53) imply that

$$U^{f^*}(T_1, T_2, v(T_1), v(T_2)) - U^{f^*}(T_1, T_2, Z_*^f(T_1), Z_*^f(T_2)) \qquad (2.57)$$

$$\geq I^{f^*}(T_1, T_2, v) - \delta - I^{f^*}(T_1, T_2, v_1^{f^*}) = -\delta + I^{f^*}(T_1, T_1 + 2^{-1}, v)$$
$$-I^{f^*}(T_1, T_1 + 2^{-1}, v_1^{f^*}) + I^{f^*}(T_2 - 2^{-1}, T_2, v) - I^{f^*}(T_2 - 2^{-1}, T_2, v_1^{f^*})$$
$$\geq -\delta + U^{f^*}(T_1, T_1 + 2^{-1}, v(T_1), v(T_1 + 2^{-1}))$$
$$-U^{f^*}(T_1, T_1 + 2^{-1}, v_1^{f^*}(T_1), v_1^{f^*}(T_1 + 2^{-1}))$$
$$+U^{f^*}(T_2 - 2^{-1}, T_2, v(T_2 - 2^{-1}), v(T_2))$$
$$-U^{f^*}(T_2 - 2^{-1}, T_2, v_1^{f^*}(T_2 - 2^{-1}), v_1^{f^*}(T_2)) \geq -\delta - 2^{-6}\epsilon_1\delta_0 r.$$

It follows from Corollary 1.3.1 that there are a.c. functions $v_i : [T_1, T_2] \to R^n$, $i = 3, 4$ such that

$$v_3(T_i) = v(T_i), \ i = 1, 2, \quad I^f(T_1, T_2, v_3) = U^f(T_1, T_2, v(T_1), v(T_2)),$$
$$(2.58)$$
$$v_4(T_i) = Z_*^f(T_i), \ i = 1, 2, \ v_4(t) = v_3(t), \ t \in [T_1 + 2^{-1}, T_2 - 2^{-1}],$$
$$I^f(S, S + 2^{-1}, v_4) = U^f(S, S + 2^{-1}, v_4(S), v_4(S + 2^{-1})), \ S = T_1, T_2 - 2^{-1}.$$

By Theorem 1.2.3, (2.9), (2.58) and (2.47),

$$|v_3(t)| \leq M(f, p), \ t \in [T_1, T_2]. \qquad (2.59)$$

By the definition of δ (see (2.45) and (2.44)), (2.58), (2.37), (2.4) and (2.59),

$$|U^f(S, S + 2^{-1}, v_3(S), v_3(S + 2^{-1})) - U^f(S, S + 2^{-1}, v_4(S), v_4(S + 2^{-1}))|$$
$$(2.60)$$
$$\leq 2^{-7}\epsilon_1\delta_0 r, \ S = T_1, T_2 - 2^{-1}.$$

Combined with (2.58) the inequality (2.60) implies that

$$U^f(T_1, T_2, v(T_1), v(T_2)) - U^f(T_1, T_2, Z_*^f(T_1), Z_*^f(T_2))$$
$$\geq I^f(T_1, T_2, v_3) - I^f(T_1, T_2, v_4)$$
$$= U^f(T_1, T_1 + 2^{-1}, v_3(T_1), v_3(T_1 + 2^{-1}))$$
$$+U^f(T_2 - 2^{-1}, T_2, v_3(T_2 - 2^{-1}), v_3(T_2))$$

$$-U^f(T_1, T_1 + 2^{-1}, v_4(T_1), v_4(T_1 + 2^{-1}))$$
$$-U^f(T_2 - 2^{-1}, T_2, v_4(T_2 - 2^{-1}), v_4(T_2))$$
$$\geq -2^{-6}\epsilon_1\delta_0 r.$$

By these relations, (2.57) and (2.56) which holds for $g = f, f^*$, the inequality (2.52) is valid with $g = f, f^*$. Combined with Assertion 4 of Theorem 1.2.1, (2.3) and (2.37) this implies that

$$U^f(T_1, T_2, v(T_1), v(T_2)) \geq U^f(T_1, T_2, Z_*^f(T_1), Z_*^f(T_2)) - \delta - 32^{-1}\epsilon_1\delta_0 r \tag{2.61}$$
$$= I^f(T_1, T_2, Z_*^f) - \delta - 32^{-1}\epsilon_1\delta_0 r$$
$$\geq U^{f^*}(T_1, T_2, Z_*^f(T_1), Z_*^f(T_2)) - \delta - 32^{-1}\epsilon_1\delta_0 r$$
$$\geq U^{f^*}(T_1, T_2, v(T_1), v(T_2)) - 2(\delta + 32^{-1}\epsilon_1\delta_0 r)$$
$$\geq I^{f^*}(T_1, T_2, v) - \delta - 2(\delta + 32^{-1}\epsilon_1\delta_0 r).$$

It follows from (2.3), (2.11), (2.47) and (2.51) that

$$I^{f^*}(T_1, T_2, v) \geq I^f(T_1, T_2, v) + \epsilon_1\delta_0 r.$$

This inequality and (2.61) imply that

$$U^f(T_1, T_2, v(T_1), v(T_2)) \geq I^f(T_1, T_2, v) + \epsilon_1\delta_0 r - \delta - 2(\delta + 32^{-1}\epsilon_1\delta_0 r)$$
$$\geq U^f(T_1, T_2, v(T_1), v(T_2)) - 3\delta + 15 \cdot 16^{-1}\epsilon_1\delta_0 r.$$

This is contradictory to (2.43). The contradiction we have reached proves the lemma.

Now we will prove an auxiliary result which generalizes Lemma 2.2.3. This result shows that the convergence property established in Lemma 2.2.3 for the integrand f^* is also valid for all integrands from a small neighborhood of f^*.

LEMMA 2.2.4 *For each $\epsilon \in (0,1)$ there exist an open neighborhood $\mathcal{U}$ of f^* in $\mathcal{M}$ with the weak topology and $\delta \in (0, \epsilon)$ such that the following property holds:*
If $g \in \mathcal{U}$, $T_1 \in [0, \infty)$, $T_2 \in [T_1 + 1, \infty)$ and if an a.c. function $v : [T_1, T_2] \to R^n$ satisfies

$$|v(T_i) - Z_*^f(T_i)| \leq \delta, \ i = 1, 2,$$

$$I^g(T_1, T_2, v) \leq U^g(T_1, T_2, v(T_1), v(T_2)) + \delta, \tag{2.62}$$

then

$$|v(t) - Z_*^f(t)| \leq \epsilon, \ t \in [T_1, T_2]. \tag{2.63}$$

Proof. Let $\epsilon \in (0,1)$. It follows from Lemma 2.2.3 that there exists $\delta \in (0, \epsilon)$ such that if $T_1 \in [0, \infty)$, $T_2 \in [T_1+1, \infty)$ and if an a.c. function $v : [T_1, T_2] \to R^n$ satisfies

$$|v(T_i) - Z_*^f(T_i)| \le 8\delta, \ i = 1, 2,$$

$$I^{f^*}(T_1, T_2, v) \le U^{f^*}(T_1, T_2, v(T_1), v(T_2)) + 8\delta, \qquad (2.64)$$

then

$$|v(t) - Z_*^f(t)| \le \epsilon, \ t \in [T_1, T_2]. \qquad (2.65)$$

Lemma 2.2.2 implies that there are an integer $N \ge 8$ and an open neighborhood $\mathcal{U}_0$ of f^* in $\mathcal{M}$ with the weak topology such that

$$\mathcal{U}_0 \subset W(f, p)$$

and for each $g \in \mathcal{U}_0$, each $T \ge 0$ and each a.c. function $v : [T, T+N] \to R^n$ the following property holds:

If

$$\sup\{|v(T)|, |v(T+N)|\} \le 2M_0(f, p) + 2, \qquad (2.66)$$

$$I^g(T, T+N, v) \le U^g(T, T+N, v(T), v(T+N)) + 2M_0(f, p) + 2,$$

then there exists an integer $i_0 \in [0, N-6]$ such that

$$|v(t) - Z_*^f(t)| \le \delta, \ t \in [i_0 + T, i_0 + T + 6]. \qquad (2.67)$$

By Proposition 1.3.6 there exist an open neighborhood $\mathcal{U}_1$ of f^* in $\mathcal{M}$ with the weak topology and a positive number S such that

$$\mathcal{U}_1 \subset \mathcal{U}_0, \ |U^g(T_1, T_2, y_1, y_2)| + 1 < S \ \text{for each } g \in \mathcal{U}_1, \ \text{each } T_1 \in [0, \infty), \qquad (2.68)$$

$T_2 \in [T_1 + 2^{-1}, T_1 + 8N + 8]$ and each $y_1, y_2 \in B(M(f, p) + 2)$, $i = 1, 2$.

It follows from Propositions 1.3.8 and 1.3.9 that there exist an open neighborhood $\mathcal{U}$ of f^* in $\mathcal{M}$ with the weak topology such that $\mathcal{U} \subset \mathcal{U}_1$ and that for each $g \in \mathcal{U}$, each $T_1 \in [0, \infty)$ and each $T_2 \in [T_1 + 4^{-1}, T_1 + 8N + 8]$ the following properties hold:

a) If an a.c. function $v : [T_1, T_2] \to R^n$ satisfies

$$\min\{I^{f^*}(T_1, T_2, v), I^g(T_1, T_2, v)\} \le 2S + 2M(f, p) + 4, \qquad (2.69)$$

then

$$|I^{f^*}(T_1, T_2, v) - I^g(T_1, T_2, v)| \le 4^{-1}\delta; \qquad (2.70)$$

b) If $y_1, y_2 \in B(M(f, p) + 2)$, $i = 1, 2$, then

$$|U^{f^*}(T_1, T_2, y_1, y_2) - U^g(T_1, T_2, y_1, y_2)| \le 4^{-1}\delta. \qquad (2.71)$$

Let $g \in \mathcal{U}$, $T_1 \in [0, \infty)$, $T_2 \in [T_1 + 1, \infty)$ and let an a.c. function $v : [T_1, T_2] \rightarrow R^n$ satisfy (2.62). We show that the inequality (2.63) is true.

We have two cases: (i) $T_2 - T_1 \leq 6N$; (ii) $T_2 - T_1 > 6N$. Consider the case (i). (2.4), (2.62) and (2.68) imply the inequality (2.69). The inequality (2.70) follows from (2.69) and property a). By property b) and (2.62),

$$|U^{f^*}(T_1, T_2, v(T_1), v(T_2)) - U^g(T_1, T_2, v(T_1), v(T_2))| \leq 4^{-1}\delta.$$

Combined with (2.62) and (2.70) this inequality implies that

$$I^{f^*}(T_1, T_2, v) \leq I^g(T_1, T_2, v)$$

$$+4^{-1}\delta \leq U^g(T_1, T_2, v(T_1), v(T_2)) + \delta + 4^{-1}\delta$$

$$\leq U^{f^*}(T_1, T_2, v(T_1), v(T_2)) + 3 \cdot 2^{-1}\delta.$$

It follows from this inequality, the inequality (2.62) and the definition of δ (see (2.64), (2.65)) that the inequality (2.63) is true.

Consider the case (ii). By Theorem 1.2.3, the definitions of $M_0(f, p)$ and $W_0(f, p)$ (see (2.7)) and the inequality (2.62),

$$|v(t)| \leq M_0(f, p) + 1, \ t \in [T_1, T_2]. \tag{2.72}$$

It follows from the choice of $\mathcal{U}_0$ and N (see (2.66) and (2.67)) and the inequalities (2.72) and (2.62) that the following property holds:

For each $\tau \in [T_1, T_2 - N]$ there exists an integer $i_\tau \in [0, N - 6]$ such that

$$|v(t) - Z_*^f(t)| \leq \delta, \ t \in [i_\tau + \tau, i_\tau + \tau + 6]. \tag{2.73}$$

It is easy to see that there exists a finite sequence $\{t_i\}_{i=0}^q \subset [T_1, T_2]$ such that

$$t_0 = T_1, \ t_q = T_2, \ t_{i+1} - t_i \in [6, N], \ i = 0, \ldots, q-2, \ t_q - t_{q-1} \in [N, 2N], \tag{2.74}$$

$$|v(t_i) - Z_*^f(t_i)| \leq \delta, \ i = 0, \ldots, q. \tag{2.75}$$

Fix an integer $i \in \{0, \ldots, q-1\}$. (2.62), (2.74), (2.75) and (2.68) imply that

$$I^g(t_i, t_{i+1}, v) \leq U^g(t_i, t_{i+1}, v(t_i), v(t_{i+1})) + \delta \leq \delta + S. \tag{2.76}$$

It follows from (2.76), (2.74), (2.72) and the properties a), b) that

$$|I^{f^*}(t_i, i_{i+1}, v) - I^g(t_i, t_{i+1}, v)| \leq 4^{-1}\delta,$$

$$|U^{f^*}(t_i, t_{i+1}, v(t_i), v(t_{i+1})) - U^g(t_i, t_{i+1}, v(t_i), v(t_{i+1}))| \leq 4^{-1}\delta,$$

$$I^{f^*}(t_i, i_{i+1}, v) \leq I^g(t_i, t_{i+1}, v) + 4^{-1}\delta$$

$$\leq U^g(t_i, t_{i+1}, v(t_i), v(t_{i+1})) + \delta + \delta/4 \leq U^{f^*}(t_i, t_{i+1}, v(t_i), v(t_{i+1})) + 3\delta/2. \tag{2.77}$$

Combined with the choice of δ (see (2.64), (2.65)) and (2.75) the inequality (2.77) implies that $|v(t) - Z_*^f(t)| \leq \epsilon$ for all $t \in [t_i, t_{i+1}]$. This completes the proof of the lemma.

We need the next lemma in order to establish the convergence property of Theorem 2.1.3. This lemma follows from Lemmas 2.2.4 and 2.2.2.

LEMMA 2.2.5 *For each $\epsilon \in (0,1)$ there exist an open neighborhood $\mathcal{U}$ of f^* in $\mathcal{M}$ with the weak topology, a number $l_* \geq 8$ and an integer $q_* \geq 4$ such that for each $g \in \mathcal{U}$, each $T_1 \in [0, \infty)$, $T_2 \in [T_1 + l_* q_*, \infty)$ and each a.c. function $v : [T_1, T_2] \to R^n$ the following property holds:*
   If

$$|v(t)| \leq 2M_0(f, p) + 2, \ t \in [T_1, T_2]$$

and

$$I^g(T_1, T_2, v) \leq U^g(T_1, T_2, v(T_1), v(T_2)) + p, \tag{2.78}$$

then there exist sequences of numbers $\{d_i\}_{i=1}^q$, $\{\bar{d}_i\}_{i=1}^q$ such that

$$q \leq q_*, \ d_i < \bar{d}_i \leq d_i + l_*, \ i = 1, \dots, q, \tag{2.79}$$

$$|v(t) - Z_*^f(t)| \leq \epsilon, \ t \in [T_1, T_2] \setminus \cup_{i=1}^q [d_i, \bar{d}_i]. \tag{2.80}$$

Proof. Let $\epsilon \in (0,1)$. It follows from Lemma 2.2.4 that there exist $\delta \in (0, \epsilon)$ and a neighborhood $\mathcal{U}_0$ of f^* in $\mathcal{M}$ with the weak topology such that the inequality (2.63) is true for each $g \in \mathcal{U}_0$, each $T_1 \in [0, \infty)$, $T_2 \in [T_1 + 1, \infty)$ and each a.c. function $v : [T_1, T_2] \to R^n$ which satisfies (2.62).

By Lemma 2.2.2 there exist a neighborhood $\mathcal{U}$ of f^* in $\mathcal{M}$ with the weak topology and an integer $N \geq 8$ such that $\mathcal{U} \subset \mathcal{U}_0$ and if $g \in \mathcal{U}$, $T \geq 0$ and if an a.c. function $v : [T, T + N] \to R^n$ satisfies (2.66) then there exists an integer $i_0 \in [0, N - 6]$ such that (2.67) is true. Fix an integer

$$q_* > 4 + \delta^{-1}p \text{ and a number } l_* \geq 2N. \tag{2.81}$$

Assume that $g \in \mathcal{U}$, $T_1 \in [0, \infty)$, $T_2 \in [T_1 + l_* q_*, \infty)$ and $v : [T_1, T_2] \to R^n$ is an a.c. function satisfying (2.78). It follows from the definition of $\mathcal{U}$ and (2.78) that for each $\tau \in [T_1, T_2 - N]$ there is an integer $i_\tau \in [0, N-6]$ such that

$$|v(t) - Z_*^f(t)| \leq \delta \text{ for all } t \in [i_\tau + \tau, i_\tau + \tau + 6].$$

This implies that there exists a sequence of numbers $t_0, \ldots, t_G$ such that

$$t_0 = T_1, \; t_G = T_2, \; t_{i+1} - t_i \in [3, N], \; i = 0, \ldots, G - 1, \qquad (2.82)$$

$$|v(t_i) - Z_*^f(t_i)| \le \delta, \; i = 1, \ldots, G - 1. \qquad (2.83)$$

Set

$$C = \{i \in \{1, \ldots, G - 2\} : \; I^g(t_i, t_{i+1}, v) > U^g(t_i, t_{i+1}, v(t_i), v(t_{i+1})) + \delta\} \qquad (2.84)$$

and denote by $\mathrm{Card}(C)$ the cardinality of C. By (2.78),

$$p \ge I^g(T_1, T_2, v) - U^g(T_1, T_2, v(T_1), v(T_2)) \ge \sum_{i \in C} [I^g(t_i, t_{i+1}, v) \qquad (2.85)$$

$$-U^g(t_i, t_{i+1}, v(t_i), v(t_{i+1}))] \ge \delta \, \mathrm{Card}(C), \qquad \mathrm{Card}(C) \le \delta^{-1} p.$$

Let $i \in \{1, \ldots, G - 2\} \setminus C$. It follows from the definition of δ, $\mathcal{U}_0$ and (2.84), (2.83), (2.82) that $|v(t) - Z_*^f(t)| \le \epsilon$ for all $t \in [t_i, t_{i+1}]$. Therefore

$$|v(t) - Z_*^f(t)| \le \epsilon,$$

$$t \in [t_i, t_{i+1}], \; i \in \{1, \ldots, G - 2\} \setminus C.$$

It is easy to see that

$$[T_1, T_2] \setminus \cup\{[t_i, t_{i+1}] : \; i \in \{1, \ldots, G - 2\} \setminus C\}$$

$$\subset \cup\{[t_i, t_{i+1}] : \; i \in C \cup \{0, G - 1\}\}$$

and by (2.82), (2.81), (2.85),

$$t_{i+1} - t_i \le N \le l_*, \; i = 0, \ldots, G - 1, \; \mathrm{Card}(C \cup \{0, G - 1\}) \le q_*.$$

This completes the proof of the lemma.

LEMMA 2.2.6 *Let $\epsilon \in (0, 1)$. Then there exist a neighborhood $\mathcal{U}$ of f^* in $\mathcal{M}$ with the weak topology, a number $l_* \ge 8$, an integer $q_* \ge 4$ such that for each $g \in \mathcal{U}$, each $T_1 \in [0, \infty)$, $T_2 \in [T_1 + l_* q_*, \infty)$ and each a.c. function $v : [T_1, T_2] \to R^n$ which satisfies one of the following relations below:*

$$(i) |v(T_i)| \le 2M^f + 2p + 2, \; i = 1, 2,$$

$$I^g(T_1, T_2, v) \le U^g(T_1, T_2, v(T_1), v(T_2)) + p;$$

$$(ii) |v(T_1)| \le 2M^f + 2p + 2, \; I^g(T_1, T_2, v) \le \sigma^g(T_1, T_2, v(T_1)) + p,$$

there are sequences of numbers $\{d_i\}_{i=1}^q$, $\{\bar{d}_i\}_{i=1}^q$ for which (2.79) and (2.80) hold.

Lemma 2.2.6, which is an extension of Lemma 2.2.5, now follows from Lemma 2.2.5, Theorem 1.2.3 and the definition of $W_0(f, p)$, $M_0(f, p)$ (see (2.6), (2.7)).

The next auxiliary result which follows from Lemmas 2.2.4 and 2.2.2 will be used in order to establish the convergence property of Theorems 2.1.1 and 2.1.2.

LEMMA 2.2.7 *Let $\epsilon \in (0, 1)$. Then there exist a neighborhood $\mathcal{U}$ of f^* in $\mathcal{M}$ with the weak topology such that, for each $g \in \mathcal{U}$ and each (g)-good function $v : [0, \infty) \to R^n$, the inequality $|v(t) - Z_*^f(t)| \leq \epsilon$ is valid for all sufficiently large t.*

Proof. It follows from Lemma 2.2.4 that there exist a number $\delta \in (0, \epsilon)$ and a neighborhood $\mathcal{U}_0$ of f^* in $\mathcal{M}$ with the weak topology such that the inequality (2.63) is true for each $g \in \mathcal{U}_0$, each $T_1 \in [0, \infty)$, $T_2 \in [T_1 + 1, \infty)$ and each a.c. function $v : [T_1, T_2] \to R^n$ which satisfies (2.62).

It follows from Lemma 2.2.2 that there exist a neighborhood $\mathcal{U}$ of f^* in $\mathcal{M}$ with the weak topology and an integer $N \geq 8$ such that

$$\mathcal{U} \subset \mathcal{U}_0 \cap W(f, p)$$

and that the following property holds:

If $g \in \mathcal{U}$, $T \geq 0$ and if an a.c. function $v : [T, T + N] \to R^n$ satisfies (2.12), then there is an integer $i_0 \in [0, N - 6]$ such that

$$|v(t) - Z_*^f(t)| \leq \delta, \ t \in [i_0 + T, i_0 + T + 6]. \tag{2.86}$$

Assume that $g \in \mathcal{U}$ and $v : [0, \infty) \to R^n$ is a (g)-good function. By the definition of W^f (see (2.5)),

$$|v(t)| \leq M^f \text{ for all large } t. \tag{2.87}$$

Since v is a (g)-good function there exists $T_0 > 0$ such that

$$I^g(t_1, t_2, v) \leq U^g(t_1, t_2, v(t_1), v(t_2)) + \delta \tag{2.88}$$

for each $t_1 \geq T_0$, $t_2 > t_1$. We may assume that $|v(t)| \leq M^f$ for all $t \in [T_0, \infty)$.

Let $T \geq T_0$. It follows from the definition of $\mathcal{U}$, N and (2.88) which holds with $t_1 = T$, $t_2 = T + N$ that there exists an integer $i_T \in [0, N - 6]$ such that

$$|v(t) - Z_*^f(t)| \leq \delta, \ t \in [i_T + T, i_T + T + 6].$$

Therefore there exists a sequence $\{T_i\}_{i=1}^\infty \subset (T_0, \infty)$ such that

$$T_{i+1} - T_i \in [6, N], \ i = 0, 1, \ldots, \ |v(T_i) - Z_*^f(T_i)| \leq \delta, \ i = 1, 2, \ldots.$$

It follows from these relations, (2.88) which holds with $t_1 = T_i$, $t_2 = T_{i+1}$ and the definition of $\mathcal{U}_0$, δ that

$$|v(t) - Z_*^f(t)| \leq \epsilon, \ t \in [T_i, T_{i+1}], \ i = 1, 2, \ldots.$$

The lemma is proved.

The next lemma plays a crucial role in the proof of Theorem 2.1.4. Its proof is based on Lemmas 2.2.4 and 2.2.2.

LEMMA 2.2.8 *Let $\epsilon \in (0, 1)$. Then there exist a neighborhood $\mathcal{U}$ of f^* in $\mathcal{M}$ with the weak topology, $\delta \in (0, \epsilon)$ and $\Delta > 1$ such that the following property holds:*

For each $g \in \mathcal{U}$, each $T_1 \in [0, \infty)$, $T_2 \in [T_1 + 2\Delta, \infty)$ and each a.c. function $v : [T_1, T_2] \to R^n$ which satisfies

$$|v(t)| \leq M_0(f, p), \ t \in [T_1, T_2],$$

$$I^g(T_1, T_2, v) \leq U^g(T_1, T_2, v(T_1), v(T_2)) + \delta, \tag{2.89}$$

there exist $\tau_1 \in [T_1, T_1 + \Delta]$ and $\tau_2 \in [T_2 - \Delta, T_2]$ such that

$$|v(t) - Z_*^f(t)| \leq \epsilon, \ t \in [\tau_1, \tau_2]. \tag{2.90}$$

Moreover if $|v(T_1) - Z_*^f(T_1)| \leq \delta$, then $\tau_1 = T_1$.

Proof. By Lemma 2.2.4 there exist $\delta \in (0, \epsilon)$ and a neighborhood $\mathcal{U}_0$ of f^* in $\mathcal{M}$ with the weak topology such that for each $g \in \mathcal{U}_0$, each $T_1 \in [0, \infty)$, $T_2 \in [T_1 + 1, \infty)$ and each a.c. function $v : [T_1, T_2] \to R^n$ which satisfies (2.62), relation (2.63) holds.

By Lemma 2.2.2 there exist a neighborhood $\mathcal{U}$ of f^* in $\mathcal{M}$ with the weak topology and an integer $N \geq 8$ such that $\mathcal{U} \subset \mathcal{U}_0$ and for each $g \in \mathcal{U}$, each $T \geq 0$ and each a.c. function $v : [T, T + N] \to R^n$ which satisfies (2.12), there is an integer $i_0 \in [0, N - 6]$ for which (2.86) holds. Set

$$\Delta = 2N. \tag{2.91}$$

Assume that $g \in \mathcal{U}$, $T_1 \in [0, \infty)$, $T_2 \in [T_1 + 2\Delta, \infty)$ and $v : [T_1, T_2] \to R^n$ is an a.c. function satisfying (2.89). It follows from the definition of $\mathcal{U}, N$ and (2.89) that for each $T \in [T_1, T_2 - N]$ there is an integer $i_0 \in [0, N - 6]$ for which (2.86) holds. Therefore there exists a sequence of numbers $\{t_i\}_{i=0}^G \subset [T_1, T_2]$ such that

$$t_0 = T_1, \ t_{i+1} - t_i \in [6, N], \ i = 0, \ldots, G - 1, \ T_2 - t_G \leq N, \tag{2.92}$$

$$|v(t_i) - Z_*^f(t_i)| \leq \delta, \ i = 1, \ldots, G. \tag{2.93}$$

It follows from the definition of δ, $\mathcal{U}_0$, (2.89), (2.93), (2.92) that

$$|v(t) - Z_*^f(t)| \leq \epsilon, \ t \in [t_i, t_{i+1}] \tag{2.94}$$

for all $i = 1, \ldots, G - 1$ and if $|v(T_1) - Z_*^f(T_1)| \leq \delta$, then (2.94) holds for $i = 0, \ldots, G - 1$. This completes the proof of the lemma.

The following lemma is an extension of Lemma 2.2.8.

LEMMA 2.2.9 *Let* $\epsilon \in (0, 1)$. *Then there exist a neighborhood* $\mathcal{U}$ *of* f^* *in* $\mathcal{M}$ *with the weak topology,* $\delta \in (0, \epsilon)$, $\Delta > 1$ *such that for each* $g \in \mathcal{U}$, *each* $T_1 \in [0, \infty)$, $T_2 \in [T_1 + 2\Delta, \infty)$ *and each a.c. function* $v : [T_1, T_2] \to R^n$ *the following properties hold:*

(i) If $|v(T_i)| \leq 2M^f + 2 + 2p$, $i = 1, 2$

and

$$I^g(T_1, T_2, v) \leq U^g(T_1, T_2, v(T_1), v(T_2)) + \delta,$$

then (2.90) holds with $\tau_1 \in [T_1, T_1 + \Delta]$ *and* $\tau_2 \in [T_2 - \Delta, T_2]$. *Moreover if* $|v(T_1) - Z_*^f(T_1)| \leq \delta$, *then* $\tau_1 = T_1$.

(ii) If $|v(T_1)| \leq 2M^f + 2 + 2p$, $I^g(T_1, T_2, v) \leq \sigma^g(T_1, T_2, v(T_1)) + \delta$, *then (2.90) holds with* $\tau_1 \in [T_1, T_1 + \Delta]$ *and* $\tau_2 \in [T_2 - \Delta, T_2]$. *Moreover if* $|v(T_1) - Z_*^f(T_1)| \leq \delta$, *then* $\tau_1 = T_1$.

Lemma 2.2.9 now follows from Lemma 2.2.8, Theorem 1.2.3 and the definition of $W_0(f, p)$, $M_0(f, p)$ (see (2.6), (2.7)).

LEMMA 2.2.10 *Let* $\epsilon \in (0, 1)$. *Then there exist a neighborhood* $\mathcal{U}$ *of* f^* *in* $\mathcal{M}$ *with the weak topology,* $\Delta > 1$ *such that for each* $g \in \mathcal{U}$ *and each* (g)-*overtaking optimal function* $v : [0, \infty) \to R^n$ *satisfying* $|v(0)| \leq 2M^f + 2 + 2p$ *the relation* $|v(t) - Z_*^f(t)| \leq \epsilon$ *holds for all* $t \in [\Delta, \infty)$.

Lemma 2.2.10 follows from the definition of W^f, M^f (see (2.5)) and Lemma 2.2.9.

2.3.   Proofs of Theorems 2.1.1-2.1.4

Construction of the set $\mathcal{F}$. Fix $z_* \in R^n$ and an integer $p \geq 1$. For each $f \in \mathcal{M}$ we define a function $Z_*^f : [0, \infty) \to R^n$, numbers M^f, $M_0(f, p)$, $M(f, p)$, a function $\phi^{M(f,p)+1}$, a number $r(f, p) > 0$ and neighborhoods W^f, $W_0(f, p)$, $W(f, p)$ of f in $\mathcal{M}$ with the weak topology as in Section 2.2 (see (2.3)-(2.10)). Set

$$E_p = \{f_r^{M(f,p)+1} : f \in \mathcal{M}, \ r \in (0, r(f, p))\}. \tag{3.1}$$

Clearly for each $f \in \mathcal{M}$ and each $r \in (0, r(f, p))$ Lemmas 2.2.2-2.2.10 hold with $f^* = f_r^{M(f,p)+1}$. By Lemma 2.2.1 E_p is everywhere dense in $\mathcal{M}$ with the strong topology.

For each $f \in \mathcal{M}$, each $r \in (0, r(f, p))$ and each integer $k \geq 1$ there exist an open neighborhood $V(f, p, r, k)$ of $f_r^{M(f,p)+1}$ in $\mathcal{M}$ with the weak topology, an integer $q(f, p, r, k) \geq 4$, numbers

$$\delta(f, p, r, k) \in (0, (4k)^{-1}), \ l(f, p, r, k) \geq 8, \ \Delta(f, p, r, k) > 1$$

such that:

(i) Lemma 2.2.6 holds with

$$f^* = f_r^{M(f,p)+1}, \ \epsilon = (4k)^{-1}, \ \mathcal{U} = V(f, p, r, k),$$

$$q_* = q(f, p, r, k), \ l_* = l(f, p, r, k);$$

(ii) Lemma 2.2.7 holds with

$$f^* = f_r^{M(f,p)+1}, \ \epsilon = (4k)^{-1}, \ \mathcal{U} = V(f, p, r, k);$$

(iii) Lemmas 2.2.9 and 2.2.10 hold with

$$f^* = f_r^{M(f,p)+1}, \ \mathcal{U} = V(f, p, r, k), \ \epsilon = (4k)^{-1},$$

$$\delta = \delta(f, p, r, k), \ \Delta = \Delta(f, p, r, k).$$

We define

$$\mathcal{F}_p = \cap_{k=1}^{\infty} \cup \{V(f, p, r, k) : \ f \in \mathcal{M}, \ r \in (0, r(f, p))\}, \quad (3.2)$$

$$\mathcal{F} = \cap_{p=1}^{\infty} \mathcal{F}_p. \quad (3.3)$$

Clearly $\mathcal{F}$ is a countable intersection of open (in the weak topology) everywhere dense (in the strong topology) subsets of $\mathcal{M}$.

Proof of Theorem 2.1.1. We prove Assertion 1. Assume that $g \in \mathcal{F}$ and $v_i : [0, \infty) \rightarrow R^n$, $i = 1, 2$ are (g)-good functions. Let $\epsilon > 0$. Fix an integer $k \geq 4\epsilon^{-1}$. There exist $f \in \mathcal{M}$ and $r \in (0, r(f, 1))$ such that $g \in V(f, 1, r, k)$. By condition (ii) and Lemma 2.2.7,

$$|v_2(t) - v_1(t)| \leq (2k)^{-1} < \epsilon \text{ for all large } t.$$

Since ϵ is an arbitrary positive number we conclude that

$$|v_2(t) - v_1(t)| \rightarrow 0 \text{ as } t \rightarrow \infty.$$

Assertion 1 is proved.

We prove Assertion 2. Let $g \in \mathcal{F}$ and let $y \in R^n$. By Theorem 1.2.1 there exists a (g)-good function $Y : [0, \infty) \to R^n$ such that $Y(0) = y$ and

$$I^g(0, T, Y) = U^g(0, T, Y(0), Y(T)) \qquad (3.4)$$

for each $T \geq 0$. We show that Y is a (g)-overtaking optimal function.

Let us assume the converse. Then there exists a number $\epsilon > 0$ and an a.c. function $Z : [0, \infty) \to R^n$ such that $Z(0) = y$ and

$$\limsup_{T \to \infty} [I^g(0, T, Y) - I^g(0, T, Z)] > \epsilon. \qquad (3.5)$$

There exists a sequence of positive numbers $\{T_i\}_{i=1}^{\infty}$ such that

$$T_i \to +\infty \text{ as } i \to \infty,$$

$$I^g(0, T_i, Y) - I^g(0, T_i, Z) > \epsilon, \ i = 1, 2, \dots. \qquad (3.6)$$

By Theorem 1.2.1, Z is a bounded (g)-good function. Therefore

$$Y(t) - Z(t) \to 0 \text{ as } t \to \infty. \qquad (3.7)$$

The functions Y and Z are (f)-good and bounded. Therefore we can choose a number

$$S > \sup\{|Z(t)|, Y(t)| : t \in [0, \infty)\}. \qquad (3.8)$$

It follows from Proposition 1.3.7 that there exists $\delta > 0$ such that the following property holds:

If $T_1 \in [0, \infty)$, $T_2 \in [T_1 + 4^{-1}, T_1 + 4]$ and if $y_1, y_2, z_1, z_2 \in R^n$ satisfy

$$|y_i|, |z_i| \leq S, \ i = 1, 2, \ |y_1 - y_2|, |z_1 - z_2| \leq \delta,$$

then

$$|U^g(T_1, T_2, y_1, z_1) - U^g(T_1, T_2, y_2, z_2| \leq 8^{-1}\epsilon. \qquad (3.9)$$

Since $|Y(t) - Z(t)| \to 0$ as $t \to \infty$ there exists $\tau > 0$ such that

$$|Z(t) - Y(t)| \leq 2^{-1}\delta, \ t \in [\tau, \infty). \qquad (3.10)$$

Fix a natural number j such that $T_j > \tau$. There exists an a.c. function $X : [0, \infty) \to R^n$ such that

$$X(t) = Z(t), t \in [0, T_j], \ X(t) = Y(t), \ t \in [T_j + 1, \infty), \qquad (3.11)$$

$$I^g(T_j, T_j + 1, X) = U^g(T_j, T_j + 1, X(T_j), X(T_j + 1)).$$

It follows from (3.11), (3.4), (3.8), (3.10) and the definition of δ that

$$X(0) = Y(0), \ X(T_j + 1) = Y(T_j + 1),$$

$$|I^g(T_j, T_j + 1, X) - I^g(T_j, T_j + 1, Y)| = |U^g(T_j, T_j + 1, X(T_j), X(T_j + 1))$$
$$-U^g(T_j, T_j + 1, Y(T_j), Y(T_j + 1))| \leq 8^{-1}\epsilon.$$

Together with (3.11) and (3.6) these relations imply that

$$I^g(0, T_j + 1, Y) - U^g(0, T_j + 1, Y(0), Y(T_j + 1))$$

$$\geq I^g(0, T_j + 1, Y) - I^g(0, T_j + 1, X) =$$

$$I^g(0, T_j, Y) - I^g(0, T_j, Z) + I^g(T_j, T_j + 1, Y) - I^g(T_j, T_j + 1, X) \geq \epsilon - 8^{-1}\epsilon.$$

This is contradictory to (3.4). The obtained contradiction proves Assertion 2.

We prove Assertion 3. Let $g \in \mathcal{F}$, $\epsilon > 0$ and $Y : [0, \infty) \to R^n$ be a (g)-overtaking optimal function. Fix an integer $k \geq 4\epsilon^{-1}$. There exist $f \in \mathcal{M}$, $r \in (0, r(f, 1))$ such that $g \in V(f, 1, r, k)$.

Assume that $h \in V(f, 1, r, k)$ and $v : [0, \infty) \to R^n$ is an (h)-good function. It follows from condition (ii) and Lemma 2.2.7 that $|v(t) - Y(t)| < (2k)^{-1}$ for all large t. Assertion 3 is proved.

Proof of Theorem 2.1.2. Let $g \in \mathcal{F}$, $M, \epsilon > 0$ and let $Y : [0, \infty) \to R^n$ be a (g)-overtaking function. Fix integers

$$p > 2M + 2|Y(0)| + 2, \ k > 4\epsilon^{-1}. \tag{3.12}$$

There exists $f \in \mathcal{M}$ and $r \in (0, r(f, p))$ such that $g \in V(f, p, r, k)$. By condition (ii) and Lemma 2.2.7 there exists a number $\tau_0 > 0$ such that

$$|Y(t) - Z_*^f(t)| \leq (4k)^{-1}, \ t \in [\tau_0, \infty). \tag{3.13}$$

Set

$$\tau = \tau_0 + \Delta(f, p, r, k) + 1. \tag{3.14}$$

Assume that $h \in V(f, p, r, k)$ and $v : [0, \infty) \to R^n$ is an (h)-overtaking optimal function such that $|v(0)| \leq M$. By condition (iii), Lemma 2.2.10 and (3.12),

$$|v(t) - Z_*^f(t)| \leq (4k)^{-1}, \ t \in [\Delta(f, p, r, k), \infty).$$

Together with (3.13), (3.14) and (3.12) this implies that $|v(t) - Y(t)| \leq \epsilon$ for all $t \in [\tau, \infty)$. The theorem is proved.

Proof of Theorem 2.1.3. Let $g \in \mathcal{F}$, $S_1, S_2, \epsilon > 0$ and $Y : [0, \infty) \to R^n$ be a (g)-overtaking optimal function. By Theorem 1.2.1, Y is a (g)-good function and $\sup\{|Y(t)| : t \in [0, \infty)\} < \infty$. Fix integers

$$p > 2S_1 + 2S_2 + 1 + \sup\{|Y(t)| : t \in [0, \infty)\}, \ k > 8\epsilon^{-1}. \tag{3.15}$$

There exist $f \in \mathcal{M}$ and $r \in (0, r(f,p))$ such that $g \in V(f,p,r,k)$. By condition (ii) and Lemma 2.2.7 there exists a number $\tau_0 > 0$ such that

$$|Y(t) - Z_*^f(t)| \le (4k)^{-1}, \ t \in [\tau_0, \infty). \tag{3.16}$$

Set

$$\mathcal{U} = V(f,p,r,k), \ L = \tau_0 + l(f,p,r,k), \ Q = q(f,p,r,k) + 1. \tag{3.17}$$

Assume that $h \in \mathcal{U}$, $T_1 \in [0, \infty)$, $T_2 \in [T_1 + LQ, \infty)$ and an a.c. function $v : [T_1, T_2] \to R^n$ satisfies one of the conditions (a), (b) of the theorem. By condition (i) and Lemma 2.2.6 there are numbers $\{d_i\}_{i=1}^q$, $\{\bar{d}_i\}_{i=1}^q$ such that

$$q \le q(f,p,r,k), \ d_i < \bar{d}_i \le d_i + l(f,p,r,k), \ i = 1, \ldots, q,$$

$$|v(t) - Z_*^f(t)| \le (4k)^{-1}, \ t \in [T_1, T_2] \setminus \cup_{i=1}^q [d_i, \bar{d}_i].$$

By these relations and (3.15), (3.16),

$$|v(t) - Y(t)| \le \epsilon, \ t \in [T_1, T_2] \setminus ([0, \tau_0] \cup_{i=1}^q [d_i, \bar{d}_i]).$$

This completes the proof of the theorem.

Proof of Theorem 2.1.4. Let $g \in \mathcal{F}$, $S, \epsilon > 0$ and let $Y : [0, \infty) \to R^n$ be a (g)-overtaking optimal function. By Theorem 1.2.1, Y is a (g)-good function and $\sup\{|Y(t)| : t \in [0, \infty)\} < \infty$. Fix integers

$$p > 2S + 1 + \sup\{|Y(t)| : t \in [0, \infty)\}, \ k > 8\epsilon^{-1}. \tag{3.18}$$

There exist $f \in \mathcal{M}$ and $r \in (0, r(f,p))$ such that $g \in V(f,p,r,k)$. By condition (ii) and Lemma 2.2.7 there exists a number τ_0 such that (3.16) holds. Set

$$\mathcal{U} = V(f,p,r,k), \ L = \tau_0 + \Delta(f,p,r,k), \ \delta = \delta(f,p,r,k). \tag{3.19}$$

Assume that $h \in \mathcal{U}$, $T_1 \in [0, \infty)$, $T_2 \in [T_1 + 2L, \infty)$ and an a.c. function $v : [T_1, T_2] \to R^n$ satisfies one of the conditions (a), (b) of the theorem. It follows from condition (iii) and Lemma 2.2.9 that

$$|v(t) - Z_*^f(t)| \le (4k)^{-1}, \ t \in [T_1 + \Delta(f,p,r,k), T_2 - \Delta(f,p,r,k)].$$

By this relation, (3.16), (3.18) and (3.19), $|v(t) - Y(t)| \le \epsilon$ for all $t \in [T_1 + L, T_2 - L]$. The theorem is proved.

2.4. Periodic variational problems

Let $a > 0$, $\mathbf{Z} = \{0, \pm 1, \pm 2, \ldots\}$ and let $\psi : [0, \infty) \to [0, \infty)$ be an increasing function such that $\psi(t) \to +\infty$ as $t \to \infty$. Denote by $\mathcal{M}^p$ the set of continuous functions $f : [0, \infty) \times R^n \times R^n \to R^1$ which satisfy the following assumptions:

A (i) $f(t, x + q, u) = f(t, x, u)$ for each $t \in [0, \infty)$, $x, u \in R^n$, $q \in \mathbf{Z}^n$;

A (ii) for each $(t, x) \in [0, \infty) \times R^n$ the function $f(t, x, \cdot) : R^n \to R^1$ is convex;

A (iii) the function f is bounded on $[0, \infty) \times R^n \times E$ for any bounded set $E \subset R^n$;

A (iv) $f(t, x, u) \geq \psi(|u|)|u| - a$ for each $(t, x, u) \in [0, \infty) \times R^n \times R^n$;

A (v) for each $\epsilon > 0$ there exist positive numbers Γ, δ such that if $t \in [0, \infty)$ and if $u_1, u_2, x_1, x_2 \in R^n$ are such that

$$|u_i| \geq \Gamma, \ i = 1, 2, \quad \max\{|x_1 - x_1|, |u_1 - u_2|\} \leq \delta,$$

then

$$|f(t, x_1, u_1) - f(t, x_2, u_2)| \leq \epsilon \max\{f(t, x_1, u_1), f(t, x_2, u_2)\};$$

A (vi) for each $M, \epsilon > 0$ there exists a positive number δ such that if $t \in [0, \infty)$ and if $u_1, u_2, x_1, x_2 \in R^n$ satisfy

$$|u_i| \leq M, \ i = 1, 2, \quad \max\{|x_1 - x_2|, |u_1 - u_2|\} \leq \delta,$$

then

$$|f(t, x_1, u_1) - f(t, x_2, u_2)| \leq \epsilon.$$

We equip the set $\mathcal{M}^p$ with the uniformity which is determined by the following base:

$$E(N, \epsilon, \lambda) = \{(f, g) \in \mathcal{M}^p \times \mathcal{M}^p : |f(t, x, u) - g(t, x, u)| \leq \epsilon$$

$$\text{for each } t \in [0, \infty) \text{ and each } x, u \in R^n \text{ satisfying } |u| \leq N,$$

$$(|f(t, x, u)| + 1)(|g(t, x, u)| + 1)^{-1} \in [\lambda^{-1}, \lambda]$$

$$\text{for each } t \in [0, \infty) \text{ and each } x, u \in R^n\},$$

where $N > 0$, $\epsilon > 0$, $\lambda > 1$.

We can show that the uniform space $\mathcal{M}^p$ is metrizable and complete. Consider functionals of the form

$$I^f(T_1, T_2, x) = \int_{T_1}^{T_2} f(t, x(t), x'(t)) dt$$

where $f \in \mathcal{M}^p$, $0 \leq T_1 < T_2 < +\infty$ and $x : [T_1, T_2] \to R^n$ is an a.c. function.

For $f \in \mathcal{M}^p$, $y, z \in R^n$ and numbers T_1, T_2 satisfying $0 \leq T_1 < T_2$ we define $U^f(T_1, T_2, y, z)$ and $\sigma^f(T_1, T_2, y)$ by (1.2) and (1.3) and set

$$\tilde{U}^f(T_1, T_2, y, z) = \inf\{U^f(T_1, T_2, y, z + m) : m \in \mathbf{Z}^n\}.$$

It is easy to see that $-\infty < U^f(T_1, T_2, y, z) < +\infty$ for each $f \in \mathcal{M}^p$, each $y, z \in R^n$ and each pair of numbers T_1, T_2 satisfying $0 \leq T_1 < T_2$.

For $f \in \mathcal{M}^p$ we use the notions of an (f)-good function and an (f)-overtaking optimal function.

The methods used in the proofs of Theorems 1.2.1-1.2.3 and Theorems 2.1.1-2.1.4 are applicable to the space $\mathcal{M}^p$. The following results are valid.

THEOREM 2.4.1 *For each $h \in \mathcal{M}^p$ and each $z \in R^n$ there exists an (h)-good function $Z^h : [0, \infty) \to R^n$ satisfying $Z^h(0) = z$ such that:*
1.

$$\tilde{U}^f(T_1, T_2, Z^f(T_1), Z^f(T_2)) = I^f(T_1, T_2, Z^f)$$

for each $f \in \mathcal{M}$, each $z \in R^n$ and each $T_1 \geq 0$, $T_2 > T_1$.

2. For each $f \in \mathcal{M}^p$, each $z \in R^n$ and each a.c. function $y : [0, \infty) \to R^n$ either

$$I^f(0, T, y) - I^f(0, T, Z^f) \to +\infty \ \text{as} \ T \to \infty$$

or

$$\sup\{|I^f(0, T, y) - I^f(0, T, Z^f)| : T \in (0, \infty)\} < \infty \qquad (4.1)$$

and if (4.1) is valid, then

$$\sup\{|y(t_1) - y(t_2)| : t_1 \in [0, \infty), \ t_2 \in [t_1, t_1 + 1]\} < \infty.$$

3. For each $f \in \mathcal{M}^p$ there exist a neighborhood $\mathcal{U}$ of f in $\mathcal{M}^p$ and a number $Q > 0$ such that

$$\sup\{|Z^g(t_1) - Z^g(t_2)| : t_1 \in [0, \infty), \ t_2 \in [t_1, t_1 + 1]\} \leq Q$$

for each $g \in \mathcal{U}$ and each $z \in R^n$.

4. For each $f \in \mathcal{M}^p$ there exist a neighborhood $\mathcal{U}$ of f in $\mathcal{M}^p$ and a number $Q > 0$ such that

$$I^g(T_1, T_2, Z^g) \leq I^g(T_1, T_2, y) + Q$$

for $g \in \mathcal{U}$, each $z \in R^n$, each $T_1 \geq 0$, $T_2 > T_1$ and each a.c. function $y : [T_1, T_2] \to R^n$.

THEOREM 2.4.2 *For each $f \in \mathcal{M}^p$ there exist a neighborhood $\mathcal{U}$ of f in $\mathcal{M}^p$ and $M > 0$ such that if $g \in \mathcal{U}$ and if $x : [0, \infty) \to R^n$ is a (g)-good function, then*

$$\limsup_{T \to \infty}(\sup\{|x(t_1) - x(t_2)| :$$

$$t_1 \in [T, \infty),\ t_2 \in [t_1, t_1 + 1]\}) < M.$$

We can show that there exists a set $\mathcal{F} \subset \mathcal{M}^p$ which is a countable intersection of open everywhere dense sets in $\mathcal{M}^p$ such that the following theorems are valid.

THEOREM 2.4.3 *1. For each $g \in \mathcal{F}$ and each pair of (g)-good functions $v_i : [0, \infty) \to R^n$, $i = 1, 2$,*

$$|v_2(t) - v_1(t) - m| \to 0 \ as\ t \to \infty$$

with $m \in \mathbf{Z}^n$.

2. For each $g \in \mathcal{F}$ and each $y \in R^n$ there exists a (g)-overtaking optimal function $Y : [0, \infty) \to R^n$ such that $Y(0) = y$.

3. Let $g \in \mathcal{F}$, $\epsilon > 0$ and $Y : [0, \infty) \to R^n$ be a (g)-overtaking optimal function. Then there exists a neighborhood $\mathcal{U}$ of g in $\mathcal{M}^p$ such that the following property holds:

If $h \in \mathcal{U}$ and if $v : [0, \infty) \to R^n$ is an (h)-good function, then there is $m \in \mathbf{Z}^n$ such that

$$|v(t) - Y(t) - m| \le \epsilon\ for\ all\ large\ t.$$

THEOREM 2.4.4 *Let $g \in \mathcal{F}$, $\epsilon > 0$ and let $Y : [0, \infty) \to R^n$ be a (g)-overtaking optimal function. Then there exists a neighborhood $\mathcal{U}$ of g in $\mathcal{M}^p$ and a number $\tau > 0$ such that the following property holds:*

If $h \in \mathcal{U}$ and if $v : [0, \infty) \to R^n$ is an (h)-overtaking optimal function, then there exists $m \in \mathbf{Z}^n$ such that

$$|v(t) - Y(t) - m| \le \epsilon,\ t \in [\tau, \infty).$$

THEOREM 2.4.5 *Let $g \in \mathcal{F}$, $S, \epsilon > 0$ and let $Y : [0, \infty) \to R^n$ be a (g)-overtaking optimal function. Then there exists a neighborhood $\mathcal{U}$ of g in $\mathcal{M}^p$, a number $L > 0$ and an integer $Q \ge 1$ such that for each $h \in \mathcal{U}$ and each pair of numbers $T_1 \in [0, \infty)$, $T_2 \in [T_1 + LQ, \infty)$ the following property holds:*

If an a.c. function $v : [T_1, T_2] \to R^n$ satisfies

$$I^h(T_1, T_2, v) \le \tilde{U}^h(T_1, T_2, v(T_1), v(T_2)) + S,$$

then there exist sequences of numbers

$$\{d_i\}_{i=1}^q, \ \{b_i\}_{i=1}^q \subset [T_1, T_2]$$

such that

$$q \leq Q, \ b_i < d_i \leq b_i + L, \ i = 1, \ldots, q$$

and that for each interval

$$\mathcal{J} \subset [T_1, T_2] \setminus \cup_{i=1}^q [b_i, d_i]$$

there is $m \in \mathbf{Z}^n$ *for which*

$$|v(t) - Y(t) - m| \leq \epsilon, \ t \in \mathcal{J}.$$

THEOREM 2.4.6 *Let* $g \in \mathcal{F}$, $\epsilon > 0$ *and let* $Y : [0, \infty) \to R^n$ *be a* (g)-*overtaking optimal function. Then there exists a neighborhood* $\mathcal{U}$ *of* g *in* $\mathcal{M}^p$ *and numbers* $\delta, L > 0$ *such that for each* $h \in \mathcal{U}$, *each pair of numbers* $T_1 \in [0, \infty)$, $T_2 \in [T_1 + 2L, \infty)$ *and each a.c. function* $v : [T_1, T_2] \to R^n$ *which satisfies* $I^h(T_1, T_2, v) \leq \tilde{U}^h(T_1, T_2, v(T_1), v(T_2)) + \delta$ *the following property holds:*

There exists $m \in \mathbf{Z}^n$ *such that*

$$|v(t) - Y(t) - m| \leq \epsilon, \ t \in [T_1 + L, T_2 - L].$$

2.5.   Spaces of smooth integrands

Consider the complete metric space $\mathcal{M}$ defined in Section 2.1. For any function $g : R^1 \times R^n \times R^n \to R^1$ denote by $\mathcal{L}(g)$ the restriction of g to $[0, \infty) \times R^{2n}$ and for an integer $k \geq 1$ denote by $C(k, \mathcal{M})$ the space of all integrands $f = f(t, x, u) \in C^k(R^{2n+1})$ such that $\mathcal{L}(f) \in \mathcal{M}$.

Let $k \geq 1$ be an integer. For $p = (p_1, \ldots, p_{2n+1}) \in \{0, \ldots, k\}^{2n+1}$ and $f \in C^k(R^{2n+1})$ we set

$$|p| = \sum_{i=1}^{2n+1} p_i, \ D^p f = \partial^{|p|} f / \partial y_1^{p_1} \ldots \partial y_{2n+1}^{p_{2n+1}}.$$

For the set $C(k, \mathcal{M})$ we consider the uniformity which is determined by the following base:

$$E(N, \epsilon, \lambda) = \{(f, g) \in C(k, \mathcal{M}) \times C(k, \mathcal{M}) :$$

$$|D^p f(t, x, u) - D^p g(t, x, u)| \leq \epsilon$$

for each $(t, x, u) \in R^{2n+1}$ satisfying $|t|, |x|, |u| \leq N$

and each $p \in \{0, \ldots k\}^{2n+1}$ such that $|p| \leq k$,

$|f(t, x, u) - g(t, x, u)| \leq \epsilon$ for each $t \in [0, \infty)$

and each $x, u \in R^n$ for which $|x|, |u| \leq N$,

$(|f(t, x, u)| + 1)(|g(t, x, u)| + 1)^{-1} \in [\lambda^{-1}, \lambda]$ for each $t \in [0, \infty)$

and each $x, u \in R^n$ such that $|x| \leq N\}$,

where $N > 0$, $\epsilon > 0$, $\lambda > 1$.

Clearly the uniform space $C(k, \mathcal{M})$ is Hausdorff and has a countable base. Therefore $C(k, \mathcal{M})$ is metrizable. It is easy to verify that the uniform space $C(k, \mathcal{M})$ is complete and the operator $\mathcal{L} : C(k, \mathcal{M}) \to \mathcal{M}$ is continuous. Denote by $\mathcal{M}_k$ the space of all functions $f \in C(k, \mathcal{M})$ which satisfy the following conditions:

$$f \in C^k(R^{2n+1}), \ \partial f / \partial u_i \in C^k(R^{2n+1}) \text{ for } i = 1, \ldots, n; \qquad (5.1)$$

the matrix $(\partial^2 f / \partial u_i \partial u_j)(t, x, u)$, $i, j = 1, \ldots, n$ is positive definite
$$(5.2)$$

for all $(t, x, u) \in R^{2n+1}$;

there exist a number $c_0 > 1$ and monotone increasing functions $\phi_i : [0, \infty) \to [0, \infty)$, $i = 0, 1, 2$ such that

$$\phi_0(t)t^{-1} \to +\infty \text{ as } t \to +\infty,$$

$$f(t, x, u) \geq \phi_0(c_0|u|) - \phi_1(|x|), \ t \in R^1, \ x, u \in R^n; \qquad (5.3)$$

$$\sup\{|\partial f / \partial x_i(t, x, u)|, \ |\partial f / \partial u_i(t, x, u)|\} \leq \phi_2(|x|)(1 + \phi_0(|u|)), \quad (5.4)$$

$$t \in R^1, \ x, u \in R^n, \ i = 1, \ldots, n.$$

Denote by $\bar{\mathcal{M}}_k$ the closure of $\mathcal{M}_k$ in $C(k, \mathcal{M})$. We consider the topological subspace $\bar{\mathcal{M}}_k \subset C(k, \mathcal{M})$ with the relative topology. In order to show that the conclusions of Theorems 2.1.1-2.1.4 also hold for a G_δ-subset of the space $\bar{\mathcal{M}}_k$ we need the following result.

PROPOSITION 2.5.1 *Let $k \geq 1$ be an integer, a number $c_0 > 1$, $\phi_i :$ $[0, \infty) \to [0, \infty)$, $i = 0, 1, 2$ be monotone increasing functions and let an integrand $f : R^{2n+1} \to R^1$ satisfy (5.1)-(5.4). Assume that $T_1 \in [0, \infty)$, $T_2 > T_1$ and an a.c. function $w : [T_1, T_2] \to R^n$ satisfies*

$$I^f(T_1, T_2, w) = U^f(T_1, T_2, w(T_1), w(T_2)) < \infty. \qquad (5.5)$$

Then $w \in C^{k+1}([T_1, T_2]; R^n)$ and

$$(d/dt)(\partial f/\partial u_i(t, w(t), w'(t))) = (\partial f/\partial x_i(t, w(t), w'(t))) \qquad (5.6)$$

for each $i \in \{1, \ldots, n\}$ and each $t \in [T_1, T_2]$.

Denote by $< x, y >$ the scalar product of $x, y \in R^n$. In the proof of Proposition 2.5.1 we need the following simple result.

LEMMA 2.5.1 *Let $n \geq 1$ be an integer, $-\infty < T_1 < T_2 < +\infty$ and let $f \in L^1([T_1, T_2]; R^n)$ have the following property:*

$$\int_{T_1}^{T_2} < f(t), g(t) > dt = 0$$

for every function $g \in L^\infty([T_1, T_2]; R^n)$ such that $\int_{T_1}^{T_2} g(t) dt = 0$.
Then there exists $d \in R^n$ such that $f(t) = d$ for almost all $t \in [T_1, T_2]$.

Proof of Proposition 2.5.1. Put

$$x = w(T_1), \ y = w(T_2).$$

In proving Proposition 2.5.1 we follow [64, Theorem 1.10.1]. For $t \in [T_1, T_2]$ we set $B(t) = (t, w(t), w'(t))$. Analogously to the proof of Theorem 1.10.1 of [64] we can show that if an a.c. function $h : [T_1, T_2] \to R^n$ satisfies

$$h(T_1) = h(T_2) = 0, \ h' \in L^\infty([T_1, T_2]; R^n), \qquad (5.7)$$

then

$$\sum_{i=1}^{n} \partial f/\partial x_i(B(t))h_i(t) + \sum_{i=1}^{n} \partial f/\partial u_i(B(t))h_i'(t) \in L^1(T_1, T_2), \qquad (5.8)$$

$$\int_{T_1}^{T_2} [\sum_{i=1}^{n} \partial f/\partial x_i(B(t))h_i(t) + \sum_{i=1}^{n} \partial f/\partial u_i(B(t))h_i'(t)] dt = 0. \qquad (5.9)$$

It follows from (5.3)-(5.5) that the function

$$t \to |\partial f/\partial x_i(B(t))| + |\partial f/\partial u_i(B(t))|, \ t \in [T_1, T_2]$$

belongs to the space $L^1(T_1, T_2)$ for $i = 1, \ldots, n$.
Consider a function $g \in L^\infty(T_1, T_2); R^n)$ such that

$$\int_{T_1}^{T_2} g(t) dt = 0 \qquad (5.10)$$

and put

$$h(t) = \int_{T_1}^{t} g(\tau)d\tau, \ t \in [T_1, T_2],$$

$$E_i(t) = \int_{T_1}^{t} (\partial f/\partial x_i)(B(\tau))d\tau, \ t \in [T_1, T_2], \ i = 1, \ldots, n.$$

Clearly h satisfies (5.7). Thus (5.9) and (5.8) are true. The Fubini theorem implies that for $i = 1, \ldots, n$,

$$\int_{T_1}^{T_2} (\partial f/\partial x_i)(B(t))h_i(t)dt = \int_{T_1}^{T_2} g_i(\tau)(E_i(T_2) - E_i(\tau))d\tau.$$

Combined with (5.9) this equality implies that

$$\int_{T_1}^{T_2} [\sum_{i=1}^{n} \partial f/\partial u_i(B(t)) + E_i(T_2) - E_i(t)]g_i(t)dt = 0.$$

We have shown that this equality holds for every $g \in L^{\infty}([T_1, T_2]; R^n)$ satisfying (5.10). Therefore Lemma 2.5.1 implies that there is $d = (d_1, \ldots, d_n) \in R^n$ such that

$$\partial f/\partial u_i(B(t)) + E_i(T_2) - E_i(t) = d_i \qquad (5.11)$$

for each $i \in \{1, \ldots, n\}$ and almost all $t \in [T_1, T_2]$.

Define a mapping $G : R^1 \times R^n \times R^n \to R^1 \times R^n \times R^n$ by

$$G(t, x, u) = (t, x, (\partial f/\partial u_i(t, x, u))_{i=1}^{n}).$$

Assume that $(t^i, x^i, u^i) \in R^{2n+1}$, $i = 1, 2$ and

$$G(t^1, x^1, u^1) = G(t^2, x^2, u^2).$$

Clearly $t^1 = t^2$, $x^1 = x^2$. We show that $u_1 = u_2$. For $\lambda \in [0, 1]$ we denote by $A(\lambda)$ the matrix

$$(\partial^2 f/\partial u_i \partial u_j)(t^1, x^1, u^1 + \lambda(u^2 - u^1)), \ i, j = 1, \ldots, n.$$

For $i = 1, \ldots, n$ we have

$$0 = \partial f/\partial u_i(t^1, x^1, u^2) - \partial f/\partial u_i(t^1, x^1, u^1)$$

$$= \int_0^1 (d/d\lambda)(\partial f/\partial u_i(t^1, x^1, u^1 + \lambda(u^2 - u^1))d\lambda$$

$$= \int_0^1 < (\partial^2 f/\partial u_i \partial u_j)(t^1, x^1, u^1 + \lambda(u^2 - u^1))_{j=1}^{n}, u_2 - u_1 > d\lambda.$$

This implies that

$$\int_0^1 A(\lambda)(u_2 - u_1)d\lambda = 0,$$

$$\int_0^1 < A(\lambda)(u_2 - u_1), u_2 - u_1 > d\lambda = 0.$$

By the definition of $A(\lambda)$ and (5.2) $u_2 = u_1$. Therefore the mapping G is injective.

By the inverse function theorem and the conditions of the proposition, $G(R^{2n+1})$ is an open subset of R^{2n+1}, there exists $G^{-1} : G(R^{2n+1}) \to R^{2n+1} \in C^1$ and

$$(G^{-1})'(y) = [G' \cdot G^{-1}(y)]^{-1}, \; y \in G(R^{2n+1}). \tag{5.12}$$

We show that

$$G(R^{2n+1}) = R^{2n+1}. \tag{5.13}$$

Let (t, x, U), $(t^i, x^i, u^i) \in R^{2n+1}$, $i = 1, 2, \ldots$,

$$G(t^i, x^i, u^i) \to (t, x, U) \text{ as } i \to \infty. \tag{5.14}$$

It is sufficient to prove that $(t, x, U) \in G(R^{2n+1})$. Clearly

$$(t^i, x^i) \to (t, x) \text{ as } i \to \infty. \tag{5.15}$$

We show that the sequence $\{u_i\}_{i=1}^\infty$ is bounded. Let us assume the converse. By (5.14) and (5.15) there exists a number $M > 0$ such that

$$|G(t^i, x^i, u^i)|, \; t^i, \; |x^i| \le M \text{ as } i = 1, 2, \ldots. \tag{5.16}$$

There exist $M_0 > 0$ such that

$$|f(\tau, y, 0)|, \; |(\partial f/\partial u_j)(\tau, y, 0)| \le M_0, \; j = 1, \ldots, n, \; |\tau| \le M, |y| \le M. \tag{5.17}$$

We may assume that

$$|u_i| \to \infty \text{ as } i \to \infty. \tag{5.18}$$

It follows from (5.2) and (5.16)-(5.18) that for any integer $i \ge 1$

$$f(t^i, x^i, 0) \ge f(t^i, x^i, u^i) - < (\partial f/\partial u_j(t^i, x^i, u^i))_{j=1}^n, u^i >, \tag{5.19}$$

$$f(t^i, x^i, u^i) \le M_0 + M|u^i|, \; \limsup_{i \to \infty} f(t^i, x^i, u^i)/|u^i| \le M.$$

On the other hand by (5.16), (5.18) and (5.3) for any integer $i \ge 1$

$$f(t^i, x^i, u^i) \ge \phi_0(c_0|u^i|) - \phi_1(M),$$

$$\limsup_{i\to\infty} f(t^i, x^i, u^i)/|u^i| \geq \limsup_{i\to\infty} \phi_0(c_0|u^i|)/|u^i| = +\infty.$$

This is contradictory to (5.19). The obtained contradiction proves the boundedness of $\{u_i\}_{i=1}^{\infty}$. We may assume that $u_i \to u \in R^n$ as $i \to \infty$. Together with (5.14), (5.15) this implies that

$$(t, x, U) = \lim_{i\to\infty} G(t^i, x^i, u^i) = G(t, x, u) \in G(R^{2n+1}).$$

Therefore (5.13) holds. By (5.11) for almost all $t \in [T_1, T_2]$,

$$(t, w(t), w'(t)) = G^{-1}(t, w(t), (d_i - E_i(T_2) - E_i(t))_{i=1}^n). \qquad (5.20)$$

It is now easy to see that the last relation holds for all $t \in [T_1, T_2]$ and $w \in C^2([T_1 T_2]; R^n)$. (5.9) implies that for each $h \in C^1([T_1, T_2]; R^n)$ satisfying (5.7),

$$\sum_{i=1}^n \int_{T_1}^{T_2} h_i(t)[\partial f/\partial x_i(B(t)) - (d/dt)(\partial f/\partial u_i(B(t)))]dt = 0.$$

This implies (5.6) for each $t \in [T_1, T_2]$ and each $i \in \{1, \ldots, n\}$. By (5.12) $G^{-1} \in C^k$. Together with (5.20) this implies that $w \in C^{k+1}([T_1, T_2]; R^n)$. This completes the proof of the proposition.

THEOREM 2.5.1 *Let $k \geq 1$ be an integer. Then there exists a G_δ-set $\mathcal{F} \subset \mathcal{M}$ which is a countable intersection of open (in the weak topology) everywhere dense (in the strong topology) sets in $\mathcal{M}$ and for which the conclusions of Theorems 2.1.1.-2.1.4 hold and a set $\mathcal{F}_k \subset \bar{\mathcal{M}}_k$ which is a countable intersection of open everywhere dense sets in $\bar{\mathcal{M}}_k$ such that $\mathcal{L}(\mathcal{F}_k) \subset \mathcal{F}$.*

Proof. Fix $z_* \in R^n$. For each $f \in \mathcal{M}$ let $Z_*^f : [0, \infty) \to R^n$ be as guaranteed by Theorem 1.2.1. For each $M > 0$ there exists a function $\psi^M \in C^\infty(R^1)$ such that

$$\psi^M(t) = 1, \; t \in [-M-1, M+1], \; \psi^M(t) = 0, \; |t| \geq M+2,$$

$$\psi^M(t) \in (0, 1),$$

$$t \in (-M-2, -M-1) \cup (M+1, M+2).$$

For $f \in \mathcal{M}$ and $M > |z_*|$ we define $\phi^M : [0, \infty) \times R^n \to R^1$ as follows:

$$\phi^M(t, x) = |x - Z_*^f(t)|^2 \psi^M(|x|), \; t \in [0, \infty), \; x \in R^n.$$

By Remark 2.2.1 the function ϕ^M satisfies Assumption B.

For each $f \in \mathcal{M}$ and each integer $p \geq 1$ we define numbers M^f, $M_0(f,p)$, $M(f,p)$, $r(f,p) > 0$ and neighborhoods W^f, $W_0(f,p)$, $W(f,p)$ of f in $\mathcal{M}$ as in Section 2.2 (see (2.3)-(2.10)). Consider the set $E_p \subset \mathcal{M}$ defined by (3.1), (2.3). For each $f \in \mathcal{M}$, each pair of integers $p, q \geq 1$ and each $r \in (0, r(f,p))$ we define an open neighborhood $V(f,p,r,q)$ of $f_r^{M(f,p)+1}$ in $\mathcal{M}$ as in Section 2.3 and define a set $\mathcal{F}$ by (5.2), (5.3). In Section 2.5, Theorems 2.2.1-2.2.4 were established for the set $\mathcal{F}$.

Let $g \in \mathcal{M}_k$, $p \geq 1$ be an integer and $r \in (0, r(\mathcal{L}(g), p))$. By Theorem 1.2.1 and Proposition 2.5.1, $Z_*^{\mathcal{L}(g)} \in C^{k+1}$. There exists $Y^g : R^1 \to R^n \in C^{k+1}$ such that

$$Y^g(t) = Z_*^{\mathcal{L}(g)}(t) \text{ for } t \in [0, \infty),$$

$$(d^{k+1}Y^g/dt^{k+1})(t) = (d^{k+1}Y^g/dt^{k+1})(0) \text{ for } t \in (-\infty, 0).$$

We define a function $g_r^p : R^{2n+1} \to R^1$ as follows:

$$g_r^p(t, x, u) = g(t, x, u) + r|x - Y^g(t)|^2 \psi^S(|x|),$$

$$t \in R^1, \ x, u \in R^n$$

where $S = M(\mathcal{L}(g), p) + 1$.

It is easy to verify that $g_r^p \in \mathcal{M}_k$,

$$g_r^p \to g \text{ as } r \to 0 \text{ in } C(k, \mathcal{M}),$$

$\mathcal{L}(g_r^p) \in E_p$ for $g \in \mathcal{M}_k$, $p \geq 1$ and $r \in (0, r(\mathcal{L}(g), p))$. For each integer $p \geq 1$ we set

$$G_p = \{g_r^p : \ g \in \mathcal{M}_k, \ r \in (0, r(\mathcal{L}(g), p))\}.$$

For each $g \in \mathcal{M}_k$, each pair of integers $p, q \geq 1$ and each

$$r \in (0, r(\mathcal{L}(g), p))$$

we set

$$U(g, p, r, q) = \mathcal{L}^{-1}(V(\mathcal{L}(g), p, r, q)) \cap \bar{\mathcal{M}}_k.$$

Evidently $U(g, p, r, q)$ is an open neighborhood of g_r^p in $\bar{\mathcal{M}}_k$. We define

$$\mathcal{F}_{kp} = \cap_{q=1}^{\infty} \cup \{U(g, p, r, q) : \ g \in \mathcal{M}_k, \ r \in (0, r(\mathcal{L}(g), p))\},$$

$$\mathcal{F}_k = \cap_{p=1}^{\infty} \mathcal{F}_{kp}.$$

Clearly $\mathcal{F}_k$ is a countable intersection of open everywhere dense sets in $\bar{\mathcal{M}}_k$ and $\mathcal{L}(\mathcal{F}_k) \subset \mathcal{F}$. This completes the proof of the theorem.

2.6.   Examples

Let $n \geq 1$. Fix a positive constant a and set $\psi(t) = t$, $t \in [0, \infty)$. Consider a complete metric space $\mathcal{M}$ of integrands $f : [0, \infty) \times R^1 \times R^1 \to R^1$ defined in Section 2.1 and a G_δ-subset $\mathcal{F} \subset \mathcal{M}$ constructed in Section 2.3. Define by $\mathcal{F}_0$ the set of all integrands $g \in \mathcal{M}$ for which the conclusions of Theorems 2.1.1-2.1.4 are valid. Clearly the set $\mathcal{F}$ is everywhere dense in $\mathcal{M}$ and $\mathcal{F} \subset \mathcal{F}_0$.

Example 6.1. Consider an integrand $f(t, x, u) = x^2 + u^2$, $t, x, u \in R^1$. It is easy to see that $f \in \mathcal{M}$. Applying the methods used in the proofs of Theorems 2.1.1-2.1.4 we can show that $f \in \mathcal{F}_0$.

Example 6.2. Fix a number $q > 0$ and consider an integrand

$$g(t, x, u) = qx^2(x - 1)^2 + u^2, \ t, x, u \in R^1.$$

It is easy to see that $g \in \mathcal{M}$ if a is large enough. Clearly the function $v_1(t) = 0$, $v_2(t) = 1$, $t \in [0, \infty)$ are (g)-overtaking optimal. Assertion 1 of Theorem 2.1.1 implies that $g \notin \mathcal{F}_0$. It is easy to verify that $f, g \in \mathcal{M}_k$ for any integer $k \geq 1$.

Chapter 3

EXTREMALS
OF AUTONOMOUS PROBLEMS

In this chapter we establish the turnpike property for autonomous variational problems with nonconvex integrands. For this class of integrands the "turnpike" is a compact subset of R^n. We consider the complete metric space of integrands $\mathcal{M}$ introduced in Section 2.1 and the subspace $\mathcal{A} \subset \mathcal{M}$ of all integrands $f \in \mathcal{M}$ which do not depend on t. We establish the existence of a set $\mathcal{F} \subset \mathcal{A}$ which is a countable intersection of open everywhere dense sets in $\mathcal{A}$ such that each $f \in \mathcal{F}$ has the turnpike property.

Moreover we show that the turnpike property holds for approximate solutions of variational problems with a generic integrand f and that the turnpike phenomenon is stable under small perturbations of a generic integrand f.

3.1. Main results

Let $a > 0$ be a constant and let $\psi : [0, \infty) \to [0, \infty)$ be an increasing function such that

$$\psi(t) \to +\infty \text{ as } t \to \infty.$$

Denote by $|\cdot|$ the Euclidean norm in R^n and denote by $\mathcal{A}$ the set of continuous functions $f : R^n \times R^n \to R^1$ which satisfy the following assumptions:

A(i) for each $x \in R^n$ the function $f(x, \cdot) : R^n \to R^1$ is convex;

A(ii) $f(x, u) \geq \max\{\psi(|x|), \psi(|u|)|u|\} - a$ for each $(x, u) \in R^n \times R^n$;

A(iii) for each $M, \epsilon > 0$ there exist $\Gamma, \delta > 0$ such that if $u_1, u_2, x_1, x_2 \in R^n$ satisfy

$$|x_i| \leq M, \ |u_i| \geq \Gamma, \ i = 1, 2, \quad \max\{|x_1 - x_2|, |u_1 - u_2|\} \leq \delta,$$

then
$$|f(x_1, u_1) - f(x_2, u_2)| \leq \epsilon \max\{f(x_1, u_1), f(x_2, u_2)\}.$$

Clearly, $\mathcal{A}$ is the set of all integrands $f \in \mathcal{M}$ which do not depend on t.

It is easy to show that an integrand $f = f(x, u) \in C^1(R^n \times R^n)$ belongs to $\mathcal{A}$ if f satisfies Assumptions A(i), A(ii) and also there exists an increasing function $\psi_0 : [0, \infty) \to [0, \infty)$ such that

$$\sup\{|\partial f/\partial x(x, u)|, |\partial f/\partial u(x, u)|\} \leq \psi_0(|x|)(1 + \psi(|u|)|u|)$$

for each $x, u \in R^n$.

We consider the topological subspace $\mathcal{A} \subset \mathcal{M}$ with the relative weak and strong topologies introduced in Section 2.1. Note that $\mathcal{A}$ is the closed subset of $\mathcal{M}$ with the weak topology.

The strong topology is induced by the uniformity which is determined by the following base:

$$E_s(\epsilon) = \{(f, g) \in \mathcal{A} \times \mathcal{A} : |f(x, u) - g(x, u)| \leq \epsilon$$

$$\text{for each } x, u \in R^n\},$$

where $\epsilon > 0$. It is easy to see that the space $\mathcal{A}$ with this uniformity is metrizable and complete.

The weak topology is induced by the uniformity which is determined by the following base:

$$E(N, \epsilon, \lambda) = \{(f, g) \in \mathcal{A} \times \mathcal{A} : |f(x, u) - g(x, u)| \leq \epsilon$$

$$\text{for each } x, u \in R^n \text{ satisfying } |x|, |u| \leq N,$$

$$(|f(x, u)| + 1)(|g(x, u)| + 1)^{-1} \in [\lambda^{-1}, \lambda]$$

$$\text{for each } x, u \in R^n \text{ satisfying } |x| \leq N\},$$

where $N > 0$, $\epsilon > 0$, $\lambda > 1$.

Clearly, the space $\mathcal{A}$ with this uniformity is metrizable and complete (see Proposition 1.3.2).

We consider functionals of the form

$$I^f(T_1, T_2, x) = \int_{T_1}^{T_2} f(x(t), x'(t)) dt \tag{1.1}$$

where $f \in \mathcal{A}$, $0 \leq T_1 < T_2 < +\infty$ and $x : [T_1, T_2] \to R^n$ is an a.c. function.

For $f \in \mathcal{A}$, $y, z \in R^n$ and numbers T_1, T_2 satisfying $0 \leq T_1 < T_2$ we set

$$U^f(T_1, T_2, y, z) = \inf\{I^f(T_1, T_2, x) : x : [T_1, T_2] \to R^n \tag{1.2}$$

is an a.c. function satisfying $x(T_1) = y$, $x(T_2) = z$},

$$\sigma^f(T_1, T_2, y) = \inf\{U^f(T_1, T_2, y, u) : u \in R^n\}. \tag{1.3}$$

It is easy to see that $-\infty < U^f(T_1, T_2, y, z) < +\infty$ for each $f \in \mathcal{A}$, each $y, z \in R^n$ and all numbers T_1, T_2 satisfying $0 \leq T_1 < T_2$.

Recall the definition of a good function given in Section 1.2 and the definition of an overtaking optimal function introduced in Section 1.1.

Let $f \in \mathcal{A}$. We say that an a.c. function $x : [0, \infty) \to R^n$ is an (f)-*good function* if for any a.c function $y : [0, \infty) \to R^n$ the function

$$T \to I^f(0, T, y) - I^f(0, T, x), \ T \in (0, \infty)$$

is bounded from below.

Let $f \in \mathcal{A}$. We say that an a.c. function $x : [0, \infty) \to R^n$ is (f)-overtaking optimal if

$$\limsup_{T \to \infty} \int_0^T [f(x(t), x'(t)) - f(y(t), y'(t))]dt \leq 0$$

for any a.c. function $y : [0, \infty) \to R^n$ satisfying $y(0) = x(0)$.

In this paper we employ the following weakened version of this criterion [12, 42, 88, 89].

Let $f \in \mathcal{A}$. We say that an a.c. function $x : [0, \infty) \to R^n$ is (f)-weakly optimal if

$$\liminf_{T \to \infty} \int_0^T [f(x(t), x'(t)) - f(y(t), y'(t))]dt \leq 0$$

for any a.c. function $y : [0, \infty) \to R^n$ satisfying $y(0) = x(0)$.

Let $f \in \mathcal{A}$. For any a.c. function $x : [0, \infty) \to R^n$ we set

$$J(x) = \liminf_{T \to \infty} T^{-1} I^f(0, T, x). \tag{1.4}$$

Of special interest is the minimal long-run average cost growth rate

$$\mu(f) = \inf\{J(x) : x : [0, \infty) \to R^n \text{ is an a.c. function}\}. \tag{1.5}$$

Clearly $-\infty < \mu(f) < +\infty$ and for every (f)-good function $x : [0, \infty) \to R^n$,

$$\mu(f) = J(x). \tag{1.6}$$

In Section 3.2 we will establish the following result.

PROPOSITION 3.1.1 *For any a.c. function $x : [0, \infty) \to R^n$ either*

$$I^f(0, T, x) - T\mu(f) \to \infty \text{ as } T \to \infty$$

or

$$\sup\{|I^f(0, T, x) - T\mu(f)| : T \in (0, \infty)\} < \infty. \qquad (1.7)$$

Moreover (1.7) holds if and only if x is an (f)-good function.

We denote $d(x, B) = \inf\{|x - y| : y \in B\}$ for $x \in R^n$, $B \subset R^n$. Denote by $\mathrm{dist}(A, B)$ the Hausdorff metric for two sets $A \subset R^n$, $B \subset R^n$ and denote by $\mathrm{Card}(A)$ the cardinality of a set A.

For every bounded a.c. function $x : [0, \infty) \to R^n$ define

$$\Omega(x) = \{y \in R^n : \text{ there exists a sequence } \{t_i\}_{i=0}^{\infty} \subset (0, \infty)$$

$$\text{for which } t_i \to \infty, \ x(t_i) \to y \text{ as } i \to \infty\}. \qquad (1.8)$$

We say that an integrand $f \in \mathcal{A}$ has the asymptotic turnpike property, or briefly (ATP), if $\Omega(v_1) = \Omega(v_2)$ for each pair of (f)-good functions $v_i : [0, \infty) \to R^n$, $i = 1, 2$.

Let $f \in \mathcal{A}$ have the asymptotic turnpike property. Put

$$H(F) = \Omega(v) \qquad (1.9)$$

where $v : [0, \infty) \to R^n$ is an (f)-good function. Clearly, $H(f)$ does not depend on v. By Theorem 1.2.1, H(f) is a compact subset of R^n. We say that $H(f)$ is the turnpike of f.

In this chapter we prove the following results.

THEOREM 3.1.1 *There exists a set $\mathcal{F} \subset \mathcal{A}$ which is a countable intersection of open (in the weak topology) everywhere dense (in the strong topology) subsets of $\mathcal{A}$ such that each $f \in \mathcal{F}$ has the asymptotic turnpike property.*

Theorem 3.1.1 describes the limit behavior of (f)-good functions for a generic $f \in \mathcal{A}$. The following result establishes the existence of an (f)-weakly optimal function for each $f \in \mathcal{A}$ which has (ATP) and for each initial state $x \in R^n$.

THEOREM 3.1.2 *Assume that $f \in \mathcal{A}$ has the asymptotic turnpike property. Then for each $x \in R^n$ there exists an (f)-weakly optimal function $X : [0, \infty) \to R^n$ satisfying $X(0) = x$.*

It follows from Theorems 3.1.1 and 3.1.2 that for a generic $f \in \mathcal{A}$ and every $x \in R^n$ there exists an (f)-weakly optimal function $X : [0, \infty) \to R^n$ satisfying $X(0) = x$.

THEOREM 3.1.3 *Assume that $f \in \mathcal{A}$ has (ATP) and $H(f)$ is the turnpike of f. Let ϵ be a positive number. Then there exist an integer $L \geq 1$*

and a neighborhood $\mathcal{U}$ of f in $\mathcal{A}$ with the weak topology such that the following property holds:

If $g \in \mathcal{U}$ and if $v : [0, \infty) \to R^n$ is a (g)-good function, then

$$\text{dist}(H(f), \{v(t) : t \in [T, T + L]\}) \leq \epsilon \text{ for all large } T.$$

THEOREM 3.1.4 *Assume that $f \in \mathcal{A}$ has (ATP) and $H(f) \subset R^n$ is the turnpike of f.*

Let $M_0, M_1, \epsilon > 0$. Then there exists a neighborhood $\mathcal{U}$ of f in $\mathcal{A}$ with the weak topology, numbers $l, S > 0$ and integers $L, _* Q \geq 1$ such that for each $g \in \mathcal{U}$, each pair of numbers $T_1 \in [0, \infty)$, $T_2 \in [T_1 + L + lQ_*, \infty)$ and each a.c. function $v : [T_1, T_2] \to R^n$ which satisfies

$$|v(T_i)| \leq M_1, \ i = 1, 2, \quad I^g(T_1, T_2, v) \leq U^g(T_1, T_2, v(T_1), v(T_2)) + M_0$$

the following properties hold:

$$|v(t)| \leq S \text{ for all } t \in [T_1, T_2];$$

there exist sequences of numbers $\{b_i\}_{i=1}^Q$, $\{c_i\}_{i=1}^Q \subset [T_1, T_2]$ such that

$$Q \leq Q_*, \ 0 \leq c_i - b_i \leq l, \ i = 1, \ldots, Q,$$

$$\text{dist}(H(f), \{v(t) : t \in [T, T + L]\}) \leq \epsilon$$

$$\text{for each } T \in [T_1, T_2 - L] \setminus \cup_{i=1}^Q [b_i, c_i].$$

Theorem 3.1.4 shows that if an integrand $f \in \mathcal{A}$ has the asymptotic turnpike property with the turnpike $H(f) \subset R^n$, then for any finite horizon problem the turnpike property holds with the set $H(f)$ being the attractor.

Let $k \geq 1$ be an integer. Denote by $\mathcal{A}_k$ the space of all integrands $f \in \mathcal{A} \cap C^k(R^{2n})$. For $p = (p_1, \ldots, p_{2n}) \in \{0, \ldots, k\}^{2n}$ and $f \in C^k(R^{2n})$ we set

$$|p| = \sum_{i=1}^{2n} p_i, \ D^p f = \partial^{|p|} f / \partial y_1^{p_1} \ldots \partial y_{2n}^{p_{2n}}.$$

For the set $\mathcal{A}_k$ we consider the uniformity which is determined by the following base:

$$E(N, \epsilon, \lambda) = \{(f, g) \in \mathcal{A}_k \times \mathcal{A}_k : |D^p f(x, u) - D^p g(x, u)| \leq \epsilon$$

$$\text{for each } (x, u) \in R^{2n} \text{ satisfying } |x|, |u| \leq N$$

$$\text{and each } p \in \{0, \ldots, k\}^{2n} \text{ such that } |p| \leq k,$$

$$|f(x, u) - g(x, u)| \leq \epsilon \text{ for each } x, u \in R^n \text{ satisfying } |x|, |u| \leq N,$$

$$(|f(x,u)| + 1)(|g(x,u)| + 1)^{-1} \in [\lambda^{-1}, \lambda]$$

$$\text{for each } x, u \in R^n \text{ such that } |x| \leq N\},$$

where $N > 0$, $\epsilon > 0$, $\lambda > 1$.

Clearly the uniform space $\mathcal{A}_k$ is Hausdorff and has a countable base. Therefore $\mathcal{A}_k$ is metrizable [37]. It is easy to verify that the uniform space $\mathcal{A}_k$ is complete. We establish the following result.

THEOREM 3.1.5 *Let* $k \geq 1$ *be an integer. Then there exists a set* $\mathcal{F}_k \subset \mathcal{A}_k$ *which is a countable intersection of open everywhere dense subsets of* $\mathcal{A}_k$ *such that each* $f \in \mathcal{F}_k$ *has the asymptotic turnpike property.*

The results of this chapter have been established in [93].

Note that the turnpike result of this chapter (Theorem 3.1.4) is weaker than the turnpike result of Chapter 2 (Theorem 2.1.4). In Chapter 2 for nonautonomous integrands we obtained that an approximate solution v on an interval $[T_1, T_2]$ is close to the turnpike Z^f for most points $t \in [T_1, T_2]$. In this chapter for autonomous integrands we obtain only that the image $v([T, T + L])$ of the interval $[T, T + L] \subset [T_1, T_2]$ is close to the turnpike $H(f)$ for most points $T \in [T_1, T_2]$. (Here $L > 0$ is a constant.) Moreover, for the nonautonomous case the set of all bad points is contained in the union of two intervals $[T_1, T_1 + L_0]$ and $[T_2 - L_0, T_2]$ where L_0 is a positive constant. In the autonomous case considered in this chapter the set of all bad points of the interval $[T_1, T_2]$ is contained in the union of a finite number of closed intervals and we can only say that the length of each of these intervals and their number are bounded by a constant which do not depend on $[T_1, T_2]$ and $v(T_1), v(T_2)$.

Chapter 3 is organized as follows. Proposition 3.1.1 is proved in Section 3.2. In Section 3.3 we will establish a weakened version of Theorem 3.1.3. The continuity of the function $U^f(T_1, T_2, x, y)$ is studied in Section 3.4. Section 3.5 contains some useful results on discrete-time control systems while Theorem 3.1.2 is proved in Section 3.6. Section 3.7 contains auxiliary results for Theorem 3.1.1 while Section 3.8 contains auxiliary results for Theorems 3.1.3 and 3.1.4. Theorem 3.1.4 is proved in Section 3.9 and Theorem 3.1.3 is proved in Section 3.10. Section 3.11 contains the proofs of Theorems 3.1.1 and 3.1.5. Certain examples are given in Section 3.12.

3.2.   Proof of Proposition 3.1.1

In the sequel we associate with any $f \in \mathcal{A}$ a related discrete-time control system. We need the following result established in [39] (see also [16]) for such discrete-time control systems.

PROPOSITION 3.2.1 *Let $K \subset R^n$ be a compact set, $v : K \times K \to R^1$ be a continuous function and define*

$$\mu(v) = \inf \left\{ \liminf_{N \to \infty} N^{-1} \sum_{i=0}^{N-1} v(z_i, z_{i+1}) : \{z_i\}_{i=0}^{\infty} \subset K \right\},$$

$$\pi^v(x)$$

$$= \inf \left\{ \liminf_{N \to \infty} N^{-1} \sum_{i=0}^{N-1} [v(z_i, z_{i+1}) - \mu(v)] : \{z_i\}_{i=0}^{\infty} \subset K, \ z_0 = x \right\},$$

$$\theta^v(x, y) = v(x, y) - \mu(v) + \pi^v(y) - \pi^v(x)$$

for $x, y \in K$. Then $\pi^v : K \to R^1$, $\theta^v : K \times K \to R^1$ are continuous functions, θ^v is nonnegative and $E(x) = \{y \in K : \theta^v(x, y) = 0\}$ is nonempty for every $x \in K$.

Proof of Proposition 3.1.1. Let $f \in \mathcal{A}$, $z \in R^n$ and let an a.c. function $Z^f : [0, \infty) \to R^n$ be as guaranteed in Theorem 1.2.1. It follows from Theorem 1.2.1, (1.4), (1.6) and assumption A(ii) that

$$\mu(f) = \liminf_{N \to \infty} N^{-1} I^f(0, N, Z^f)$$

where N is an integer. Combined with Theorem 1.2.1 this equality implies that

$$\mu(f) = \liminf_{N \to \infty} N^{-1} \sum_{i=0}^{N-1} U^f(i, i+1, Z^f(i), Z^f(i+1)). \tag{2.1}$$

It follows from Theorems 1.2.1 and 1.2.2 that there is a number

$$Q > \sup\{|Z^f(t)| : \ t \in [0, \infty)\} \tag{2.2}$$

such that

$$\limsup_{t \to \infty} |x(t)| < Q \tag{2.3}$$

for each (f)-good function $x : [0, \infty) \to R^n$. Put

$$B_Q = \{x \in R^n : |x| \le Q\},$$

$$v(x, y) = U^f(0, 1, x, y) \ (x, y \in B_Q). \tag{2.4}$$

It is easy to see that for each $x, y \in R^n$ and each integer $i \ge 0$,

$$U^f(i, i+1, x, y) = U^f(0, 1, x, y). \tag{2.5}$$

Proposition 1.3.7 implies that the function $v : B_Q \times B_Q \to R^1$ is continuous. We show that

$$\sup\left\{ \left| \sum_{i=0}^{N-1} [v(Z^f(i), Z^f(i+1)) - \mu(f)] \right| : N = 1, 2, \ldots \right\} < \infty. \quad (2.6)$$

Fix an integer $N \geq 2$. By Proposition 3.2.1 there exists a sequence $\{x_j\}_{j=0}^N \subset B_Q$ such that

$$x_0 = z, \ x_N = Z^f(N), \ \theta^v(x_i, x_{i+1}) = 0, \ (0 \leq i < N-1). \quad (2.7)$$

Proposition 3.2.1 implies that

$$\sum_{i=0}^{N-1} [v(x_i, x_{i+1}) - \mu(v)] = \pi^v(z) - \pi^v(x_{N-1}) + v(x_{N-1}, x_N) - \mu(v). \quad (2.8)$$

By (2.2) and Proposition 3.2.1,

$$\sum_{i=0}^{N-1} [v(Z^f(i), Z^f(i+1)) - \mu(v)] \geq \pi^v(z) - \pi^v(Z^f(N)). \quad (2.9)$$

On the other hand it follows from Assertion 4 of Theorem 1.2.1, (2.4), (2.5), (2.7), and Corollary 1.3.1 that

$$\sum_{i=0}^{N-1} v(Z^f(i), Z^f(i+1)) = \sum_{i=0}^{N-1} U^f(i, i+1, Z^f(i), Z^f(i+1)) = I^f(0, N, Z^f)$$

$$\quad (2.10)$$

$$\leq \sum_{i=0}^{N-1} U^f(i, i+1, x_i, x_{i+1}) = \sum_{i=0}^{N-1} v(x_i, x_{i+1}).$$

Combining (2.8)-(2.10) we obtain that for any integer $N \geq 2$,

$$\left| \sum_{i=0}^{N-1} [v(Z^f(i), Z^f(i+1)) - \mu(v)] \right| \leq 2 \sup\{|\pi^v(y)| : y \in B_Q\} \quad (2.11)$$

$$+ 2 \sup\{|v(x, y)| : x, y \in B_Q\}.$$

By (2.1), (2.4), (2.5) and (2.11),

$$\mu(f) = \mu(v). \quad (2.12)$$

Combined with (2.11) this equality implies (2.6). We show that

$$\sup\{|I^f(0, T, Z^f) - T\mu(f)| : T \in (0, \infty)\} < \infty. \quad (2.13)$$

Let $T > 0$. There exists an integer $N \geq 0$ such that $N \leq T < N + 1$. By Assumption A(ii),

$$I^f(T, N + 1, Z^f) \geq -a. \tag{2.14}$$

On the other hand it follows from (2.2), (2.4), (2.5), Theorem 1.2.1 and Assumption A(ii) that

$$I^f(T, N + 1, Z^f) = I^f(N, N + 1, Z^f) - I^f(N, T, Z^f)$$

$$\leq v(Z^f(N), Z^f(N + 1)) + a.$$

By these relations, Theorem 1.2.1, (2.4), (2.5) and (2.14) ,

$$|I^f(0, T, Z^f) - T\mu(f)| \leq |I^f(0, N + 1, Z^f) - (N + 1)\mu(f)|$$

$$+ |I^f(T, N + 1, Z^f) - (N + 1 - T)\mu(f)|$$

$$\leq |\sum_{i=0}^{N}[v(Z^f(i), Z^f(i+1)) - \mu(f)]| + |\mu(f)| + a + \sup\{|v(x, y)| : x, y \in B_Q\}.$$

Together with (2.6) this implies (2.13). Proposition 3.1.1 now follows from (2.13) and Theorem 1.2.1.

3.3. Weakened version of Theorem 3.1.3

The proof of Theorem 3.1.3 is difficult and it is based on a number of auxiliary results. Now we are ready to prove its weakened version which establishes the convergence property of Theorem 3.1.3 when $g = f$. We begin this section with the following useful property of good functions.

PROPOSITION 3.3.1 *Let $g \in \mathcal{A}$ and let $y : [0, \infty) \rightarrow R^n$ be a (g)-good function. Then for each $\epsilon > 0$ there exists $T_0 > 0$ such that the following property holds:*
If $T \geq T_0$ and $\bar{T} > T$, then

$$I^g(T, \bar{T}, y) \leq U^g(T, \bar{T}, y(T), y(\bar{T})) + \epsilon.$$

Proof. Let us assume the converse. Then there exist $\epsilon > 0$ and sequences $\{T_i\}_{i=1}^{\infty}, \{\bar{T}_i\}_{i=1}^{\infty} \subset (0, \infty)$ such that for each natural number i,

$$T_i < \bar{T}_i < T_{i+1},$$

$$I^g(T_i, \bar{T}_i, y) > U^g(T_i, \bar{T}_i, y(T_i), y(\bar{T}_i)) + \epsilon.$$

Put $\bar{T}_0 = 0$. It follows from Corollary 1.3.1 that there is an a.c. function $x : [0, \infty) \to R^n$ such that for each natural number i,

$$x(t) = y(t), \ t \in [\bar{T}_i, T_{i+1}],$$

$$I^g(T_i, \bar{T}_i, x) = U^g(T_i, \bar{T}_i, x(T_i), x(\bar{T}_i)).$$

It is not difficult to verify that

$$I^g(0, \bar{T}_i, y) - I^g(0, \bar{T}_i, x) \to \infty \text{ as } i \to \infty.$$

Therefore y is not a (g)-good function. The contradiction we have reached proves the proposition.

We will use the next auxiliary result in the proof of the weakened version of Theorem 3.1.3 in order to show that a certain function is good.

PROPOSITION 3.3.2 *Let $f \in A$ and let $x : [0, \infty) \to R^n$ be an a.c. function such that*

$$\sup\{|x(t)| : \ t \in [0, \infty)\} < \infty$$

and

$$\sup\{I^f(0, i, x) - U^f(0, i, x(0), x(i)), \ i = 1, 2, \ldots\} < \infty.$$

Then x is an (f)-good function.

Proof. Choose positive numbers S_0 and S_1 such that

$$|x(t)| \leq S_0 \text{ for all } t \in [0, \infty)$$

and

$$I^f(0, i, x) \leq U^f(0, i, x(0), x(i)) + S_1, \ i = 1, 2, \ldots.$$

Fix $z \in R^n$ satisfying $|z| \leq S_0$ and let an a.c. function $Z^f : [0, \infty) \to R^n$ be as guaranteed in Theorem 1.2.1. It follows from Theorems 1.2.1 and 1.2.2 that there exists a number

$$Q > \sup\{|Z^f(t)| : \ t \in [0, \infty)\} + S_0$$

such that for each (f)-good function $y : [0, \infty) \to R^n$,

$$\limsup_{t \to \infty} |y(t)| < Q.$$

Put

$$B_Q = \{y \in R^n : \ |y| \leq Q\},$$

$$v(y_1, y_2) = U^f(0, 1, y_1, y_2), \quad y_1, y_2 \in B_Q.$$

By Proposition 1.3.7 the function $v : B_Q \times B_Q \to R^1$ is continuous. It was shown in Section 3.2 that $\mu(f) = \mu(v)$.

Fix an integer $N \geq 2$. By Proposition 3.2.1 there exists a sequence

$$\{y_i\}_{i=0}^N \subset B_Q$$

such that

$$y_0 = x(0), \ \theta^v(y_i, y_{i+1}) = 0 \ (0 \leq i < N - 1), y_N = x(N).$$

It follows from these relations, the conditions of the proposition, Proposition 3.2.1 and Corollary 1.3.1 that

$$\sum_{i=0}^{N-1} [v(y_i, y_{i+1}) - \mu(v)] = \pi^v(x(0)) - \pi^v(y_{N-1}) + v(y_{N-1}, x(N)) - \mu(v),$$

$$I^f(0, N, x) - N\mu(f) \leq U^f(0, N, x(0), x(N)) + S_1 - N\mu(v)$$

$$\leq \sum_{i=0}^{N-1} [v(y_i, y_{i+1}) - \mu(v)] + S_1$$

$$\leq S_1 + 2\sup\{|\pi^v(h)| : h \in B_Q\} + 2\sup\{|v(h_1, h_2)| : h_1, h_2 \in B_Q\}.$$

It follows from these relations and Proposition 3.1.1 that x is an (f)-good function. The proposition is proved.

The following result is a weakened version of Theorem 3.1.3.

THEOREM 3.3.1 *Assume that $f \in \mathcal{A}$ has the asymptotic turnpike property with the turnpike $H(f) \subset R^n$. Then for each $\epsilon > 0$ there exists an integer $L \geq 1$ such that the following property holds:*
If $v : [0, \infty) \to R^n$ is an (f)-good function, then

$$dist(H(f), \{v(t) : t \in [T, T + L]\}) \leq \epsilon \text{ for all large } T.$$

Proof. Let ϵ be a positive number. Assume that the assertion of the theorem does not hold. Then for every integer $N \geq 1$ there exists an (f)-good function $x_N : [0, \infty) \to R^n$ such that

$$\limsup_{T \to \infty} \ dist(H(f), \{x_N(t) : t \in [T, T + N]\}) \geq \epsilon. \tag{3.1}$$

It follows from Theorem 1.2.2 that there exists $M_0 > 0$ such that

$$\limsup_{t \to \infty} |x(t)| < M_0 \tag{3.2}$$

for each (f)-good function $x : [0, \infty) \to R^n$.

By the definition of M_0, Propositions 3.1.1 and 3.3.1 we may assume that

$$|x_N(t)| < M_0, \ (t \in [0, \infty), \ N = 1, 2, \ldots) \tag{3.3}$$

and that for each integer $N \geq 1$ and each pair of numbers $T_1 \geq 0$, $T_2 > T_1$,

$$I^f(T_1, T_2, x_N) \leq U^f(T_1, T_2, x_N(T_1), x_N(T_2)) + N^{-1}. \tag{3.4}$$

It is not difficult to see that if $x : [0, \infty) \to R^n$ is an (f)-good function and γ is the positive number, then

$$d(x(t), H(f)) \leq \gamma \text{ for all large } t.$$

Therefore we may assume that

$$d(x_N(t), H(f)) \leq 4^{-1}\epsilon \ (t \in [0, \infty), \ N = 1, 2, \ldots). \tag{3.5}$$

Let $N \geq 4$ be an integer. By (3.1) there exists $T_N \in [0, \infty)$ such that

$$\text{dist}(H(f), \{x_N(t) : \ t \in [T_N, T_N + N]\}) \geq 7 \cdot 8^{-1}\epsilon. \tag{3.6}$$

By (3.5) and (3.6) there exists $h_N \in H(f)$ for which

$$d(h_N, \{x_N(t) : \ t \in [T_N, T_N + N]\}) \leq 2^{-1}\epsilon. \tag{3.7}$$

Fix an integer $j_N \geq 0$ for which $j_N \leq T_N < j_N + 1$ and put

$$v_N(t) = x_N(t + 1 + j_N) \quad (t \in [0, N - 2]). \tag{3.8}$$

It follows from the inequalities (3.3), (3.4) and (3.7) that

$$|v_N(t)| < M_0 \quad (t \in [0, N - 2]), \tag{3.9}$$

$$I^f(T_1, T_2, v_N) \leq U^f(T_1, T_2, v_N(T_1), v_N(T_2)) + N^{-1} \tag{3.10}$$

for each $T_1 \in [0, N - 2)$, $T_2 \in (T_1, N - 2]$ and that

$$d(h_N, \{v_N(t) : \ t \in [0, N - 2]\}) \geq 2^{-1}\epsilon. \tag{3.11}$$

Fix an integer $j \geq 1$. By (3.9), (3.10) and Proposition 1.3.7,

$$\sup\{I^f(0, j, v_N) : \ N \text{ is an integer}, \ N > 4 + j\}$$

$$\leq \sup\{U^f(0, j, v_N(0), v_N(j)) : \ N \text{ is an integer}, \ N \geq 4 + j\} + 1 < \infty.$$

By this inequality and Proposition 1.3.5 there exist an a.c. function $v_* : [0, \infty) \to R^n$ and a subsequence $\{v_{N_i}\}_{i=1}^{\infty}$ such that for any integer $j \geq 1$,

$$v_{N_k}(t) \to v_*(t) \text{ as } k \to \infty \text{ uniformly in } [0, j], \tag{3.12}$$

$$v'_{N_k} \to v'_* \text{ as } k \to \infty \text{ weakly in } L^1(R^n; (0, j))$$

and

$$I^f(0, j, v_*) \leq \liminf_{k \to \infty} I^f(0, j, v_{N_k}). \tag{3.13}$$

(3.9) and (3.12) imply that $|v_*(t)| \leq M_0$ for all $t \in [0, \infty)$. It follows from Proposition 1.3.7, (3.12), (3.13) and (3.10) that for any integer $j \geq 1$,

$$I^f(0, j, v_*) = U^f(0, j, v_*(0), v_*(j)).$$

Combined with Proposition 3.3.2 this equality implies that v_* is an (f)-good function.

We may assume without loss of generality that there exists

$$h_* = \lim_{k \to \infty} h_{N_k} \in H(f). \tag{3.14}$$

Since v_* is an (f)-good function we have

$$\Omega(v_*) = H(f).$$

Therefore there exists a sequence $\{t_i\}_{i=1}^{\infty} \subset (0, \infty)$ such that $t_i \to \infty$ and $v_*(t_i) \to h_*$ as $i \to \infty$.

On the other hand it follows from (3.11), (3.12), (3.14) that

$$d(h_*, \{v_*(t) : t \in [0, \infty)\}) \geq 2^{-1}\epsilon.$$

The contradiction we have reached proves the theorem.

3.4. Continuity of the function $U^f(T_1, T_2, x, y)$

THEOREM 3.4.1 *Assume that $f \in \mathcal{A}$. Then the mapping*

$$(T_1, T_2, x, y) \to U^f(T_1, T_2, x, y)$$

is continuous for $T_1 \in [0, \infty)$, $T_2 \in (T_1, \infty)$, $x, y \in R^n$.

Proof. Lower semicontinuity. Let $x, y \in R^n$, $T_1 \in [0, \infty)$, $T_2 \in (T_1, \infty)$, $\{x_i\}_{i=1}^{\infty}$, $\{y_i\}_{i=1}^{\infty} \subset R^n$, $\{T_{1i}\}_{i=1}^{\infty}$, $\{T_{2i}\}_{i=1}^{\infty} \subset [0, \infty)$, $T_{2i} > T_{1i}$ $(i = 1, 2, \ldots)$,

$$x_i \to x, \ y_i \to y, \ T_{1i} \to T_1, \ T_{2i} \to T_2 \text{ as } i \to \infty. \tag{4.1}$$

By Corollary 1.3.1 for each integer $i \geq 1$ there exists an a.c. function $z_i : [T_{1i}, T_{2i}] \to R^n$ such that

$$z_i(T_{1i}) = x_i, \ z_i(T_{2i}) = y_i, \ I^f(T_{1i}, T_{2i}, z_i) = U^f(T_{1i}, T_{2i}, x_i, y_i). \tag{4.2}$$

By Proposition 1.3.6 and (4.1),

$$\sup\{|U^f(T_{1i}, T_{2i}, x_i, y_i)| : \ i = 1, 2, \ldots\} < \infty. \tag{4.3}$$

We may assume that there exists $\lim_{i\to\infty} U^f(T_{1i}, T_{2i}, x_i, y_i)$. Fix $\delta \in (0, 1)$ and set

$$T_1(\delta) = \sup\{0, T_1 - \delta\}. \tag{4.4}$$

There exists a natural number $N(\delta)$ such that

$$T_1 + \delta \geq T_{1i} \geq T_1(\delta), \ T_2 - \delta \leq T_{2i} \leq T_2 + \delta \text{ for all integers } i \geq N(\delta). \tag{4.5}$$

Set

$$\gamma = \sup\{|f(h, 0)| : \ h \in R^n, \ |h| \leq \sup\{|x_i| + |y_i| : \ i = 1, 2, \ldots\}\}. \tag{4.6}$$

For every integer $i \geq N(\delta)$ we define an a.c. function $z_{\delta i} : [T_1(\delta), T_2 + \delta] \to R^n$ as follows:

$$z_{\delta 1}(t) = z_i(T_{1i}), \ t \in [T_1(\delta), T_{1i}], \ z_{\delta i}(t) = z_i(t), \ t \in [T_{1i}, T_{2i}],$$

$$z_{\delta i}(t) = z_i(T_{2i}), \ t \in [T_{2i}, T_2 + \delta]. \tag{4.7}$$

It follows from the relations (4.2), (4.3), (4.5) and (4.7) that the sequence $\{I^f(T_1(\delta), T_2 + \delta, z_{\delta i})\}_{i=N(\delta)}^{\infty}$ is bounded. By Proposition 1.3.5 there exist an a.c. function $z_\delta : [T_1(\delta), T_2 + \delta] \to R^n$ and a subsequence $\{z_{\delta i_k}\}_{k=1}^{\infty}$ $(i_1 \geq N(\delta))$ such that

$$z_{\delta i_k}(t) \to z_\delta(t) \text{ as } k \to \infty \text{ uniformly in } [T_1(\delta), T_2 + \delta],$$

$$z'_{\delta i_k} \to z'_\delta \text{ as } k \to \infty \text{ weakly in } L^1(R^n; (T_1, T_2)),$$

$$I^f(T_1(\delta), T_2 + \delta, z_\delta) \leq \liminf_{k\to\infty} I^f(T_1(\delta), T_2 + \delta, z_{\delta i_k}). \tag{4.8}$$

By (4.6)-(4.8) and (4.2) for any integer $k \geq 1$,

$$I^f(T_1(\delta), T_2 + \delta, z_{\delta i_k}) = I^f(T_1(\delta), T_{1i_k}, z_{\delta i_k}) + I^f(T_{1i_k}, T_{2i_k}, z_{\delta i_k})$$

$$+ I^f(T_{2i_k}, T_2 + \delta, z_{\delta i_k}) \leq (T_{1i_k} - T_1(\delta))\gamma + U^f(T_{1i_k}, T_{2i_k}, x_{i_k}, y_{i_k})$$

$$+ (T_2 + \delta - T_{2i_k})\gamma \leq U^f(T_{1i_k}, T_{2i_k}, x_{i_k}, y_{i_k}) + 4\gamma\delta. \tag{4.9}$$

By (4.8) and (4.9),

$$I^f(T_1(\delta), T_2 + \delta, z_\delta) \leq \lim_{i\to\infty} U^f(T_{1i}, T_{2i}, x_i, y_i) + 4\gamma\delta.$$

By this relation, (4.4) and Assumption A(ii),

$$I^f(T_1, T_2, z_\delta) \leq \lim_{i\to\infty} U^f(T_{1i}, T_{2i}, x_i, y_i) + 4\gamma\delta + 2a\delta.$$

To complete the proof of the lower semicontinuity it is sufficient to show that

$$Z_\delta(T_1) = x, \ z_\delta(T_2) = y. \tag{4.10}$$

By (4.8),

$$z_{\delta i_k}(T_1) \to z_\delta(T_1), \ z_{\delta i_k}(T_2) \to z_\delta(T_2) \text{ as } k \to \infty. \tag{4.11}$$

It follows from (4.1), (4.3), (4.9) and Proposition 1.3.4 that

$$z_{\delta i_k}(T_1) - z_{\delta i_k}(T_{1 i_k}) \to 0, \ z_{\delta i_k}(T_2) - z_{\delta i_k}(T_{2 i_k}) \to 0 \text{ as } k \to \infty.$$

Together with (4.11), (4.7), (4.2) and (4.1) this relation implies (4.10). The lower semicontinuity is proved.

Upper semicontinuity. Let $x, y \in R^n$, $T_1 \in [0, \infty)$, $T_2 \in (T_1, \infty)$,

$$\{x_i\}_{i=1}^\infty, \ \{y_i\}_{i=1}^\infty \subset R^n,$$

$$\{T_{1i}\}_{i=1}^\infty, \ \{T_{2i}\}_{i=1}^\infty \subset [0, \infty),$$

$$T_{2i} > T_{1i} \ (i = 1, 2, \ldots),$$

$$x_i \to x, \ y_i \to y, \ T_{1i} \to T_1, \ T_{2i} \to T_2 \text{ as } i \to \infty. \tag{4.12}$$

By Corollary 1.3.1 there exists an a.c. function $z : [T_1, T_2] \to R^n$ such that

$$z(T_1) = x, \ x(T_2) = y,$$
$$I^f(T_1, T_2, z) = U^f(T_1, T_2, x, y). \tag{4.13}$$

Fix $\delta \in (0, 1)$ and define γ, $T_1(\delta)$ by (4.6) and (4.4). There is an integer $N(\delta) \geq 1$ satisfying (4.5). Define an a.c. function $z_\delta : [T_1(\delta), T_2 + \delta] \to R^n$ as follows:

$$z_\delta(t) = x \ (t \in [T_1(\delta), T_1]), \ z_\delta(t) = z(t) \ (t \in [T_1, T_2]),$$

$$z_\delta(t) = y \ (t \in [T_2, T_2 + \delta]). \tag{4.14}$$

For an integer $i \geq N(\delta)$ we set

$$b_i = (T_{2i} - T_{1i})^{-1}[y_i + z_\delta(T_{1i}) - x_i - z_\delta(T_{2i})], \ a_i = x_i - z_\delta(T_{1i}) - b_i T_{1i},$$

$$z_{\delta i}(t) = z_\delta(t) + a_i + b_i t \ (t \in [T_1(\delta), T_2 + \delta]). \tag{4.15}$$

Clearly

$$z_{\delta i}(T_{1i}) = x_i, \ z_{\delta i}(T_{2i}) = y_i, \ (i \text{ is an integer}, \ i \geq N(\delta)), \tag{4.16}$$

$$a_i \to 0, \ b_i \to 0 \text{ as } i \to \infty. \tag{4.17}$$

We will show that

$$I^f(T_1(\delta), T_2 + \delta, z_{\delta i})$$
$$\to I^f(T_1(\delta), T_2 + \delta, z_\delta) \text{ as } i \to \infty. \tag{4.18}$$

It is easy to see that $|I^f(T_1(\delta), T_2 + \delta, z_\delta)| < \infty$. Set

$$S_0 = \sup\{|z_\delta(t)| : t \in [T_1(\delta), T_2 + \delta]\}. \tag{4.19}$$

Fix $\epsilon > 0$. There exists a number $\Delta \in (0, 1)$ such that

$$\Delta(|I^f(T_1(\delta), T_2 + \delta, z_\delta)| + a(T_2 - T_1 + 2\delta)) < 8^{-1}\epsilon \tag{4.20}$$

(recall a in Assumption A(ii)).

By Proposition 1.3.1 there exist $\Gamma_0 > 0$, $\delta_0 \in (0, 1)$ such that the following property holds:

If $u_1, u_2, h_1, h_2 \in R^n$ satisfy

$$|h_i| \le S_0 + 4, \ |u_i| \ge \Gamma_0 \ (i = 1, 2), \ |u_1 - u_2|, \ |h_1 - h_2| \le \delta_0, \tag{4.21}$$

then

$$|f(h_1, u_1) - f(h_2, u_2)| \le \Delta \min\{f(h_1, u_1), f(h_2, u_2)\}. \tag{4.22}$$

Since the function f is continuous there exists $\delta_1 \in (0, \delta_0)$ such that the following property holds:

If $h_1, h_2, u_1, u_2 \in R^n$ satisfy

$$|h_i|, |u_i| \le \Gamma_0 + S_0 + 6, \ i = 1, 2, \ \max\{|h_1 - h_2|, |u_1 - u_2|\} \le \delta_1, \tag{4.23}$$

then

$$|f(h_1, u_1) - f(h_2, u_2)| \le [8(T_2 - T_1 + 2)]^{-1}\epsilon. \tag{4.24}$$

It follows from (4.17) that there is an integer $N_1 > N(\delta)$ such that

$$|b_i| \le 2^{-1}\delta_1, \ |a_i + b_i t| \le 2^{-1}\delta_1 \ (t \in [T_1(\delta), T_2 + \delta]) \tag{4.25}$$

for any integer $i \ge N_1$.

Assume that an integer $i \ge N_1$ and estimate

$$I^f(T_1(\delta), T_2 + \delta, z_{\delta i}) - I^f(T_1(\delta), T_2 + \delta, z_\delta).$$

Put

$$E_1 = \{t \in [T_1(\delta), T_2 + \delta] : |z'_\delta(t)| \ge \Gamma_0 + 1\}, \ E_2 = [T_1(\delta), T_2 + \delta] \setminus E_1. \tag{4.26}$$

Clearly

$$|I^f(T_1(\delta), T_2 + \delta, z_{\delta i}) - I^f(T_1(\delta), T_2 + \delta, z_\delta)| = \sigma_1 + \sigma_2 \tag{4.27}$$

with

$$\sigma_i = \int_{E_i} |f(z_\delta(t), z'_\delta(t)) - f(z_{\delta i}(t), z'_{\delta i}(t))| dt, \ i = 1, 2. \qquad (4.28)$$

By (4.26), (4.19), (4.25), (4.15) and the definition of Γ_0, δ_0 (see (4.21), (4.22)) for each $t \in E_1$,

$$|f(z_\delta(t), z'_\delta(t)) - f(z_{\delta i}(t), z'_{\delta i}(t))| \leq \Delta f(z_\delta(t), z'_\delta(t)).$$

Combined with (4.20) and Assumption A(ii) this inequality implies that

$$\sigma_1 \leq \Delta \int_{E_1} f(z_\delta(t), z'_\delta(t)) dt$$

$$\leq \Delta (I^f(T_1(\delta), T_2 + \delta, z_\delta) + a(T_2 - T_1 + 2\delta)) \leq 8^{-1}\epsilon. \qquad (4.29)$$

It follows from (4.26), (4.19), (4.25), (4.15) and the definition of δ_1 (see (4.23), (4.24)) that

$$|f(z_\delta(t), z'_\delta(t)) - f(z_{\delta i}(t), z'_{\delta i}(t))| \leq [8(T_2 - T_1 + 2)]^{-1}\epsilon$$

for each $t \in E_2$ and

$$\sigma_2 \leq 8^{-1}\epsilon. \qquad (4.30)$$

By (4.27), (4.29) and (4.30) for each integer $i \geq N_1$,

$$|I^f(T_1(\delta), T_2 + \delta, z_\delta) - I^f(T_1(\delta), T_2 + \delta, z_{\delta i})| \leq \epsilon.$$

Therefore we have proved (4.18).

By (4.16), (4.4), (4.5), Assumption A(ii) and (4.18) for any integer $i \geq N(\delta)$,

$$U^f(T_{1i}, T_{2i}, x_i, y_i) \leq I^f(T_{1i}, T_{2i}, z_{\delta i}) \leq I^f(T_1(\delta).T_2 + \delta, z_{\delta i}) + 4\delta a$$

$$\rightarrow I^f(T_1(\delta), T_2 + \delta, z_\delta) + 4a\delta \text{ as } i \to \infty.$$

Together with (4.14), (4.4), (4.6) and (4.13) this implies that

$$\limsup_{i \to \infty} U^f(T_{1i}, T_{2i}, x_i, y_i) \leq I^f(T_1(\delta), T_2 + \delta, z_\delta) + 4a\delta$$

$$\leq I^f(T_1, T_2, z) + 4a\delta + 2\delta\gamma \leq 4a\delta + 2\delta\gamma + U^f(T_1, T_2, x, y).$$

This completes the proof of the upper semicontinuity. The theorem is proved.

3.5.   Discrete-time control systems

In the sequel we associate with any $f \in \mathcal{A}$ a related discrete-time control system. In this section we establish some useful properties of such systems.

Consider a continuous function $v : R^n \times R^n \to R^1$ satisfying

$$v(x, y) \to \infty \text{ as } |x| + |y| \to \infty.$$

We have the following result [39].

PROPOSITION 3.5.1 *Given a compact set $C \subset R^n$ there is a ball $B \subset R^n$ such that for every sequence $\{z_k\}_{k=0}^{\infty} \subset R^n$ not included in B with $z_0 \in C$ there exists a sequence $\{s_k\}_{k=1}^{\infty} \subset B$ with $s_0 = z_0$ such that*

$$\sum_{k=0}^{N} v(s_k, s_{k+1}) < \sum_{k=0}^{N} v(z_k, z_{k+1}) \text{ for all large } N.$$

Let $x \in R^n$. Define

$$\mu(v) = \inf \left\{ \liminf_{N \to \infty} N^{-1} \sum_{k=0}^{N-1} v(z_k, z_{k+1}) : \{z_k\}_{k=0}^{\infty} \subset R^n, \ z_0 = x \right\}. \tag{5.1}$$

Clearly $\mu(v) \in R^1$ and is independent of x. For $x, y \in R^n$ we set

$$\pi^v(x) = \inf \left\{ \liminf_{N \to \infty} \sum_{k=0}^{N-1} [v(z_k, z_{k+1}) - \mu(v)] : \{z_k\}_{k=0}^{\infty} \subset R^n, \ z_0 = x \right\}, \tag{5.2}$$

$$\theta^v(x, y) = v(x, y) - \mu(v) + \pi^v(y) - \pi^v(x). \tag{5.3}$$

For a compact set $C \subset R^n$ we define $v^C : C \times C \to R^1$ by

$$v^C(x, y) = v(x, y), \ x, y \in C.$$

Propositions 3.2.1 and 3.5.1 imply the following result.

PROPOSITION 3.5.2 *1. Given a compact set $C \subset R^n$ there is a ball $B \supset C$ such that*

$$\mu(v^B) = \mu(v), \ \pi^v(x) = \pi^{v^B}(x), \ \theta^v(x, y) = \theta^{v^B}(x, y) \quad (x, y \in C).$$

$$\pi^{v^B}(x) \geq \pi^v(x) \ (x \in B), \quad \theta^{v^B}(x, y) \geq \theta^v(x, y) \ (x \in C, y \in B);$$

for every $x \in C$ there exists $y \in B$ satisfying $\theta^v(x, y) = \theta^{v^B}(x, y) = 0$.
2. π^v, θ^v are continuous functions, θ^v is nonnegative.

PROPOSITION 3.5.3 $\pi^v(x) \to \infty$ *as* $|x| \to \infty$.

Proof. There is a number $c_0 > 0$ such that

$$\inf\{v(z, y) : z, y \in R^n, \ |z| + |y| \geq c_0\} \geq 4|\mu(v)| + 4.$$

Let

$$x \in R^n, \ \{z_k\}_{k=0}^{\infty} \subset R^n, \ z_0 = x,$$

$$\liminf_{N \to \infty} \sum_{k=0}^{N} [v(z_k, z_{k+1}) - \mu(v)] \leq \pi^v(x) + 1.$$

It is easy to see that the set

$$E := \{k \in \{1, 2, \ldots\} : \ |z_k| \leq c_0\}$$

is nonempty. Denote by m the minimal element of the set E. We have

$$\pi^v(x) + 1 \geq \liminf_{N \to \infty} \sum_{k=0}^{N} [v(z_k, z_{k+1}) - \mu(v)]$$

$$\geq \sum_{k=0}^{m-1} [v(z_k, z_{k+1}) - \mu(v)]$$

$$+\pi^v(z_m) \geq v(x, z_1) - \mu(v) + \inf\{\pi^v(y) : \ y \in R^n, \ |y| \leq c_0\} \to \infty$$

as $|x| \to \infty$. This completes the proof of the proposition.

PROPOSITION 3.5.4 *Let* $\{x_k\}_{k=1}^{\infty} \subset R^n$. *Then either*

$$\sum_{k=0}^{N} [v(x_k, x_{k+1}) - \mu(v)] \to \infty \ \text{as} \ N \to \infty$$

or the sequences

$$\{x_k\}_{k=1}^{\infty} \ \text{and} \ \{\sum_{k=0}^{N} [v(x_k, x_{k+1}) - \mu(v)]\}_{N=0}^{\infty}$$

are bounded.

Proof. Assume that the sequence

$$\{\sum_{k=0}^{N} [v(x_k, x_{k+1}) - \mu(v)]\}_{N=1}^{\infty}$$

does not tend to $+\infty$. Then for every integer $k \geq 1$,

$$\infty > \liminf_{N \to \infty} \sum_{q=0}^{N}[v(x_q, x_{q+1}) - \mu(v)] \geq \sum_{q=0}^{k-1}[v(x_q, x_{q+1}) - \mu(v)] + \pi^v(x_k)$$

$$\geq \pi^v(x_0) + \sum_{q=0}^{k-1}\theta^v(x_q, x_{q+1}).$$

Hence $\sum_{q=0}^{\infty}\theta^v(x_q, x_{q+1}) < \infty$. The validity of the proposition follows from this inequality and Proposition 3.5.3.

3.6.   Proof of Theorem 3.1.2

In this section we associate with any $f \in \mathcal{A}$ a related discrete-time control system. There is a simple correspondence between solutions of variational problems with the integrand f and solutions of this control system.

Let $f \in \mathcal{A}$. It is easy to see that $U^f(t, t+T, x, y) = U^f(0, T, x, y)$ for each $x, y \in R^n$, $T, t \in (0, \infty)$. For every $T > 0$ define a function $v_T^f : R^n \times R^n \to R^1$ as follows:

$$v_T^f(x, y) = U^f(0, T, x, y) \ (x, y \in R^n). \tag{6.1}$$

Theorem 3.4.1 and Proposition 1.3.3 imply that for any $T > 0$ the function v_T^f is continuous and $v_T^f(x, y) \to \infty$ as $|x| + |y| \to \infty$. We define

$$\mu_T^f = T^{-1}\mu(u), \ \pi_T^f = \pi^u, \ \theta_T^f = \theta^u \tag{6.2}$$

where $u = v_T^f$ (see (5.1-5.3)).

We consider a discrete-time control system studied in Section 3.5 with the cost function $v_T^f : R^n \times R^n \to R^1$, where $T > 0$. Theorem 3.1.2 will be proved in the following way. We fix $T > 0$ and obtain an optimal solution $\{x_i\}_{i=0}^{\infty} \subset R^n$ of the discrete-time control system. Then the function $v : [0, \infty) \to R^n$ which satisfies

$$v(iT) = x_i, \ I^f(iT, (i+1)T, v) = U^f(0, 1, x_i, x_{i+1}), \ i = 0, 1, \ldots$$

will be an (f)-weakly optimal function.

We begin with the following auxiliary result which establishes an important relation between the infinite horizon variational problem with the integrand f and the infinite horizon discrete-time control system with the cost function v_T^f.

PROPOSITION 3.6.1 $\mu_\tau^f = \mu(f)$ *for every* $\tau > 0$ *(recall* $\mu(f)$ *in (1.5)).*

Proof. Let $\tau > 0$. Consider any (f)-good function $x : [0, \infty) \to R^n$. By Proposition 3.1.1 the function $I^f(0, T, x) - T\mu(f)$, $T \in (0, \infty)$ is bounded. Then relations (5.1) and (6.2) imply that

$$\mu_\tau^f \leq \tau^{-1} \liminf_{N \to \infty} N^{-1} \sum_{i=0}^{N-1} v_\tau^f(x(i\tau), x((i+1)\tau))$$

$$\leq \tau^{-1} \liminf_{N \to \infty} N^{-1} \sum_{i=0}^{N-1} U^f(i\tau, (i+1)\tau, x(i\tau), x((i+1)\tau))$$

$$\leq \tau^{-1} \liminf_{N \to \infty} N^{-1} I^f(0, N\tau, x) \leq \mu(f).$$

By Proposition 3.5.2 there exists a sequence $\{y_i\}_{i=0}^\infty \subset R^n$ such that $\theta_\tau^f(y_i, y_{i+1}) = 0$, $i = 0, 1, \ldots$. Propositions 3.5.3 and 3.5.4 imply that

$$\sup \left\{ |\sum_{i=0}^{N-1} [v_\tau^f(y_i, y_{i+1}) - \tau\mu_\tau^f]| : N = 0, 1, \ldots \right\} < \infty. \qquad (6.3)$$

By Corollary 1.3.1 there exists an a.c. function $x : [0, \infty) \to R^n$ such that

$$x(i\tau) = y_i, \; I^f(i\tau, (i+1)\tau, x) = U^f(i\tau, (i+1)\tau, y_i, y_{i+1})$$

$$= v_\tau^f(y_i, y_{i+1}) \quad (i = 0, 1, \ldots).$$

It follows from these relations, (1.4), (1.5) and (6.3) that

$$\mu(f) \leq \liminf_{T \to \infty} T^{-1} I^f(0, T, x) \leq \liminf_{N \to \infty} (\tau N)^{-1} \sum_{i=0}^{N-1} v_\tau^f(y_i, y_{i+1}) = \mu_\tau^f.$$

Combined with (6.3) this inequality completes the proof of the proposition.

Let $m(ds)$ be the Lebesgue measure. The following result was established in [44].

LEMMA 3.6.1 *Let* $\{d_k\}_{k=1}^\infty$ *be an increasing sequence of positive numbers such that* $d_k \to \infty$ *as* $k \to \infty$. *Consider numbers* $T > 0$ *with the property that for every* $p \geq 1$,

$$\inf_{k,i} \{d_k - iT : k \geq p, \; i \geq 0, \; d_k \geq iT\} = 0. \qquad (6.4)$$

Then there is a set $D \subset [0, \infty)$ *with* $m([0, \infty) \setminus D) = 0$ *such that every* $T \in D$ *satisfies (6.4) for every integer* $p \geq 1$.

By a simple modification of the proof of Proposition 4.4 in [44] we can establish the following result which is a continuous-time version of Proposition 3.2.1.

THEOREM 3.6.1 *There exist a continuous function* $\pi^f : R^n \to R^1$, *a continuous nonnegative function* $(T, x, y) \to \bar{\theta}_T^f(x, y) \in R^1$ *defined for* $T > 0$, $x, y \in R^n$, *and a set* $D \subset [0, \infty)$ *with* $m([0, \infty) \setminus D) = 0$ *such that*

$$\pi^f(x) = \pi_T^f(x) \text{ for every } x \in R^n \text{ and every } T \in D;$$

$$\pi^f(x) \leq \pi_T^f(x) \text{ for every } x \in R^n \text{ and every } T > 0;$$

$$U^f(0, T, x, y) = T\mu(f) + \pi^f(x) - \pi^f(y) + \bar{\theta}_T^f(x, y)$$

$$\text{for each } x, y \in R^n \text{ and each } T > 0;$$

$$\text{for every } T > 0 \text{ and every } x \in R^n \text{ there is}$$

$$y \in R^n \text{ satisfying } \bar{\theta}_T^f(x, y) = 0.$$

For each $f \in \mathcal{A}$, each pair of numbers $T_1 \geq 0$, $T_2 > T_1$ and each a.c. function $v : [T_1, T_2] \to R^n$ put

$$\sigma^f(T_1, T_2, v) = I^f(T_1, T_2, v) - (T_2 - T_1)\mu(f) + \pi^f(v(T_2)) - \pi^f(v(T_1)),$$
$$(6.5)$$
$$\Phi^f(T_1, T_2, v) = I^f(T_1, T_2, v) - (T_2 - T_1)\mu(f).$$

THEOREM 3.6.2 *For every* $x \in R^n$,

$$\pi^f(x) = \inf\{\liminf_{T \to \infty}[I^f(0, T, v) - \mu(f)T] : v : [0, \infty)$$

$$\to R^n \text{ is an a.c. function satisfying } v(0) = x\}.$$

Proof. Let $x \in R^n$ and $v : [0, \infty) \to R^n$ be an a.c. function such that $v(0) = x$. By Theorem 3.6.1 and Proposition 3.6.1 we need to show that

$$\pi^f(x) \leq \liminf_{T \to \infty} \Phi^f(0, T, v).$$

By Proposition 3.1.1 we may assume that v is an (f)-good function and that the function

$$T \to \Phi^f(0, T, v), \ T \in (0, \infty)$$

is bounded. Theorem 1.2.2 implies that the function $v : [0, \infty) \to R^n$ is bounded. There exists a sequence of positive numbers $\{d_k\}_{k=1}^\infty$ such that $d_{k+1} - d_k \geq 1$ ($k = 1, 2, \ldots$) and

$$\lim_{k \to \infty} \Phi^f(0, d_k, v) = \liminf_{T \to \infty} \Phi^f(0, T, v). \quad (6.6)$$

By Lemma 3.6.1 there is a set $D \subset (0, \infty)$ with $m([0, \infty) \setminus D) = 0$ such that every $T \in D$ satisfies (6.5) for every integer $p \geq 1$.

Let $T \in D$. It follows from the definition of D and (6.5) that there exists a sequence of positive integers $M_j \to \infty$ as $j \to \infty$ $(j = 1, 2, \ldots)$ and a subsequence $\{d_{k_j}\}_{j=1}^\infty$ such that

$$d_{k_j - 1} < M_j T \leq d_{k_j} \ (j = 1, 2, \ldots), \ d_{k_j} - M_j T \to 0 \text{ as } j \to \infty. \quad (6.7)$$

It follows from Assumption A(ii) and (6.7) that for any natural number j

$$\Phi^f(0, TM_j, v) - \Phi^f(0, d_{k_j}, v) \leq -I^f(TM_j, d_{k_j}, v)$$
$$+ |\mu(f)|(d_{k_j} - TM_j) \leq (d_{k_j} - TM_j)(|\mu(f)| + a) \to 0 \text{ as } j \to \infty.$$

It follows from this relation, Proposition 3.6.1 and Theorem 3.6.1 that

$$\pi^f(x) \leq \pi_T^f(x) \leq \liminf_{N \to \infty} \sum_{i=0}^N [U^f(iT, (i+1)T, v(iT), v((i+1)T)) - T\mu(f)]$$

$$\leq \liminf_{j \to \infty} \Phi^f(0, TM_j, v) \leq \liminf_{j \to \infty} \Phi^f(0, d_{k_j}, v).$$

The theorem now follows from this relation and (6.6).

Theorems 3.6.1 and 3.6.2 are important tools in our study of autonomous variational problems. It is not difficult to see that a minimization problem with the functional $I^f(T_1, T_2, v)$ is equivalent to the corresponding minimization problem with the functional $\sigma^f(T_1, T_2, v)$. In the sequel we prefer to work with the functional $\sigma^f(T_1, T_2, v)$ which is always nonnegative. The next theorem establishes the existence of an a.c. function $v : [0, \infty) \to R^n$ such that $\sigma^f(0, T, v) = 0$ for all $T > 0$. We will show (see Theorem 3.6.4) that if the integrand f has the asymptotic turnpike property, then any such function is (f)-weakly optimal.

THEOREM 3.6.3 *For every* $x \in R^n$ *there exists an* (f)-*good function* $v : [0, \infty) \to R^n$ *such that* $v(0) = x$ *and for each* $T_1 \in [0, \infty)$ *and each* $T_2 \in (T_1, \infty)$,
$$\sigma^f(T_1, T_2, v) = 0.$$

Proof. Let $x \in R^n$. Choose a number $T > 0$ such that $\pi_T^f(y) = \pi^f(y)$ for all $y \in R^n$. It follows from Proposition 3.5.2 that there exists a sequence $\{x_i\}_{i=0}^\infty \subset R^n$ such that

$$x_0 = x, \ \theta_T^f(x_i, x_{i+1}) = 0 \ (i = 0, 1, \ldots). \quad (6.8)$$

Propositions 3.5.3 and 3.5.4 imply that the sequence $\{x_i\}_{i=0}^\infty$ is bounded. By Corollary 1.3.1 there is an a.c. function $v : [0, \infty) \to R^n$ such that for every integer $i \geq 0$,

$$v(iT) = x_i, \ I^f(iT, (i+1)T, v) = U^f(iT, (i+1)T, v(iT), v((i+1)T)). \tag{6.9}$$

Since the sequence $\{x_i\}_{i=0}^{\infty}$ is bounded it follows from (6.9) and Propositions 1.3.6 and 1.3.3 that the function $v : [0, \infty) \to R^n$ is also bounded. It follows from Propositions 3.1.1 and 3.6.1, (6.8) and (6.9) that v is an (f)-good function. By Theorem 3.6.1,

$$\sigma^f(\tau_1, \tau_2, v) \geq 0 \text{ for each } \tau_1 \geq 0, \ \tau_2 > \tau_1. \tag{6.10}$$

It follows from (6.8) and (6.9) and the definition of T that for any integer $i \geq 0$,

$$\sigma^f(iT, (i+1)T, v) = 0. \tag{6.11}$$

The theorem now follows from (6.10) and the boundedness of v.

THEOREM 3.6.4 *Assume that the integrand f has the asymptotic turnpike property with the turnpike $H(f) \subset R^n$. Let $v : [0, \infty) \to R^n$ be an a.c. function such that $\sigma^f(T_1, T_2, v) = 0$ for each $T_1 \geq 0$, $T_2 > T_1$. Then v is an (f)-weakly optimal function. Moreover, there exists a sequence of numbers $t_j \to \infty$ as $j \to \infty$ such that*

$$\limsup_{j \to \infty} [I^f(0, t_j, v) - I^f(0, t_j, w)] \leq 0$$

for each a.c. function $w : [0, \infty) \to R^n$ satisfying $v(0) = w(0)$.

Proof. There exists $x_* \in H(f)$ such that $\pi^f(x_*) \geq \pi^f(z)$ for all $z \in H(f)$. It follows from Theorem 3.6.1, Propositions 3.1.1 and 3.5.3 that v is an (f)-good function. To prove the theorem it is sufficient to note that there exists a sequence of numbers $\{t_j\}_{j=1}^{\infty}$ such that $t_j \to \infty$, $v(t_j) \to x_*$ as $j \to \infty$.

Theorem 3.1.2 now follows from Theorems 3.6.3 and 3.6.4.

3.7.   Preliminary lemmas for Theorem 3.1.1

In this section we show that the set of all integrands $f \in \mathcal{A}$ which have the asymptotic turnpike property is an everywhere dense subset of $\mathcal{A}$ with the strong topology.

The following result was established in [5, Chapter 2, section 3].

PROPOSITION 3.7.1 *Let Ω be a closed subset of R^q. Then there exists a bounded nonnegative function $\phi \in C^{\infty}(R^q)$ such that $\Omega = \{x \in R^q :$*

$\phi(x) = 0\}$ *and for each sequence of nonnegative integers* $p_1, p_2, \ldots, p_q$ *the function* $\partial^{|p|}\phi/\partial x_1^{p_1} \ldots \partial x_q^{p_q} : R^q \to R^1$ *is bounded where* $|p| = \sum_{i=1}^q p_i$.

Let $f \in \mathcal{A}$. For any $r \in (0,1)$ we construct $f_r \in \mathcal{A}$ which has the asymptotic turnpike property such that $f_r \to f$ as $r \to 0^+$ in the strong topology.

By Theorem 1.2.2 there exists a number $M > 0$ such that

$$\limsup_{t \to \infty} |v(t)| < M \tag{7.1}$$

for each (f)-good function $v : [0, \infty) \to R^n$. Define

$$\mathcal{D}(f) = \{\Omega(v) : v \text{ is an } (f)\text{-good function}\}. \tag{7.2}$$

LEMMA 3.7.1 *There exists* $H^* \in \mathcal{D}(f)$ *such that for every* $D \in \mathcal{D}(f) \setminus \{H^*\}$

$$D \setminus H^* \neq \emptyset.$$

Proof. Let $D_1, D_2 \in \mathcal{D}(f)$. We will say that $D_1 \leq D_2$ if and only if $D_1 \subset D_2$. We will show that there exists a minimal element of the ordered set $\mathcal{D}(f)$.

Consider a nonempty set $E \subset \mathcal{D}(f)$ such that for each $D_1, D_2 \in E$ one of the relations below holds: $D_1 \leq D_2$; $D_2 \leq D_1$. By Zorn's lemma it is sufficient to show that there is $\tilde{D} \in \mathcal{D}(f)$ such that $\tilde{D} \subset D$ for any $D \in E$. Set

$$\bar{D} = \cap_{D \in E} D.$$

Clearly $\bar{D} \neq \emptyset$. We will show that there exists an (f)-good function $v : [0, \infty) \to R^n$ such that $\Omega(v) \subset \bar{D}$.

For every integer $p \geq 1$ there exists $D_p \in E$ such that

$$\text{dist}(\bar{D}, D_p) \leq p^{-1} \tag{7.3}$$

and there exists an (f)-good function $v_p : [0, \infty) \to R^n$ such that

$$\Omega(v_p) = D_p. \tag{7.4}$$

By Proposition 3.3.1 we may assume without loss of generality that

$$I^f(T_1, T_2, v_p) \leq U^f(T_1, T_2, v_p(T_1), v_p(T_2)) + 1 \tag{7.5}$$

for each $p \geq 1$ and each $T_1 \in [0, \infty)$, $T_2 \in (T_1, \infty)$ and

$$d(v_p(t), D_p) \leq p^{-1} \quad (t \in [0, \infty), \ p = 1, 2, \ldots). \tag{7.6}$$

It follows from (7.5), (7.6), Theorem 3.4.1 and Proposition 1.3.5 that there exist a subsequence $\{v_{p_j}\}_{j=1}^{\infty}$ and an a.c. function $u : [0, \infty) \to R^n$ such that for every integer $N \geq 1$,

$$v_{p_j}(t) \to u(t) \text{ as } j \to \infty \text{ uniformly in } [0, N] \text{ and} \tag{7.7}$$

$$I^f(0, N, u) \leq U^f(0, N, u(0), u(N)) + 1.$$

It follows from (7.3) and (7.6) that

$$u(t) \in \bar{D} \text{ for all } t \in [0, \infty). \tag{7.8}$$

By Proposition 3.3.2, (7.7) and (7.8), u is an (f)-good function. This completes the proof of the lemma.

LEMMA 3.7.2 *Let* $u : [0, \infty) \to R^n$ *be an* (f)-*good function. Then there exists an* (f)-*good function* $v : [0, \infty) \to \Omega(u)$ *such that for each* $T_1 \in [0, \infty)$, $T_2 \in (T_1, \infty)$,

$$I^f(T_1, T_2, v) = U^f(T_1, T_2, v(T_1), v(T_2)).$$

Proof. We may assume that

$$d(u(t), \Omega(u)) \leq 1 \text{ for all } t \in [0, \infty). \tag{7.9}$$

By Proposition 3.3.1 we may assume without loss of generality that

$$I^f(T_1, T_2, u) \leq U^f(T_1, T_2, u(T_1), u(T_2)) + 1 \tag{7.10}$$

for each $T_1 \in [0, \infty)$, $T_2 \in (T_2, \infty)$. For every integer $p \geq 1$ we set

$$v_p(t) = u(t + p) \quad (t \in [0, \infty)). \tag{7.11}$$

By (7.9), (7.10), Theorem 3.4.1 and Proposition 1.3.5, there exist a subsequence $\{v_{p_j}\}_{j=1}^{\infty}$ and an a.c. function $v : [0, \infty) \to R^n$ such that for every integer $N \geq 1$,

$$v_{p_j}(t) \to v(t) \text{ as } j \to \infty \text{ uniformly in } [0, N] \text{ and}$$

$$I^f(0, N, v) \leq \liminf_{j \to \infty} I^f(0, N, v_{p_j}). \tag{7.12}$$

Together with (7.11) this implies that

$$v(t) \in \Omega(u) \quad (t \in [0, \infty)). \tag{7.13}$$

It follows from (7.11), (7.12), Proposition 3.3.1 and Theorem 3.4.1 that

$$I^f(0, N, v) = U^f(0, N, v(0), v(N)) \text{ for any integer } N \geq 1.$$

Together with (7.13) and Proposition 3.3.2 this implies that v is an (f)-good function. The lemma is proved.

By Lemma 3.7.1 there exists $H^* \in \mathcal{D}(f)$ such that

$$D \setminus H^* \neq \emptyset$$

for every $D \in \mathcal{D}(f) \setminus \{H^*\}$. By Lemma 3.7.2 there exists an (f)-good function $w_f : [0, \infty) \to H^*$ such that for each number $T > 0$,

$$I^f(0, T, w_f) = U^f(0, T, w_f(0), w_f(T)). \tag{7.14}$$

Clearly

$$\Omega(w_f) = H^*. \tag{7.15}$$

LEMMA 3.7.3 *Suppose that* $\phi : R^n \to [0, \infty)$ *is a continuous bounded function such that*

$$H^* \subset \{x \in R^n : \phi(x) = 0\} \subset H^* \cup \{x \in R^n : |x| \geq M + 3\}. \tag{7.16}$$

For $r \in (0, 1]$ *we set*

$$f_r(x, u) = f(x, u) + r\phi(x) \quad (x, u \in R^n). \tag{7.17}$$

Then $f_r \in \mathcal{A}$ *for all* $r \in (0, 1]$ *and the following property holds:*
For any neighborhood $\mathcal{U}$ *of* f *in* $\mathcal{A}$ *with the strong topology there exists a number* $r_0 \in (0, 1)$ *such that* $f_r \in \mathcal{U}$ *for every* $r \in (0, r_0)$.

The next auxiliary result shows that for any $r \in (0, 1]$ the integrand f_r has the asymptotic turnpike property.

LEMMA 3.7.4 *Suppose that* $\phi : R^n \to [0, \infty)$ *is a continuous bounded function which satisfies (7.16),* $r \in (0, 1]$ *and a function* $f_r : R^n \times R^n \to R^1$ *is defined by (7.17). Then* $\Omega(v) = H^*$ *for each* (f_r)-*good function* $v : [0, \infty) \to R^n$.

Proof. Let us assume the converse. Then there exists an (f_r)-good function $v : [0, \infty) \to R^n$ such that

$$\Omega(v) \neq H^*. \tag{7.18}$$

It is easy to see that $\mu(f_r) = \mu(f)$ and w_f is an (f_r)-good function. Then Proposition 3.1.1 implies that v is an (f)-good function and

$$\sup_{T>0}\{|\Phi^{f_r}(0, T, v)|\} < \infty, \quad \sup_{T>0}\{|\Phi^{f}(0, T, v)|\} < \infty. \tag{7.19}$$

It follows from (7.18) and the definition of H^* that there exists a sequence $\{T_i\}_{i=1}^\infty \subset (0, \infty)$ such that

$$T_i \to \infty, \; v(T_i) \to z \in \Omega(v) \setminus H^* \text{ as } i \to \infty. \tag{7.20}$$

By the definition of M (see (7.1)) and Proposition 3.3.1 we may assume that

$$|v(t)| < M \; (t \in [0, \infty)),$$
$$I^f(T_1, T_2, v) \leq U^f(T_1, T_2, v(T_1), v(T_2)) + 1 \tag{7.21}$$
$$\text{for each } T_1 \in [0, \infty), \; T_2 \in (T_1, \infty).$$

We may assume without loss of generality that

$$T_{i+1} - T_i \geq 10, \; d(v(T_i), H^*) \geq \gamma, \; i = 1, 2, \ldots \tag{7.22}$$

with a constant $\gamma > 0$. It follows from Propositions 1.3.4 and 1.3.6 and (7.21) that there exists $\delta \in (0, 1)$ such that the following property holds:
   If $t_1, t_2 \in [0, \infty)$ satisfy $|t_1 - t_2| \leq \delta$, then $|v(t_1) - v(t_2)| \leq 8^{-1}\gamma$.
   By the definition of δ, (7.21) and (7.22),

$$d(v(t), H^*) \geq 2^{-1}\gamma$$

for every integer $i \geq 1$ and every $t \in [T_i, T_i + \delta]$.
   Combined with (7.21) and (7.16) this implies that there exists a constant $\gamma_1 > 0$ such that

$$\phi(v(t)) \geq \gamma_1$$

for every integer $i \geq 1$ and every $t \in [T_i, T_i + \delta]$. By this property, (7.19) and (7.17),

$$\Phi^{fr}(0, T_N, v) \geq \Phi^f(0, T_N, v) + r\gamma_1\delta(N - 1) \to \infty \text{ as } N \to \infty.$$

This is contradictory to (7.19). The obtained conradiction proves the lemma.

Lemmas 3.7.3 and 3.7.4 imply the following result.

LEMMA 3.7.5 *There exists an everywhere dense subset E of the space $\mathcal{A}$ with the strong topology such that each $f \in E$ has the asymptotic turnpike property.*

Propositions 3.7.1 and Lemmas 3.7.3 and 3.7.4 imply the following result.

LEMMA 3.7.6 *Let $k \geq 1$ be an integer. Then there exists an everywhere dense subset $E_k \subset \mathcal{A}_k$ such that each $f \in E_k$ has the asymptotic turnpike property.*

3.8. Preliminary lemmas for Theorems 3.1.3 and 3.1.4

Assume that $f \in \mathcal{A}$ has the asymptotic turnpike property with the turnpike $H(f) \subset R^n$. Then $\Omega(v) = H(f)$ for every (f)-good function $v : [0, \infty) \to R^n$.

We begin with the following auxiliary result which shows that an approximate solution of a variational problem with the integrand f defined on a finite interval is close to the turnpike $H(f)$ at some points if the length of the interval is large enough.

LEMMA 3.8.1 *Let $\epsilon_0 \in (0, 1)$, $K_0, M_0 > 0$ and let l be a positive integer such that for each (f)-good function $x : [0, \infty) \to R^n$,*

$$dist(H(f), \{x(t) : t \in [T, T + l]\}) \leq 8^{-1}\epsilon_0 \qquad (8.1)$$

for all large T (the existence of l follows from Theorem 3.3.1). Then there exists an integer $N \geq 10$ such that the following property holds: If an a.c. function $x : [0, Nl] \to R^n$ satisfies $|x(0)|, |x(Nl)| \leq K_0$ and

$$I^f(0, Nl, x) \leq U^f(0, Nl, x(0), x(Nl)) + M_0,$$

then there exists an integer $i_0 \in [0, N - 8]$ such that

$$dist(H(f), \{x(t) : t \in [T, T + l]\}) \leq \epsilon_0 \text{ for all } T \in [i_0 l, (i_0 + 7)l].$$

Proof. Assume that the lemma is wrong. Then for every integer $N \geq 10$ there exists an a.c. function $x_N : [0, N] \to R^n$ satisfying

$$|x_N(0)|, |x_N(Nl)| \leq K_0, \ I^f(0, Nl, x_N) \leq U^f(0, Nl, x_N(0), x_N(Nl)) + M_0 \qquad (8.2)$$

such that the following property holds:
If an integer $i \in [0, N - 8]$, then there is $T(i) \in [il, (i + 7)l]$ for which

$$dist(H(f), \{x(t) : t \in [T(i), T(i) + l]\}) > \epsilon_0. \qquad (8.3)$$

By Theorem 1.2.3,

$$\sup\{|x_N(t)| : t \in [0, Nl], \ N \text{ is an integer}, N \geq 10\} < \infty. \qquad (8.4)$$

It follows from (8.2), (8.4) and Theorem 3.4.1 that for any integer $j \geq 1$,

$$\sup\{I^f(0, jl, x_N) : N \text{ is an integer}, N \geq \max\{j, 10\}\} < \infty. \qquad (8.5)$$

By (8.5) and Proposition 1.3.5 there exist a subsequence $\{x_{N_i}\}_{i=1}^{\infty}$ and an a.c. function $x : [0, \infty) \to R^n$ such that for each integer $q \geq 1$,

$$x_{N_i}(t) \to x(t) \text{ as } i \to \infty \text{ uniformly in } [0, ql],$$

$$x'_{N_i} \to x' \text{ as } i \to \infty \text{ weakly} \tag{8.6}$$

in $L^1(R^n; (0, ql))$, $I^f(0, ql, x) \leq \liminf\limits_{i \to \infty} I^f(0, ql, x_{N_i})$.

(8.2), (8.6) and Theorem 3.4.1 imply that

$$I^f(0, ql, x) \leq U^f(0, ql, x(0), x(ql)) + M_0 \text{ for any integer } q \geq 1. \tag{8.7}$$

It follows from Proposition 3.3.2, (8.7), (8.4) and (8.6) that $x : [0, \infty) \to R^n$ is an (f)-good function. By the definition of l there exists an integer $Q \geq 1$ such that (8.1) holds for all $T \in [Ql, \infty)$. By (8.6) there exists an integer $p \geq 1$ such that

$$N_p \geq 2Q + 20, \ |x_{N_p}(t) - x(t)| \leq 8^{-1}\epsilon_0, \ t \in [0, (2Q + 20)l]. \tag{8.8}$$

It follows from (8.1) which holds for every $T \in [Ql, \infty)$ and (8.8) that

$$\text{dist}(H(f), \{x_{N_p}(t) : t \in [T, T+l]\}) \leq 4^{-1}\epsilon_0 \text{ for each } T \in [Ql, Ql + 10l].$$

This is contradictory to the definition of x_{N_p} (see (8.3)). The obtained contradiction proves the lemma.

The following auxiliary result is an extension of Lemma 3.8.1. It shows that the property established in Lemma 3.8.1 for the variational problems with the integrand f also holds for variational problems with integrands which belong to a small neighborhood of f.

LEMMA 3.8.2 *Let $\epsilon_0 \in (0, 1)$, $K_0, M_0 > 0$ and let l be a positive integer such that each (f)-good function $x : [0, \infty) \to R^n$ satisfies (8.1) for all large T. Then there exist an integer $N \geq 10$ and a neighborhood $\mathcal{U}$ of f in $\mathcal{A}$ with the weak topology such that the following property holds:*
If $g \in \mathcal{U}$, $S \in [0, \infty)$ and if an a.c. function $x : [S, S + Nl] \to R^n$ satisfies

$$|x(S)|, |x(S + Nl)| \leq K_0, \ I^g(S, S + Nl, x)$$
$$\leq U^g(S, S + Nl, x(S), x(S + Nl)) + M_0, \tag{8.9}$$

then there exists an integer $i_0 \in [0, N - 8]$ such that

$$\text{dist}(H(f), \{x(t) : t \in [T, T+l]\}) \leq \epsilon_0 \tag{8.10}$$

for all $T \in [S + i_0 l, S + (i_0 + 7)l]$.

Proof. It follows from Lemma 3.8.1 that there exists an integer $N \geq 10$ such that the following property holds:
If an a.c. function $x : [0, Nl] \to R^n$ satisfies

$$|x(0)|, |x(Nl)| \leq K_0, \ I^f(0, Nl, x) \leq U^f(0, Nl, x(0), x(Nl)) + M_0 + 8, \tag{8.11}$$

then there exists an integer $i_0 \in [0, N - 8]$ such that (8.10) holds for all $T \in [i_0 l, (i_0 + 7)l]$.

Put

$$\tilde{S} = \sup\{|U^f(0, Nl, y, z)| : y, z \in R^n, |y|, |z| \leq 2K_0 + 1\}. \qquad (8.12)$$

By Theorem 3.4.1, $\tilde{S}$ is finite. It follows from Proposition 1.3.9 that there exists a neighborhood $\mathcal{U}_1$ of f in $\mathcal{A}$ with the weak topology such that the following property holds:

If $g \in \mathcal{U}_1$, $S \in [0, \infty)$ and if $y, z \in R^n$ satisfy $|y|, |z| \leq 2K_0 + 1$, then

$$|U^f(S, S + Nl, y, z) - U^g(S, S + Nl, y, z)| \leq 8^{-1}. \qquad (8.13)$$

It follows from Proposition 1.3.8 that there is a neighborhood $\mathcal{U}$ of f in $\mathcal{A}$ with the weak topology such that $\mathcal{U} \subset \mathcal{U}_1$ and that the following property holds:

If $g \in \mathcal{U}$, $S \in [0, \infty)$ and if an a.c. function $x : [S, S + Nl] \to R^n$ satisfies

$$\min\{I^f(S, S + Nl, x), I^g(S, S + Nl, x)\} \leq 4 + M_0 + \tilde{S}, \qquad (8.14)$$

then

$$|I^f(S, S + Nl, x) - I^g(S, S + Nl, x)| \leq 8^{-1}. \qquad (8.15)$$

Assume that $g \in \mathcal{U}$, $S \in [0, \infty)$ and an a.c. function $x : [S, S + Nl] \to R^n$ satisfies (8.9). It follows from (8.9) and the choice of $\mathcal{U}_1$ that

$$|U^f(S, S+Nl, x(S), x(S+Nl)) - U^g(S, S+Nl, x(S), x(S+Nl))| \leq 8^{-1}. \qquad (8.16)$$

By (8.16), (8.12) and (8.9),

$$I^g(S, S + Nl, x)$$

$$\leq U^f(S, S + Nl, x(S), x(S + Nl)) + M_0 + 8^{-1} \leq \tilde{S} + M_0 + 8^{-1}. \qquad (8.17)$$

By the choice of $\mathcal{U}$ and the inequality (8.17),

$$|I^g(S, S + Nl, x) - I^f(S, S + Nl, x)| \leq 8^{-1}. \qquad (8.18)$$

(8.17) and (8.18) imply that

$$I^f(S, S + Nl, x) \leq U^f(S, S + Nl, x(S), x(S + Nl)) + M_0 + 4^{-1}.$$

By the definition of N there exists an integer $i_0 \in [0, N - 8]$ such that (8.10) holds for all $T \in [S + i_0 l, S + (i_0 + 7)l]$. The lemma is proved.

In general for given $x_1, x_2 \in R^n$ we cannot guarantee the existence of $T > 0$ and $v : [0, T] \to R^n$ such that $v(0) = x_1$, $v(T) = x_2$ and $\sigma^f(0, T, v)$

is small. The following lemma shows that such T and v exist if x_1 and x_2 are close to the turnpike $H(f)$.

LEMMA 3.8.3 *For each $\epsilon > 0$ there exists $\delta > 0$ such that the following property holds:*
If $x_1, x_2 \in R^n$ satisfy $d(x_i, H(f)) \leq \delta$, $i = 1, 2$, then there exists an a.c. function $v : [0, T] \to R^n$ such that

$$T \geq 1, \ v(0) = x_1, \ v(T) = x_2, \tag{8.19}$$

$$\sigma^f(0, T, v) \leq \epsilon. \tag{8.20}$$

Proof. Let $\epsilon > 0$. It follows from Theorems 3.4.1 and 3.6.1 that there exists $\delta_0 \in (0, \min\{1, \epsilon\})$ for which the following property holds:
If $y_1, y_2, z_1, z_2 \in R^n$ satisfy

$$d(y_i, H(f)) \leq 1, \ d(z_i, H(f)) \leq 1, \ i = 1, 2, \tag{8.21}$$

$$|y_i - z_i| \leq \delta_0, \ i = 1, 2, \tag{8.22}$$

then

$$|U^f(0, 1, y_1, y_2) - U^f(0, 1, z_1, z_2)| \leq 8^{-1}\epsilon,$$
$$|\pi^f(y_i) - \pi^f(z_i)| \leq 8^{-1}\epsilon, \ i = 1, 2. \tag{8.23}$$

It follows from Theorem 3.3.1 that there exists a natural number G such that for each (f)-good function $v : [0, \infty) \to R^n$ the inequality

$$\mathrm{dist}(H(f), \{v(t) : \ t \in [T, T + G]\}) \leq 4^{-1}\delta_0 \tag{8.24}$$

is valid for all large T. Theorem 3.6.3 implies that there exists an (f)-good function $w : [0, \infty) \to R^n$ such that $w(0) = 0$ and that for each $T_1 \geq 0$, $T_2 > T_1$,

$$\sigma^f(T_1, T_2, w) = 0. \tag{8.25}$$

By the choice of G there exists $T_0 > 0$ such that for any $T \geq T_0$,

$$\mathrm{dist}(H(f), \{w(t) : \ t \in [T, T + G]\}) \leq 4^{-1}\delta_0. \tag{8.26}$$

Put

$$\delta = 8^{-1}\delta_0. \tag{8.27}$$

Let $x_1, x_2 \in R^n$, $d(x_i, H(f)) \leq \delta$, $i = 1, 2$. There exist $h_1, h_2 \in H(f)$ such that

$$|x_i - h_i| \leq \delta, \ i = 1, 2. \tag{8.28}$$

It follows from (8.26), which holds for any $T \geq T_0$, that there exist $T_1 \in [T_0, T_0 + G]$ and $T_2 \in [T_0 + 10G, T_0 + 11G]$ such that

$$|w(T_i) - h_i| \leq 4^{-1}\delta_0, \ i = 1, 2. \tag{8.29}$$

There exists an a.c. function $v : [0, T_2 - T_1] \to R^n$ such that

$$v(0) = x_1, \; v(t) = w(T + T_1) \quad (t \in [1, T_2 - T_1 - 1]), \; v(T_2 - T_1) = x_2,$$

$$I^f(0, 1, v) = U^f(0, 1, v(0), v(1)),$$

$$I^f(T_2 - T_1 - 1, T_2 - T_1, v)$$

$$= U^f(T_2 - T_1 - 1, T_2 - T_1, v(T_2 - T_1 - 1), v(T_2 - T_1)). \quad (8.30)$$

By the definition of δ_0 (see (8.21)-(8.23)), (8.30), (8.28), (8.27) and (8.29),

$$|U^f(0, 1, v(0), v(1)) - U^f(T_1, T_1 + 1, w(T_1), w(T_1 + 1))| \le 8^{-1}\epsilon,$$

$$|U^f(T_2 - T_1 - 1, T_2 - T_1, v(T_2 - T_1 - 1), v(T_2 - T_1))$$
$$- U^f(T_2 - 1, T_2, w(T_2 - 1), w(T_2))| \le 8^{-1}\epsilon. \quad (8.31)$$

(8.30) implies that

$$\sigma^f(0, T_2 - T_1, v) = \sigma^f(0, 1, v) + \sigma^f(1, T_2 - T_1, v) + \sigma^f(T_2 - T_1 - 1, T_2 - T_1, v).$$
$$(8.32)$$

It follows from (8.30) and (8.25) that

$$\sigma^f(1, T_2 - T_1 - 1, v) = 0. \quad (8.33)$$

By the definition of δ_0 (see (8.21)-(8.23)), (8.28), (8.27), (8.29) and (8.31),
$$|\pi^f(w(T_i)) - \pi^f(x_i)| \le 8^{-1}\epsilon, \; i = 1, 2.$$

Combined with (8.30), (8.31) and (8.25) this inequality implies that

$$|\sigma^f(0, 1, v)| \le |U^f(0, 1, v(0), v(1)) - U^f(T_1, T_1 + 1, w(T_1), w(T_1 + 1))|$$

$$+ |\pi^f(w(T_1)) - \pi^f(x_1)| + |U^f(T_1, T_1 + 1, w(T_1), w(T_1 + 1)) - \pi^f(w(T_1))$$

$$+ \pi^f(w(T_1 + 1)) - \mu(f)| \le 4^{-1}\epsilon,$$

$$|\sigma^f(T_2 - T_1 - 1, T_2 - T_1, v)|$$

$$\le |U^f(T_2 - T_1 - 1, T_2 - T_1, v(T_2 - T_1 - 1), v(T_2 - T_1))$$

$$- U^f(T_2 - 1, T_2, w(T_2 - 1), w(T_2))| + |\pi^f(w(T_2)) - \pi^f(v(T_2 - T_1))|$$

$$+ |U^f(T_2 - 1, T_2, w(T_2 - 1), w(T_2)) - \pi^f(w(T_2 - 1))$$

$$+ \pi^f(w(T_2)) - \mu(f)| \le 4^{-1}\epsilon.$$

It follows from these inequalities, (8.32) and (8.33) that

$$0 \le \sigma^f(0, T_2 - T_1, v) \le \epsilon.$$

This completes the proof of the lemma.

The next lemma is an important tool in the proof of Theorem 3.1.4. It establishes the turnpike property for approximate solutions $v : [0, T] \rightarrow R^n$ such that $v(0)$ and $v(T)$ are close to the turnpike and $\sigma^f(0, T, v)$ is small enough.

LEMMA 3.8.4 *Let $\epsilon \in (0, 1)$ and let L be a positive integer such that for each (f)-good function $v : [0, \infty) \rightarrow R^n$ the inequality*

$$dist(H(f), \{v(t) : t \in [T, T + L]\}) \leq \epsilon \tag{8.34}$$

is valid for all large T (the existence of L follows from Theorem 3.3.1). Then there exists $\delta > 0$ such that the following property holds: If $T \in [L, \infty)$ and if an a.c. function $v : [0, T] \rightarrow R^n$ satisfies

$$d(v(0), H(f)) \leq \delta, \; d(v(T), H(f)) \leq \delta, \tag{8.35}$$

$$I^f(0, T, v) - T\mu(f) - \pi^f(v(0)) + \pi^f(v(T)) \leq \delta, \tag{8.36}$$

then for every $S \in [0, T - L]$,

$$dist(H(f), \{v(t) : t \in [S, S + L]\}) \leq \epsilon. \tag{8.37}$$

Proof. It follows from Lemma 3.8.3 that for every integer $k \geq 1$ there exists

$$\delta_k \in (0, 2^{-k}) \tag{8.38}$$

such that the following property holds:
  If $x_1, x_2 \in R^n$ satisfy

$$d(x_i, H(f)) \leq \delta_k, \; i = 1, 2, \tag{8.39}$$

then there is an a.c. function $v : [0, T] \rightarrow R^n$ such that

$$T \geq 1, \; v(0) = x_1, \; v(T) = x_2,$$

$$\sigma^f(0, T, v) \leq 2^{-k}. \tag{8.40}$$

Assume that the lemma is wrong. Then for every integer $k \geq 1$ there exist $T_k \geq L$, an a.c. function $v_k : [0, T_k] \rightarrow R^n$ and $S_k \in [0, T_k - L]$ such that

$$d(v_k(s), H(f)) \leq \delta_k, \; s = 0, \; T_k, \tag{8.41}$$

$$\sigma^f(0, T_k, v_k) \leq \delta_k, \tag{8.42}$$

$$dist(H(f), \{v_k(t) : t \in [S_k, S_k + L]\}) > \epsilon. \tag{8.43}$$

By the choice of $\{\delta_k\}_{t=1}^{\infty}$ (see (8.38)-(8.40)) and (8.41) for each natural number k there exists an a.c. function $u_k : [0, \tau_k] \to R^n$ such that

$$\tau_k \geq 1, \ u_k(0) = v_k(T_k), \ u_k(\tau_k) = v_{k+1}(0), \qquad (8.44)$$

$$\sigma^f(0, \tau_k, u_k) \leq 2^{-k}.$$

By (8.44) there exists an a.c. function $v : [0, \infty) \to R^n$ such that

$$v(t) = v_1(t) \ (t \in [0, T_1]), \ v(t) = u_1(t - T_1) \ (t \in [T_1, T_1 + \tau_1]),$$

$$v(t) = v_k \left(t - \sum_{i=1}^{k-1}(T_i + \tau_i) \right), \ t \in \left[\sum_{i=1}^{k-1}(T_i + \tau_i), \sum_{i=1}^{k-1}(T_i + \tau_i) + T_k \right],$$

$$v(t) = u_k \left(t - \sum_{i=1}^{k-1}(T_i + \tau_i) - T_k \right),$$

$$t \in \left[\sum_{i=1}^{k-1}(T_i + \tau_i) + T_k, \sum_{i=1}^{k}(T_i + \tau_i) \right], \ k = 2, 3, \ldots . \qquad (8.45)$$

Put $T_0, \tau_0 = 0$. It follows from (8.38), (8.42), (8.44) and (8.45) that for any natural number $k \geq 2$,

$$\sigma^f \left(0, \sum_{i=0}^{k}(T_i + \tau_i), v \right) = \sum_{q=1}^{k} \sigma^f \left(\sum_{i=0}^{q-1}(T_i + \tau_i), \sum_{i=0}^{q-1}(T_i + \tau_i) + T_q, v \right)$$

$$+ \sum_{q=1}^{k} \left(\sigma^f \left(\sum_{i=0}^{q-1}(T_i + \tau_i) + T_q, \sum_{i=0}^{q}(T_i + \tau_i), v \right) \right)$$

$$= \sum_{q=1}^{k} [\sigma^f(0, T_q, v_q) + \sigma^f(0, \tau_q, u_q)] \leq \sum_{q=1}^{k}(\delta_q + 2^{-q}) \leq 4.$$

It follows from this relation, Proposition 3.1.1, (8.45), (8.44) and (8.41) that v is an (f)-good function. By the definition of L, (8.34) holds for all large T.

On the other hand by (8.45) and (8.43) for every integer $k \geq 2$,

$$\text{dist} \left(H(f), \left\{ v(t) : t \in \left[\sum_{i=1}^{k-1}(T_i + \tau_i) + S_k, \sum_{i=1}^{k-1}(T_i + \tau_i) + S_k + L \right] \right\} \right)$$

$$> \epsilon.$$

The obtained contradiction proves the lemma.

3.9.   Proof of Theorem 3.1.4

Assume that $f \in \mathcal{A}$ has the asymptotic turnpike property with the turnpike $H(f) \subset R^n$. Let M_0, M_1, $\epsilon > 0$. We may assume without loss of generality that

$$M_1 > 4 + \sup\{|y| : y \in H(f)\}, \ \epsilon < 1. \tag{9.1}$$

It follows from Theorem 3.3.1 that there is a natural number L such that if $v : [0, \infty) \to R^n$ is an (f)-good function, then for all large T,

$$\mathrm{dist}(H(f), \{v(t) : t \in [T, T+L]\}) \le 4^{-1}\epsilon. \tag{9.2}$$

It follows from Theorem 1.2.3 that there exist a number $S > M_0 + M_1$ and a neighborhood $\mathcal{U}_0$ of f in $\mathcal{A}$ with the weak topology such that the following property holds:

If $g \in \mathcal{U}_0$, $T_1 \in [0, \infty)$, $T_2 \in [T_1 + 1, \infty)$ and if an a.c. function $v : [T_1, T_2] \to R^n$ satisfies

$$|v(T_i)| \le M_1, \ i = 1, 2, \ I^g(T_1, T_2, v) \le U^g(T_1, T_2, v(T_1), v(T_2)) + M_0, \tag{9.3}$$

then

$$|v(t)| \le S \ t \in [T_1, T_2]. \tag{9.4}$$

It follows from Lemma 3.8.4 that there exists a number

$$\delta_0 \in (0, 8^{-1}\epsilon) \tag{9.5}$$

such that if $T \in [L, \infty)$ and if an a.c. function $v : [0, T] \to R^n$ satisfies

$$d(v(\tau), H(f)) \le \delta_0, \ \tau = 0, T, \tag{9.6}$$

$$\sigma^f(0, T, v) \le \delta_0,$$

then for every $\tau \in [0, T - L]$,

$$\mathrm{dist}(H(f), \{v(t) : t \in [\tau, \tau + L]\}) \le 4^{-1}\epsilon. \tag{9.7}$$

Choose an integer

$$Q_* > 20 + 2(M_0 + 4 + 2\sup\{|U_f(0, 1, x, y)| : x, y \in R^n, |x|, |y| \le S\} \tag{9.8}$$

$$+ 2\sup\{|\pi^f(x)| : x \in R^n, |x| \le S\} + 2|\mu(f)|)\delta_0^{-1}.$$

By Theorems 3.4.1 and 3.6.1 there exists

$$\delta_1 \in (0, 8^{-1}\delta_0) \tag{9.9}$$

such that the following property holds:
    If $y_1, y_2, z_1, z_2 \in R^n$ satisfy

$$d(y_i, H(f)) \leq 4, \ d(z_i, H(f)) \leq 4, \ |y_i - z_i| \leq 8\delta_1, \ i = 1, 2, \qquad (9.10)$$

then

$$|U^f(0, 1, y_1, y_2) - U^f(0, 1, z_1, z_2)| \leq (16Q_*)^{-1}\delta_0 \qquad (9.11)$$

and

$$|\pi^f(y_i) - \pi^f(z_i)| \leq (16Q_*)^{-1}\delta_0, \ i = 1, 2.$$

It follows from Theorem 3.3.1 that there is a natural number L_1 such that if $v : [0, \infty) \to R^n$ is an (f)-good function, then for all large T

$$\text{dist}(H(f), \{v(t) : t \in [T, T + L_1]\}) \leq 8^{-1}\delta_1. \qquad (9.12)$$

We may assume without loss of generality that

$$L_1 \geq 8L. \qquad (9.13)$$

It follows from Lemma 3.8.2 that there exist a natural number $N \geq 10$ and a neighborhood $\mathcal{U}_1$ of f in $\mathcal{A}$ with the weak topology such that

$$\mathcal{U}_1 \subset \mathcal{U}_0 \qquad (9.14)$$

and that the following property holds:
    If $g \in \mathcal{U}_1$, $T \in [0, \infty)$, and if an a.c. function $x : [T, T + NL_1] \to R^n$ satisfies

$$|x(T)|, |x(T + NL_1)| \leq S,$$

$$I^g(T, T+NL_1, x) \leq U^g(T, T+NL_1, x(T), x(T+NL_1))+M_0+4, \qquad (9.15)$$

then there exists an integer $i_0 \in [0, N - 8]$ such that for all $\tau \in [T + i_0 L_1, T + (i_0 + 7)L_1]$,

$$\text{dist}(H(f), \{x(t) : t \in [\tau, \tau + L_1]\}) \leq \delta_1. \qquad (9.16)$$

Put

$$D_0 = \sup\{|\pi^f(x)| : x \in R^n, \ d(x, H(f)) \leq 4\}, \ l = 50NL_1. \qquad (9.17)$$

It follows from Propositions 1.3.8 and 1.3.9 that there exists a neighborhood $\mathcal{U}$ of f in $\mathcal{A}$ with the weak topology such that

$$\mathcal{U} \subset \mathcal{U}_1 \qquad (9.18)$$

and that the following properties hold:

(i) If $g \in \mathcal{U}$, $T_1 \in [0, \infty)$, $T_2 \in [T_1 + 1, T_1 + 100N(L_1 + l)]$ and if an a.c. function $v : [T_1, T_2] \to R^n$ satisfies

$$\min\{I^f(T_1, T_2, v), I^g(T_1, T_2, v)\} \leq 4D_0 + 8 + 200NL_1|\mu(f)|, \quad (9.19)$$

$$+2\sup\{|U^f(0, 1, x, y)| : x, y \in R^n, |x|, |y| \leq S\},$$

then

$$|I^f(T_1, T_2, v) - I^g(T_1, T_2, v)| \leq (64Q_*)^{-1}\delta_1; \quad (9.20)$$

(ii) if $g \in \mathcal{U}$, $T_1 \in [0, \infty)$, $T_2 \in [T_1 + 1, T_1 + 100N(L_1 + l)]$ and if $y, z \in R^n$ satisfy $|y|, |z| \leq 2S + 8$, then

$$|U^f(T_1, T_2, y, z) - U^g(T_1, T_2, y, z)| \leq (64Q_*)^{-1}\delta_1. \quad (9.21)$$

Let $g \in \mathcal{U}$, $T_1 \in [0, \infty)$, $T_2 \in [T_1 + L + lQ_*, \infty)$ and let $v : [T_1, T_2] \to R^n$ be an a.c. function such that

$$|v(T_i)| \leq M_1, \ i = 1, 2, \ I^g(T_1, T_2, v) \leq U^g(T_1, T_2, v(T_1), v(T_2)) + M_0. \quad (9.22)$$

By the definition of S and $\mathcal{U}_0$, (9.4) holds. Put

$$\mathcal{E} = \{h \in R^1 : T_1 + 10NL_1 \leq h \leq T_2 - 10NL_1 \text{ and }$$

$$\text{dist}(H(f), \{v(t) : t \in [h, h + L]\}) > \epsilon\}. \quad (9.23)$$

If $\mathcal{E} = \emptyset$, then Theorem 3.1.4 is valid. Hence we may assume that $\mathcal{E} \neq \emptyset$. We set

$$\tilde{h} = \inf\{h : h \in \mathcal{E}\}$$

and choose $h_1 \in \mathcal{E}$ satisfying $h_1 \leq \tilde{h} + 4^{-1}$. It follows from the definition of $\mathcal{U}_1$, N (see (9.14)-(9.16)) that there are integers $i_1, i_2 \in \{0, \ldots, N-8\}$ such that

$$\text{dist}(H(f), \{v(t) : t \in [T, T + L_1]\}) \leq \delta_1 \quad (9.24)$$

for any number $T \in [h_1 - (2N - i_1)L_1, h_1 - (2N - i_1 - 7)L_1] \cup [h_1 + (N + i_2)L_1, h_1 + (N + i_2 + 7)L_1]$. Set

$$b_1 = h_1 - 2NL_1 + i_1L_1, \ c_1 = h_1 + NL_i + i_2L_1.$$

By induction we define a sequence of intervals $[b_q, c_q]$ $(q \geq 1)$ such that:

(B) $NL_1 \leq c_q - b_q \leq 4NL_1$, $b_q \geq c_{q-1}$ if $q \geq 2$,

$$[b_q, c_q - NL_1] \cap \mathcal{E} \neq \emptyset, \ \mathcal{E} \setminus \cup_{j=1}^q [b_j, c_j] \subset (c_q, T_2];$$

(C) for $h \in \{b_q, c_q\}$, (9.24) holds for every $T \in [h, h + 7L_1]$. (9.25)

Clearly for $q = 1$ properties (B) and (C) hold.

Let k be a natural number. Assume that a sequence of intervals $\{[b_q, c_q]\}_{q=1}^{k}$ was defined and properties (B), (C) hold for $q = 1, \ldots, k$. If

$$\mathcal{E} \setminus \cup_{q=1}^{k}[b_q, c_q] = \emptyset,$$

then the construction of the sequence is completed and $[b_k, c_k]$ is its last element.

Let us assume the converse. Choose $h_2 \in \mathcal{E} \setminus \cup_{q=1}^{k}[b_q, c_q]$ such that

$$h_2 \leq 4^{-1} + \inf\{h : h \in \mathcal{E} \setminus \cup_{q=1}^{k}[b_q, c_q]\}.$$

It follows from the definition of $\mathcal{U}_1$, N that there are integers $j_1, j_2 \in [0, N-8]$ such that (9.24) holds for every

$$T \in [h_2 - (2N - j_1)L_1, h_2 - (2N - j_1 - 7)L_1]$$

$$\cup [h_2 + (N + j_2)L_1, h_2 + (N + j_2 + 7)L_1].$$

Set $c_{k+1} = h_2 + NL_1 + j_2 L_1$, $b_{k+1} = \max\{c_k, h_2 - 2NL_1 + j_1 L_1\}$. It is easy to verify that properties (B), (C) hold with $q = k + 1$.

Evidently the construction of the sequence will be completed in a finite number of steps. Let $[b_Q, c_Q]$ be the last element of the sequence. Clearly

$$\mathcal{E} \subset \cup_{q=1}^{Q}[b_q, c_q]. \tag{9.26}$$

It follows from Theorem 3.6.3 that there exists an (f)-good function $w : [0, \infty) \to R^n$ such that $w(0) = 0$ and

$$\sigma^f(T_1, T_2, w) = 0 \tag{9.27}$$

for each $T_1 \in [0, \infty)$, $T_2 \in (T_1, \infty)$. By the choice of L_1 (see (9.12)) there exists a positive number $\tau_0 > 0$ such that

$$\text{dist}(H(f), \{w(t) : t \in [T, T + L_1]\}) \leq 8^{-1}\delta_1 \tag{9.28}$$

for each $T \geq \tau_0$.

Let $i \in \{1, \ldots, Q\}$. By properties (B) and (C), (9.13), (9.9), (9.24), the choice of δ_0 (see (9.5)-(9.7)) and the definition of $\mathcal{E}$,

$$\sigma^f(b_i, c_i, v) > \delta_0. \tag{9.29}$$

By properties (B) and (C), (9.4), (9.7), (9.29) and the definition of $\mathcal{U}$ (see (9.19) and (9.20)),

$$\sigma^g(b_i, c_i, v) > \delta_0 - 64^{-1}\delta_0. \tag{9.30}$$

It follows from (9.24), (9.28) and property (C) that there exist $t(i,1) \in [\tau_0, \infty)$ and $t(i,2) \in [t(i,1) + 2L_1, t(i,1) + 3L_1]$ such that

$$|w(t(i,1)) - v(b_i)|, \ |w(t(i,2)) - v(c_i)| \le 2\delta_1. \qquad (9.31)$$

Put

$$t(i,3) = t(i,2) - t(i,1). \qquad (9.32)$$

By Corollary 1.3.1 there exists an a.c. function $v_i : [t(i,1), t(i,2)] \to R^n$ such that

$$v_i(t(i,1)) = v(b_i),$$

$$v_i(t) = w(t) \ (t \in [t(i,1) + 1, t(i,2) - 1]), \ v_i(t(i,2)) = v(c_i),$$

$$I^f(t(i,1), t(i,1) + 1, v_i) = U^f(t(i,1), t(i,1) + 1, v_i(t(i,1)), \ v_i(t(i,1) + 1)),$$

$$I^f(t(i,2) - 1, t(i,2), v_i) = U^f(t(i,2) - 1, t(i,2), v_i(t(i,2) - 1), v_i(t(i,2))). \qquad (9.33)$$

By (9.33), (9.31), (9.27), (9.28) and the choice of δ_1 (see (9.9)-(9.11)),

$$\sigma^f(t(i,1), t(i,2), v_i) = \sigma^f(t(i,1), t(i,1) + 1, v_i) + \sigma^f(t(i,2) - 1, t(i,2), v_i)$$

$$\le |U^f(t(i,1), t(i,1) + 1, v_i(t(i,1)), v_i(t(i,1) + 1))$$

$$-U^f(t(i,1), t(i,1) + 1, w(t(i,1)), w(t(i,1) + 1))|$$

$$+|\pi^f(w(t(i,1))) - \pi^f(v_i(t,1))|$$

$$+|U^f(t(i,2) - 1, t(i,2), v_i(t(i,2) - 1), v_i(t(i,2)))$$

$$-U^f(t(i,2) - 1, t(i,2), w(t(i,2) - 1), w(t(i,2)))|$$

$$+|\pi^f(w(t(i,2))) - \pi^f(v_i(t(i,2)))| \le (4Q_*)^{-1}\delta_0. \qquad (9.34)$$

Property (C), (9.34), (9.33) and (9.17) imply that

$$I^f(t(i,1), t(i,2), v_i) \le 3L_1|\mu(f)| + 2D_0 + 1. \qquad (9.35)$$

By induction we define sequences of numbers $\{\bar{b}_i\}_{i=1}^Q$, $\{\bar{c}_i\}_{i=1}^Q$ as follows:

$$\bar{b}_1 = b_1, \ \bar{c}_i = \bar{b}_i + t(i,3) \ (i = 1, \dots, Q), \qquad (9.36)$$

$$\bar{b}_{i+1} = b_{i+1} + \bar{c}_i - c_i \ (i \text{ is an integer}, \ i \le Q - 1).$$

Define an a.c. function $u : [T_1, \bar{c}_Q] \to R^n$ by

$$u(t) = v(t) \ (t \in [T_1, b_1]),$$

$$u(t) = v_i(t - \bar{b}_i + t(i,1)) \ (t \in [\bar{b}_i, \bar{c}_i], \ i = 1, \dots, Q),$$

$$u(t) = v(t - \bar{c}_i + c_i) \ (t \in [\bar{c}_i, \bar{b}_{i+1}], \ i \le Q - 1). \qquad (9.37)$$

(9.23) and property (B) imply that

$$c_Q \leq T_2 - 6NL_1,$$

$$\bar{c}_Q = c_Q - \sum_{i=1}^{Q}(c_i - b_i - t(i,3))$$

$$\in [c_Q - 4QNL_1, c_Q - (N-3)QL_1]. \tag{9.38}$$

By (9.34), (9.35), (9.36) and the definitions of u and $\mathcal{U}$ (see (9.19)-(9.21)) for $i = 1, \ldots, Q$,

$$\sigma^g(\bar{b}_i, \bar{c}_i, u) \leq \sigma^f(t(i,1), t(i,2), v_i) + (64Q^*)^{-1}\delta_1$$

$$\leq (64Q_*)^{-1}\delta_1 + (4Q_*)^{-1}\delta_0. \tag{9.39}$$

There exists an a.c. function $u_0 : [T_1, T_2] \to R^n$ such that

$$u_0(t) = u(t) \ (t \in [T_1, \bar{c}_Q]), \ u_0(t) = w(t - \bar{c}_Q + t(Q,2)) \ (t \in [\bar{c}_Q + 1, c_Q - 1]),$$

$$u_0(t) = v(t) \ (t \in [c_Q, T_2]),$$

$$I^f(h, h+1, u_0) = U^f(h, h+1, u_0(h), u_0(h+1)) \ (h \in \{\bar{c}_Q, c_Q - 1\}). \tag{9.40}$$

It follows from (9.40), (9.37), (9.33), (9.28), (9.4) and (9.1) that

$$|u_0(t)| \leq S \ (t \in \{\bar{c}_Q, \bar{c}_Q + 1, c_Q - 1, c_Q\}). \tag{9.41}$$

By this relation, the definition of $\mathcal{U}$ (see (9.19)-(9.21) and (9.40))

$$I^g(h, h + 1, u_0)$$

$$\leq \sup\{|U^f(0, 1, x, y)| : \ x, y \in R^n, \ |x|, |y| \leq S\}$$

$$+(64Q_*)^{-1}\delta_1 \ (h \in \{\bar{c}_Q, c_Q - 1\}). \tag{9.42}$$

By (9.17), (9.27), (9.28), (9.40) and the choice of $\mathcal{U}$ (see (9.19) and (9.20)) if $t_1 \in [\bar{c}_Q + 1, \infty)$ and if $t_2 \in (-\infty, c_Q - 1]$ satisfy $1 \leq t_2 - t_1 \leq 100NL_1$, then

$$|\sigma^g(t_1, t_2, u_0)| \leq (64Q_*)^{-1}\delta_1. \tag{9.43}$$

(9.40), (9.37) and (9.42) imply that

$$I^g(T_1, T_2, u_0) - I^g(T_1, T_2, v)$$

$$= I^g(b_1, c_Q, u_0) - I^g(b_1, c_Q, v) = I^g(c_Q - 1, c_Q, u_0)$$

$$+I^g(\bar{c}_Q + 1, c_Q - 1, u_0) + I^g(\bar{c}_Q, \bar{c}_Q + 1, u_0) + \sum_{i=1}^{Q}[I^g(\bar{b}_i, \bar{c}_i, u) - I^g(b_i, c_i, v)]$$

$$\leq \sum_{i=1}^{Q}[I^g(\bar{b}_i, \bar{c}_i, u) - I^g(b_i, c_i, v)] + I^g(\bar{c}_Q + 1, c_Q - 1, u_0)$$

$$+2\sup\{|U^f(0, 1, x, y)| : x, y \in R^n, |x|, |y| \leq S\} + (32Q_*)^{-1}\delta_1.$$

It follows from this relation, (9.39), (9.33), (9.37), (9.36) and (9.30) that

$$I^g(T_1, T_2, u_0) - I^g(T_1, T_2, v) \leq I^g(\bar{c}_Q + 1, c_Q - 1, u_0)$$

$$+2\sup\{|U^f(0, 1, x, y)| : x, y \in R^n, |x|, |y| \leq S\} + (32Q_*)^{-1}\delta_1$$

$$+\sum_{i=1}^{Q}[\delta_1(64Q_*)^{-1}+(4Q_*)^{-1}\delta_0+\mu(f)(t(i,3)-c_i+b_i)-\delta_0+64^{-1}\delta_0]. \quad (9.44)$$

Put

$$t_i = \bar{c}_Q + 1 + iQ^{-1}(c_Q - \bar{c}_Q - 2), \ i = 0, \ldots, Q.$$

By (9.27), (9.38), (9.40), (9.41) and (9.43),

$$I^g(\bar{c}_Q + 1, c_Q - 1, u_0) \leq \sum_{i=0}^{Q-1}[I^g(t_i, t_{i+1}, u_0) - \pi^f(u_0(t_i)) + \pi^f(u_0(t_{i+1}))]$$

$$+2\sup\{|\pi^f(x)| : x \in R^n, |x| \leq S\} \leq 2\sup\{|\pi^f(x)| : x \in R^n, |x| \leq S\}$$

$$+\mu(f)(c_Q - \bar{c}_Q - 2) + Q\delta_1(64Q_*)^{-1}.$$

By this relation, (9.44), (9.22), (9.40), (9.37) and (9.38),

$$-M_0 \leq I^g(T_1, T_2, u_0) - I^g(T_1, T_2, v) \leq 2\sup\{|\pi^f(x)| : x \in R^n, |x| \leq S\}$$

$$+2\sup\{|U^f(0, 1, x, y)| : x, y \in R^n, |x|, |y| \leq S\} + (32Q_*)^{-1}\delta_1$$

$$+Q[\delta_1(64Q_*)^{-1} + \delta_0(4Q_*)^{-1} + \delta_0(64)^{-1} - \delta_0] + Q\delta_1(64Q_*)^{-1} + 2|\mu(f)|.$$

It follows from this relation, (9.8) and (9.9) that $Q + 20 \leq Q_*$. This completes the proof of the theorem.

3.10.   Proof of Theorem 3.1.3

Assume that $f \in \mathcal{A}$ has the asymptotic turnpike property with the turnpike $H(f) \subset R^n$. It follows from Theorem 1.2.2 that there exist a neighborhood $\mathcal{U}_1$ of f in $\mathcal{A}$ with the weak topology and a number $M_1 > 0$ such that

$$\limsup_{t\to\infty} |v(t)| < M_1 \qquad (10.1)$$

for each $g \in \mathcal{U}_1$ and each (g)-good function $v : [0, \infty) \to R^n$. It follows from Theorem 3.1.4 that there exist a number $l > 0$, integers $L, Q_* \geq 1$ and a neighborhood $\mathcal{U}$ of f in $\mathcal{A}$ with the weak topology such that $\mathcal{U} \subset \mathcal{U}_1$ and the following property holds:

If $g \in \mathcal{U}$, $T_1 \in [0, \infty)$, $T_2 \in [T_1 + L, +lQ_*, \infty)$ and if an a.c. function $v : [T_1, T_2] \to R^n$ satisfies

$$|v(T_i)| \leq M_1, \ i = 1, 2, \ I^g(T_1, T_2, v) \leq U^g(T_1, T_2, v(T_1), v(T_2)) + 1,$$
(10.2)

then there exist sequences of numbers $\{b_i\}_{i=1}^Q$, $\{c_i\}_{i=1}^Q \subset [T_1, T_2]$ such that

$$Q \leq Q_*, \ 0 \leq c_i - b_i \leq l \ (i = 1, \ldots, Q),$$
(10.3)

and for every $T \in [T_1, T_2 - L] \setminus \cup_{i=1}^Q [b_i, c_i]$,

$$\mathrm{dist}(H(f), \{v(t) : t \in [T, T + L]\}) \leq \epsilon.$$
(10.4)

Let $g \in \mathcal{U}$ and $v : [0, \infty) \to R^n$ be a (g)-good function. It is sufficient to show that (10.4) holds for all large T.

Let us assume the converse. Then there exists a sequence $\{t_i\}_{i=1}^\infty \subset (0, \infty)$ such that

$$t_{i+1} - t_i \geq 4l + 4, \ \mathrm{dist}(H(f), \{v(t) : t \in [t_i, t_i + L]\}) > \epsilon, \ i = 1, 2, \ldots.$$
(10.5)

By the definition of $\mathcal{U}_1$, M_1 (see (10.1)) and Proposition 3.3.1 we may assume that

$$|v(t)| < M_1 \ (t \in [0, \infty)),$$
(10.6)

$$I^g(T_1, T_2, v) \leq U^g(T_1, T_2, v(T_1), v(T_2)) + 1$$
(10.7)

for each $T_1 \in [0, \infty)$, $T_2 \in (T_1, \infty)$. Fix an integer

$$\tau > (L + l)Q_* + t_{2Q_*+2} + 4L + 4.$$
(10.8)

It follows from (10.6), (10.7), (10.8) and the definition of $\mathcal{U}$, l, L, Q_* that there exist sequences of numbers $\{b_i\}_{i=1}^Q$, $\{c_i\}_{i=1}^Q \subset [0, \tau]$ such that (10.3) holds and (10.4) is valid for every $T \in [0, \tau - L] \setminus \cup_{i=1}^Q [b_i, c_i]$. Together with (10.5) and (10.8) this implies that for every $i \in \{1, \ldots, 2Q_* + 2\}$ there exists $p(i) \in \{1, \ldots, Q\}$ for which $t_i \in [b_{p(i)}, c_{p(i)}]$. By (10.3) there exist $i, j \in \{1, \ldots, 2Q_* + 2\}$ such that

$$i < j, \ p(i) = p(j), \ t_j - t_i \leq c_{p(i)} - b_{p(i)} \leq l.$$

This is contradictory to (10.5). The obtained contradiction proves the theorem.

3.11.   Proofs of Theorems 3.1.1 and 3.1.5

Proof of Theorem 3.1.1. By Lemma 3.7.5 there exists an everywhere dense subset E of $\mathcal{A}$ with the strong topology such that each $f \in E$ has the asymptotic turnpike property with the turnpike $H(f) \subset R^n$.

Let $f \in E$ and $p \geq 1$ be an integer. It follows from Theorem 3.1.3 that there exist an integer $L(f,p) \geq 1$ and an open neighborhood $\mathcal{U}(f,p)$ of f in $\mathcal{A}$ with the weak topology such that if $g \in \mathcal{U}(f,p)$ and if $v : [0,\infty) \to R^n$ is a (g)-good function, then

$$\mathrm{dist}(H(f), \{v(t) : t \in [T, T+L(f,p)]\}) \leq p^{-1} \text{ for all large } T.$$

Put

$$\mathcal{F} = \cap_{p=1}^{\infty} \cup \{\mathcal{U}(f,p) : f \in E\}.$$

Let $g \in \mathcal{F}$ and $v_i : [0,\infty) \to R^n$ $(i = 1,2)$ be (g)-good functions. Let $\epsilon \in (0,1)$. There exist an integer $p \geq 1$ and $f \in E$ such that

$$p > 8\epsilon^{-1}, \ g \in \mathcal{U}(f,p).$$

It follows from the definition of $\mathcal{U}(f,p)$ and $L(f,p)$ that

$$\mathrm{dist}(H(f), \Omega(v_i)) \leq p^{-1}, \ i = 1,2, \ \mathrm{dist}(\Omega(v_1), \Omega(v_2)) \leq 2p^{-1} < \epsilon.$$

This completes the proof of the theorem.

By a simple modification of the above proof (basically by replacing Lemma 3.7.5 by Lemma 3.7.6) we can easily establish Theorem 3.1.5.

3.12.   Examples

Fix a positive constant a and set $\psi(t) = t$, $t \in [0,\infty)$. Consider a complete metric space $\mathcal{A}$ of integrands $f : R^n \times R^n \to R^1$ defined in Section 3.1.

Example 1. Consider an integrand $f(x,u) = |x|^2 + |u|^2$, $x, u \in R^n$. It is easy to see that $f \in \mathcal{A}$. Applying the methods used in the proofs of Theorems 3.1.1-3.1.4 we can show that $\Omega(v) = \{0\}$ for every (f)-good function $v : [0,\infty) \to R^n$.

Example 2. Fix a number $q > 0$ and consider an integrand $g(x,u) = q|x|^2|x-e|^2 + |u|^2$, $x, u \in R^n$ where $e = (1,1,\dots,1) \in R^n$. It is easy to see that $g \in \mathcal{A}$ if a is large enough. Clearly the function $v_1(t) = 0$, $v_2(t) = e$, $t \in [0,\infty)$ are (g)-good and g does not have the asymptotic turnpike property (see Theorem 3.1.1).

Chapter 4

INFINITE HORIZON
AUTONOMOUS PROBLEMS

In this chapter we study the structure of weakly optimal solutions of infinite-horizon autonomous variational problems with vector-valued functions and with integrands which belong to the space $\mathcal{A}$ introduced in Section 3.1. The results of this chapter have been established in [97].

4.1. Main results

Denote by $|\cdot|$ the Euclidean norm in R^n. Let $a > 0$ be a constant and let $\psi : [0, \infty) \to [0, \infty)$ be an increasing function such that $\psi(t) \to +\infty$ as $t \to \infty$.

We consider the space $\mathcal{A}$ introduced in Section 3.1 which consists of all continuous functions $f : R^n \times R^n \to R^1$ which satisfy the following assumptions:

A(i) for each $x \in R^n$ the function $f(x, \cdot) : R^n \to R^1$ is convex;

A(ii) $f(x, u) \geq \max\{\psi(|x|), \psi(|u|)|u|\} - a$ for each $(x, u) \in R^n \times R^n$;

A(iii) for each $M, \epsilon > 0$ there exist $\Gamma, \delta > 0$ such that if $u_1, u_2, x_1, x_2 \in R^n$ satisfy

$$|x_i| \leq M, \ |u_i| \geq \Gamma, \ i = 1, 2, \quad \max\{|x_1 - x_2|, |u_1 - u_2|\} \leq \delta,$$

then

$$|f(x_1, u_1) - f(x_2, u_2)| \leq \epsilon \max\{f(x_1, u_1), f(x_2, u_2)\}.$$

It is easy to show that an integrand $f = f(x, u) \in C^1(R^n \times R^n)$ belongs to $\mathcal{A}$ if f satisfies Assumptions A(i), A(ii), and also there exists an increasing function $\psi_0 : [0, \infty) \to [0, \infty)$ such that for each $x, u \in R^n$,

$$\sup\{|\partial f / \partial x(x, u)|, \ |\partial f / \partial u(x, u)|\} \leq \psi_0(|x|)(1 + \psi(|u|)|u|).$$

For the set $\mathcal{A}$ we consider the uniformity introduced in Section 3.1. This uniformity is determined by the following base:

$$E(N, \epsilon, \lambda) = \{(f, g) \in \mathcal{A} \times \mathcal{A} : |f(x, u) - g(x, u)| \leq \epsilon$$

$$\text{for each } x, u \in R^n \text{ satisfying } |x|, |u| \leq N,$$

$$(|f(x, u)| + 1)(|g(x, u)| + 1)^{-1} \in [\lambda^{-1}, \lambda]$$

$$\text{for each } x, u \in R^n \text{ satisfying } |x| \leq N\},$$

where $N > 0$, $\epsilon > 0$, $\lambda > 1$ [37].

It is known that the uniform space $\mathcal{A}$ is metrizable and complete (see Section 3.1). This uniformity induces a topology in $\mathcal{A}$ which is called the weak topology.

We consider functionals of the form

$$I^f(T_1, T_2, x) = \int_{T_1}^{T_2} f(x(t), x'(t))dt \tag{1.1}$$

where $f \in \mathcal{A}$, $0 \leq T_1 < T_2 < +\infty$ and $x : [T_1, T_2] \to R^n$ is an a.c. function.

For $f \in \mathcal{A}$, $y, z \in R^n$ and numbers T_1, T_2 satisfying $0 \leq T_1 < T_2$ we set

$$U^f(T_1, T_2, y, z) = \inf\{I^f(T_1, T_2, x) : x : [T_1, T_2] \to R^n \tag{1.2}$$

$$\text{is an a.c. function satisfying } x(T_1) = y, \ x(T_2) = z\}.$$

It is easy to see that $-\infty < U^f(T_1, T_2, y, z) < +\infty$ for each $f \in \mathcal{A}$, each $y, z \in R^n$ and all numbers T_1, T_2 satisfying $0 \leq T_1 < T_2$.

Let $f \in \mathcal{A}$. For any a.c. function $x : [0, \infty) \to R^n$ we set

$$J(x) = \liminf_{T \to \infty} T^{-1} I^f(0, T, x). \tag{1.3}$$

Of special interest is the minimal long-run average cost growth rate

$$\mu(f) = \inf\{J(x) : x : [0, \infty) \to R^n \text{ is an a.c. function}\}. \tag{1.4}$$

Clearly $-\infty < \mu(f) < +\infty$.

Recall the definition of a good function given in Section 1.2 and the definition of a weakly optimal function introduced in Section 3.1.

An a.c. function $x : [0, \infty) \to R^n$ is called an (f)-*good function* if the function $\phi_x^f : T \to I^f(0, T, x) - \mu(f)T$, $T \in (0, \infty)$ is bounded.

An a.c. function $x : [0, \infty) \to R^n$ is called (f)-weakly optimal if for any a.c. function $y : [0, \infty) \to R^n$ satisfying $y(0) = x(0)$,

$$\liminf_{T \to \infty} \int_0^T [f(x(t), x'(t)) - f(y(t), y'(t))]dt \leq 0.$$

Proposition 3.1.1 and Theorem 1.2.2 imply the following result.

PROPOSITION 4.1.1 *For any a.c. function* $x : [0, \infty) \to R^n$ *either*

$$I^f(0, T, x) - T\mu(f) \to \infty \text{ as } T \to \infty$$

or

$$\sup\{|I^f(0, T, x) - T\mu(f)| : T \in (0, \infty)\} < \infty.$$

Moreover any (f)-*good function* $x : [0, \infty) \to R^n$ *is bounded.*

We denote $d(x, B) = \inf\{|x-y| : y \in B\}$ for $x \in R^n$, $B \subset R^n$. Denote by $\mathrm{dist}(A, B)$ the Hausdorff metric for two sets $A \subset R^n$, $B \subset R^n$. For every bounded a.c. function $x : [0, \infty) \to R^n$ define

$$\Omega(x) = \{y \in R^n : \text{ there exists a sequence } \{t_i\}_{i=0}^{\infty} \subset (0, \infty)$$

$$\text{for which } t_i \to \infty, \ x(t_i) \to y \text{ as } i \to \infty\}. \tag{1.5}$$

We say that an integrand $f \in A$ has the asymptotic turnpike property if $\Omega(v_1) = \Omega(v_2)$ for each pair of (g)-good functions $v_i : [0, \infty) \to R^n$, $i = 1, 2$.

Let $f \in A$ have the asymptotic turnpike property. Put

$$H(F) = \Omega(v)$$

where $v : [0, \infty) \to R^n$ is an (f)-good function. Clearly, $H(f)$ does not depend on v. By Theorem 1.2.1, H(f) is a compact subset of R^n. We say that $H(f)$ is the turnpike of f.

Consider any $f \in A$. In analyzing the infinite-horizon variational problem with the integrand f we study the function $U^f(T_1, T_2, y, z)$ $(T_2 > T_1 \geq 0, \ y, z \in R^n)$ defined by (1.2). By Theorems 3.6.1 and 3.6.2,

$$U^f(0, T, x, y) = T\mu(f) + \pi^f(x) - \pi^f(y) + \bar{\theta}_T^f(x, y), \ x, y \in R^n, \ T \in (0, \infty), \tag{1.6}$$

where $\pi^f : R^n \to R^1$ is a continuous function and $(T, x, y) \to \bar{\theta}_T^f(x, y) \in R^1$ is a continuous function defined for $T > 0$, $x, y \in R^n$,

$$\pi^f(x) = \inf\{\liminf_{T \to \infty}[I^f(0, T, v) - \mu(f)T] : v : [0, \infty)$$

$$\to R^n \text{ is an a.c. function satisfying } v(0) = x\}, \ x \in R^n, \tag{1.7}$$

and for every $T > 0$ and every $x \in R^n$ there is $y \in R^n$ satisfying $\bar{\theta}_T^f(x, y) = 0$.

Assume that the integrand f has the asymptotic turnpike property with the turnpike $H(f) \subset R^n$. (By Theorem 3.1.1 a generic $f \in \mathcal{A}$ has the asymptotic turnpike property.)

By Theorems 3.6.3 and 3.6.4 for every $x \in R^n$ there exists an (f)-good function $v : [0, \infty) \to R^n$ such that $v(0) = x$ and the relation

$$I^f(T_1, T_2, v) = (T_2 - T_1)\mu(f) + \pi^f(v(T_1)) - \pi^f(v(T_2)) \qquad (1.8)$$

holds for each $T_1 \in [0, \infty)$, $T_2 \in (T_1, \infty)$, and moreover, each a.c. function $v : [0, \infty) \to R^n$ such that (1.8) holds for each $T_1 \in [0, \infty)$, $T_2 \in (T_1, \infty)$ is an (f)-weakly optimal function.

For each function $f \in \mathcal{A}$ denote by $A(f)$ the set of all a.c. functions $v : [0, \infty) \to R^n$ which satisfy (1.8) for each $T_1 \in [0, \infty)$, $T_2 \in (T_1, \infty)$.

If $f \in \mathcal{A}$ has the turnpike property, then by Theorem 3.6.4 any function $v \in A(f)$ is an (f)-weakly optimal function.

We establish the following results.

THEOREM 4.1.1 *Assume that $f \in \mathcal{A}$ has the asymptotic turnpike property with the turnpike $H(f) \subset R^n$. Then f is a continuity point of the mapping*

$$g \to (\mu(g), \pi^g) \in R^1 \times C(R^n), \ g \in \mathcal{A},$$

where $C(R^n)$ is the space of all continuous functions $\phi : R^n \to R^1$ with the topology of the uniform convergence on bounded subsets and $\mathcal{A}$ is equipped with the weak topology.

Theorems 4.1.2 and 4.1.3 establish turnpike properties of good functions which satisfy (1.8) for a generic $f \in \mathcal{A}$.

THEOREM 4.1.2 *Assume that $f \in \mathcal{A}$ has the asymptotic turnpike property with the turnpike $H(f) \subset R^n$. Let ϵ be a positive number. Then there exist numbers $L, \delta > 0$ and a neighborhood $\mathcal{U}$ of f in $\mathcal{A}$ with the weak topology such that the following property holds:*
If $g \in \mathcal{U}$ and if $v \in A(g)$ satisfies $d(v(0), H(f)) \leq \delta$, then

$$dist(H(f), \{v(t) : t \in [T, T + L]\}) \leq \epsilon \qquad (1.9)$$

for each $T \geq 0$.

Theorem 4.1.2 shows that if f has the asymptotic turnpike property, g belongs to a small neighborhood of f, $v \in A(g)$ and $v(0)$ is close to the turnpike $H(f)$, then v is close to the turnpike $H(f)$ at any point $t \geq 0$.

The following result shows that the convergence property established in Theorem 4.1.2 holds for all $T \geq Q$ where Q is a positive constant, if we do not assume that $v(0)$ is close to the turnpike $H(f)$.

THEOREM 4.1.3 *Assume that $f \in \mathcal{A}$ has the asymptotic turnpike property with the turnpike $H(f) \subset R^n$. Let ϵ and K be positive numbers. Then there exist numbers $L, Q > 0$ and a neighborhood $\mathcal{U}$ of f in $\mathcal{A}$ with the weak topology such that if $g \in \mathcal{U}$ and if $v \in A(g)$ satisfies $|v(0)| \leq K$, then the inequality (1.9) is valid for all $T \in [Q, \infty)$.*

For each $f \in \mathcal{A}$ denote by $B(f)$ the set of all a.c. functions $v : R^1 \to R^n$ such that (1.8) holds for each $T_1 \in R^1$, $T_2 \in (T_1, \infty)$, and $\liminf_{t \to -\infty} |v(t)| < \infty$.

THEOREM 4.1.4 *Assume that $f \in \mathcal{A}$ has the asymptotic turnpike property with the turnpike $H(f) \subset R^n$. Then the following properties hold:*
(1) for each $h \in H(f)$ there exists $v \in B(f)$ satisfying $v(0) = h$;
(2) for each $v \in B(f)$ the relation $v(t) \in H(f)$ holds for all $t \in R^1$;
(3) for each $\epsilon > 0$ there exists $L > 0$ such that (1.9) holds for each $v \in B(f)$ and each $T \in R^1$.

Chapter 4 is organized as follows. Theorems 4.1.1-4.1.3 are proved in Section 4.2. Section 4.3 contains the proof of Theorem 4.1.4.

4.2. Proofs of Theorems 4.1.1-4.1.3

We preface the proofs of Theorems 4.1.1-4.1.3 by a number of auxiliary results.

Theorem 3.6.1, Proposition 3.5.3 and equation (6.2) in Section 3.6 imply the following result.

PROPOSITION 4.2.1 *Let $f \in \mathcal{A}$. Then $\pi^f(x) \to \infty$ as $|x| \to \infty$.*

Assume that $f \in \mathcal{A}$ has the asymptotic turnpike property with the turnpike $H(f) \subset R^n$.

For each $g \in \mathcal{A}$, each $r_1 \geq 0$, $r_2 > r_1$ and each a.c. function $v : [r_1, r_2] \to R^n$ set

$$\sigma^g(r_1, r_2, v) = I^g(r_1, r_2, v) - \pi^f(v(r_1)) + \pi^f(v(r_2)),$$

$$\tilde{\sigma}^g(r_1, r_2, v) = I^g(r_1, r_2, v) - \pi^g(v(r_1)) + \pi^g(v(r_2)),$$

$$\Phi^g(r_1, r_2, v) = I^g(r_1, r_2, v) - U^g(r_1, r_2, v(r_1), v(r_2)). \qquad (2.1)$$

Lemma 3.8.2 and Theorem 1.2.3 imply the following result which will be used in the proof of Theorem 3.1.3.

LEMMA 4.2.1 *Let $\epsilon_0 \in (0,1)$, $K_0, M_0 > 0$ and let l be a positive integer such that each (f)-good function $x : [0, \infty) \to R^n$ satisfies*

$$dist(H(f), \{x(t) :\ t \in [T, T+l]\}) \leq 8^{-1}\epsilon_0$$

for all large T. Then there exist an integer $N \geq 10$ and a neighborhood $\mathcal{U}$ of f in $\mathcal{A}$ with the weak topology and a number $M_1 > 0$ such that the following property holds:

If $g \in \mathcal{U}$, $T_1 \geq 0$, $T_2 \geq T_1 + Nl$ and if an a.c. function $x : [T_1, T_2] \to R^n$ satisfies

$$|x(T_i)| \leq K_0,\ i = 1, 2,\ \Phi^g(T_1, T_2, x) \leq M_0, \tag{2.2}$$

then

$$|x(t)| \leq M_1 \text{ for all } t \in [T_1, T_2];$$

for each $S \in [T_1, T_2 - Nl]$ there exists an integer $i_0 \in [0, N-8]$ such that for all $T \in [S + i_0 l, S + (i_0 + 7)l]$,

$$dist(H(f), \{x(t) :\ t \in [T, T+l]\}) \leq \epsilon_0.$$

Put

$$D_f = \sup\{|h| :\ h \in H(f)\}. \tag{2.3}$$

For each $g \in \mathcal{A}$ denote by $A(g)$ the set of all a.c. functions $v : [0, \infty) \to R^n$ such that

$$\tilde{\sigma}^g(T_1, T_2, v) = (T_2 - T_1)\mu(g) \tag{2.4}$$

for each $T_1 \in [0, \infty)$, $T_2 \in (T_1, \infty)$.

For $K, \tau > 0$ and $g \in \mathcal{A}$ put

$$l(g, K, \tau) = \inf\{U^g(0, \tau, x, y) - \pi^f(x) + \pi^f(y) :\ x, y \in R^n, |x|, |y| \leq K\}. \tag{2.5}$$

It follows from the representation formula (see (1.6), (1.7)) and Theorem 3.6.3 that

$$l(f, K, \tau) = \mu(f)\tau,\ \tau > 0,\ K > D_f. \tag{2.6}$$

Equations (2.5), (2.6) and Proposition 1.3.9 imply the following result.

LEMMA 4.2.2 *Let $K > D_f$, $0 < \tau_1 < \tau_2$, $\delta > 0$. Then there exists a neighborhood $\mathcal{U}$ of f in $\mathcal{A}$ with the weak topology such that for each $g \in \mathcal{U}$ and each $\tau \in [\tau_1, \tau_2]$,*

$$|l(g, K, \tau) - \tau\mu(f)| \leq \delta.$$

LEMMA 4.2.3 *For each $h \in H(f)$ there exists an (f)-good function $v : [0, \infty) \to H(f)$ such that $v(0) = h$ and $v \in A(f)$.*

Proof. Let $h \in H(f)$. By Proposition 4.2.1 there exists an (f)-good function $u \in A(f)$. We may assume that

$$d(u(t), H(f)) \leq 1 \text{ for all } t \in [0, \infty). \tag{2.7}$$

There exists a sequence of positive numbers $\{T_p\}_{p=0}^{\infty}$ such that

$$T_{p+1} \geq T_p + 1, \; p = 0, 1, \ldots, \; u(T_p) \to h \text{ as } p \to \infty. \tag{2.8}$$

For every integer $p \geq 1$ we set

$$v_p(t) = u(t + T_p), \; t \in [0, \infty). \tag{2.9}$$

By Proposition 1.3.5, (2.9) and (2.7) there exist a subsequence $\{v_{p_j}\}_{j=1}^{\infty}$ and an a.c. function $v : [0, \infty) \to R^n$ such that for every integer $N \geq 1$,

$$v_{p_j}(t) \to v(t) \text{ as } j \to \infty \text{ uniformly in } [0, N], \tag{2.10}$$

$$I^f(0, N, v) \leq \liminf_{j \to \infty} I^f(0, N, v_{p_j}). \tag{2.11}$$

Since f has the asymptotic turnpike property with the turnpike $H(f)$ we have $\Omega(v) = H(f)$. It follows from (2.9) and (2.10) that $v(t) \in H(f)$ for all $t \in [0, \infty)$. By (2.8)-(2.10), $v(0) = h$. Since $u \in A(f)$ it follows from (2.9)-(2.11) that for each integer $N \geq 1$,

$$I^f(0, N, v) \leq \liminf_{j \to \infty} I^f(T_{p_j}, T_{p_j} + N, u)$$

$$\leq \liminf_{j \to \infty} [N\mu(f) + \pi^f(u(T_{p_j})) - \pi^f(u(T_{p_j} + N))]$$

$$= N\mu(f) + \pi^f(v(0)) - \pi^f(v(N)).$$

Together with the representation formula (see (1.6), (1.7)) this implies that $v \in A(f)$. The lemma is proved.

In the next two lemmas we study the existence of a function $x : [0, T] \to R^n$ such that $\sigma^g(0, T, x) \leq l(g, K, T) + \epsilon$ where ϵ is a given small positive number, g belongs to a small neighborhood of f in $\mathcal{A}$ with the weak topology and $x(0)$, $x(T)$ belong to the turnpike $H(f)$.

LEMMA 4.2.4 *For each $\epsilon \in (0, 1)$ and each $K > D_f + 1$ there exists a neighborhood $\mathcal{U}$ of f in $\mathcal{A}$ with the weak topology such that the following property holds:*

If $g \in \mathcal{U}$ and $T \geq 1$, then there exists an a.c. function $x : [0, T] \to R^n$ for which $x(0), x(T) \in H(f)$ and

$$\sigma^g(0, T, x) \leq l(g, K, T) + \epsilon. \tag{2.12}$$

Proof. Let $\epsilon \in (0,1)$ and $K > D_f + 1$. By Lemma 4.2.3 there exists an (f)-good function

$$v_0 : [0,\infty) \to H(f) \text{ such that } v_0 \in A(f). \tag{2.13}$$

By Theorem 1.2.3 there exist a number $M_1 > 0$ and a neighborhood $\mathcal{U}_1$ of f in $\mathcal{A}$ with the weak topology such that the following property holds:
    If $g \in \mathcal{U}_1$, $T_1 \in [0,\infty)$, $T_2 \in [T_1 + 8^{-1}, \infty)$ and if an a.c. function $x : [T_1, T_2] \to R^n$ satisfies

$$|x(T_i)| \le K, \ i = 1,2, \ \Phi^g(T_1, T_2, x) \le 1, \tag{2.14}$$

then

$$|x(t)| \le M_1, \ t \in [T_1, T_2]. \tag{2.15}$$

It follows from Theorem 3.4.1 that there exists a number

$$\delta \in (0, 8^{-1}\epsilon) \tag{2.16}$$

such that the following property holds:
    If $y_1, y_2, z_1, z_2 \in R^n$ satisfy

$$|y_i|, |z_i| \le M_1 + 2K + 1, \ i = 1,2, \ |y_i - z_i| \le 4\delta, \ i = 1,2, \tag{2.17}$$

then

$$|U^f(0,1,y_1,y_2) - U^f(0,1,z_1,z_2)| \le 2^{-6}\epsilon,$$
$$|\pi^f(y_i) - \pi^f(z_i)| \le 2^{-6}\epsilon, \ i = 1,2. \tag{2.18}$$

By Theorem 3.3.1 there exists an integer $L \ge 1$ such that if $v : [0,\infty) \to R^n$ is an (f)-good function, then for all large T,

$$\text{dist}(H(f), \{v(t) : t \in [T, T+L]\}) \le 8^{-1}\delta. \tag{2.19}$$

It follows from the choice of L and Lemma 4.2.1 that there exist a neighborhood $\mathcal{U}_2$ of f in $\mathcal{A}$ with the weak topology and an integer $N_1 \ge 10$ such that the following property holds:
    If $g \in \mathcal{U}_2$, $T_1 \ge 0$, $T_2 \ge T_1 + N_1 L$ and if an a.c. function $x : [T_1, T_2] \to R^n$ satisfies (2.14), then for each $S \in [T_1, T_2 - N_1 L]$ there exists an integer $i_0 \in [0, N_1 - 8]$ such that for all $T \in [S + i_0 L, S + (i_0 + 7)L]$

$$\text{dist}(H(f), \{x(t) : t \in [T, T+L]\}) \le \delta. \tag{2.20}$$

Since f has the asymptotic turnpike property we have $\Omega(v_0) = H(f)$. Then it follows from (2.13) that there exist integers

$$N_2 \ge 4N_1 L + 4, \ N_3 \ge 8 \tag{2.21}$$

such that

$$\text{dist}(H(f), \{v_0(t) : t \in [0, N_2 L]\}) \leq 8^{-1}\delta,$$

$$\text{dist}(H(f), \{v_0(t) : t \in [8(N_2+1)L, 8(N_2+1)L+N_3L]\}) \leq 8^{-1}\delta. \quad (2.22)$$

Fix an integer

$$N_0 \geq 8L(N_1 + N_2 + N_3 + 4). \quad (2.23)$$

It follows from Proposition 1.3.9 that there exists a neighborhood $\mathcal{U}_3$ of f in $\mathcal{A}$ with the weak topology such that

$$|U^f(0, 1, y, z) - U^g(0, 1, y, z)| \leq 2^{-6}\epsilon \quad (2.24)$$

for each $g \in \mathcal{U}_3$ and each $z, y \in R^n$ which satisfy $|y|, |z| \leq 4M_1 + 4K + 4$.

It follows from Proposition 1.3.8 that there exists a neighborhood $\mathcal{U}_4$ of f in $\mathcal{A}$ with the weak topology such that the following property holds:

If $g \in \mathcal{U}_4$, $T_1 \geq 0$, $T_2 \in [T_1 + 1, T_1 + N_0 + 1]$ and if and a.c. function $x : [T_1, T_2] \to R^n$ satisfies

$$\min\{I^f(T_1, T_2, x), \ I^g(T_1, T_2, x)\}$$

$$\leq N_0(|\mu(f)| + 1) + 2\sup\{|\pi^f(h)| : \ h \in H(f)\}, \quad (2.25)$$

then

$$|I^f(T_1, T_2, x) - I^g(T_1, T_2, x)| \leq 2^{-6}\epsilon. \quad (2.26)$$

It follows from Lemma 4.2.2 that there exists a neighborhood $\mathcal{U}_5$ of f in $\mathcal{A}$ with the weak topology such that

$$|l(g, K, \tau) - \mu(f)\tau| \leq 2^{-6}\epsilon \quad (2.27)$$

for each $g \in \mathcal{U}_5$ and each $\tau \in [1, 8(N_0 + 1)]$. Put

$$\mathcal{U} = \cap_{i=1}^{5} \mathcal{U}_i. \quad (2.28)$$

Let $g \in \mathcal{U}$ and a number $T \geq 1$. There are two cases: (i) $T \leq N_0$; (ii) $T > N_0$. Consider the case (i). Put

$$x(t) = v_0(t), \ t \in [0, T]. \quad (2.29)$$

By the definition of $A(f)$ (see (2.4)) and (2.13),

$$I^f(0, T, x) \leq N_0 |\mu(f)| + 2\sup\{|\pi^f(h)| : \ h \in H(f)\}.$$

It follows from this inequality and the choice of $\mathcal{U}_4$ that

$$|I^f(0, T, x) - I^g(0, T, x)| \leq 2^{-6}\epsilon. \quad (2.30)$$

It follows from the choice of $\mathcal{U}_5$ that

$$|l(g, K, T) - \mu(f)T| \leq 2^{-6}\epsilon. \qquad (2.31)$$

Combining (2.30), (2.29) and (2.13) we obtain that

$$\sigma^g(0, T, x) \leq \sigma^f(0, T, v_0) + 2^{-6}\epsilon$$

$$\leq \mu(f)T + 2^{-6}\epsilon \leq l(g, K, T) + 2^{-5}\epsilon.$$

Therefore in the case (i) the assertion of the lemma is valid.
Consider the case (ii). Then

$$T > N_0. \qquad (2.32)$$

There exists an a.c. function $y : [0, T] \to R^n$ such that

$$|y(0)|, |y(T)| \leq K, \ \sigma^g(0, T, y) \leq l(g, K, T) + 16^{-1}\delta. \qquad (2.33)$$

We have

$$\Phi^g(0, T, y) \leq 16^{-1}\delta.$$

It follows from this inequality, (2.23), (2.32), (2.33), and the choice of $N_1, \mathcal{U}_2$ (see (2.20)) that there exist integers $i_1, i_2 \in [0, N_1 - 8]$ such that

$$\text{dist}(H(f), \{y(t) : t \in [S, S + L]\}) \leq \delta \qquad (2.34)$$

for each

$$S \in [i_1 L, (i_1 + 7)L] \cup [T - 2N_1 L + i_2 L, T - 2N_1 L + (i_2 + 7)L]. \qquad (2.35)$$

By (2.34), (2.35) and (2.22) there exist

$$t_1 \in [8(N_2 + 1)L, 8(N_2 + 1)L + N_3 L], \ t_2 \in [0, N_2 L] \qquad (2.36)$$

such that

$$|y(i_1 L + 1) - v_0(t_1)| \leq \delta + 4^{-1}\delta, \ |y(T - 2N_1 L + i_2 L + 1) - v_0(t_2)| \leq \delta + 4^{-1}\delta. \qquad (2.37)$$

By Corollary 1.3.1, (2.13), (2.36), (2.21), (2.32), and (2.23) there exists an a.c. function $x : [0, T] \to R^n$ such that

$$x(t) = v_0(t + t_1 - i_i L - 1), \ t \in [0, i_1 L + 1),$$

$$x(t) = y(t), \ t \in [i_1 L + 2, T - 2N_1 L + i_2 L],$$

$$x(t) = v_0(t + t_2 - (T - 2N_1 L + i_2 L + 1)), \ t \in [T - 2N_1 L + i_2 L + 1, T], \qquad (2.38)$$

$$I^g(\tau, \tau + 1, x) = U^g(0, 1, x(\tau), x(\tau + 1)), \ \tau = i_1 L + 1, \ T - 2N_1 L + i_2 L. \qquad (2.39)$$

Put

$$\tau_0 = 0, \ \tau_1 = i_1 L + 1, \ \tau_2 = i_1 L + 2, \ \tau_3 = T - 2N_1 L + i_2 L,$$

$$\tau_4 = T - 2N_1 L + i_2 L + 1, \ \tau_5 = T. \tag{2.40}$$

(2.39) and (2.40) imply that

$$\sigma^g(0, T, x) - \sigma^g(0, T, y) = \sum_{i=0}^{1} [\sigma^g(\tau_i, \tau_{i+1}, x) - \sigma^g(\tau_i, \tau_{i+1}, y)]$$

$$+ \sum_{i=3}^{4} [\sigma^g(\tau_i, \tau_{i+1}, x) - \sigma^g(\tau_i, \tau_{i+1}, y)]. \tag{2.41}$$

Analogously to the case (i) we can show that

$$\sigma^g(t_1 - i_1 L - 1, t_1, v_0) \le l(g, K, i_1 L + 1) + 2^{-5}\epsilon,$$

$$\sigma^g(t_2, t_2 + 2N_1 L + i_2 L + 1, v_0) \le l(g, K, 2N_1 L + i_2 L + 1) + 2^{-5}\epsilon.$$

By these inequalities and (2.38)-(2.40),

$$\sigma^g(\tau_i, \tau_{i+1}, x) \le l(g, K, \tau_{i+1} - \tau_i) + 2^{-5}\epsilon, \ i = 0, 4. \tag{2.42}$$

It follows from (2.13), (2.16), (2.33), (2.37) and (2.40) that

$$\sigma^g(\tau_i, \tau_{i+1}, y) \ge l(g, K, \tau_{i+1} - \tau_i), \ i = 0, 4. \tag{2.43}$$

(2.38)-(2.40) imply that for $i = 1, 3$,

$$\sigma^g(\tau_i, \tau_{i+1}, x) - \sigma^g(\tau_i, \tau_{i+1}, y)$$

$$\le U^g(0, 1, x(\tau_i), x(\tau_{i+1})) - \pi^f(x(\tau_i)) + \pi^f(x(\tau_{i+1}))$$

$$- [U^g(0, 1, y(\tau_i), y(\tau_{i+1})) - \pi^f(y(\tau_i)) + \pi^f(y(\tau_{i+1}))]. \tag{2.44}$$

By the choice of $\mathcal{U}_1$ and M_1 (see (2.14), (2.15)) and (2.33),

$$|y(t)| \le M_1, \ t \in [0, T]. \tag{2.45}$$

It follows from (2.13), (2.37), (2.38), (2.40), and the choice of δ (see (2.16)-(2.18)) that for $i = 1, 3$,

$$|\pi^f(x(\tau_{i+1})) - \pi^f(x(\tau_i)) - (\pi^f(y(\tau_{i+1})) - \pi^f(y(\tau_i)))| \le 2^{-6}\epsilon. \tag{2.46}$$

By (2.13), (2.37), (2.38), (2.40), (2.45), the choice of $\mathcal{U}_3$ (see (2.24)) and the choice of δ (see (2.16)-(2.18)) for $i = 1, 3$

$$|U^g(0, 1, x(\tau_i), x(\tau_{i+1})) - U^g(0, 1, y(\tau_i), y(\tau_{i+1}))|$$

$$\leq 2^{-5}\epsilon + |U^f(0, 1, x(\tau_i), x(\tau_{i+1})) - U^f(0, 1, y(\tau_i), y(\tau_{i+1}))| \leq 2^{-4}\epsilon.$$

Combined with (2.44) and (2.46) this inequality implies that for $i = 1, 3$,

$$\sigma^g(\tau_i, \tau_{i+1}, x) - \sigma^g(\tau_i, \tau_{i+1}, y) \leq 2^{-3}\epsilon.$$

This inequality and (2.41)-(2.43) imply that

$$\sigma^g(0, T, x) - \sigma^g(0, T, y) \leq 2^{-1}\epsilon.$$

Combined with (2.16), (2.33) and (2.39) this inequality implies (2.12). This completes the proof of the lemma.

The following auxiliary result generalizes Lemma 4.2.4. It shows the existence of $x : [0, T] \to R^n$ satisfying (2.48) and $x(0) = h$ for any $h \in H(f)$ while Lemma 4.2.4 establishes the existence of such x only for a certain $h \in H(f)$.

LEMMA 4.2.5 *For each $\epsilon \in (0, 1)$ and each $K > D_f + 1$ there exists a neighborhood $\mathcal{U}$ of f in $\mathcal{A}$ with the weak topology such that the following property holds:*

For each $g \in \mathcal{U}$, each $h \in H(f)$ and each number $T \geq 1$ there exists an a.c. function $x : [0, T] \to R^n$ such that

$$x(0) = h, \ x(T) \in H(f), \tag{2.47}$$

$$\sigma^g(0, T, x) \leq l(g, M_1, T) + \epsilon. \tag{2.48}$$

Proof. Let $\epsilon \in (0, 1)$ and $K > D_f + 1$. It follows from Theorem 1.2.3 that there exist a number $M_1 > 2K + 1$ and a neighborhood $\mathcal{U}_1$ of f in $\mathcal{A}$ with the weak topology such that the following property holds:

If $g \in \mathcal{U}_1$, $T_1 \in [0, \infty)$, $T_2 \in [T_1 + 8^{-1}, \infty)$, and if an a.c. function $x : [T_1, T_2] \to R^n$ satisfies

$$|x(T_i)| \leq K, \ i = 1, 2, \ \Phi^g(T_1, T_2, x) \leq 1, \tag{2.49}$$

then

$$|x(t)| \leq M_1, \ t \in [T_1, T_2]. \tag{2.50}$$

Lemma 4.2.4 implies that there exists a neighborhood $\mathcal{U}_2$ of f in $\mathcal{A}$ with the weak topology such that for each $g \in \mathcal{U}_2$ and each number $T \geq 1$ there exists an a.c. function $x : [0, T] \to R^n$ such that $x(0), x(T) \in H(f)$,

$$\sigma^g(0, T, x) \leq l(g, M_1, T) + 2^{-6}\epsilon. \tag{2.51}$$

It follows from Theorem 3.4.1 that there exists a number

$$\delta \in (0, 8^{-1}\epsilon) \tag{2.52}$$

such that if $y_1, y_2, z_1, z_2 \in R^n$ satisfy

$$|y_i|, |z_i| \leq 2M_1 + 4, \ i = 1, 2, \ |y_i - z_i| \leq 4\delta, \ i = 1, 2, \qquad (2.53)$$

then

$$|U^f(0, 1, y_1, y_2) - U^f(0, 1, z_1, z_2)| \leq 2^{-6}\epsilon,$$
$$|\pi^f(y_i) - \pi^f(z_i)| \leq 2^{-6}\epsilon, \ i = 1, 2. \qquad (2.54)$$

It follows from Proposition 1.3.9 that there exists a neighborhood $\mathcal{U}_3$ of f in $\mathcal{A}$ with the weak topology such that

$$|U^f(0, 1, y, z) - U^g(0, 1, y, z)| \leq 2^{-6}\epsilon \qquad (2.55)$$

for each $g \in \mathcal{U}_3$ and each $y, z \in R^n$ which satisfy $|y|, |z| \leq 2M_1 + 4$.

By Lemma 4.2.3 there exists an (f)-good function

$$v_0 : [0, \infty) \to H(f) \text{ such that } v_0 \in A(f). \qquad (2.56)$$

Since the integrand f has the asymptotic turnpike property with the turnpike $H(f)$ we have $\Omega(v_0) = H(f)$. Then it follows from (2.56) that there exist integers $N_1, N_2 \geq 8$ for which

$$\text{dist}(H(f), \{v_0(t) : \ t \in [4, N_1 + 4]\}) \leq 8^{-1}\delta,$$

$$\text{dist}(H(f), \{v_0(t) : \ t \in [2N_1 + 16, 2N_1 + 16 + N_2]\}) \leq 8^{-1}\delta. \qquad (2.57)$$

Fix an integer

$$N_0 \geq 8(N_1 + N_2 + 20). \qquad (2.58)$$

It follows from Proposition 1.3.8 that there exists a neighborhood $\mathcal{U}_4$ of f in $\mathcal{A}$ with the weak topology such that the following property holds:

If $g \in \mathcal{U}_4$, $T_1 \geq 0$, $T_2 \in [T_1 + 1, T_1 + N_0]$ and if an a.c. function $x : [T_1, T_2] \to R^n$ satisfies

$$\min\{I^f(T_1, T_2, x), \ I^g(T_1, T_2, x)\}$$

$$\leq N_0|\mu(f)| + 2\sup\{|\pi^f(z)| : \ z \in H(f)\}, \qquad (2.59)$$

then

$$|I^f(T_1, T_2, x) - I^g(T_1, T_2, x)| \leq 2^{-6}\epsilon. \qquad (2.60)$$

By Lemma 4.2.2 there exists a neighborhood $\mathcal{U}_5$ of f in $\mathcal{A}$ with the weak topology such that

$$|l(g, M_1, T) - \mu(f)T| \leq 2^{-6}\epsilon \qquad (2.61)$$

for each $g \in \mathcal{U}_5$ and each $T \in [1, 2N_0 + 8]$. Put

$$\mathcal{U} = \cap_{i=1}^{5}\mathcal{U}_i. \qquad (2.62)$$

Let $g \in \mathcal{U}$, $h \in H(f)$, and a number $T \geq 1$. There are two cases: (i) $T \leq N_0$; (ii) $T > N_0$. Consider the case (i). By Lemma 4.2.3 there exists an (f)-good function

$$x : [0, \infty) \to H(f) \text{ such that } x \in A(f), \ x(0) = h.$$

Analogously to the case (i) in the proof of Lemma 4.2.4 we can show that the relation (2.48) holds. Therefore in the case (i) the assertion of the lemma is valid.

Consider the case (ii). Then

$$T > N_0. \tag{2.63}$$

By (2.58), (2.63) and the choice of $\mathcal{U}_2$ (see (2.51)) there exists an a.c. function $u : [0, T - 4(N_1 + N_2 + 5)] \to R^n$ such that

$$u(0), u(T - 4(N_1 + N_2 + 5)) \in H(f)), \tag{2.64}$$

$$\sigma^g(0, T - 4(N_1 + N_2 + 5), u) \leq l(g, M_1, T - 4(N_1 + N_2 + 5)) + 2^{-6}\epsilon. \tag{2.65}$$

(2.57) and (2.64) imply that there exist

$$t_1, t_3 \in [4, N_1 + 4], \ t_2 \in [2N_1 + 16, 2N_1 + 16 + N_2] \tag{2.66}$$

for which

$$|h - v_0(t_1)| \leq 4^{-1}\delta, \ |u(0) - v_0(t_2)| \leq 4^{-1}\delta,$$

$$|u(T - 4(N_1 + N_2 + 5)) - v_0(t_3)| \leq 4^{-1}\delta. \tag{2.67}$$

By Corollary 1.3.1, (2.58), (2.63) and (2.66) there exists an a.c. function $x : [0, T] \to R^n$ such that

$$x(0) = h, \ x(t) = v_0(t + t_1), \ t \in [1, t_2 - t_1 - 1],$$

$$x(t) = u(t - t_2 + t_1),$$

$$t \in [t_2 - t_1, T - 4(N_1 + N_2 + 5) + t_2 - t_1],$$

$$x(t) = v_0(t + t_3 - (T - 4(N_1 + N_2 + 5) + t_2 - t_1)),$$

$$t \in [T - 4(N_1 + N_2 + 5) + t_2 - t_1 + 1, T], \tag{2.68}$$

$$\Phi^g(\tau, \tau + 1, x) = 0, \ \tau = 0, \ t_2 - t_1 - 1, \ T - 4(N_1 + N_2 + 5) + t_2 - t_1. \tag{2.69}$$

Clearly (2.47) holds. We show that (2.48) holds.

Put

$$\tau_0 = 0, \ \tau_1 = 1, \ \tau_2 = t_2 - t_1 - 1, \ \tau_3 = t_2 - t_1,$$

$$\tau_4 = T - 4(N_1 + N_2 + 5) + t_2 - t_1,$$

$$\tau_5 = T - 4(N_1 + N_2 + 5) + t_2 - t_1 + 1, \ \tau_6 = T. \tag{2.70}$$

There exists an a.c. function $y : [0, T] \to R^n$ such that

$$|y(0)|, |y(T)| \le K, \ \sigma^g(0, T, y) \le l(g, K, T) + 2^{-6}\epsilon. \tag{2.71}$$

By the choice of $\mathcal{U}_1$ and M_1 (see (2.49), (2.50)) and (2.71),

$$|y(t)| \le M_1, \ t \in [0, T]. \tag{2.72}$$

It follows from (2.69) and (2.70) that

$$\sigma^g(0, T, x) - \sigma^g(0, T, y) = \sum_{i=0}^{5} [\sigma^g(\tau_i, \tau_{i+1}, x) - \sigma^g(\tau_i, \tau_{i+1}, y)]. \tag{2.73}$$

(2.72) and (2.69) imply that

$$\sigma^g(\tau_i, \tau_{i+1}, y) \ge l(g, M_1, \tau_{i+1} - \tau_i), \ i = 0, \dots, 5. \tag{2.74}$$

By (2.65), (2.68), (2.69) and (2.70),

$$\sigma^g(\tau_3, \tau_4, x) \le l(g, M_1, \tau_4 - \tau_3) + 2^{-6}\epsilon. \tag{2.75}$$

Analogously to (2.42) (see the proof of Lemma 4.2.4) we can show that

$$\sigma^g(\tau_i, \tau_{i+1}, x) \le l(g, M_1, \tau_{i+1} - \tau_i) + 2^{-5}\epsilon, \ i = 1, 5. \tag{2.76}$$

It follows from (2.56), (2.64), (2.68)-(2.70) and the choice of $\mathcal{U}_3$ (see (2.55)) that for $i = 0, 2, 4$,

$$\sigma^g(\tau_i, \tau_{i+1}, x) = U^g(0, 1, x(\tau_i), x(\tau_{i+1})) - \pi^f(x(\tau_i)) + \pi^f(x(\tau_{i+1}))$$
$$\le U^f(0, 1, x(\tau_i), x(\tau_{i+1})) - \pi^f(x(\tau_i)) + \pi^f(x(\tau_{i+1})) + 2^{-6}\epsilon. \tag{2.77}$$

Put

$$\gamma_0 = t_1, \ \gamma_2 = t_2 - 1, \ \gamma_4 = t_3. \tag{2.78}$$

Equations (2.78), (2.70), (2.68) and (2.67) imply that for $i = 0, 2, 4$,

$$|x(\tau_i) - v_0(\gamma_i)|, \ |x(\tau_i + 1) - v_0(\gamma_i + 1)| \le 4^{-1}\delta. \tag{2.79}$$

By (2.56), (2.70), (2.77), (2.79) and the choice of δ (see (2.52)-(2.54)) for $i = 0, 2, 4$,

$$\sigma^g(\tau_i, \tau_{i+1}, x) \le U^f(0, 1, x(\tau_i), x(\tau_i + 1)) - \pi^f(x(\tau_i)) + \pi^f(x(\tau_i + 1)) + 2^{-6}\epsilon$$
$$\le U^f(0, 1, v_0(\gamma_i), v_0(\gamma_i + 1)) - \pi^f(v_0(\gamma_i)) + \pi^f(v_0(\gamma_i + 1))$$
$$+ 2^{-4}\epsilon = \mu(f) + 2^{-4}\epsilon.$$

It follows from this relation and the choice of $\mathcal{U}_5$ that for $i = 0, 2, 4$,

$$\sigma^g(\tau_i, \tau_{i+1}, x) \leq l(g, M_1, \tau_{i+1} - \tau_i) + 2^{-4}\epsilon + 2^{-6}\epsilon.$$

Combining this relation and (2.73)-(2.76) we obtain that

$$\sigma^g(0, T, x) - \sigma^g(0, T, y) \leq 2^{-6}\epsilon + 2^{-4}\epsilon + 3(2^{-6}\epsilon + 2^{-4}\epsilon) \leq 2^{-1}\epsilon.$$

Combined with (2.69) and (2.71) this implies (2.48). This completes the proof of the lemma.

The next lemma establishes the turnpike property for a function $v : [0, T] \rightarrow R^n$ such that $v(0), v(T)$ are close to the turnpike $H(f)$ and $\sigma^g(0, T, v) \leq l(g, K, T) + \delta$, where δ is small, T is large and g belongs to a small neighborhood of f in $\mathcal{A}$.

LEMMA 4.2.6 *Let $\epsilon \in (0, 1)$, $K > D_f + 1$ and L be a positive integer such that for each (f)-good function $v : [0, \infty) \rightarrow R^n$,*

$$dist(H(f), \{v(t) : t \in [T, T + L]\}) \leq 8^{-1}\epsilon \qquad (2.80)$$

for all large T (the existence of L follows from Theorem 3.3.1).
Then there exists a neighborhood $\mathcal{U}$ of f in $\mathcal{A}$ with the weak topology and $\delta \in (0, 1)$ such that the following property holds:
If $g \in \mathcal{U}$, $T \in [L, \infty)$ and if an a.c. function $v : [0, T] \rightarrow R^n$ satisfies

$$d(v(0), H(f)) \leq \delta, \ d(v(T), H(f)) \leq \delta, \qquad (2.81)$$

$$\sigma^g(0, T, v) \leq l(g, K, T) + \delta, \qquad (2.82)$$

then for every $S \in [0, T - L]$,

$$dist(H(f), \{v(t) : t \in [S, S + L]\}) \leq \epsilon. \qquad (2.83)$$

Proof. It follows from Lemma 3.8.4 that there exists

$$\delta_1 \in (0, \epsilon) \qquad (2.84)$$

such that the following property holds:
If $T \in [L, \infty)$ and if an a.c. function $v : [0, T] \rightarrow R^n$ satisfies

$$d(v(0), H(f)) \leq \delta_1, \ d(v(T), H(f)) \leq \delta_1, \ \sigma^f(0, T, v) - T\mu(f) \leq \delta_1, \qquad (2.85)$$

then the inequality (2.83) holds for every $S \in [0, T - L]$.
Fix a number

$$\delta \in (0, 8^{-1}\delta_1). \qquad (2.86)$$

It follows from Theorem 3.3.1 that there exists an integer $L_1 \geq 1$ such that for each (f)-good function $v : [0, \infty) \to R^n$,

$$\text{dist}(H(f), \{v(t) : t \in [T, T + L_1]\}) \leq 8^{-1}\delta \qquad (2.87)$$

for all large T. We may assume that

$$L_1 \geq 10L + 24. \qquad (2.88)$$

It follows from Lemma 4.2.1 that there exist a neighborhood $\mathcal{U}_1$ of f in $\mathcal{A}$ with the weak topology and an integer $N_1 \geq 10$ such that the following property holds:

If $g \in \mathcal{U}_1$, $T_1 \geq 0$, $T_2 \geq T_1 + N_1 L_1$ and if an a.c. function $x : [T_1, T_2] \to R^n$ satisfies

$$|x(T_i)| \leq K, \; i = 1, 2, \; \Phi^g(T_1, T_2, x) \leq 1, \qquad (2.89)$$

then for each $S \in [T_1, T_2 - N_1 L_1]$ there exists an integer $i_0 \in [0, N_1 - 8]$ such that

$$\text{dist}(H(f), \{x(t) : t \in [T, T + L_1]\}) \leq \delta \qquad (2.90)$$

for all $T \in [S + i_0 L_1, S + (i_0 + 7)L_1]$.

It follows from Lemma 4.2.5 that there exists a neighborhood $\mathcal{U}_2$ of f in $\mathcal{A}$ with the weak topology such that if $g \in \mathcal{U}_2$, $T \geq 1$ and if $h \in H(f)$, then there exists an a.c. function $x : [0, T] \to R^n$ such that

$$x(0) = h, \; x(T) \in H(f),$$

$$\sigma^g(0, T, x) \leq l(g, K, T) + 8^{-1}\delta. \qquad (2.91)$$

Choose an integer

$$N_0 \geq 100 L_1 N_1. \qquad (2.92)$$

By Lemma 4.2.2 there exists a neighborhood $\mathcal{U}_3$ of f in $\mathcal{A}$ with the weak topology such that

$$|l(g, K, \tau) - \mu(f)\tau| \leq 4^{-1}\delta \qquad (2.93)$$

for each $g \in \mathcal{U}_3$ and each $\tau \in [1, N_0]$.

It follows from Proposition 1.3.8 that there exists a neighborhood $\mathcal{U}_4$ of f in $\mathcal{A}$ with the weak topology such that the following property holds:

If $g \in \mathcal{U}_4$, $T_1, T_2 \geq 0$ satisfy $T_2 - T_1 \in [1, 2N_0 + 1]$ and if an a.c. function $x : [T_1, T_2] \to R^n$ satisfies

$$\min\{I^f(T_1, T_2, x), \; I^g(T_1, T_2, x)\}$$

$$\leq 2N_0|\mu(f)| + 4 + 2\sup\{|\pi^f(h)| : h \in R^n, \; |h| \leq K + 2\}, \qquad (2.94)$$

then

$$|I^f(T_1, T_2, x) - I^g(T_1, T_2, x)| \le \delta. \tag{2.95}$$

Put

$$\mathcal{U} = \cap_{i=1}^4 \mathcal{U}_i. \tag{2.96}$$

Assume that $g \in \mathcal{U}$, $T \ge L$ and an a.c. function $v : [0, T] \to R^n$ satisfies (2.81) and (2.82). There are two cases: (i) $T \le N_0$; (ii) $T > N_0$. Consider case (i). By the choice of $\mathcal{U}_3$ (see (2.93)) and (2.82),

$$\sigma^g(0, T, v) \le |\mu(f)|T + 2\delta. \tag{2.97}$$

It follows from the choice of $\mathcal{U}_4$ (see (2.94), (2.95)) and (2.81) that

$$|I^f(0, T, v) - I^g(0, T, v)| \le \delta.$$

Combined with (2.86), (2.81) and (2.97) this inequality implies (2.85). It follows from (2.85) and the definition of δ_1 (see (2.84)) that (2.83) holds for all $S \in [0, T - L]$. Therefore in case (i) the assertion of the lemma is valid.

Consider case (ii). Then

$$T > N_0. \tag{2.98}$$

It follows from (2.81), (2.82), (2.98), (2.91) and the definition of N_1 and $\mathcal{U}_1$ (see (2.89), (2.90)) that there exists a sequence of numbers $\{T_i\}_{i=0}^q$ such that

$$T_0 = 0, \ T_q = T, \ T_{i+1} - T_i \in [2L_1, 2(2N_1 - 6)L_1], \ i = 0, \dots, q-1, \tag{2.99}$$

$$d(v(T_i), H(f)) \le \delta, \ i = 0, \dots, q. \tag{2.100}$$

For each a.c. function $y : [0, T] \to R^n$ and each $r \in [0, T]$ we set

$$\sigma^g(r, r, y) = 0. \tag{2.101}$$

We set

$$l(g, K, 0) = 0. \tag{2.102}$$

Let integers $j, p \in [0, q]$, $j < p$. We estimate

$$\sigma^g(T_j, T_p, v) - l(g, K, T_p - T_j).$$

It follows from (2.99)-(2.102) and the choice of $\mathcal{U}_2$ (see (2.91)) that there exists an a.c. function $y : [0, T] \to R^n$ such that

$$y(T_i) \in H(f), \ i = 0, j, p, q, \ \sigma^g(0, T_j, y) \le l(g, K, T_j) + 8^{-1}\delta,$$

$$\sigma^g(T_j, T_p, y) \le l(g, K, T_p - T_j) + 8^{-1}\delta,$$

$$\sigma^g(T_p, T_q, y) \le l(g, K, T_q - T_p) + 8^{-1}\delta.$$

These relations, (2.82) and (2.99)-(2.102) imply that

$$\delta \ge \sigma^g(0, T, v) - \sigma^g(0, T, y)$$

$$= [\sigma^g(0, T_j, v) - \sigma^g(0, T_j, y)] + [\sigma^g(T_j, T_p, v) - \sigma^g(T_j, T_p, y)]$$

$$+ [\sigma^g(T_p, T_q, v) - \sigma^g(T_p, T_q, y)]$$

$$\ge \sigma^g(T_j, T_p, v) - \sigma^g(T_j, T_p, y) - 4^{-1}\delta,$$

$$\sigma^g(T_j, T_p, v) \le \delta + 4^{-1}\delta + l(g, K, T_p - T_j) + 8^{-1}\delta. \qquad (2.103)$$

We have shown that (2.103) holds for each pair of integers $j, p \in [0, q]$ satisfying $j < p$.

Let $s \in [0, T - L]$. By (2.99) and (2.88) there exist integers $j, p \in [0, q]$ such that

$$j < p, \ S \in [T_j, T_p - L], \ T_p - T_j \in [2L_1, 8N_1L_1]. \qquad (2.104)$$

Evidently (2.103) holds. It follows from (2.92), (2.100), (2.103), (2.104) and the choice of $\mathcal{U}_3$ (see (2.93)) that

$$\sigma^g(T_j, T_p, v) \le \mu(f)(T_p - T_j) + 2\delta.$$

It follows from this inequality, the choice of $\mathcal{U}_4$ (see (2.94), (2.95)), (2.92), (2.99) and (2.104) that

$$|I^f(T_j, T_p, v) - I^g(T_j, T_p, v)| \le \delta,$$

$$\sigma^f(T_j, T_p, v) \le \mu(f)(T_p - T_j) + 3\delta.$$

By these inequalities, (2.99), (2.88), (2.86) and the choice of δ_1 (see (2.84)), (2.83) holds. This completes the proof of the lemma.

The proof of the next result is based on Lemmas 4.2.5 and 4.2.6.

LEMMA 4.2.7 *Let* $\epsilon \in (0, 1)$, $K > D_f + 4$ *and* L *be a positive integer such that for each* (f)-*good function* $v : [0, \infty) \to R^n$,

$$dist(H(f), \{v(t) : \ t \in [T, T + L]\}) \le 8^{-1}\epsilon$$

for all large T *(the existence of* L *follows from Theorem 3.3.1).*

Then there exists a neighborhood $\mathcal{U}$ *of* f *in* $\mathcal{A}$ *with the weak topology such that for each* $g \in \mathcal{U}$ *and each* $h \in H(f)$ *there exists an a.c. function* $x : [0, \infty) \to R^n$ *such that* $x(0) = h$;
for all $S \in [0, \infty)$,

$$dist(H(f), \{x(t) : \ t \in [S, S + L]\}) \le \epsilon; \qquad (2.105)$$

for each $T \geq 1$

$$\sigma^g(0, T, v) \leq l(g, K, T) + \epsilon; \tag{2.106}$$

$$\sigma^g(t_1, t_2, x) \leq l(g, K, t_2 - t_1) + \epsilon \tag{2.107}$$

for each $t_1 \geq 1$, $t_2 \geq t_1 + 1$.

Proof. It follows from Lemma 4.2.6 that there exist a neighborhood $\mathcal{U}_1$ of f in $\mathcal{A}$ with the weak topology and

$$\delta_1 \in (0, 8^{-1}\epsilon) \tag{2.108}$$

such that the following property holds:

If $g \in \mathcal{U}_1$, $T \in [L, \infty)$, and if an a.c. function $v : [0, T] \to R^n$ satisfies

$$d(v(0), H(f)) \leq \delta_1, \ d(v(T), H(f)) \leq \delta_1, \tag{2.109}$$

$$\sigma^g(0, T, v) \leq l(g, K, T) + \delta_1,$$

then the inequality (2.105) holds for every $S \in [0, T - L]$.

By Lemma 4.2.5 there exists a neighborhood $\mathcal{U}_2$ of f in $\mathcal{A}$ with the weak topology such that for each $g \in \mathcal{U}_2$, each $h \in H(f)$ and each number $T \geq 1$ there exists an a.c. function $v : [0, T] \to R^n$ such that

$$v(0) = h, \ v(T) \in H(f),$$

$$\sigma^g(0, T, v) \leq l(g, K, T) + 8^{-1}\delta_1. \tag{2.110}$$

Put

$$\mathcal{U} = \mathcal{U}_1 \cap \mathcal{U}_2. \tag{2.111}$$

Assume that $g \in \mathcal{U}$ and $h \in H(f)$. By the definition of $\mathcal{U}_2$ (see (2.110)) for each $N \geq 1$ there exists an a.c. function $x_N : [0, N] \to R^n$ such that (2.110) holds with $T = N$, $v = x_N$. It follows from the definition of $\mathcal{U}_1$ and δ_1 (see (2.109)) that for each integer $N \geq L$ and each number $S \in [0, N - L]$,

$$\mathrm{dist}(H(f), \{x_N(t) : t \in [S, S + L]\}) \leq \epsilon. \tag{2.112}$$

Let $N \geq L$ be an integer. Assume that an integer $q \in \{2, 3, 4\}$, $\{t_i\}_{i=0}^{q} \subset [0, N]$, $t_0 = 0$, $t_q = N$, $t_{i+1} - t_i \geq 1$, $i = 0, \ldots, q - 1$. Equation (2.112), which holds for each $S \in [0, N - L]$, implies that

$$\sigma^g(t_i, t_{i+1}, x_N) \geq l(g, K, t_{i+1} - t_i), \ i = 0, \ldots, q - 1. \tag{2.113}$$

By the definition of $\mathcal{U}_2$ (see (2.110)) there exists an a.c. function $y : [0, N] \to R^n$ for which

$$y(t_i) \in H(f), \ i = 0, \ldots, q, \ \sigma^g(t_i, t_{i+1}, y)$$

$$\le l(g, K, t_{i+1} - t_i) + 8^{-1}\delta_1, \ i = 0, \ldots, q - 1.$$

Together with (2.110) which holds with $T = N$, $v = x_N$ and (2.113) this implies that for each $j \in [0, q - 1]$,

$$8^{-1}\delta_1 \ge \sigma^g(0, N, x_N) - \sigma^g(0, N, y) = \sum_{i=0}^{q-1}[\sigma^g(t_i, t_{i+1}, x_N) - \sigma^g(t_i, t_{i+1}, y)]$$

$$\ge \sigma^g(t_j, t_{j+1}, x_N) - l(g, K, t_{j+1} - t_j) - 8^{-1}\delta_1 q.$$

This implies that for each integer $N \ge L+3$ and each $\tau_1 \in \{0\}\cup[1, N-2]$, $\tau_2 \in [\tau_1 + 1, N - 1]$,

$$\sigma^g(\tau_1, \tau_2, x_N) \le l(g, K, \tau_2 - \tau_1) + 3 \cdot 4^{-1}\delta_1. \tag{2.114}$$

By (2.114), (2.112), which holds for each $N \ge L$ and each number $S \in [0, N - L]$, and by Proposition 1.3.5 there exist a subsequence $\{x_{N_p}\}_{p=1}^{\infty}$ and an a.c. function $x : [0, \infty) \to R^n$ such that for each integer $N \ge 1$,

$$x_{Np}(t) \to x(t) \text{ as } p \to \infty \text{ uniformly in } [0, N] \tag{2.115}$$

and

$$I^g(T_1, T_2, x) \le \liminf_{p\to\infty} I^g(T_1, T_2, x_{N_p}) \tag{2.116}$$

for each $T_1 \in [1, \infty) \cup \{0\}$, $T_2 \ge T_1 + 1$.

Equation (2.110), which holds with $T = N$, $v = x_N$ and (2.115) imply that $x(0) = h$. Equations (2.112) and (2.115) imply (2.105) for all $S \in [0, \infty)$. Equations (2.114)-(2.116) and (2.108) imply (2.106) for each $T \ge 1$ and (2.107) for each $t_1 \ge 1$, $t_2 \ge t_1 + 1$. This completes the proof of the lemma.

LEMMA 4.2.8 $\sup\{\pi^f(h) : h \in H(f)\} = 0$.

Proof. There exists $h_0 \in H(f)$ for which

$$\pi^f(h_0) \ge \pi^f(h), \ h \in H(f). \tag{2.117}$$

Let $v : [0, \infty) \to R^n$ be an a.c. function and let

$$v(0) = h_0. \tag{2.118}$$

We will show that $\liminf_{T\to\infty}[I^f(0, T, v) - T\mu(f)] \ge 0$.

By Proposition 4.1.1 we may assume that v is an (f)-good function. Then $\Omega(v) = H(f)$. It follows from this relation, the representation formula (see (1.6)), (2.117), and (2.118) that

$$\liminf_{T\to\infty}[I^f(0, T, v) - T\mu(f)] \ge \liminf_{T\to\infty}[\pi^f(v(0)) - \pi^f(v(T))] \ge 0.$$

This implies that $\pi^f(h_0) \geq 0$. By Theorem 3.6.3 there exists an (f)-good function $u \in A(f)$ satisfying $u(0) = h_0$. It is easy to see that $\Omega(v) = H(f)$,

$$\liminf_{T\to\infty}[I^f(0,T,u) - T\mu(f)] = \liminf_{T\to\infty}[\pi^f(u(0)) - \pi^f(u(T))] = 0.$$

This completes the proof of the lemma.

The next lemma establishes the continuity of the function $f \to \mu(f)$, $f \in \mathcal{A}$.

LEMMA 4.2.9 *Let $\epsilon \in (0,1)$. Then there exists a neighborhood $\mathcal{U}$ of f in $\mathcal{A}$ with the weak topology such that $|\mu(f) - \mu(g)| \leq \epsilon$ for each $g \in \mathcal{U}$.*

Proof. By Theorem 1.2.2 there exist a neighborhood $\mathcal{U}_1$ of f in $\mathcal{A}$ with the weak topology and a number $M_0 > 0$ such that for each $g \in \mathcal{U}_1$ and each (g)-good function $x : [0,\infty) \to R^n$,

$$\limsup_{t\to\infty}|x(t)| < M_0. \tag{2.119}$$

Set

$$M_1 = \sup\{|U^f(0,1,x,y)| : x,y \in R^n, |x|,|y| \leq 2M_0 + 2\}. \tag{2.120}$$

It follows from Proposition 1.3.8 that there exists a neighborhood $\mathcal{U}_2$ of f in $\mathcal{A}$ with the weak topology such that the following property holds: If $g \in \mathcal{U}_2$, $T \geq 0$, and if an a.c. function $y : [T, T+1] \to R^n$ satisfies

$$\min\{I^f(T,T+1,y), I^g(T,T+1,y)\} \leq 2M_1 + 4,$$

then

$$|I^f(T,T+1,y) - I^g(T,T+1,y)| \leq \epsilon/8. \tag{2.121}$$

It follows from Proposition 1.3.9 that there exists a neighborhood $\mathcal{U}_3$ of f in $\mathcal{A}$ with the weak topology such that the following property holds: If $g \in \mathcal{U}_3$ and if $x,y \in R^n$ satisfy $|y|,|x| \leq 2M_0 + 2$, then

$$|U^f(0,1,x,y) - U^g(0,1,x,y)| \leq \epsilon/16. \tag{2.122}$$

Put

$$\mathcal{U} = \cap_{i=1}^3 \mathcal{U}_i. \tag{2.123}$$

Let $g_1, g_2 \in \mathcal{U}$ and let $x : [0,\infty) \to R^n$ be a (g_1)-good function. We have

$$\sup\{|I^{g_1}(0,T,x) - T\mu(g_1)| : T \in (0,\infty)\} < \infty. \tag{2.124}$$

By the definition of $\mathcal{U}_1$, M_0 (see (2.119)) we may assume that

$$|x(t)| \leq M_0, \ t \in [0, \infty). \tag{2.125}$$

By Proposition 4.1.1 we may assume without loss of generality that

$$\Phi^{g_1}(T, T+1, x) \leq 4^{-1} \text{ for all } t \in [0, \infty).$$

By this inequality, (2.120), (2.125) and the choice of $\mathcal{U}_3$ (see (2.122)),

$$I^{g_1}(T, T+1, x) \leq M_1 + 2^{-1} \text{ for all } T \in [0, \infty).$$

It follows from this inequality and the choice of $\mathcal{U}_2$ that for each $T \in [0, \infty)$,

$$|I^f(T, T+1, x) - I^{g_1}(T, T+1, x)| \leq 8^{-1}\epsilon,$$

$$I^f(T, T+1, x) \leq M_1 + 1,$$

$$|I^f(T, T+1, x) - I^{g_2}(T, T+1, x)| \leq 8^{-1}\epsilon,$$

$$|I^{g_1}(T, T+1, x) - I^{g_2}(T, T+1, x)| \leq 4^{-1}\epsilon. \tag{2.126}$$

Equations (2.124) and (2.126) imply that

$$\sup\{|I^{g_2}(0, N, x) - 4^{-1}\epsilon N - N\mu(g)| : \ N = 1, 2, \ldots\} < \infty.$$

Together with Proposition 4.1.1 this implies that $\mu(g_2) \leq \mu(g_1) + 4^{-1}\epsilon$. This completes the proof of the lemma.

LEMMA 4.2.10 *Let $K > D_f + 1$. Then there exists a neighborhood $\mathcal{U}$ of f in $\mathcal{A}$ with the weak topology such that $l(g, K, \tau) \leq \tau\mu(g)$ for each $g \in \mathcal{U}$ and each $\tau > 0$.*

Proof. It follows from Theorem 3.1.3 that there exists a neighborhood $\mathcal{U}$ of f in $\mathcal{A}$ with the weak topology such that if $g \in \mathcal{U}$ and if $x : [0, \infty) \to R^n$ is an (g)-good function, then

$$\limsup_{t \to \infty} |x(t)| < K.$$

Let $g \in \mathcal{U}$, $\tau > 0$ and let $x : [0, \infty) \to R^n$ be a (g)-good function. We may assume that

$$|x(t)| \leq K, \ t \in [0, \infty). \tag{2.127}$$

By (2.127) for each integer $N \geq 1$,

$$\sigma^g(0, N\tau, x) = \sum_{i=0}^{N-1} [\sigma^g(i\tau, (i+1)\tau, x)] \geq Nl(g, K, \tau),$$

$$I^g(0, N\tau, x) \geq Nl(g, K, \tau) + 2\sup\{|\pi^f(z)| : z \in R^n, |z| \leq K\}.$$

Since x is a (g)-good function we have

$$\sup\{Nl(g, K, \tau) - N\tau\mu(g) : N = 1, 2, \ldots\} < \infty.$$

This completes the proof of the lemma.

There exists

$$h_* \in H(f) \text{ such that } \pi^f(h_*) \geq \pi^f(h), \ h \in H(f). \tag{2.128}$$

The next auxiliary result is an extension of Lemma 4.2.7. It shows the existence of a good function which in addition to the properties established in Lemma 4.2.7 has some other important properties. The proof is based on a number of results established in the book including Lemma 4.2.7.

LEMMA 4.2.11 *For each* $\epsilon \in (0, 1)$ *and each* $K > D_f + 4$ *there exist an integer* $L \geq 1$ *and a neighborhood* $\mathcal{U}$ *of* f *in* $\mathcal{A}$ *with the weak topology such that if* $g \in \mathcal{U}$ *and if* $h \in H(f)$, *then there exists a* (g)-*good function* $v : [0, \infty) \to R^n$ *such that*

$$v(0) = h; \tag{2.129}$$

$$dist(H(f), \{v(t) : t \in [T, T + L]\}) \leq \epsilon \tag{2.130}$$

for all $T \in [0, \infty)$;

$$\sigma^g(T_1, T_2, v) \leq l(g, K, T_2 - T_1) + \epsilon \tag{2.131}$$

for each $T_1 \in \{0\} \cup [1, \infty)$, $T_2 \geq T_1 + 1$;

$$|\sigma^g(0, T, v) - T\mu(g)| \leq \epsilon \tag{2.132}$$

for each $T \in [1, \infty)$;

$$|\liminf_{T \to \infty}[I^g(0, T, v) - T\mu(g)] - \pi^f(h)| \leq 2\epsilon. \tag{2.133}$$

Proof. Let $\epsilon \in (0, 1)$ and $K > D_f + 4$. It follows from (2.128) and Lemma 4.2.3 that there exists an (f)-good function $v_0 : [0, \infty) \to H(f)$ such that

$$v_0 \in A(f), \ v_0(0) = h_*. \tag{2.134}$$

It follows from Theorem 3.4.1 that there exists a positive number

$$\delta < 2^{-8}\epsilon \tag{2.135}$$

such that the following property holds:

If $y_1, y_2, z_1, z_2 \in R^n$ satisfy

$$|y_i|, |z_i| \leq 2K + 8, \ i = 1, 2, \ |y_i - z_i| \leq 4\delta, \ i = 1, 2,$$

then

$$|U^f(0, 1, y_1, y_2) - U^f(0, 1, z_1, z_2)| \leq 2^{-8}\epsilon,$$
$$|\pi^f(y_i) - \pi^f(z_i)| \leq 2^{-8}\epsilon, \ i = 1, 2. \tag{2.136}$$

It follows from Proposition 1.3.9 that there exists a neighborhood $\mathcal{U}_1$ of f in $\mathcal{A}$ with the weak topology such that the following property holds: If $g \in \mathcal{U}_1$ and if $y, z \in R^n$ satisfy $|y|, |z| \leq 2K + 8$, then

$$|U^f(0, 1, y, z) - U^g(0, 1, y, z)| \leq 2^{-4}\delta. \tag{2.137}$$

Since f has the asymptotic turnpike property with the turnpike $H(f)$ we have $\Omega(v_0) = H(f)$. Therefore there exist integers $N_1, N_2 \geq 10$ for which

$$\text{dist}(H(f), \{v_0(t) : \ t \in [0, N_1]\}) \leq 2^{-4}\delta,$$
$$\text{dist}(H(f), \{v_0(t) : \ t \in [4N_1, 4N_1 + N_2]\}) \leq 2^{-4}\delta. \tag{2.138}$$

By Theorem 3.3.1 there exists an integer $L \geq 1$ such that for each (f)-good function $v : [0, \infty) \to R^n$,

$$\text{dist}(H(f), \{v(t) : \ t \in [T, T + L]\}) \leq 2^{-4}\delta \tag{2.139}$$

for all large T.

It follows from Lemma 4.2.7 that there exists a neighborhood $\mathcal{U}_2$ of f in $\mathcal{A}$ with the weak topology such that if $g \in \mathcal{U}_2$ and if $h \in H(f)$, then there exists an a.c. function $v : [0, \infty) \to R^n$ satisfying (2.129) and such that for each $T \in [0, \infty)$,

$$\text{dist}(H(f), \{v(t) : \ t \in [T, T + L]\}) \leq \delta, \tag{2.140}$$

$$\sigma^g(T_1, T_2, v) \leq l(g, K, T_2 - T_1) + \delta \tag{2.141}$$

for each $T_1 \in \{0\} \cup [1, \infty)$, $T_2 \geq T_1 + 1$.

Lemma 4.2.10 implies that there exists a neighborhood $\mathcal{U}_3$ of f in $\mathcal{A}$ with the weak topology such that

$$l(g, K, T) \leq T\mu(g) \tag{2.142}$$

for each $g \in \mathcal{U}_3$ and each $T > 0$.

It follows from Lemma 4.2.9 that there exists a neighborhood $\mathcal{U}_4$ of f in $\mathcal{A}$ with the weak topology such that

$$|\mu(f) - \mu(g)| \leq 2^{-8}\delta(8N_1 + 8N_2)^{-1} \tag{2.143}$$

for each $g \in \mathcal{U}_4$.

By Proposition 1.3.8 there exists a neighborhood $\mathcal{U}_5$ of f in $\mathcal{A}$ with the weak topology such that the following property holds:

If $g \in \mathcal{U}_5$, $T_1 \geq 0$, $T_2 \in [T_1+1, T_1+8(N_1+N_2)]$, and if an a.c. function $v : [T_1, T_2] \to R^n$ satisfies

$$\min\{I^f(T_1, T_2, v), \ I^g(T_1, T_2, v)\}$$

$$\leq 8(N_1+N_2)|\mu(f)|+4+2\sup\{|\pi^f(z)| : \ z \in R^n, \ |z| \leq 2K+2\}, \quad (2.144)$$

then

$$|I^f(T_1, T_2, v) - I^g(T_1, T_2, v)| \leq 2^{-8}\delta. \quad (2.145)$$

Put

$$\mathcal{U} = \cap_{i=1}^{5}\mathcal{U}_i. \quad (2.146)$$

Let $g \in \mathcal{U}$ and $h \in H(f)$. It follows from the definition of $\mathcal{U}_2$ that there exists an a.c. function $v : [0, \infty) \to R^n$ such that (2.129) holds, relation (2.140) holds for each $T \in [0, \infty)$, and relation (2.141) holds for each $T_1 \in \{0\} \cup [1, \infty)$, $T_2 \geq T_1 + 1$. By (2.141), which holds for $T_1 = 0$ and each $T_2 \geq 1$, and the definition of $\mathcal{U}_3$ (see (2.142)) v is a (g)-good function.

Fix a number $T \geq 1$. We will establish (2.132). It follows from (2.140) and (2.138) that there exist

$$t_1 \in [0, N_1], \ t_2 \in [4N_1, 4N_1 + N_2] \quad (2.147)$$

for which

$$|v(T) - v_0(t_1)| \leq \delta + 8^{-1}\delta, \ |h - v_0(t_2)| \leq \delta + 8^{-1}\delta. \quad (2.148)$$

By Corollary 1.3.1 there exists an a.c. function $x : [0, T+t_2-t_1] \to R^n$ such that

$$x(t) = v(t), \ t \in [0, T], \ x(t) = v_0(t + t_1 - T),$$

$$t \in [T + 1, T + t_2 - t_1 - 1],$$

$$x(T + t_2 - t_1) = h, \ I^g(S, S + 1, x) = U^g(0, 1, x(S), x(S + 1)),$$

$$S = T, \ T + t_2 - t_1 - 1. \quad (2.149)$$

Equations (2.149) and (2.129) imply that

$$I^g(0, T + t_2 - t_1, x) \geq \mu(g)(T + t_2 - t_1). \quad (2.150)$$

By (2.149),

$$\sigma^g(T, T + t_2 - t_1, x) = U^g(0, 1, x(T), x(T + 1)) - \pi^f(x(T))$$

$$+\pi^f(x(T+1)) + I^g(t_1+1, t_2-1, v_0)$$
$$-\pi^f(v_0(t_1+1)) + \pi^f(v_0(t_2-1))$$
$$+U^g(0, 1, x(T+t_2-t_1-1), x(T+t_2-t_1))$$
$$-\pi^f(x(T+t_2-t_1-1)) + \pi^f(x(T+t_2-t_1)). \qquad (2.151)$$

Analogously to the proof of the case (i) in Lemma 4.2.4 (see (2.30) and (2.31)) we can show by using (2.134), (2.147), and the definition of $\mathcal{U}_5$ (see (2.145)) that

$$\sigma^g(t_1+1, t_2-1, v_0) \le 2^{-8}\delta + \mu(f)(t_2-t_1-2). \qquad (2.152)$$

Set

$$S_1 = T, \; S_2 = T + t_2 - t_1 - 1, \; r_1 = t_1, \; r_2 = t_2 - 1. \qquad (2.153)$$

It follows from this relation, (2.149), the definition of $\mathcal{U}_1$, δ (see (2.135)-(2.137)), (2.140), which holds for each $T \ge 0$ and (2.134) that for $i = 1, 2$,

$$U^g(0, 1, x(S_i), x(S_i+1)) - \pi^f(x(S_i)) + \pi^f(x(S_i+1))$$
$$\le U^f(0, 1, x(S_i), x(S_i+1)) - \pi^f(x(S_i)) + \pi^f(x(S_i+1)) + 2^{-4}\delta$$
$$\le U^f(0, 1, v_0(r_i), v_0(r_i+1)) - \pi^f(v_0(r_i)) + \pi^f(v_0(r_i+1)) + 2^{-6}\epsilon$$
$$\le \mu(f) + 2^{-6}\epsilon.$$

It follows from this relation, (2.147), (2.151)-(2.153) and the choice of $\mathcal{U}_4$ (see (2.143)) that

$$\sigma^g(T, T+t_2-t_1, x) \le 2\mu(f) + 2^{-5}\epsilon + \mu(f)(t_2-t_1-2) + 2^{-8}\epsilon$$
$$\le 2^{-5}\epsilon + \mu(g)(t_2-t_1) + 2^{-7}\epsilon. \qquad (2.154)$$

By (2.129), (2.149), (2.150) and (2.154),

$$\sigma^g(0, T, v) \ge \mu(g)T - 2^{-4}\epsilon.$$

This relation, (2.141), which holds with $T_1 = 0$, $T_2 = T$ and (2.142) imply (2.132). Therefore we have shown that (2.132) holds for each $T \in [1, \infty)$. Together with (2.129), this implies that

$$\epsilon \ge |\liminf_{T \to \infty}[I^g(0, T, v) - T\mu(g)] - \liminf_{T \to \infty}[\pi^f(h) - \pi^f(v(T))]|. \qquad (2.155)$$

By (2.140), which holds for each $T \in [0, \infty)$, and the definition of δ (see (2.135), (2.136))

$$|\liminf_{T \to \infty}[\pi^f(h) - \pi^f(v(T))] - [\pi^f(h) - \sup\{\pi^f(z) : z \in H(f)\}]| \le 2^{-8}\epsilon.$$

Equation (2.133) now follows from this relation, (2.155), and Lemma 4.2.8. The lemma is proved.

The following lemma is an auxiliary result for Theorem 4.1.1.

LEMMA 4.2.12 *For each $\epsilon \in (0,1)$ there exist $\delta \in (0, \epsilon)$ and a neighborhood $\mathcal{U}$ of f in $\mathcal{A}$ with the weak topology such that the followiing property holds:*

If $g \in \mathcal{U}$, $h \in H(f)$ and if $y \in R^n$ satisfies $|y - h| \leq \delta$, then

$$|\pi^g(y) - \pi^f(y)| \leq \epsilon.$$

Proof. Let $\epsilon \in (0,1)$. Fix

$$K > D_f + 4. \tag{2.156}$$

It follows from Theorem 3.4.1 that there exists a positive number

$$\delta < 8^{-1}\epsilon \tag{2.157}$$

such that the following property holds:
If $x_1, x_2, y_1, y_2 \in R^n$ satisfy

$$|x_i|, |y_i| \leq K + 2, \ i = 1, 2, \ |x_i - y_i| \leq 8\delta, \ i = 1, 2, \tag{2.158}$$

then

$$|U^f(0, 1, x_1, x_2) - U^f(0, 1, y_1, y_2)| \leq 2^{-8}\epsilon,$$
$$|\pi^f(x_i) - \pi^f(y_i)| \leq 2^{-8}\epsilon, \ i = 1, 2. \tag{2.159}$$

It follows from Proposition 1.3.9 that there exists a neighborhood $\mathcal{U}_1$ of f in $\mathcal{A}$ with the weak topology such that the following property holds:
If $g \in \mathcal{U}_1$ and if $y_1, y_2 \in R^n$ satisfy $|y_i| \leq 2K + 2$, $i = 1, 2$, then

$$|U^f(0, 1, y_1, y_2) - U^g(0, 1, y_1, y_2)| \leq 2^{-8}\epsilon. \tag{2.160}$$

It follows from Lemma 4.2.11 that there exist an integer $L \geq 1$ and a neighborhood $\mathcal{U}_2$ of f in $\mathcal{A}$ with the weak topology such that if $g \in \mathcal{U}_2$ and if $h \in H(f)$, then there exists a (g)-good function $v : [0, \infty) \to R^n$ such that $v(0) = h$;

$$\text{dist}(H(f), \{v(t) : \ t \in [T, T + L]\}) \leq 2^{-6}\delta \tag{2.161}$$

for all $T \in [0, \infty)$;

$$\sigma^g(T_1, T_2, v) \leq l(g, K, T_2 - T_1)) + 2^{-6}\delta \tag{2.162}$$

for each $T_1 \in \{0\} \cup [1, \infty)$, $T_2 \geq T_1 + 1$;

$$|\sigma^g(0, T, v) - T\mu(g)| \leq 2^{-6}\delta \tag{2.163}$$

for each $T \in [1, \infty)$;

$$|\liminf_{T \to \infty}[I^g(0, T, v) - T\mu(g)] - \pi^f(h)| \leq 2^{-6}\delta. \qquad (2.164)$$

Lemmas 4.2.10 and 4.2.9 imply that there exists a neighborhood $\mathcal{U}_3$ of f in $\mathcal{A}$ with the weak topology such that

$$|\mu(g) - \mu(f)| \leq 2^{-6}\epsilon, \; l(g, K, T) \leq T\mu(g) \qquad (2.165)$$

for each $g \in \mathcal{U}_3$ and each $T \in (0, \infty)$.

It follows from Theorem 3.1.3 that there exist an integer $L_1 \geq 1$ and a neighborhood $\mathcal{U}_4$ of f in $\mathcal{A}$ with the weak topology such that the following property holds:

If $g \in \mathcal{U}_4$ and if $v : [0, \infty) \to R^n$ is a (g)-good function, then

$$\text{dist}(H(f), \{v(t) : t \in [T, T + L_1]\}) \leq 2^{-8}\delta \qquad (2.166)$$

for all large T.

Put

$$\mathcal{U} = \cap_{i=1}^4 \mathcal{U}_i. \qquad (2.167)$$

Assume that

$$g \in \mathcal{U}, \; h \in H(f), \; y \in R^n, \; |y - h| \leq \delta. \qquad (2.168)$$

By the definition of $\mathcal{U}_2$, L and (2.168) there exists a (g)-good function $v : [0, \infty) \to R^n$ such that $v(0) = h$, (2.161) holds for each $T \geq 0$, (2.162) holds for each $T_1 \in \{0\} \cup [1, \infty)$ and each $T_2 \geq T_1 + 1$, and (2.163) holds for each $T \in [1, \infty)$. Together with (2.165) this implies that

$$0 \leq T\mu(g) - l(g, K, T) \leq 2^{-5}\delta \text{ for all } T \in [1, \infty). \qquad (2.169)$$

Consider any (g)-good function $u : [0, \infty) \to R^n$ for which $u(0) = y$. By the definition of $\mathcal{U}_4$, L_1, (2.168) holds with $v = u$ for all large T. Together with (2.169) and (2.168), this implies that

$$\liminf_{T \to \infty}[I^g(0, T, u) - T\mu(g)] \geq \liminf_{T \to \infty}[I^g(0, T, u) - l(g, K, T)] - 2^{-5}\delta$$

$$\geq \limsup_{T \to \infty}[\pi^f(y) - \pi^f(u(T))] - 2^{-5}\delta. \qquad (2.170)$$

It follows from (2.166), which holds with $v = u$ for all large T; (2.170); (2.168); the definition of δ (see (2.157)-(2.159)); and Lemma 3.8.2 that

$$\liminf_{T \to \infty}[\pi^f(y) - \pi^f(u(T))] \geq \pi^f(h) - 2^{-8}\epsilon - \limsup_{T \to \infty} \pi^f(u(T))$$

$$\geq \pi^f(h) - 2^{-8}\epsilon - \sup\{\pi^f(z) : z \in H(f)\} - 2^{-8}\epsilon$$

$$\geq \pi^f(h) - 2^{-7}\epsilon.$$

Together with (2.170) and (2.157) this implies that

$$\liminf_{T\to\infty}[I^g(0,T,u) - T\mu(g)] \geq \pi^f(h) - \epsilon.$$

Therefore

$$\pi^g(y) \geq \pi^f(h) - \epsilon. \tag{2.171}$$

We will show that $\pi^g(y) \leq \pi^f(h) + \epsilon$. By Lemma 4.2.3 there exists an (f)-good function

$$v_0 : [0,\infty) \to H(f) \text{ such that } v_0 \in A(f), \ v_0 = h. \tag{2.172}$$

By the definition of $\mathcal{U}_2$ and L and (2.172) there exists a (g)-good function $v_1 : [0,\infty) \to R^n$ such that

$$v_1(0) = v_0(1). \tag{2.173}$$

(2.161) holds with $v = v_1$ for each $T \in [0,\infty)$; (2.162) holds with $v = v_1$ for each $T_1 \in \{0\} \cup [1,\infty)$, $T_2 \geq T_1 + 1$; (2.163) holds with $v = v_1$ for each $T \in [1,\infty)$; and (2.164) holds with $v = v_1$, $h = v_0(1)$.

By Corollary 1.3.1 there exists an a.c. function $w : [0,\infty) \to R^n$ such that

$$w(0) = y, \ w(t) = v_1(t-1), \ t \in [1,\infty), \ I^g(0,1,w) = U^g(0,1,w(0),w(1)). \tag{2.174}$$

Equations (2.173), (2.174) and (2.164), which holds with $v = v_1$, $h = v_0(1)$, imply that

$$\pi^g(y) \leq \liminf_{T\to\infty}[I^g(0,T,w) - T\mu(g)] = U^g(0,1,y,v_0(1)) - \mu(g)$$

$$+ \liminf_{T\to\infty}[I^g(0,T,v_1) - T\mu(g)]$$

$$= U^g(0,1,y,v_0(1)) - \mu(g) + \pi^f(v_0(1)) + 2^{-6}\delta. \tag{2.175}$$

It follows from (2.168), (2.156), (2.172), the definition of $\mathcal{U}_1$ (see (2.160)), and the definition of δ (see (2.157)-(2.159)) that

$$U^g(0,1,y,v_0(1)) \leq U^f(0,1,y,v_0(1))$$

$$+2^{-8}\epsilon \leq 2^{-8}\epsilon + U^f(0,1,v_0(0),v_0(1)) + 2^{-8}\epsilon$$

$$= 2^{-7}\epsilon + \pi^f(h) - \pi^f(v_0(1)) + \mu(f).$$

Together with (2.175) and (2.165) this implies that $\pi^g(y) \leq 2^{-7}\epsilon + 2^{-5}\epsilon + \pi^f(h) \leq \pi^f(h) + \epsilon$. This completes the proof of the lemma.

There exists $h_f \in H(f)$ such that

$$\pi^f(h_f) \geq \pi^f(h), \ h \in H(f). \tag{2.176}$$

The next lemma establishes a useful formula for calculation of $\pi^g(x)$ where $x \in R^n$ and g belongs to a small neighborhood of f.

LEMMA 4.2.13 *Let* $\epsilon \in (0,1)$, $K > D_f + 4$. *Then there exist a neighborhood* $\mathcal{U}$ *of* f *in* $\mathcal{A}$ *with the weak topology and integers* $Q_1 \geq 8$, $Q_2 \geq 8 + Q_1$ *such that for each* $g \in \mathcal{U}$ *and each* $x \in R^n$ *satisfying* $|x| \leq K$,

$$\pi^g(x) = \inf\{\liminf_{T \to \infty}[I^g(0, T, v) - T\mu(g)] :$$

$$v : [0, \infty) \to R^n \text{ is an a.c. function,}$$

$$v(0) = x, \ \inf\{|v(t) - h_f| : t \in [Q_1, Q_2]\} \leq \epsilon\}. \tag{2.177}$$

Proof. It follows from Theorem 3.1.3 that there exist an integer $L \geq 1$ and a neighborhood $\mathcal{U}_1$ of f in $\mathcal{A}$ with the weak topology such that the following property holds:

If $g \in \mathcal{U}_1$ and if $v : [0, \infty) \to R^n$ is a (g)-good function, then for all large T,

$$\mathrm{dist}(H(f), \{v(t) : t \in [T, T + L]\}) \leq 16^{-1}\epsilon. \tag{2.178}$$

It follows from Lemma 4.2.1 and the choice of L that there exist an integer $N \geq 10$ and a neighborhood $\mathcal{U}_2$ of f in $\mathcal{A}$ with the weak topology such that the following property holds:

If $g \in \mathcal{U}_2$, $T_1 \geq 0$, $T_2 \geq T_1 + NL$, and if an a.c. function $v : [T_1, T_2] \to R^n$ satisfies

$$|v(T_i)| \leq K, \ i = 1, 2, \ \Phi^g(T_1, T_2, v) \leq 4, \tag{2.179}$$

then for each $S \in [T_1, T_2 - NL]$ there exists an integer $i_0 \in [0, N - 8]$ such that for all $T \in [S + i_0 L, S + (i_0 + 7)L]$,

$$\mathrm{dist}(H(f), \{v(t) : t \in [T, T + L]\}) \leq 2^{-1}\epsilon. \tag{2.180}$$

Put

$$\mathcal{U} = \mathcal{U}_1 \cap \mathcal{U}_2, \ Q_1 = NL, \ Q_2 = 2NL. \tag{2.181}$$

Assume that

$$g \in \mathcal{U}, \ x \in R^n, \ |x| \leq K.$$

Denote by E the set of all (g)-good functions $v : [0, \infty) \to R^n$ for which

$$v(0) = x, \ \liminf_{T \to \infty}[I^g(0, T, v) - T\mu(g)] \leq \pi^g(x) + 1. \tag{2.182}$$

It is easy to see that

$$\pi^g(x) = \inf\{\liminf_{T\to\infty}[I^g(0,T,v) - T\mu(g)] : v \in E\}. \tag{2.183}$$

Consider any $v \in E$. By the definition of $\mathcal{U}_1$, L

$$|v(t)| \le K \text{ for all large } t. \tag{2.184}$$

By (2.182)
$$\Phi^g(0,T,v) \le 2 \text{ for all } T \in [1,\infty).$$

It follows from this relation, (2.184), (2.181), and the definition of $N, \mathcal{U}_2$ (see (2.179), (2.180)) that $\inf\{|v(t) - h_f| : t \in [Q_1, Q_2]\} \le \epsilon$. This completes the proof of the lemma.

Proof of Theorem 4.1.1. By Lemma 4.2.9, f is a continuity point of the mapping $g \to \mu(g)$, $g \in \mathcal{A}$. We will show that f is a continuity point of the mapping $g \to \pi^g$, $g \in \mathcal{A}$.

Assume that $\epsilon \in (0,1)$, $K > D_f + 4$. Lemma 4.2.12 implies that there exist a neighborhood $\mathcal{U}_1$ of f in $\mathcal{A}$ with the weak topology and a positive number

$$\delta < 16^{-1}\epsilon \tag{2.185}$$

such that

$$|\pi^g(y) - \pi^f(h)| \le 16^{-1}\epsilon \tag{2.186}$$

for each $g \in \mathcal{U}_1$, each $h \in H(f)$, and each $y \in R^n$ satisfying $|y - h| \le \delta$.

It follows from Lemma 4.2.13 that there exist a neighborhood $\mathcal{U}_2$ of f in $\mathcal{A}$ with the weak topology and integers $Q_1 \ge 8$, $Q_2 \ge 8 + Q_1$ such that if $g \in \mathcal{U}_2$ and if $x \in R^n$ satisfies $|x| \le K$, then relation (2.177) is valid with $\epsilon = 8^{-1}\delta$.

It follows from Lemma 4.2.9 and Proposition 1.3.9 that there exists a neighborhood $\mathcal{U}_3$ of f in $\mathcal{A}$ with the weak topology such that the following property holds:

If $g \in \mathcal{U}_3$, $\tau \in [1, 2Q]$, and if $x, y \in R^n$ satisfy $|x|, |y| \le 2K + 2$, then

$$|\mu(g) - \mu(f)| \le (16(Q_1 + Q_2))^{-1}\epsilon, \tag{2.187}$$

$$|U^f(0,\tau,x,y) - U^g(0,\tau,x,y)| \le 16^{-1}\epsilon. \tag{2.188}$$

Put
$$\mathcal{U} = \cap_{i=1}^3 \mathcal{U}_i.$$

By the choice of $\mathcal{U}_2$, Q_1, Q_2, and (2.177), if $g \in \mathcal{U}$ and if $x \in R^n$ satisfies $|x| \le K$, then

$$\pi^g(x) = \inf\{U^g(0,T,x,y) - T\mu(g) + \pi^g(y) :$$

$$T \in [Q_1, Q_2], \ y \in R^n, \ |y - h_f| \leq 8^{-1}\delta\}.$$

By this relation, (2.187), (2.188), and the definition of $\mathcal{U}_1$, δ (see (2.185) and (2.186)), for each $g \in \mathcal{U}$ and each $x \in R^n$ satisfying $|x| \leq K$,

$$|\pi^g(x) - \inf\{U^f(0, T, x, y) - T\mu(f) + \pi^f(h_f) :$$

$$T \in [Q_1, Q_2], \ y \in R^n, \ |y - h_f| \leq 8^{-1}\delta\}|$$

$$\leq 16^{-1}\epsilon + 8^{-1}\epsilon + 16^{-1}\epsilon \leq 4^{-1}\epsilon.$$

This implies that for each $g_1, g_2 \in \mathcal{U}$ and each $x \in R^n$ satisfying $|x| \leq K$,

$$|\pi^{g_1}(x) - \pi^{g_2}(x)| \leq 2^{-1}\epsilon.$$

Therefore f is a continuity point of the mapping $g \to \pi^g$, $g \in \mathcal{A}$. The theorem is proved.

Proof of Theorem 4.1.2. Let $\epsilon \in (0, 1)$. Theorem 3.3.1 implies that there exists a natural number L such that if $v : [0, \infty) \to R^n$ is an (f)-good function, then for all large T,

$$\text{dist}(H(f), \{v(t) : t \in [T, T + L]\}) \leq 2^{-8}\epsilon. \tag{2.189}$$

It follows from Lemma 4.2.6 that there exist a neighborhood $\mathcal{U}_1$ of f in $\mathcal{A}$ with the weak topology and a positive number

$$\delta_0 < 8^{-1}\epsilon \tag{2.190}$$

such that the following property holds:

If $g \in \mathcal{U}_1$, $T \in [L, \infty)$ and if an a.c. function $v : [0, T] \to R^n$ satisfies

$$d(v(0), H(f)) \leq \delta_0, \ d(v(T), H(f)) \leq \delta_0, \tag{2.191}$$

$$\sigma^g(0, T, v) \leq l(g, D_f + 4, T) + \delta_0,$$

then for every $S \in [0, T - L]$,

$$\text{dist}(H(f), \{v(t) : t \in [S, S + L]\}) \leq \epsilon. \tag{2.192}$$

Lemma 4.2.5 implies that there exists a neighborhood $\mathcal{U}_2$ of f in $\mathcal{A}$ with the weak topology such that the following property holds:

If $g \in \mathcal{U}_2$, $h \in H(f)$ and if a number $T \geq 1$, then there exists an a.c. function $v : [0, T] \to R^n$ such that

$$v(0) = h, \ v(T) \in H(f),$$

$$\sigma^g(0, T, x) \leq l(g, D_f + 4, T) + 8^{-1}\delta_0. \tag{2.193}$$

It is easy to see that there exists a positive number

$$\delta < 2^{-6}\delta_0 \tag{2.194}$$

such that

$$|\pi^f(x) - \pi^f(y)| \leq 2^{-4}\delta_0 \tag{2.195}$$

for each $x, y \in R^n$ satisfying $|x - y| \leq \delta$ and $|x|, |y| \leq D_f + 4$.

By Theorem 4.1.1 there exists a neighborhood $\mathcal{U}_3$ of f in $\mathcal{A}$ with the weak topology such that

$$|\pi^f(x) - \pi^g(x)| \leq 2^{-4}\delta_0 \tag{2.196}$$

for each $g \in \mathcal{U}_3$ and each $x \in R^n$ satisfying $|x| \leq D_f + 4$.

It follows from Theorem 3.1.3 that there are a neighborhood $\mathcal{U}_4$ of f in $\mathcal{A}$ with the weak topology and an integer $L_0 \geq 1$ such that for each $g \in \mathcal{U}_4$ and each (g)-good function $v : [0, \infty) \to R^n$,

$$\text{dist}(H(f), \{v(t) : t \in [T, T + L_0]\}) \leq \delta \tag{2.197}$$

for all large T.

We may assume that $L_0 \geq L$. Put

$$\mathcal{U} = \cap_{i=1}^4 \mathcal{U}_i.$$

Assume that

$$g \in \mathcal{U}, \ v \in A(g), \ d(v(0), H(f)) \leq \delta. \tag{2.198}$$

It follows from (2.198) and Proposition 4.2.1 that v is a (g)-good function. By the definition of $\mathcal{U}_4$ and L_0 there exists a number T_0 such that (2.197) holds for each $T \geq T_0$.

Let $T \geq T_0 + L$. There exists

$$\tau \in [T + L_0, T + 2L_0] \text{ such that } |v(\tau) - h_f| \leq \delta \tag{2.199}$$

(recall h_f in (2.176)). We show that

$$\sigma^g(0, \tau, v) \leq l(g, D_f + 4, \tau) + \delta_0. \tag{2.200}$$

By (2.198), (2.199) and the choice of $\mathcal{U}_3$ (see (2.196)),

$$l(g, D_f + 4, \tau) \leq \sigma^g(0, \tau, v)$$

$$\leq \tilde{\sigma}^g(0, \tau, v) + 2^{-3}\delta_0 = \tau\mu(g) + 2^{-3}\delta_0. \tag{2.201}$$

It follows from the choice of $\mathcal{U}_2$ (see (2.193)) that there exists an a.c. function $u : [0, \tau] \to R^n$ for which

$$u(0), u(\tau) \in H(f),$$

$$\sigma^g(0, \tau, u) \le l(g, D_f + 4, \tau) + 8^{-1}\delta_0.$$

By these relations, the representation formula (1.6) and the choice of $\mathcal{U}_3$ (see (2.196)),

$$l(g, D_f + 4, \tau) + 8^{-1}\delta_0 \ge \tilde{\sigma}^g(0, \tau, u)$$

$$-2^{-3}\delta_0 \ge \tau\mu(g) - 2^{-3}\delta_0.$$

Together with (2.201) this implies (2.200). By (2.198)-(2.200) and the definition of $\mathcal{U}_1$, (2.192) holds for all $S \in [0, \tau - L]$. This completes the proof of the theorem.

Proof of Theorem 4.1.3. Let $\epsilon \in (0, 1)$ and $K > D_f + 4$. Theorem 4.1.2 implies that there exist numbers $\delta \in (0, \epsilon)$, $L > 0$, and a neighborhood $\mathcal{U}_1$ of f in $\mathcal{A}$ with the weak topology such that the following property holds:
If $g \in \mathcal{U}_1$, $v \in A(g)$ satisfies $d(v(0), H(f)) \le \delta$ and if $T \in [0, \infty)$, then

$$\text{dist}(H(f), \{v(t) : t \in [T, T + L]\}) \le \epsilon. \tag{2.202}$$

It follows from Theorem 3.1.3 that there exist a neighborhood $\mathcal{U}_2$ of f in $\mathcal{A}$ with the weak topology and a natural number L_0 such that the following property holds:
For each $g \in \mathcal{U}_2$ and each (g)-good function $v : [0, \infty) \to R^n$,

$$\text{dist}(H(f), \{v(t) : t \in [T, T + L_0]\}) \le 8^{-1}\delta \tag{2.203}$$

for all large T.

By Lemma 4.2.1 there exist a neighborhood $\mathcal{U}_3$ of f in $\mathcal{A}$ with the weak topology and a natural number $N \ge 10$ such that the following property holds:
If $g \in \mathcal{U}_3$, $T_1 \ge 0$, $T_2 \ge T_1 + NL_0$ and if an a.c. function $v : [T_1, T_2] \to R^n$ satisfies

$$|v(T_i)| \le K + 8, \ i = 1, 2, \ \Phi^g(T_1, T_2, v) \le 4, \tag{2.204}$$

then for each $S \in [T_1, T_2 - NL_0]$, there exists an integer $i_0 \in [0, N - 8]$ such that

$$\text{dist}(H(f), \{v(t) : t \in [T, T + L_0]\}) \le \delta \tag{2.205}$$

for all $T \in [S + i_0L_0, S + (i_0 + 7)L_0]$.
Put

$$\mathcal{U} = \cap_{i=1}^3 \mathcal{U}_i, \ Q = NL_0. \tag{2.206}$$

Assume that

$$g \in \mathcal{U}, \ v \in A(g), \ |v(0)| \le K. \tag{2.207}$$

Equation (2.207) and Proposition 4.2.1 imply that v is a (g)-good function. Therefore by the definition of $\mathcal{U}_2$,

$$|v(t)| \leq K + 1 \text{ for all large } t.$$

It follows from this relation, (2.207), (2.206), and the definition of $\mathcal{U}_3$, N that there exists $\tau \in [0, Q]$ for which $d(v(\tau), H(f)) \leq \delta$. By this relation, (2.207), and the definition of $\mathcal{U}_1$, δ, L, the relation (2.202) holds for each $T \geq \tau$. This completes the proof of the theorem.

4.3.   Proof of Theorem 4.1.4

LEMMA 4.3.1 *Assume that $f \in \mathcal{A}$, $x : [0, \infty) \to R^n$ is an (f)-good function and $h \in \Omega(x)$. Then there exists an a.c. function $v : R^1 \to \Omega(x)$ such that $v \in B(f)$, $v(0) = h$.*

Proof. Theorem 1.2.2 implies that the function x is bounded. It is not difficult to see that the following property holds:
(a) for each $\epsilon > 0$ there exists $T(\epsilon) > 0$ such that

$$\tilde{\sigma}^f(T_1, T_2, x) - (T_2 - T_1)\mu(f) \leq \epsilon$$

for each $T_1 \geq T(\epsilon)$, $T_2 > T_1$,
    There exists a sequence of numbers $\{T_p\}_{p=0}^{\infty}$ such that

$$T_{p+1} \geq T_p + 1, \ p = 0, 1, \ldots, \ x(T_p) \to h \text{ as } p \to \infty. \tag{3.1}$$

For every integer $p \geq 1$ we set

$$v_p(t) = x(t + T_p), \ t \in [-T_p, \infty). \tag{3.2}$$

By Proposition 1.3.5, the boundedness of x, (3.1), and (3.2) there exist a subsequence $\{v_{p_j}\}_{j=1}^{\infty}$ and an a.c. function $v : R^1 \to R^n$ such that for each integer $N \geq 1$,

$$v_{p_j}(t) \to v(t) \text{ as } j \to \infty \text{ uniformly in } [-N, N],$$

$$I^f(-N, N, v) \leq \liminf_{j \to \infty} I^f(-N, N, v_{p_j}). \tag{3.3}$$

Equations (3.1)-(3.3) imply that $v(0) = h$ and $v(t) \in \Omega(x)$, $t \in R^1$. It follows from property (a), (3.3), and (3.2) that $v \in B(f)$. The lemma is proven.

    Theorems 1.2.2 and 1.2.3 and Proposition 4.2.1 imply the following result.

LEMMA 4.3.2 *Let $f \in \mathcal{A}$. Then each function $v \in B(f)$ is bounded.*

Assertion (1) of Theorem 4.1.4 follows from Lemma 4.3.1. Assertion (2) of Theorem 4.1.4 follows from Lemma 4.3.2 and Theorem 4.1.3. Assertion (3) of Theorem 4.1.4 follows from Assertion (2) and Theorem 4.1.4.

Lemma 4.3.1 implies the following result.

PROPOSITION 4.3.1 *Assume that $f \in \mathcal{A}$ and there exists a compact set $H(f) \subset R^n$ such that for each $v \in B(f)$ the following relations hold:*

$$v(t) \in H(f), \ t \in R^1,$$

$$\{y \in R^n : \quad \text{there exists a sequence } \{t_i\}_{i=0}^{\infty} \subset [0, \infty) \text{ for which}$$

$$t_i \to \infty, \ v(t_i) \to y \text{ as } i \to \infty\} = H(f).$$

Then the integrand f has the asymptotic turnpike property with the turnpike $H(f)$.

Chapter 5

TURNPIKE FOR AUTONOMOUS PROBLEMS

In this chapter we continue to study the turnpike property for autonomous variational problems. For a class of smooth nonconvex integrands we improve the results of Chapter 3. We establish the turnpike property for a generic nonconvex integrand $f(x, u)$. We show that for a generic f, any small $\epsilon > 0$ and an extremal $v : [0, T] \to R^n$ of the variational problem with large enough T, fixed end points and the integrand f, for each $\tau \in [L_1, T - L_1]$ the set $\{v(t) : t \in [\tau, \tau + L_2]\}$ is equal to a set $H(f)$ up to ϵ in the Hausdorff metric. Here $H(f) \subset R^n$ is a compact set depending only on the integrand f and $L_1 > L_2 > 0$ are constants which depend only on ϵ and $|v(0)|, |v(T)|$.

5.1.   Main results

In this chapter we analyse the structure of optimal solutions of the variational problem

$$\int_0^T f(z(t), z'(t))dt \to \min, \ z(0 = x, \ z(T) = y, \qquad (P)$$

$$z : [0, T] \to R^n \text{ is an absolutely continuous function}$$

where $T > 0$, $x, y \in R^n$ and $f : R^{2n} \to R^1$ is an integrand.

We say that an integrand $f = f(x, u) \in C(R^{2n})$ has the *turnpike property* if there exists a compact set $H(f) \subset R^n$ such that for each bounded set $K \subset R^n$ and each $\epsilon > 0$ there exist numbers $L_1 > L_2 > 0$ such that for each $T \geq 2L_1$, each $x, y \in K$ and an optimal solution $v : [0, T] \to R^n$ for the variational problem (P) the relation

$$\text{dist}(H(f), \{v(t) : t \in [\tau, \tau + L_2]\}) \leq \epsilon$$

holds for each $\tau \in [L_1, T - L_1]$. (Here $\mathrm{dist}(\cdot, \cdot)$ is the Hausdorff metric).

Our goal is to show that the turnpike property is a general phenomenon which holds for a class of autonomous variational problems with vector-valued functions. We consider the complete metric space of integrands $\bar{\mathcal{N}}_k$ (k is a nonnegative integer) described below and establish the existence of a set $\mathcal{F} \subset \bar{\mathcal{N}}_k$ which is a countable intersection of open everywhere dense sets in $\bar{\mathcal{N}}_k$ and such that each integrand $f \in \mathcal{F}$ has the turnpike property.

Moreover we show that the turnpike property holds for approximate solutions of variational problems with a generic integrand f and that the turnpike phenomenon is stable under small perturbations of a generic integrand f.

In Chapter 3 we studied the weak version of this turnpike property for optimal solutions of the variational problem (P) with $x, y \in R^n$, large enough T and a generic integrand f belonging to the space of functions $\mathcal{A}$.

In this weak version of the turnpike property established in Chapter 3 for an optimal solution of the problem (P) with $x, y \in R^n$, large enough T and a generic integrand $f \in \mathcal{A}$ the relation

$$\mathrm{dist}(H(f), \{v(t) : t \in [\tau, \tau + L_2]\}) \leq \epsilon$$

with L_2 which depends on ϵ and $|x|, |y|$ and a compact set $H(f) \subset R^n$ depending only on the integrand f, holds for each $\tau \in [0, T] \setminus E$ where $E \subset [0, T]$ is a measurable subset such that the Lebesgue measure of E does not exceed a constant which depends on ϵ and on $|x|, |y|$.

The turnpike property which is established in this chapter guarantees that we may take $E = [0, L_1] \cup [T - L_1, T]$ where $L_1 > 0$ is a constant which depends on ϵ and $|x|, |y|$. The results of this chapter have been established in [100].

Denote by $|\cdot|$ the Euclidean norm in R^n. Let $a > 0$ be a constant and let $\psi : [0, \infty) \to [0, \infty)$ be an increasing function such that $\psi(t) \to \infty$ as $t \to \infty$. Consider the space of functions $\mathcal{A}$ introduced in Section 3.1. Recall that $\mathcal{A}$ is the set of continuous functions $f : R^n \times R^n \to R^1$ which satisfy the following assumptions:

A(i) for each $x \in R^n$ the function $f(x, \cdot) : R^n \to R^1$ is convex;

A(ii) $f(x, y) \geq \max\{\psi(|x|), \psi(|u|)|u|\} - a$ for each $(x, u) \in R^n \times R^n$;

A(iii) for each $M, \epsilon > 0$ there exist $\Gamma, \delta > 0$ such that

$$|f(x_1, u_1) - f(x_2, u_2)| \leq \epsilon \max\{f(x_1, u_1), f(x_2, u_2)\}$$

for each $u_1, u_2, x_1, x_2 \in R^n$ which satisfy

$$|x_i| \leq M, \ |u_i| \geq \Gamma, \ i = 1, 2 \ \text{and} \ \max\{|x_1 - x_2|, |u_1 - u_2|\} \leq \delta.$$

For the set $\mathcal{A}$ we consider the uniformity introduced in Section 3.1. This uniformity is determined by the following base:

$$E(N, \epsilon, \lambda) = \{(f, g) \in \mathcal{A} \times \mathcal{A} :$$

$$|f(x, u) - g(x, u)| \leq \epsilon \ (u, x \in R^n, \ |x|, |u| \leq N),$$

$$(|f(x, u)| + 1)(|g(x, u)| + 1)^{-1} \in [\lambda^{-1}, \lambda] \ (x, u \in R^n, \ |x| \leq N)\}$$

where $N > 0$, $\epsilon > 0$ and $\lambda > 1$ [37].

It is known that the uniform space $\mathcal{A}$ is metrizable and complete (see Section 3.1). We consider functionals of the form

$$I^f(T_1, T_2, x) = \int_{T_1}^{T_2} f(x(t), x'(t))dt \tag{1.1}$$

where $f \in \mathcal{A}$, $0 \leq T_1 < T_2 < +\infty$ and $x : [T_1, T_2] \to R^n$ is an absolutely continuous (a.c.) function.

For $f \in \mathcal{A}$, $y, z \in R^n$ and numbers T_1, T_2 satisfying $0 \leq T_1 < T_2$ we set

$$U^f(T_1, T_2, y, z) = \inf\{I^f(T_1, T_2, x) : \ x : [T_1, T_2] \to R^n$$

$$\text{is an a.c. function satisfying } x(T_1) = y, \ x(T_2) = z\}. \tag{1.2}$$

It is easy to see that $-\infty < U^f(T_1, T_2, y, z) < \infty$ for each $f \in \mathcal{A}$, each $y, z \in R^n$ and all numbers T_1, T_2 satisfying $0 \leq T_1 < T_2$.

Let $f \in \mathcal{A}$. For any a.c. function $x : [0, \infty) \to R^n$ we set

$$J(x) = \liminf_{T \to \infty} T^{-1} I^f(0, T, x). \tag{1.3}$$

Of special interest is the *minimal long-run average cost growth rate*

$$\mu(f) = \inf\{J(x) : \ x : [0, \infty) \to R^n \text{ is an a.c. function}\}. \tag{1.4}$$

Clearly $-\infty < \mu(f) < +\infty$.

By Theorems 3.6.1 and 3.6.2,

$$U^f(0, T, x, y) = T\mu(f) + \pi^f(x) - \pi^f(y) + \theta_T^f(x, y), \tag{1.5}$$

$$x, y \in R^n, \ T \in (0, \infty),$$

where $\pi^f : R^n \to R^1$ is a continuous function and $(T, x, y) \to \theta_T^f(x, y) \in R^1$ is a continuous nonnegative function defined for $T > 0$, $x, y \in R^n$,

$$\pi^f(x) = \inf\{\liminf_{T \to \infty}[I^f(0, T, v) - \mu(f)T] : \ v : [0, \infty) \to R^n \tag{1.6}$$

$$\text{is an a.c. function satisfying } v(0) = x\}, \ x \in R^n,$$

and for every $T > 0$, every $x \in R^n$ there is $y \in R^n$ satisfying $\theta_T^f(x, y) = 0$.

Let $f \in \mathcal{A}$. An a.c. function $x : [0, \infty) \to R^n$ is called an (f)-good function if the function $\phi_x^f : T \to I^f(0, T, x) - \mu(f)T$, $T \in (0, \infty)$ is bounded. By Theorem 3.6.3 for each $f \in \mathcal{A}$ and each $z \in R^n$ there exists an (f)-good function $v : [0, \infty) \to R^n$ satisfying $v(0) = z$.

We denote $d(x, B) = \inf\{|x - y| : y \in B\}$ for $x \in R^n$, $B \subset R^n$. Denote by $\mathrm{dist}(A, B)$ the Hausdorff metric for two sets $A \subset R^n$, $B \subset R^n$. For every bounded a.c. function $x : [0, \infty) \to R^n$ define

$$\Omega(x) = \{y \in R^n : \quad \text{there exists a sequence } \{t_i\}_{i=0}^{\infty} \subset (0, \infty)$$

$$\text{for which } t_i \to \infty, \ x(t_i) \to y \text{ as } i \to \infty\}. \tag{1.7}$$

We say that an integrand $f \in \mathcal{A}$ has the asymptotic turnpike property, or briefly (ATP), if $\Omega(v_2) = \Omega(v_2)$ for all (f)-good functions $v_i : [0, \infty) \to R^n$, $i = 1, 2$.

By Theorem 3.1.1 there exists a set $\mathcal{F} \subset \mathcal{A}$ which is a countable intersection of open everywhere dense subsets of $\mathcal{A}$ and such that each $f \in \mathcal{F}$ has (ATP).

Assume that an integrand $f \in \mathcal{A}$ has the asymptotic turnpike property. Then Proposition 4.1.1 implies that there exists a compact set $H(f) \subset R^n$ such that $\Omega(v) = H(f)$ for each (f)-good function $v : [0, \infty) \to R^n$. We say that the set $H(f)$ is the turnpike of f.

Denote by $\mathcal{N}$ the set of all functions $f \in C^1(R^{2n})$ satisfying assumptions which ensure that each solution of (P) belongs to $C^2([0, T]; R^n)$:

$$\partial f / \partial u_i \in C^1(R^{2n}) \text{ for } i = 1, \ldots, n;$$

the matrix $(\partial^2 f / \partial u_i \partial u_j)(x, u)$, $i, j = 1, \ldots, n$ is positive definite for all $(x, u) \in R^{2n}$;

$$f(x, u) \geq \max\{\psi(|x|), \psi(|u|)|u|\} - a \text{ for all } x, u \in R^n \times R^n;$$

there exist a number $c_0 > 1$ and monotone increasing functions $\phi_i : [0, \infty) \to [0, \infty)$, $i = 0, 1, 2$, such that

$$\phi_0(t)t^{-1} \to \infty \text{ as } t \to \infty, \ f(x, u) \geq \phi_0(c_0|u|) - \phi_1(|x|), \ x, u \in R^n;$$

$$\max\{|\partial f / \partial x_i(x, u)|, |\partial f / \partial u_i(x, u)|\} \leq \phi_2(|x|)(1 + \phi_0(|u|)),$$

$$x, u \in R^n, \ i = 1, \ldots, n.$$

It is easy to see that $\mathcal{N} \subset \mathcal{A}$. We will establish the following result.

THEOREM 5.1.1 *Assume that an integrand $f \in \mathcal{N}$ has the asymptotic turnpike property and $\epsilon, K > 0$. Then there exists a neighborhood $\mathcal{U}$ of f*

in $\mathcal{A}$ *and numbers* $M > K$, $l_0 > l > 0$, $\delta > 0$ *such that for each* $g \in \mathcal{U}$, *each* $T \geq 2l_0$ *and each a.c. function* $v : [0, T] \to R^n$ *which satisfies*

$$|v(0)|, |v(T)| \leq K, \ I^g(0, T, v) \leq U^g(0, T, v(0), v(T)) + \delta$$

the relation $|v(t)| \leq M$ *holds for all* $t \in [0, T]$ *and*

$$dist(H(f), \{v(t) : t \in [\tau, \tau + l]\}) \leq \epsilon \qquad (1.8)$$

for each $\tau \in [l_0, T - l_0]$. *Moreover if* $d(v(0), H(f)) \leq \delta$, *then (1.8) holds for each* $\tau \in [0, T - l_0]$ *and if* $d(v(T), H(f)) \leq \delta$, *then (1.8) holds for each* $\tau \in [l_0, T - l]$.

Let $k \geq 1$ be an integer. Denote by $\mathcal{A}_k$ the set of all integrands $f \in \mathcal{A} \cap C^k(R^{2n})$. For $p = (p_1, \ldots, p_{2n}) \in \{0, \ldots, k\}^{2n}$ and $f \in C^k(R^{2n})$ we set

$$|p| = \sum_{i=1}^{2n} p_i, \ D^p f = \partial^{|p|} f / \partial y_1^{p_1} \ldots \partial y_{2n}^{p_{2n}}.$$

For the set $\mathcal{A}_k$ we consider the uniformity which is determined by the following base:

$$E(N, \epsilon, \lambda) = \{(f, g) \in \mathcal{A}_k \times \mathcal{A}_k : |D^p f(x, u) - D^p g(x, u)| \leq \epsilon$$

for each $u, x \in R^n$ satisfying $|x|, |u| \leq N$

and each $p \in \{0, \ldots, k\}^{2n}$ satisfying $|p| \leq k$,

$$|f(x, u) - g(x, u)| \leq \epsilon$$

for each $u, x \in R^n$ satisfying $|x|, |u| \leq N$,

$$(|f(x, u)| + 1)(|g(x, u)| + 1)^{-1} \in [\lambda^{-1}, \lambda]$$

for each $x, u \in R^n$ satisfying $|x| \leq N\}$

where $N > 0$, $\epsilon > 0$, $\lambda > 1$. It is easy to verify that the uniform space $\mathcal{A}_k$ is metrizable and complete.

For each integer $k \geq 1$ we define $\mathcal{N}_k = \mathcal{N} \cap \mathcal{A}_k$. Set

$$\mathcal{A}_0 = \mathcal{A}, \ \mathcal{N}_0 = \mathcal{N}.$$

Let $k \geq 0$ be an integer. Denote by $\bar{\mathcal{N}}_k$ the closure of $\mathcal{N}_k$ in $\mathcal{A}_k$ and consider the topological subspace $\bar{\mathcal{N}}_k \subset \mathcal{A}_k$ with the relative topology. We will establish the following result.

THEOREM 5.1.2 *Let* $q \geq 0$ *be an integer. Then there exists a set* $\mathcal{F}_q \subset \bar{\mathcal{N}}_q$ *which is a countable intersection of open everywhere dense subsets*

*of $\bar{N}_q$ and such that each $f \in \mathcal{F}_q$ has the asymptotic turnpike property
and the following property:*

*For each $\epsilon, K > 0$ there exist a neighborhood $\mathcal{U}$ of f in $\mathcal{A}$ and numbers
$M > K$, $l_0 > l > 0$, $\delta > 0$ such that for each $g \in \mathcal{U}$, each $T \geq 2l_0$ and
each a.c. function $v : [0, T] \to R^n$ which satisfies*

$$|v(0)|, |v(T)| \leq K, \ I^g(0, T, v) \leq U^g(0, T, v(0), v(T)) + \delta$$

the relation $|v(t)| \leq M$ holds for all $t \in [0, T]$ and

$$dist(H(f), \{v(t) : t \in [\tau, \tau + l]\}) \leq \epsilon \qquad (1.9)$$

*for each $\tau \in [l_0, T - l_0]$. Moreover if $d(v(0), H(f)) \leq \delta$, then (1.9) holds
for each $\tau \in [0, T - l_0]$ and if $d(v(T), H(f)) \leq \delta$, then (1.9) holds for
each $\tau \in [l_0, T - l]$.*

Chapter 5 is organized as follows. Theorem 5.1.1 will be proved in Section 5.2. Section 5.3 contains the proof of Theorem 5.1.2 while Section 5.4 contains an example.

5.2.   Proof of Theorem 5.1.1

Assume that $f \in \mathcal{N}$ has the asymptotic turnpike property with the turnpike $H(f) \subset R^n$.

For each a.c. function $u : [\tau_1, \tau_2] \to R^n$ where $\tau_1 \geq 0$, $\tau_2 > \tau_1$ and each $r_1, r_2 \in [\tau_1, \tau_2]$ satisfying $r_1 < r_2$ we set

$$\sigma(r_1, r_2, u) = I^f(r_1, r_2, u) - \pi^f(u(r_1)) + \pi^f(u(r_2)) - (r_2 - r_1)\mu(f),$$

$$\Phi(r_1, r_2, u) = I^f(r_1, r_2, u) - U^f(r_1, r_2, u(r_1), u(r_2)). \qquad (2.1)$$

LEMMA 5.2.1 *Assume that $h \in H(f)$. Then there exists an (f)-good
function $v : [0, \infty) \to H(f)$ such that $v(0) = h$ and $\sigma^f(T_1, T_2, x) = 0$ for
each $T_1 \geq 0, T_2 > T_1$.*

Proof. Consider any (f)-good function $w : [0, \infty) \to R^n$. Since the integrand f has the asymptotic turnpike property with the turnpike $H(f)$ we have

$$\Omega(w) = H(f).$$

Theorem 1.2.2 implies that

$$\sup\{|w(t)| : t \in [0, \infty)\} < \infty.$$

It is easy to see that the following property holds:

(a) For each $\epsilon > 0$ there exists $T(\epsilon) > 0$ such that $\sigma^f(T_1, T_2, w) \leq \epsilon$ for each $T_1 \geq T(\epsilon)$, $T_2 > T_1$.

There exists a sequence of numbers $\{T_p\}_{p=0}^{\infty} \subset [0, \infty)$ such that

$$T_{p+1} \geq T_p + 1, \ p = 0, 1, \ldots, \ w(T_p) \rightarrow h \text{ as } p \rightarrow \infty. \quad (2.2)$$

For every integer $p \geq 1$ we set

$$v_p(t) = w(t + T_p), \ t \in [0, \infty). \quad (2.3)$$

By Proposition 1.3.5, the boundedness of w, (2.3) and property (a) there exist a subsequence $\{v_{p_j}\}_{j=1}^{\infty}$ and an a.c. function $v : [0, \infty) \rightarrow R^n$ such that for each integer $N \geq 1$,

$$v_{p_j}(t) \rightarrow v(t) \text{ as } j \rightarrow \infty \text{ uniformly in } [0, N],$$

$$I^f(0, N, v) \leq \liminf_{j \to \infty} I^f(0, N, v_{p_j}). \quad (2.4)$$

(2.2)-(2.4) imply that $v(0) = h$ and $v(t) \in H(f)$, $t \in [0, \infty)$. It follows from property (a), (2.3) and (2.4) that $\sigma^f(T_1, T_2, w) = 0$ for each $T_1 \geq 0$, $T_2 > T_1$. The lemma is proved.

Lemma 5.2.1 implies that there is an (f)-good function $v_* : [0, \infty) \rightarrow H(f)$ such that for each $T_1 \geq 0$, $T_2 > T_1$,

$$I^f(T_1, T_2, v_*) = \mu(f)(T_2 - T_1) + \pi^f(v_*(T_1)) - \pi^f(v_*(T_2)). \quad (2.5)$$

By Proposition 2.5.1,

$$v_* \in C^2([0, \infty); R^n). \quad (2.6)$$

LEMMA 5.2.2 *The function* $\pi^f \cdot v_* \in C^1([0, \infty); R^1)$.

Proof. It follows from (2.5) that

$$\pi^f(v_*(T)) = -I^f(0, T, v_*) + \mu(f)T + \pi^f(v_*(0))$$

for each $T \geq 0$. Combined with (2.6) this equality implies the assertion of the lemma.

For each $\tau \in [0, \infty)$ we put

$$P_\tau(t) = t(v_*(\tau) - v_*(1)), \ t \in R^1, \ \psi(\tau) = I^f(0, 1, v_* + P_\tau). \quad (2.7)$$

LEMMA 5.2.3 $\psi \in C^1([0, \infty); R^1)$.

Proof. For $\lambda, t \in [0, \infty)$ put

$$B(\lambda, t) = (v_*(t) + P_\lambda(t), v_*'(t) + P_\lambda'(t)). \quad (2.8)$$

Let $\tau, h \in [0, \infty)$, $\tau \neq h$ and $t \in [0, 1]$. By (2.7), (2.8) there is $\lambda_h(t) \in [\min\{h, \tau\}, \max\{h, \tau\}]$ such that

$$(h - \tau)^{-1}[f(B(h, t)) - f(B(\tau, t))] = \partial f / \partial x(B(\lambda_h(t), t))tv'_*(\lambda_h(t))$$

$$+ \partial f / \partial u(B(\lambda_h(t), t))v'_*(\lambda_h(t))$$

$$\to \partial f / \partial x(B(\tau, t))v'_*(\tau)t + \partial f / \partial u(B(\tau, t))v'_*(\tau)$$

as $h \to \tau$ uniformly for all $t \in [0, 1]$. This implies that

$$\psi \in C^1([0, \infty); R^1).$$

The lemma is proved.

The next auxiliary result plays a crucial role in the proof of Theorem 5.1.1.

LEMMA 5.2.4 *For each $\epsilon > 0$ there exists a number $q \geq 8$ such that the following property holds:*

If $h_1, h_2 \in H(f)$, then there exists an a.c. function $v : [0, q] \to R^n$ such that

$$v(0) = h_1, \ v(q) = h_2, \tag{2.9}$$

$$\sigma^f(0, q, v) \leq \epsilon. \tag{2.10}$$

Proof. Define a function $\phi : [0, \infty) \to R^1$ by

$$\phi(t) = \psi(t) - \mu(f) - \pi^f(v_*(0)) + \pi^f(v_*(t)), \ t \in [0, \infty). \tag{2.11}$$

It follows from (2.11), (2.7), Lemmas 5.2.2 and 5.2.3, (2.5) and the representation formula (see (1.5), (1.6)) that

$$\phi \in C^1([0, \infty); R^1), \ \phi(1) = 0, \ \phi(t) \geq 0, \ t \in [0, \infty). \tag{2.12}$$

It follows from Theorem 3.4.1 that there exists a sequence of positive numbers $\{\delta_i\}_{i=0}^{\infty}$ such that

$$\delta_0 \in (0, 8^{-1}\epsilon), \ \delta_{i+1} < \delta_i, \ i = 0, 1, \ldots \tag{2.13}$$

and that the following property holds:

If an integer $i \geq 0$ and if $x_1, x_2, y_1, y_2 \in H(f)$ satisfy $|x_j - y_j| \leq \delta_i$, $j = 1, 2$, then

$$|U^f(0, 1, x_1, x_2) - U^f(0, 1, y_1, y_2)| \leq 2^{-i-8}\epsilon,$$

$$|\pi^f(x_j) - \pi^f(y_j)| \leq 2^{-i-8}\epsilon, \ j = 1, 2. \tag{2.14}$$

By the definition of v_* and Theorem 3.3.1 there exists an integer $L \geq 10$ such that

$$\text{dist}(H(f), \{v_*(t) : t \in [T, T + L]\}) \leq 4^{-1}\delta_0 \qquad (2.15)$$

for all $T \in [0, \infty)$.

Since $\Omega(v_*) = H(f)$ and $v_*(0) \in H(f)$ there exists a sequence of numbers $\{T_p\}_{p=1}^{\infty}$ such that

$$T_p \geq 2L + 8, \ |v_*(0) - v_*(T_p)| \leq 2^{-8}\delta_p, \ p = 1, 2, \ldots, \qquad (2.16)$$

Fix a positive number ϵ_0 for which

$$\epsilon_0 < 2^{-8}L^{-1}\epsilon. \qquad (2.17)$$

It follows from (2.12) that there exists a positive number Δ such that

$$\Delta < 2^{-8}, \ |\phi'(t)| \leq 2^{-1}\epsilon_0, \ t \in [1 - \Delta, 1 + \Delta]. \qquad (2.18)$$

Choose an integer

$$N > 64(L + 1)\Delta^{-1} \qquad (2.19)$$

and set

$$q = \sum_{i=1}^{N} T_i + 8L + 8. \qquad (2.20)$$

Let $h_1, h_2 \in H(f)$. We will construct an a.c. function $v : [0, q] \to R^n$ satisfying (2.9) and (2.10). It follows from (2.15), which holds for each $T \in [0, \infty)$, that there exist numbers t_1, t_2 such that

$$t_1 \in [0, L], \ t_2 \in [8, L + 8], \ |h_j - v_*(t_j)| \leq 4^{-1}\delta_0, \ j = 1, 2. \qquad (2.21)$$

Set

$$\Delta_0 = (N - 1)^{-1}(8L + 8 - (t_2 - t_1)). \qquad (2.22)$$

(2.22), (2.21), (2.19) and (2.18) imply that

$$\Delta_0 \in (0, \Delta). \qquad (2.23)$$

By Proposition 1.3.5, (2.16) and (2.21) there exists an a.c. function $w_0 : [0, T_1 - t_1] \to R^n$ such that

$$w_0(0) = h_1, \ w_0(t) = v_*(t_1 + t), \ t \in [1, T_1 - t_1 - 1], \ w_0(T_1 - t_1) = v_*(0), \qquad (2.24)$$

$$\Phi^f(\tau, \tau + 1, w_0) = 0, \ \tau = 0, T_1 - t_1 - 1. \qquad (2.25)$$

It follows from (2.5), (2.16), (2.24), (2.25) and the definitions of $\{\delta_j\}_{j=0}^{\infty}$ that

$$\sigma(0, T_1 - t_0, w_0) = \sigma(0, 1, w_0) + \sigma(T_1 - t_1 - 1, T_1 - t_1, w_0) \qquad (2.26)$$

$$= U^f(0, 1, h_1, v_*(t_1 + 1)) - \pi^f(h_1) + \pi^f(v_*(t_1 + 1)) - \mu(f)$$

$$+ U^f(0, 1, v_*(T_1 - 1), v_*(0)) - \pi^f(v_*(T_1 - 1)) + \pi^f(v_*(0)) - \mu(f)$$

$$\leq 4 \cdot 2^{-8}\epsilon + U^f(0, 1, v_*(t_1), v_*(t_1 + 1)) - \pi^f(v_*(t_1))$$

$$+ \pi^f(v_*(t_1 + 1)) - \mu(f) + U^f(0, 1, v_*(T_1 - 1), v_*(T_1))$$

$$- \pi^f(v_*(T_1 - 1)) + \pi^f(v_*(T_1)) - \mu(f) \leq 2^{-6}\epsilon.$$

Let $k \geq 1$ be an integer. It follows from Proposition 1.3.5, (2.7), (2.16), (2.18) and (2.23) that there exists an a.c. function $w_k : [0, \Delta_0 + T_{k+1}] \to R^n$ such that

$$w_k(t) = v_*(t) + P_{1-\Delta_0}(t), \ t \in [0, 1],$$

$$w_k(t) = v_*(t - \Delta_0), \ t \in [1, \Delta_0 + T_{k+1} - 1],$$

$$w_k(\Delta_0 + T_{k+1}) = v_*(0),$$

$$\Phi^f(\Delta_0 + T_{k+1} - 1, \Delta_0 + T_{k+1}, w_k) = 0. \tag{2.27}$$

Equations (2.27) and (2.7) imply that

$$w_k(0) = v_*(0). \tag{2.28}$$

We will estimate $\sigma(0, T_{k+1} + \Delta_0, w_k)$. By (2.5), (2.24) and (2.27),

$$\sigma(0, T_{k+1} + \Delta_0, w_k) = \sigma(0, 1, w_k) + \sigma(T_{k+1} + \Delta_0 - 1, T_{k+1} + \Delta_0, w_k). \tag{2.29}$$

It follows from (2.7), (2.11), (2.24) and (2.27) that

$$\sigma(0, 1, w_k) = \phi(1 - \Delta_0). \tag{2.30}$$

By (2.30), (2.23), (2.18) and (2.12)

$$\sigma(0, 1, w_k) \leq 2^{-1}\Delta_0\epsilon_0. \tag{2.31}$$

By (2.27), (2.24), (2.16), the definition of v_*, (2.5) and the definition of $\{\delta_i\}_{i=0}^{\infty}$ (see (2.13) and (2.14)),

$$\sigma(T_{k+1} + \Delta_0 - 1, T_{k+1} + \Delta_0, w_k)$$

$$= U^f(0, 1, v_*(T_{k+1} - 1), v_*(0)) - \pi^f(v_*(T_{k+1} - 1)) + \pi^f(v_*(0)) - \mu(f)$$

$$\leq U^f(0, 1, v_*(T_{k+1} - 1), v_*(T_{k+1})) - \pi^f(v_*(T_{k+1} - 1))$$

$$+ \pi^f(v_*(T_{k+1})) - \mu(f) + 2 \cdot 2^{-k-9}\epsilon = 2^{-k-8}\epsilon. \tag{2.32}$$

Combining (2.29), (2.31) and (2.32) we obtain that

$$\sigma(0, T_{k+1} + \Delta_0, w_k) \leq 2^{-1}\epsilon_0\Delta_0 + 2^{-k-8}\epsilon. \tag{2.33}$$

It follows from Proposition 1.3.5 that there exists an a.c. function $u_0 : [0, t_2] \to R^n$ such that

$$u_0(t) = v_*(t), \ t \in [0, t_2 - 1], \ u_0(t_2) = h_2, \tag{2.34}$$

$$\Phi^f(t_2 - 1, t_2, u_0) = 0.$$

By (2.5), (2.21), (2.24), (2.34), the definition of $\{\delta_i\}_{i=0}^\infty$ (see (2.13), (2.14)) and the definition of v_*,

$$\sigma(0, t_2, u_0) = \sigma(t_2 - 1, t_2, u_0)$$
$$= U^f(0, 1, v_*(t_2 - 1), h_2) - \pi^f(v_*(t_2 - 1)) + \pi^f(h_2) - \mu(f)$$
$$\leq U^f(0, 1, v_*(t_2 - 1), v_*(t_2)) - \pi^f(v_*(t_2 - 1))$$
$$+ \pi^f(v_*(t_2)) - \mu(f) + 2^{-7}\epsilon \leq 2^{-7}. \tag{2.35}$$

It follows from (2.20) and (2.22) that

$$T_1 - t_1 + \sum_{k=1}^{N-1}(\Delta_0 + T_{k+1}) + t_2 = q. \tag{2.36}$$

By (2.36), (2.25), (2.27), (2.28) and (2.34) there exists an a.c. function $v : [0, q] \to R^n$ such that

$$v(t) = w_0(t), \ t \in [0, T_1 - t_1],$$

$$v(t) = w_k\left(t - \left(\sum_{i=1}^k T_i + (k-1)\Delta_0 - t_1\right)\right), \tag{2.37}$$

$$t \in \left[\sum_{i=1}^k T_i + (k-1)\Delta_0 - t_1, \sum_{i=1}^{k+1} T_i + k\Delta_0 - t_1\right], \ k = 1, \ldots, N-1,$$

$$v(t) = u_0\left(t - \left(\sum_{i=1}^N T_i + (N-1)\Delta_0 - t_1\right)\right),$$

$$t \in \left[\sum_{i=1}^N T_i + (N-1)\Delta_0 - t_1, q\right].$$

(2.37), (2.25), (2.36) and (2.34) imply that

$$v(0) = h_1, \ v(q) = h_2.$$

By (2.17), (2.21), (2.22), (2.24), (2.26), (2.33), (2.35) and (2.37),

$$\sigma^f(0, q, v) = \sigma(0, T_1 - t_1, w_0) + \sum_{k=1}^{N-1} \sigma(0, T_{k+1} + \Delta_0, w_k) + \sigma(0, t_2, u_0)$$

$$\leq 2^{-6}\epsilon + \sum_{k=1}^{N-1}(2^{-1}\epsilon_0\Delta_0 + 2^{-k-8}\epsilon) + 2^{-7}\epsilon \leq 2^{-5}\epsilon + 2^{-1}(N-1)\epsilon_0\Delta_0$$

$$\leq 2^{-5}\epsilon + 2^{-1}(9L + 16)\epsilon_0 \leq 2^{-1}\epsilon.$$

This completes the proof of the lemma.

Proof of Theorem 5.1.1. Let $\epsilon, K > 0$. We may assume that

$$\epsilon < 1, \ K > \sup\{|h| : h \in H(f)\} + 4.$$

It follows from Theorem 1.2.3 that there exist a number $M > K$ and a neighborhood $\mathcal{U}_1$ of f in $\mathcal{A}$ such that the following property holds:
  If $g \in \mathcal{U}_1$, $T_1 \geq 0$, $T_2 \geq T_1 + 1$ and if an a.c. function $v : [T_1, T_2] \to R^n$ satisfies

$$|v(T_i)| \leq 2K + 4, \ i = 1, 2, \ \Phi^g(T_1, T_2, v) \leq 2, \tag{2.38}$$

then

$$|v(t)| \leq M, \ t \in [T_1, T_2]. \tag{2.39}$$

Theorem 3.4.1 implies that there exists

$$\delta_1 \in (0, 8^{-1}\epsilon) \tag{2.40}$$

such that the following property holds:
  If $x_1, x_2, y_1, y_2 \in R^n$ satisfy

$$|x_i|, |y_i| \leq 2M + 4 + 2\sup\{|h| : h \in H(f)\}, \ |x_i - y_i| \leq 4\delta_1, \ i = 1, 2, \tag{2.41}$$

then

$$|U^f(0, 1, x_1, x_2) - U^f(0, 1, y_1, y_2)| \leq 2^{-8}\epsilon,$$

$$|\pi^f(x_i) - \pi^f(y_i)| \leq 2^{-8}\epsilon, \ i = 1, 2. \tag{2.42}$$

By Theorem 3.3.1 there exists an integer $l \geq 1$ such that for each (f)-good function $v : [0, \infty) \to R^n$ the inequality

$$\text{dist}(H(f), \{v(t) : t \in [T, T + l]\}) \leq \epsilon \tag{2.43}$$

is valid for all large T. By Lemma 3.8.4 there exists a positive number

$$\delta_0 < 2^{-1}\delta_1 \tag{2.44}$$

such that the following property holds:
  If $T \in [l, \infty)$ and if an a.c. function $v : [0, T] \to R^n$ satisfies

$$d(v(0), H(f)) \leq \delta_0, \ d(v(T), H(f)) \leq \delta_0, \tag{2.45}$$

$$\sigma^f(0, T, v) \le \delta_0, \tag{2.46}$$

then for every $S \in [0, T - l]$,

$$\text{dist}(H(f), \{v(t) : t \in [S, S + l]\}) \le \epsilon. \tag{2.47}$$

By Theorem 3.4.1 there exists a positive number

$$\delta < 32^{-1}\delta_0 \tag{2.48}$$

such that the following property holds:

If $x_1, x_2, y_1, y_2 \in R^n$ satisfy

$$|x_i|, |y_i| \le 2M + 4 + 2\sup\{|h| : h \in H(f)\}, \ |x_i - y_i| \le 4\delta, \ i = 1, 2, \tag{2.49}$$

then

$$|U^f(0, 1, x_1, x_2) - U^f(0, 1, y_1, y_2)| \le 2^{-8}\delta_0,$$
$$|\pi^f(x_i) - \pi^f(y_i)| \le 2^{-8}\delta_0, \ i = 1, 2. \tag{2.50}$$

By Theorem 3.3.1 there exists an integer $L \ge 1$ such that for each (f)-good function $v : [0, \infty) \to R^n$,

$$\text{dist}(H(f), \{v(t) : t \in [T, T + L]\}) \le 8^{-1}\delta \tag{2.51}$$

for all large T.

It follows from Lemma 3.8.2 that there exist a neighborhood $\mathcal{U}_2$ of f in $\mathcal{A}$ and a natural number $N \ge 10$ such that the following property holds:

If $g \in \mathcal{U}_2$, $S \in [0, \infty)$ and if an a.c. function $x : [S, S + NL] \to R^n$ satisfies

$$|x(S)|, \ |x(S + NL)| \le 2M + 2,$$
$$\Phi^g(S, S + NL, x) \le 4, \tag{2.52}$$

then there exists an integer $i_0 \in [0, N - 8]$ such that the inequality

$$\text{dist}(H(f), \{x(t) : t \in [T, T + L]\}) \le \delta \tag{2.53}$$

is true for all $T \in [S + i_0 L, S + (i_0 + 7)L]$.

By Lemma 5.2.4 there exists a number $q \ge 8$ such that for each $h_1, h_2 \in H(f)$ there exists an a.c. function $v : [0, q] \to R^n$ which satisfies

$$v(0) = h_1, \ v(q) = h_2, \ \sigma^f(0, q, v) \le 8^{-1}\delta. \tag{2.54}$$

It follows from Proposition 1.3.8 that there exists a neighborhood $\mathcal{U}_3$ of f in $\mathcal{A}$ such that the following property holds:

If $g \in \mathcal{U}_3$, $T_1 \geq 0$, $T_2 \in [T_1 + 8^{-1}, T_1 + 6N(q + l + L)]$ and if an a.c. function $x : [T_1, T_2] \to R^n$ satisfies

$$\max\{I^f(T_1, T_2, x), I^g(T_1, T_2, x)\}$$

$$\leq 4 + 2\sup\{|\pi^f(h)| : \ h \in R^n, \ |h| \leq \sup\{|z| : z \in H(f)\} + 4\}$$

$$+6|\mu(f)|N(q + l + L), \tag{2.55}$$

then

$$|I^f(T_1, T_2, x) - I^g(T_1, T_2, x)| \leq 4^{-1}\delta. \tag{2.56}$$

By Proposition 1.3.9 there exists a neighborhood $\mathcal{U}_4$ of f in $\mathcal{A}$ such that

$$|U^f(0, 1, x_1, x_2) - U^g(0, 1, x_1, x_2)| \leq 2^{-8}\delta \tag{2.57}$$

for each $g \in \mathcal{U}_4$ and each $x_1, x_2 \in R^n$ satisfying

$$|x_1|, |x_2| \leq 2M + 4 + 2\sup\{|z| : \ z \in H(f)\}.$$

Put

$$l_0 = 2l + 2q + 2NL + 6, \tag{2.58}$$

$$\mathcal{U} = \cap_{i=1}^4 \mathcal{U}_i. \tag{2.59}$$

Assume that $g \in \mathcal{U}$, $T \geq 2l_0$ and an a.c. function $v : [0, T] \to R^n$ satisfies

$$|v(0)|, |v(T)| \leq K, \ \Phi^g(0, T, v) \leq \delta. \tag{2.60}$$

It follows from the definition of $\mathcal{U}_1$ (see (2.38), (2.39)) and (2.60) that

$$|v(t)| \leq M, \ t \in [0, T]. \tag{2.61}$$

Assume that there exist numbers $S_1, S_2 \in [0, T]$ such that

$$d(v(S_i), H(f)) \leq \delta, \ i = 1, 2, \ S_2 - S_1 \in [1 + l + q, 5N(L + l + q)]. \tag{2.62}$$

We will show that for each $\tau \in [S_1, S_2 - l]$,

$$\mathrm{dist}(H(f), \{v(t) : \ t \in [\tau, \tau + l]\}) \leq \epsilon. \tag{2.63}$$

Let us assume the converse. Then there exists a number τ such that

$$\tau \in [S_1, S_2 - l], \ \mathrm{dist}(H(f), \{v(t) : \ t \in [\tau, \tau + l]\}) > \epsilon. \tag{2.64}$$

By (2.48), (2.62), (2.64) and the definition of δ_0 (see (2.44)-(2.47)),

$$\sigma^f(S_1, S_2, v) > \delta_0. \tag{2.65}$$

We show that

$$I^g(S_1, S_2, v) - (S_2 - S_1)\mu(f) - \pi^f(v(S_1)) + \pi^f(v(S_2)) > \delta_0/2. \quad (2.66)$$

Let us assume the converse. Then (2.62) implies that

$$I^g(S_1, S_2, v)$$

$$\le 2 \sup\{|\pi^f(z)| : z \in R^n, \ d(z, H(f)) \le 1\}$$
$$+ |\mu(f)|(S_2 - S_1) + 1.$$

By this inequality, (2.62) and the choice of $\mathcal{U}_3$ (see (2.55), (2.56)),

$$|I^f(S_1, S_2, v) - I^g(S_1, S_2, v)| \le 4^{-1}\delta.$$

Combined with (2.65) this inequality implies (2.66). The contradiction we have reached proves (2.66).

By (2.62) there exist $h_1, h_2 \in H(f)$ such that

$$|v(S_i) - h_i| \le \delta, \ i = 1, 2. \quad (2.67)$$

It follows from Lemma 5.2.1 that there exists an (f)-good function

$$w_0 : [0, \infty) \to H(f) \text{ such that } w_0(0) = h_1 \quad (2.68)$$

and

$$\sigma^f(t_1, t_2, w_0) = 0 \quad (2.69)$$

for each $t_1 \ge 0$, $t_2 > t_1$. By the choice of q (see (2.54)), (2.62) and (2.68) there exists an a.c. function $w_1 : [0, q] \to R^n$ such that

$$w_1(0) = w_0(S_2 - S_1 - q), \ w_1(q) = h_2,$$

$$\sigma^f(0, q, w_1) \le 8^{-1}\delta. \quad (2.70)$$

It follows from Proposition 1.3.5, (2.62), (2.68) and (2.70) that there exists an a.c. function $u : [0, T] \to R^n$ such that

$$u(t) = v(t), \ t \in [0, S_1] \cup [S_2, T],$$

$$u(t) = w_0(t - S_1), \ t \in [S_1 + 1, S_2 - q],$$
$$u(t) = w_1(t - (S_2 - q)), \ t \in [S_2 - q, S_2 - 1],$$
$$I^g(r, r + 1, u) = U^g(0, 1, u(r), u(r + 1)), \ r = S_1, S_2 - 1. \quad (2.71)$$

For each a.c. function $y : [a, b] \to R^n$ where $a \ge 0$, $b > a$ and each $r_1, r_2 \in [a, b]$ satisfying $r_1 \le r_2$ we set

$$\sigma^g(r_1, r_2, y) = I^g(r_1, r_2, y) - (r_2 - r_1)\mu(f) - \pi^f(y(r_1)) + \pi^f(y(r_2)). \quad (2.72)$$

(2.60), (2.71) and (2.72) imply that

$$\delta \geq I^g(0, T, v) - I^g(0, T, u)$$

$$= \sigma^g(0, T, v) - \sigma^g(0, T, u) = \sigma^g(S_1, S_2, v) - \sigma^g(S_1, S_2, u). \qquad (2.73)$$

It follows from (2.70), (2.68) and the choice of M (see (2.38), (2.39)), that

$$|w_1(t)| \leq M, \ t \in [0, q]. \qquad (2.74)$$

By (2.70) there exists an a.c. function $\tilde{w} : [S_1, S_2] \to R^n$ such that

$$\tilde{w}(t) = w_0(t - S_1), \ t \in [S_1, S_2 - q], \qquad (2.75)$$

$$\tilde{w}(t) = w_1(t - (S_2 - q)), \ t \in [S_2 - q, S_2].$$

It follows from (2.73), (2.72), (2.66), (2.71) and (2.75) that

$$\delta \geq 2^{-1}\delta_0 - \sigma^g(S_1, S_2, u) = 2^{-1}\delta_0 - \sigma^g(S_1, S_2, \tilde{w}) \qquad (2.76)$$

$$+[\sigma^g(S_1, S_1 + 1, \tilde{w}) - \sigma^g(S_1, S_1 + 1, u)]$$

$$+[\sigma^g(S_2 - 1, S_2, \tilde{w}) - \sigma^g(S_2 - 1, S_2, u)].$$

We will estimate $\sigma^g(S_1, S_2, \tilde{w})$ and $\sigma^g(h, h + 1, \tilde{w}) - \sigma^g(h, h + 1, u)$, $h = S_1, S_2 - 1$.

Let $h \in \{S_1, S_2 - 1\}$. It follows from (2.70), (2.75), (2.71), (2.74), (2.68), (2.61), (2.62) and (2.67) that

$$|\tilde{w}(h)|, |\tilde{w}(h + 1)|, |u(h)|, |u(h + 1)| \leq M + \sup\{|z| : \ z \in H(f)\},$$

$$|\tilde{w}(h) - u(h)|, |\tilde{w}(h + 1) - u(h + 1)| \leq \delta.$$

By these inequalities, (2.71), (2.72), the choice of $\mathcal{U}_4$ (see (2.57)) and δ (see (2.49), (2.50), (2.48)),

$$\sigma^g(h, h + 1, \tilde{w}) - \sigma^g(h, h + 1, u)$$

$$\geq U^g(0, 1, \tilde{w}(h), \tilde{w}(h + 1)) - \pi^f(\tilde{w}(h)) + \pi^f(\tilde{w}(h + 1))$$

$$-[U^g(0, 1, u(h), u(h + 1)) - \pi^f(u(h)) + \pi^f(u(h + 1))]$$

$$\geq U^f(0, 1, \tilde{w}(h), \tilde{w}(h + 1)) - \pi^f(\tilde{w}(h)) + \pi^f(\tilde{w}(h + 1))$$

$$-[U^f(0, 1, u(h), u(h + 1)) - \pi^f(u(h)) + \pi^f(u(h + 1))] - 2^{-7}\delta$$

$$\geq -2^{-6}\delta_0, \ h \in \{S_1, S_2 - 1\}. \qquad (2.77)$$

We will estimate $\sigma^g(S_1, S_2, \tilde{w})$. It follows from (2.69), (2.70) and (2.75) that

$$I^f(S_1, S_2, \tilde{w}) - (S_2 - S_1)\mu(f) - \pi^f(\tilde{w}(S_1)) + \pi^f(\tilde{w}(S_2)) \leq 8^{-1}\delta. \quad (2.78)$$

By this inequality, (2.62), (2.75), (2.68), (2.70) and the definition of $\mathcal{U}_3$ (see (2.55), (2.56)),

$$|I^f(S_1, S_2, \tilde{w}) - I^g(S_1, S_2, \tilde{w})| \le 4^{-1}\delta.$$

Combined with (2.78) and (2.72) this inequality implies that

$$\sigma^g(S_1, S_2, \tilde{w}) \le 3 \cdot 8^{-1}\epsilon.$$

By this inequality, (2.76) and (2.77),

$$\delta \ge 2^{-1}\delta_0 - 3 \cdot 8^{-1}\delta - 2^{-5}\delta_0.$$

This is contradictory to (2.48). The obtained contradiction proves that (2.63) holds for each $\tau \in [S_1, S_2 - l]$. Therefore we have shown that the following property holds:

Property D. For each $S_1, S_2 \in [0, T]$ which satisfy (2.62) relation (2.63) holds for each $\tau \in [S_1, S_2 - l]$.

It follows from (2.60), (2.61) and the definition of $\mathcal{U}_2$, N (see (2.52), (2.53)) that for each $r_0 \in [0, T - (1 + l + q + L(N + 2))]$ there exists a number r_1 such that

$$r_1 - r_0 \in [1 + l + q + 2L, 1 + l + q + L(N + 2)], \ d(v(r_1), H(f)) \le \delta.$$

This implies that there exists a finite sequence of numbers $\{S_i\}_{i=1}^Q \subset [0, T]$ such that

$$S_0 = 0, \ S_{i+1} - S_i \in [1+l+q+2L, 1+l+q+L(N+2)], \ i = 0, \dots, Q-1,$$

$$T - S_Q \le 1 + l + q + L(N + 2), \ d(v(S_i), H(f)) \le \delta, \ i = 1, \dots, Q.$$

The assertion of the theorem follows from these relations and Property D.

5.3.   Proof of Theorem 5.1.2

LEMMA 5.3.1 *Let an integrand* $f \in \mathcal{N}$ *have the asymptotic turnpike property and let* ϵ *be a positive number. Then there exists a neighborhood* $\mathcal{U}$ *of* f *in* $\mathcal{A}$ *such that*

$$dist(\Omega(v), H(f)) \le \epsilon$$

for each $g \in \mathcal{U}$ *and each* (g)-*good function* $v : [0, \infty) \to R^n$.

Proof. It follows from Theorem 1.2.2 that there exist a positive number K and a neighborhood $\mathcal{U}_1$ of f in $\mathcal{A}$ such that

$$\limsup_{t\to\infty} |v(t)| < K$$

for each $g \in \mathcal{U}_1$ and each (g)-good function $v : [0, \infty) \to R^n$.

By Theorem 5.1.1 there exist a neighborhood $\mathcal{U}$ of f in $\mathcal{A}$ which satisfies $\mathcal{U} \subset \mathcal{U}_1$ and numbers $l_0 > l > 0$, $\delta > 0$ such that the following property holds:

If $g \in \mathcal{U}$, $T \geq 2l_0$ and if an a.c. function $v : [0, T] \to R^n$ satisfies

$$|v(0)|, |v(T)| \leq K, \ I^g(0, T, v) \leq U^g(0, T, v(0), v(T)) + \delta,$$

then for each $\tau \in [l_0, T - l_0]$,

$$\mathrm{dist}(H(f), \{v(t) : \ t \in [\tau, \tau + l]\}) \leq \epsilon.$$

Assume that $g \in \mathcal{U}$ and $v : [0, \infty) \to R^n$ is a (g)-good function. By Proposition 4.1.1 and the choice of $\mathcal{U}$, $\mathcal{U}_1$, K there exists a number $T_0 \geq 0$ such that

$$|v(t)| \leq K, \ t \in [T_0, \infty),$$

$$I^g(t_1, t_2, v) \leq U^g(t_1, t_2, v(t_1), v(t_2)) + \delta \text{ for each } t_1 \geq T_0, \ t_2 > t_1.$$

It follows from these relations and the definition of $\mathcal{U}$, l_0, l, δ that

$$\mathrm{dist}(H(f), \Omega(v)) \leq \epsilon.$$

The lemma is proved.

Construction of the set $\mathcal{F}_q$. Suppose that q is a nonnegative integer. By Lemmas 3.7.1, 3.7.3 and 3.7.4 and Proposition 3.7.1 there exists a set $E_q \subset \mathcal{N}_q$ which is an everywhere dense subset of $\bar{\mathcal{N}}_q$ and such that each integrand $f \in E_q$ has the asymptotic turnpike property with the turnpike $H(f) \subset R^n$. Therefore $\Omega(v) = H(f)$ for each $f \in E_q$ and each (f)-good function $v : [0, \infty) \to R^n$.

It follows from Theorem 5.1.1 and Lemma 5.3.1 that for each $f \in E_q$ and each integer $p \geq 1$ there exist an open neighborhood $\mathcal{U}(f, p)$ of f in $\mathcal{A}$ and numbers $M(f, p) > p$, $l_0(f, p) > l(f, p) > 0$, $\delta(f, p) \in (0, p^{-1})$ such that:

$$\mathrm{dist}(H(f), \Omega(v)) \leq 4^{-1}\delta(f, p)$$

for each $g \in \mathcal{U}(f, p)$ and each (g)-good function $v : [0, \infty) \to R^n$;

if $g \in \mathcal{U}(f, p)$, $T \geq 2l_0(f, p)$ and if an a.c. function $v : [0, T] \to R^n$ satisfies

$$|v(0)|, |v(T)| \leq p, \ I^g(0, T, v) \leq U^g(0, T, v(0), v(T)) + \delta(f, p), \quad (3.1)$$

then
$$|v(t)| \leq M(f, p), \ t \in [0, T]$$

and the following properties hold:

(i) for each $\tau \in [l_0(f, p), T - l_0(f, p)]$,

$$\text{dist}(H(f), \{v(t) : t \in [\tau, \tau + l(f, p)]\}) \leq p^{-1}; \tag{3.2}$$

(ii) if $d(v(0), H(f)) \leq \delta(f, p)$, then (3.2) holds for each $\tau \in [0, T - l_0(f, p)]$;

(iii) if
$$d(v(T), H(f)) \leq \delta(f, p),$$

then (3.2) holds for each $\tau \in [l_0(f, p), T - l(f, p)]$.

We define

$$\mathcal{F}_q = [\cap_{p=1}^{\infty} \cup \{\mathcal{U}(f, p) : f \in E_q\}] \cap \bar{\mathcal{N}}_q. \tag{3.3}$$

Clearly $\mathcal{F}_q$ is a countable intersection of open everywhere dense subsets of $\bar{\mathcal{N}}_q$.

Assume that $f \in \mathcal{F}_q$, $\epsilon, K > 0$. Fix a natural number p such that

$$p > 2K + 4 + 8\epsilon^{-1}. \tag{3.4}$$

There exists $G \in E_q$ such that

$$f \in \mathcal{U}(G, p). \tag{3.5}$$

It follows from (3.4), (3.5) and the definition of $\mathcal{U}(G, p)$, $\delta(G, p)$ that for each (f)-good function $v : [0, \infty) \to R^n$,

$$\text{dist}(H(G), \Omega(v)) \leq 4^{-1}\delta(G, p) < (4p)^{-1} < 8^{-1}\epsilon. \tag{3.6}$$

This implies that for each (f)-good function $v_i : [0, \infty) \to R^n$, $i = 1, 2$,

$$\text{dist}(\Omega(v_1), \Omega(v_2)) \leq \epsilon.$$

Since ϵ is any positive number we conclude that f has the asymptotic turnpike property and there exists a compact set $H(f) \subset R^n$ such that $\Omega(w) = H(f)$ for each (f)-good function $w : [0, \infty) \to R^n$. It follows from (3.6) that

$$\text{dist}(H(G), H(f)) \leq 4^{-1}\delta(G, p). \tag{3.7}$$

Set
$$\mathcal{U} = \mathcal{U}(G, p), \ M = M(G, p),$$
$$l_0 = l_0(G, p), \ l = l(G, p), \ \delta = 8^{-1}\delta(G, p). \tag{3.8}$$

Assume that $g \in \mathcal{U}$, $T \geq 2l_0$ and an a.c. function $v : [0, T] \rightarrow R^n$ satisfies

$$|v(0)|, |v(T)| \leq K, \ I^g(0, T, v) \leq U^g(0, T, v(0), v(T)) + \delta. \qquad (3.9)$$

It follows from (3.9), (3.8), (3.6), (3.4) and the definition of $\mathcal{U}(G, p)$, $M(G, p)$, $l_0(G, p)$, $l(G, p)$, $\delta(G, p)$ that

$$|v(t)| \leq M, \ t \in [0, T], \qquad (3.10)$$

and properties (i)-(iii) hold with $f = G$. Together with (3.7), (3.8) and (3.4) this implies that

$$\mathrm{dist}(H(f), \{v(t) : \ t \in [\tau, \tau + l]\}) \leq \epsilon \qquad (3.11)$$

for each $\tau \in [l_0, T - l_0]$; if $d(v(0), H(f)) \leq \delta$, then (3.11) holds for each $\tau \in [0, T - l_0]$. If $d(v(T), H(f)) \leq \delta$, then (3.11) holds for each $\tau \in [l_0, T - l]$. This completes the proof of the theorem.

5.4.   Examples

Fix a constant $a > 0$ and set $\psi(t) = t$, $t \in [0, \infty)$. Consider the complete metric space $\mathcal{A}$ of integrands $f : R^n \times R^n \rightarrow R^1$ defined in Section 5.1.

Example 1. Consider an integrand $f(x, u) = |x|^2 + |u|^2$, $x, u \in R^n$. It is easy to see that $f \in \mathcal{N}_q$ for each integer $q \geq 0$ if the constant a is large enough. We can show (see Section 3.12) that $\Omega(v) = \{0\}$ for every (f)-good function $v : [0, \infty) \rightarrow R^n$. Therefore Theorem 5.1.1 holds with the integrand f.

Example 2. Fix a number $q > 0$ and consider an integrand $g(x, u) = q|x|^2|x - e|^2 + |u|^2$, $x, u \in R^n$, where $e = (1, 1, \ldots, 1)$. It is easy to see that $g \in \mathcal{N}$ if the constant a is large enough. Clearly f does not have the turnpike property.

Chapter 6

LINEAR PERIODIC
CONTROL SYSTEMS

In this chapter we study the existence and asymptotic behavior of overtaking optimal trajectories for linear control systems with periodic convex integrands $f : [0, \infty) \times R^n \times R^m \to R^1$. We extend the results obtained by Artstein and Leizarowitz for tracking periodic problems with quadratic integrands [3] and establish the existence and uniqueness of optimal trajectories on an infinite horizon. The asymptotic behavior of finite time optimizers is examined.

6.1. Main results

We consider a linear control system defined by

$$x'(t) = Ax(t) + Bu(t), \ x(0) = x_0, \tag{1.1}$$

where A and B are given matrices of dimensions $n \times n$ and $n \times m$, $x(t) \in R^n$, $u(t) \in R^m$ and the admissible controls are measurable functions.

We assume that the linear system (1.1) is controllable and that the integrand f is a Borel measurable function.

The performance of the above control system is measured on any finite interval $[T_1, T_2]$ by the integral functional

$$I(T_1, T_2, x, u) = \int_{T_1}^{T_2} f(t, x(t), u(t))dt. \tag{1.2}$$

Artstein and Leizarowitz [3] analyzed the existence and structure of solutions of the linear system (1.1) with an integrand

$$f(t, x, u) = (x - \Gamma(t))'Q(x - \Gamma(t)) + u'Pu \ (t \in [0, \infty), \ x \in R^n, \ u \in R^m), \tag{1.3}$$

where P is a given positive definite symmetric matrix, Q is a positive semidefinite symmetric matrix, the pair (A, Q) is observable and Γ : $[0, \infty) \to R^n$ is a measurable function satisfying

$$\Gamma(t + T) = \Gamma(t) \ (t \in [0, \infty))$$

for some constant $T > 0$. Artstein and Leizarowitz [3] showed the existence of a unique solution for the infinite horizon tracking of the periodic trajectory Γ and established a turnpike property for finite time optimizers. Their methods are based on explicit expressions for optimal solutions to tracking on finite intervals. In this chapter we discuss an another approach which was developed in [111] to extend the results of [3] to a cost function $f : [0, \infty) \times R^n \times R^n \to R^1$ which satisfies

Assumption A

(i) $f(t + \tau, x, u) = f(t, x, u) \ (t \in [0, \infty), \ x \in R^n, \ u \in R^m)$ for some constant $\tau > 0$.

(ii) For any $t \in [0, \infty)$ the function $f(t, \cdot, \cdot) : R^n \times R^m \to R^1$ is strictly convex.

(iii) The function f is bounded on any bounded subset of $[0, \infty) \times R^n \times R^m$;

(iv) $f(t, x, u) \to \infty$ as $|x| \to \infty$ uniformly in $(t, u) \in [0, \infty) \times R^m$;

(v) $f(t, x, u)|u|^{-1} \to \infty$ as $|u| \to \infty$ uniformly in $(t, x) \in [0, \infty) \times R^n$.

Assumption A implies that f is bounded below on $[0, \infty) \times R^n \times R^m$.

We denote by $|\cdot|$ the Euclidean norm in R^n. For each $r > 0$ and each $x \in R^n$ set

$$\mathcal{B}(r) = \{y \in R^n : |y| \le r\}, \ \mathcal{B}(x, r) = \{y \in R^n : |x - y| \le r\}.$$

In this chapter we assume that the integrand f satisfies Assumption A and prove the following results.

PROPOSITION 6.1.1 *There exists a trajectory-control pair* $x^* : [0, \tau] \to R^n$, $u^* : [0, \tau] \to R^m$ *which is the unique solution of the following variational problem:*

$$minimize \ I(0, \tau, x, u) \ subject \ to \ x(0) = x(\tau).$$

We show that the periodic trajectory $\{x^*(t) : t \in [0, \infty)\}$ is a turnpike for optimal solutions of our control problem.

Put

$$\mu = \tau^{-1} I(0, \tau, x^*, u^*). \tag{1.4}$$

The scalar μ is the minimal long-run average cost growth rate.

The following results were obtained in [111].

THEOREM 6.1.1 *For any trajectory-control pair* $x : [0, \infty) \to R^n$, $u :$
$[0, \infty) \to R^m$ *either*

(i) $I(0, T, x, u) - T\mu \to \infty$ *as* $T \to \infty$

or

(ii) $\sup\{|I(0, T, x, u) - T\mu| : T > 0\} < \infty$.

Moreover, in the case relation (ii) holds, then

$$\sup\{|x(i\tau + t) - x^*(t)| : t \in [0, \tau]\} \to 0 \ as \ i \to \infty$$

over the integers.

We say that a trajectory-control pair $x : [0, \infty) \to R^n$, $u : [0, \infty) \to$
R^m is *good* if

$$\sup\{|I(0, T, x, u) - T\mu| : T > 0\} < \infty.$$

The second statement of Theorem 6.1.1 describes the asymptotic behavior of good trajectory-control pairs.

We say that a trajectory-conrol pair $\tilde{x} : [0, \infty) \to R^n$, $\tilde{u} : [0, \infty) \to R^m$
is *overtaking optimal* if

$$\limsup_{T \to \infty} [I(0, T, \tilde{x}, \tilde{u}) - I(0, T, x, u)] \leq 0$$

for each trajectory-control pair $x : [0, \infty) \to R^n$, $u : [0, \infty) \to R^m$
satisfying $x(0) = \tilde{x}(0)$.

THEOREM 6.1.2 *Let* $x_0 \in R^n$. *Then there exists an overtaking optimal*
trajectory-control pair $\tilde{x} : [0, \infty) \to R^n$, $\tilde{u} : [0, \infty) \to R^m$ *satisfying*
$\tilde{x}(0) = x_0$. *Moreover, if a trajectory-control pair* $x : [0, \infty) \to R^n$,
$u : [0, \infty) \to R^m$ *satisfies* $x(0) = x_0$, *then there are a time* T_0 *and* $\epsilon > 0$
such that

$$I(0, T, x, u) \geq I(0, T, \tilde{x}, \tilde{u}) + \epsilon$$

for all $T \geq T_0$.

Theorem 6.1.3 describes the limit behavior of overtaking optimal trajectories.

THEOREM 6.1.3 *Let* M, ϵ *be positive numbers. Then there exists a nat-*
ural number N *such that for any overtaking optimal trajectory-control*
pair $x : [0, \infty) \to R^n$, $u : [0, \infty) \to R^m$ *which satisfies* $|x(0)| \leq M$ *the*
relation

$$\sup\{|x(i\tau + t) - x^*(t)| : t \in [0, \tau]\} \leq \epsilon \qquad (1.5)$$

holds for all integers $i \geq N$. *Moreover, there exists a positive number* δ
such that for any overtaking optimal trajectory-control pair $x : [0, \infty) \to$

R^n, $u : [0, \infty) \to R^m$ *satisfying* $|x(0) - x^*(0)| \leq \delta$, *the relation (1.5)* *holds for all integers* $i \geq 0$.

For each $z \in R^n$ and $T > 0$ we set

$$\Delta(z, T) = \inf\{I(0, T, x, u) : x : [0, T] \to R^n, u : [0, T] \to R^m$$

is a trajectory-control pair satisfying $x(0) = z\}$. (1.6)

We will see that $-\infty < \Delta(z, T) < \infty$.

Theorem 6.1.4 establishes the turnpike property for optimal trajectories with the turnpike $x^*(\cdot)$.

THEOREM 6.1.4 *Let* $M, \epsilon > 0$. *Then there exists an integer* $N \geq 1$ *and a positive number* δ *such that the following property holds:*

For each $T > 2N\tau$ *and each trajectory-control pair* $x : [0, T] \to R^n$, $u : [0, T] \to R^m$ *which satisfies*

$$|x(0)| \leq M, (1.7)$$

$$I(0, T, x, u) \leq \Delta(x(0), T) + \delta, (1.8)$$

the inequality (1.5) holds for all integers $i \in [N, \tau^{-1}T - N]$. *Moreover if* $|x(0) - x^*(0)| \leq \delta$, *then the inequality (1.5) holds for all integers* $i \in [0, \tau^{-1}T - N]$.

The chapter is organized as follows. Section 6.2 contains auxiliary results. In Section 6.3 we discuss discrete-time optimal control problems related to the continuous-time optimal control problems. Theorem 6.1.1 is proved in Section 6.4. Section 6.5 contains the proof of Theorem 6.1.2. Section 6.6 contains the proof of Theorem 6.1.3 while Theorem 6.1.4 is proved in Section 6.7.

6.2.   Preliminary results

We have the following result [41].

PROPOSITION 6.2.1 *For every* $\tilde{y}, \tilde{z} \in R^n$ *and every* $T > 0$ *there exists a solution* $x(\cdot), y(\cdot)$ *of the following system:*

$$x' = Ax + BB^t y, \quad y' = x - A^t y$$

with the boundary conditions $x(0) = \tilde{y}$, $x(T) = \tilde{z}$ *(where* B^t *denotes the transpose of* B*).*

Assumption A and Proposition 6.2.1 imply that

$$-\infty < \Delta(z, T) < +\infty \tag{2.1}$$

for each $z \in R^n$ and $T > 0$.

For each $y, z \in R^n$ define

$$v(y, z) = \min I(0, \tau, x, u) \tag{2.2}$$

$$\text{subject to } x(0) = y, \ x(\tau) = z; \tag{2.3}$$

here $x(\cdot)$ is the response to $u(\cdot)$.

It follows from Proposition 6.2.1 and Assumption A that the function v is convex and

$$-\infty < v(y, z) < \infty \text{ for each } y, z \in R^n. \tag{2.4}$$

PROPOSITION 6.2.2 *Let* $M_1 > 0$ *and* $0 < \tau_0 < \tau_1$. *Then there exists a positive number* M_2 *such that the following property holds:*

If $T \in [\tau_0, \tau_1]$ *and if an a.c. trajectory-control pair* $x : [0, T] \to R^n$, $u : [0, T] \to R^m$ *satisfies*

$$I(0, T, x, u) \le M_1, \tag{2.5}$$

then

$$|x(t)| \in \mathcal{B}(M_2), \ t \in [0, T]. \tag{2.6}$$

Proof. We may assume without loss of generality that the function f is nonnegative. Fix a positive number

$$\delta < \min\{8^{-1}\tau_0, (2\|A\| + 2)^{-1}\}. \tag{2.7}$$

It follows from Assumption A that there exist a number $c_0 > 1$ such that

$$f(t, x, u) \ge 8|u|(\|B\| + 1) \text{ for each } (t, x, u) \in [0, \infty) \times R^n \times R^m$$

$$\text{satisfying } |u| \ge c_0 \tag{2.8}$$

and a positive number h_0 such that

$$f(t, x, u) \ge 4M_1\delta^{-1} \text{ for each } (t, x, u) \in [0, \infty) \times R^n \times R^m$$

$$\text{satisfying } |x| \ge h_0. \tag{2.9}$$

Fix a number

$$M_2 > 2 + 2h_0 + 2\|B\|\delta c_0 + 2M_1. \tag{2.10}$$

Let $T \in [\tau_0, \tau_1]$ and $x : [0, T] \to R^n$, $u : [0, T] \to R^m$ be a trajectory-control pair which satisfies (2.5). We show that (2.6) is fulfilled.

Assume the contrary. Then there exists $t_0 \in [0, T]$ such that

$$|x(t_0)| > M_2. \tag{2.11}$$

It follows from (2.5), (2.7) and (2.9) that there exists $t_1 \in [0, T]$ for which

$$x(t_1) \in \mathcal{B}(h_0), \quad |t_1 - t_0| \leq \delta. \tag{2.12}$$

There exists a number t_2 such that

$$\min\{t_0, t_1\} \leq t_2 \leq \max\{t_0, t_1\},$$

$$|x(t_2)| = \max\{|x(t)| : t \in [\min\{t_0, t_1\}, \max\{t_0, t_1\}]\}. \tag{2.13}$$

(1.1), (2.12) and (2.13) imply that

$$|x(t_1) - x(t_2)| = \left| \int_{t_1}^{t_2} x'(t) dt \right| \leq ||A|| \left| \int_{t_1}^{t_2} |x(t)| dt \right| + ||B|| \left| \int_{t_1}^{t_2} |u(t)| dt \right|$$

$$\leq ||A|| |x(t_2)| \delta + ||B|| \left| \int_{t_1}^{t_2} |u(t)| dt \right|. \tag{2.14}$$

It follows from the inequalities (2.8), (2.5) and (2.12) that

$$\left| \int_{t_1}^{t_2} |u(t)| dt \right| \leq c_0 |t_1 - t_2| + (8||B|| + 8)^{-1} I(0, T, x, u)$$

$$\leq \delta c_0 + (8||B|| + 8)^{-1} M_1.$$

It follows from this inequality, (2.7) and (2.12)-(2.14) that

$$|x(t_1) - x(t_2)| \leq 2^{-1} |x(t_2)| + ||B|| \delta c_0 + M_1.$$

Combined with (2.11), (2.12) and (2.13) this inequality implies that

$$2^{-1} M_2 - h_0 \leq ||B|| \delta c_0 + M_1.$$

This relation is contradictory to (2.10). The contradiction we have reached proves the proposition.

PROPOSITION 6.2.3 *Let* $M_1, \epsilon, \tau_0, \tau_1 > 0$ *and let* $\tau_0 < \tau_1$. *Then there exists a number* $\delta > 0$ *such that the following property holds:*

If $T \in [\tau_0, \tau_1]$ *and if a trajectory-control pair* $x : [0, T] \to R^n$, $u : [0, T] \to R^m$ *satisfies (2.5), then*

$$|x(t_1) - x(t_2)| \leq \epsilon$$

for each $t_1, t_2 \in [0, T]$ such that $|t_1 - t_2| \leq \delta$.

Proof. Let a number $M_2 > 0$ be as guaranteed in Proposition 6.2.2. We may assume without loss of generality that the function f is non-negative. Fix a large positive number c_1. It follows from Assumption A that there exists a number $c_2 > 0$ such that the following property holds:

If $(t, x, u) \in [0, \infty) \times R^n \times R^m$ satisfies $|u| \geq c_2$, then

$$f(t, x, u) \geq c_1 |u|.$$

Let $T \in [\tau_0, \tau_1]$ and $x : [0, T] \to R^n$, $u : 0, T] \to R^m$ be a trajectory-control pair satisfying (2.5). Then (2.6) holds. Let $t_1, t_2 \in [0, T]$, $t_1 < t_2$. By (2.5) and the choice of c_2,

$$\int_{t_1}^{t_2} |u(t)| dt \leq c_2 (t_2 - t_1) + c_1^{-1} M_1.$$

It follows from this inequality, (2.6) and (1.1) that

$$|x(t_1) - x(t_2)| = \left| \int_{t_1}^{t_2} x'(t) dt \right| \leq ||A|| M_2 (t_2 - t_1) + ||B|| \int_{t_1}^{t_2} |u(t)| dt$$

$$\leq ||A|| M_2 (t_2 - t_1) + ||B|| c_2 (t_2 - t_1) + ||B|| c_1^{-1} M_1.$$

This relation implies the validity of the proposition.

PROPOSITION 6.2.4 *Let M_1 and T be positive numbers and let $\mathcal{F}$ be the set of all trajectory-control pairs $x : [0, T] \to R^n$, $u : [0, T] \to R^m$ satisfying (2.5). Then for every sequence $\{(x_i, u_i)\}_{i=1}^{\infty} \subset \mathcal{F}$ there exist a subsequence $\{(x_{i_k}, u_{i_k})\}_{k=1}^{\infty}$ and $(x, u) \in \mathcal{F}$ such that $x_{i_k}(t) \to x(t)$ as $k \to \infty$ uniformly in $[0, T]$, $x'_{i_k} \to x'$ as $k \to \infty$ weakly in $L^1(R^n; (0, T))$, and $u_{i_k} \to u$ as $k \to \infty$ weakly in $L^1(R^m; (0, T))$.*

Proof. It follows from Proposition 6.2.2 that there is a positive number M_2 such that the inequality (2.6) is valid for each trajectory-control pair $(x, u) \in \mathcal{F}$. Assume that $\{(x_i, u_i)\}_{i=1}^{\infty} \subset \mathcal{F}$ so that

$$I(0, T, x_i, u_i) \leq M_1, \ x_i(t) \in \mathcal{B}(M_2) \ (t \in [0, T], \ i = 1, 2, \ldots). \quad (2.15)$$

The boundedness below of f, (2.15) and Assumption A imply that the sequence of functions $\{u_i\}$ is equiabsolutely integrable on $[0, T]$. Therefore there exists a subsequence $\{(x_{i_k}, u_{i_k})\}_{k=1}^{\infty}$ and $u \in L^1(R^m; (0, T))$ such that

$$u_{i_k} \to u \text{ as } k \to \infty \text{ weakly in } L^1(R^m; (0, T)). \quad (2.16)$$

By (2.15), (2.16) and (1.1) we may assume that there exists a function $h \in L^1(R^n; (0, T))$ such that

$$x'_{i_k} \to h \text{ as } k \to \infty \text{ weakly in } L^1(R^n; (0, T)). \tag{2.17}$$

We may also assume that $\lim_{k \to \infty} x_{i_k}(0)$ exists, and for $t \in (0, T]$ we put

$$x(t) = \lim_{k \to \infty} x_{i_k}(0) + \int_0^t h(s)ds. \tag{2.18}$$

It is easy to see that $x_{i_k}(t) \to x(t)$ as $k \to \infty$ for any $t \in [0, T]$. By Proposition 6.2.3 $x_{i_k}(t) \to x(t)$ as $k \to \infty$ uniformly in $[0, T]$.

We need to show that $(x, u) \in \mathcal{F}$. It follows from Assumption A that the set $\mathcal{F}$ is convex. Therefore there exists a sequence $\{(y_i, v_i)\}_{i=1}^\infty \subset \mathcal{F}$ for which

$$y_i(t) \to x(t) \text{ as } i \to \infty \text{ uniformly in } [0, T] \tag{2.19}$$

and

$$y'_i(t) \to h(t), \ v_i(t) \to u(t) \text{ as } i \to \infty \text{ a.e. in } [0, T]. \tag{2.20}$$

(1.1), (2.18), (2.19) and (2.20) imply that (x, u) is a trajectory-control pair. By Assumption A and Fatou's lemma, (x, u) satisfies (2.5). This completes the proof of the proposition.

COROLLARY 6.2.1 *Let $x_1, x_2 \in R^n$. Then there is a unique trajectory-control pair $x : [0, \tau] \to R^n$, $u : [o, \tau] \to R^m$ such that*

$$x(0) = x_1, \ x(\tau) = x_2 \text{ and } I(0, \tau, x, u) = v(x_1, x_2).$$

Corollary 6.2.1 and Assumption A imply that the function v is strictly convex. It follows from Proposition 6.2.2 that

$$v(y, z) \to \infty \text{ as } |y| + |z| \to \infty. \tag{2.21}$$

Put
$$D = \{(z, z) : \ z \in R^n\}$$

and denote by (z^*, z^*) a unique minimum of v on D. Corollary 6.2.1 now implies Proposition 6.1.1 with

$$x^*(0) = z^*, \ \mu = \tau^{-1} v(z^*, z^*). \tag{2.22}$$

PROPOSITION 6.2.5 *There exists $p \in R^n$ such that the function $\theta : R^n \times R^n \to R^1$ defined by*

$$\theta(y, z) = v(y, z) - v(z^*, z^*) - p'(y - z), \ y, z \in R^n \tag{2.23}$$

is strictly convex and has the following property:
$$\theta(y, z) > 0 \text{ if } (y, z) \neq (z^*, z^*), \text{ and } \theta(z^*, z^*) = 0.$$

Proof. For any function $\phi : \mathcal{B} \to R^1$ set

$$\text{epi}(\phi) = \{(r, x) : x \in \mathcal{B}, r \geq \phi(x)\}.$$

Let

$$\Omega = (\text{epi}(v)) \cup \{(z, z, v(z^*, z^*)) : z \in R^n\}$$

and let $w(y, z)$ be the function whose epigraph is $\text{conv}(\Omega)$ (namely, the closed convex hull of Ω). It is not difficult to see that the function $w : R^n \times R^n \to R^1$ is convex,

$$-\infty < w(y, z) \leq v(y, z) \ (y, z \in R^n)$$

and

$$w(z, z) \geq v(z^*, z^*) \text{ for all } z \in R^n.$$

There exists a subgradient vector (p_1, p_2) of w at (z^*, z^*) for which

$$w(y, z) \geq w(z^*, z^*) + (p_1, p_2)'((y - z) - (z^*, z^*)) \ (y, z \in R^n).$$

Since $w(z, z) = v(z^*, z^*)$ for each $z \in R^n$ we have $p_2 = -p_1$. This equality implies that

$$v(y, z) \geq w(y, z) \geq v(z^*, z^*) + p_1'(y - z) \ (y, z \in R^n).$$

This completes the proof of the proposition.

Proposition 6.2.4 and the uniqueness in Corollary 6.2.1 imply the following result.

PROPOSITION 6.2.6 *Let $\epsilon > 0$. Then there exists $\delta > 0$ such that if a trajectory-control pair $x : [0, \tau] \to R^n$, $u : [0, \tau] \to R^n$ satisfies*

$$x(0), x(\tau) \in \mathcal{B}(z^*, \delta), \ I(0, \tau, x, u) \leq v(z^*, z^*) + \delta, \tag{2.24}$$

then

$$|x(t) - x^*(t)| \leq \epsilon \ (t \in [0, \tau]).$$

PROPOSITION 6.2.7 *For each positive number ϵ there exists $\delta > 0$ such that the following property holds:*
If a trajectory-control pair $x : [0, \tau] \to R^n$, $u : [0, \tau] \to R^m$ satisfies (2.24), then for all $T \in (0, \tau]$,

$$|I(0, T, x, u) - I(0, T, x^*, u^*)| \leq \epsilon.$$

Proof. Assume the converse. Then there exist $\epsilon > 0$, a sequence of numbers $\{T_i\}_{i=0}^{\infty} \subset (0, \tau]$ and a sequence of trajectory-control pairs

$$x_i : [0, \tau] \to R^n, \ u_i : [0, \tau] \to R^m, \ i = 0, 1, \ldots,$$

such that

$$|x_i(0) - z^*| + |x_i(\tau) - z^*| \to 0,$$

$$I(0, \tau, x_i, u_i) - v(z^*, z^*) \to 0 \text{ as } i \to \infty, \tag{2.25}$$

$$|I(0, T_i, x_i, u_i) - I(0, T_i, x^*, u^*)| \geq \epsilon, \ i = 0, 1, 2, \ldots. \tag{2.26}$$

We may assume that the limit

$$T = \lim_{i \to \infty} T_i \tag{2.27}$$

exists and one of the relations below holds:

(a) $I(0, T_i, x_i, u_i) - I(0, T_i, x^*, u^*) \geq \epsilon$ for all natural numbers i;
(b) $I(0, T_i, x_i, u_i) - I(0, T_i, x^*, u^*) \leq -\epsilon$ for all natural numbers i.

If relation (a) holds we define

$$E = [T, \tau], \ E_i = [T_i, \tau], \ i = 0, 1, 2, \ldots. \tag{2.28}$$

Otherwise we set

$$E = [0, T], \ E_i = [0, T_i], \ i = 0, 1, 2, \ldots. \tag{2.29}$$

In view of (2.25) we may assume that

$$\int_{E_i} f(t, x_i(t), u_i(t))dt - \int_{E_i} f(t, x^*(t), u^*(t))dt$$

$$\leq -2^{-1}\epsilon \text{ for all integers } i \geq 1. \tag{2.30}$$

By Assumption A and (2.25) the sequence of the functions $\{u_i\}_{i=0}^{\infty}$ is equiabsolutely integrable on $[0, \tau]$. In view of Corollary 6.2.1, Proposition 6.2.4 and (2.25) we may assume without loss of generality that

$$x_i(t) \to x^*(t) \text{ as } i \to \infty \text{ uniformly in } [0, \tau],$$

$$x_i' \to (x^*)' \text{ as } i \to \infty \text{ weakly in } L^1(R^n; (0, \tau))$$

and $u_i \to u^*$ as $i \to \infty$ weakly in $L^1(R^m; (0, \tau))$.

Since the function $f(t, \cdot, \cdot)$ is convex for any $t \geq 0$ we conclude that there exists a sequence of trajectory-control pairs

$$y_i : [0, \tau] \to R^n, \ w_i : [0, \tau] \to R^m, \ i = 0, 1, 2, \ldots$$

such that

$$|y_i(0) - z^*| + |y_i(\tau) - z^*| \to 0,$$

$$I(0, \tau, y_i, w_i) - v(z^*, z^*) \to 0 \text{ as } i \to \infty, \tag{2.31}$$

$$y_i(t) \to x^*(t) \text{ as } i \to \infty \text{ uniformly in } [0, \tau],$$

$$\int_0^\tau |w_i(t) - u^*(t)| dt \to 0 \text{ as } i \to \infty, \tag{2.32}$$

$$\int_{E_i} f(t, y_i(t), w_i(t)) dt - \int_{E_i} f(t, x^*(t), u^*(t)) dt$$

$$\leq -2^{-1}\epsilon \text{ for all integers } i \geq 1. \tag{2.33}$$

We may assume by extracting a subsequence and re-indexing that

$$w_i(t) \to u^*(t) \text{ as } i \to \infty \text{ a.e. in } [0, \tau]. \tag{2.34}$$

For a set $F \subset [0, \tau]$ we define $\lambda_F : [0, \tau] \to R^1$ as follows:

$$\lambda_F(t) = 1 \ (t \in F), \ \lambda_F(t) = 0 \ (t \in [0, \tau] \setminus F).$$

By Fatou's lemma, (2.32)-(2.34) and Assumption A,

$$\int_0^\tau f(t, x^*(t), u^*(t)) \lambda_E(t) dt \leq \limsup_{i \to \infty} \int_0^\tau f(t, y_i(t), w_i(t)) \lambda_{E_i}(t) dt$$

$$\leq \limsup_{i \to \infty} \int_{E_i} f(t, x^*(t), u^*(t)) dt - 2^{-1}\epsilon = \int_E f(t, x^*(t), u^*(t)) dt - 2^{-1}\epsilon.$$

The contradiction obtained proves the proposition.

6.3. Discrete-time control systems

Let K be a compact metric space and let $w : K \times K \to R^1$ be bounded and lower semicontinuous. We define

$$a(w) = \sup\{w(x, y) : x, y \in K\}, \ b(w) = \inf\{w(x, y) : x, y \in K\},$$

$$\mu(w) = \inf\left\{\liminf_{N \to \infty} N^{-1} \sum_{i=0}^{N-1} w(z_i, z_{i+1}) : \{z_i\}_{i=0}^\infty \subset K\right\}. \tag{3.1}$$

The following two results were established in [39] when K was a compact set in R^n but their proofs remain in force when K is any compact metric space.

PROPOSITION 6.3.1 *1. For each natural number N and each sequence* $\{z_i\}_{i=0}^N \subset K,$

$$\sum_{i=0}^{N-1} [w(z_i, z_{i+1}) - \mu(w)] \geq b(w) - a(w).$$

2. For every sequence $\{z_i\}_{i=0}^\infty \subset K,$ *either*

$$\sum_{i=0}^{N-1} [w(z_i, z_{i+1}) - \mu(w)] \to \infty \text{ as } N \to \infty$$

or

$$\sup \left\{ |\sum_{i=0}^{N-1} [w(z_i, z_{i+1}) - \mu(w)]| : \; N = 1, 2, \dots \right\} < \infty.$$

3. For every initial value z_0 there is a sequence $\{z_i\}_{i=0}^\infty \subset K$ *such that*

$$\left| \sum_{i=0}^{N-1} [w(z_i, z_{i+1}) - \mu(w)] \right| \leq 4|a(v) - b(v)| \; (N = 1, 2, \dots).$$

PROPOSITION 6.3.2 *Let $w : K \times K \to R^1$ be a continuous function. Define*

$$\pi^w(x) = \inf \left\{ \liminf_{N \to \infty} \sum_{i=0}^{N-1} [w(z_i, z_{i+1}) - \mu(v)] : \; \{z_i\}_{i=0}^\infty \subset K, \; z_0 = x \right\},$$

$$\theta^w(x, y) = w(x, y) - \mu(w) + \pi^w(y) - \pi^w(x)$$

for $x, y \in K$. Then π^w and θ^w are continuous functions,

$$\theta^w(x, y) \geq 0 \text{ for all } x, y \in K$$

and for every $x \in K$ there is $y \in K$ for which $\theta^w(x, y) = 0$.

Consider a continuous function $w : R^n \times R^n \to R^1$ satisfying

$$w(x, y) \to \infty \text{ as } |x| + |y| \to \infty.$$

Let $x \in R^n$. Define

$$\mu(w) = \inf\{\liminf_{N \to \infty} N^{-1} \sum_{k=0}^{N-1} w(z_k, z_{k+1}) : \; \{z_k\}_{k=0}^\infty \subset R^n, \; z_0 = x\}.$$

$$(3.2)$$

By Propositions 6.3.1 and 3.5.1, $\mu(w)$ is independent of the initial point x_0. For $x, y \in R^n$ we set

$$\pi^w(x) = \inf \left\{ \liminf_{N \to \infty} \sum_{i=0}^{N-1} [w(z_i, z_{i+1}) - \mu(v)] : \{z_i\}_{i=0}^\infty \subset R^n, \ z_0 = x \right\},$$

$$(3.3)$$

$$\theta^w(x, y) = w(x, y) - \mu(w) + \pi^w(y) - \pi^w(x). \qquad (3.4)$$

We have the following result.

PROPOSITION 6.3.3 *Let* $M_1, M_2 > 0$,

$$\inf\{w(x, y) : \ x, y \in R^n, \ |x| + |y| \geq M_1\} > |w(0, 0)| + 1. \qquad (3.5)$$

Then there exists a natural number $N > 2$ *such that for each natural number* $q \geq N$ *and any sequence* $\{x_k\}_{k=0}^q \subset R^n$ *the following properties hold:*
  1. *If*

$$\{k \in \{0, \ldots, q\} : \ x_k \in \mathcal{B}(M_1)\} = \{0, q\}$$

and $y_0 = x_0$, $y_q = x_q$, $y_k = 0$ $(k = 1, \ldots, q - 1)$, *then*

$$\sum_{k=0}^{q-1} [w(x_k.x_{k+1}) - w(y_k.y_{k+1})] \geq M_2; \qquad (3.6)$$

  2. *If*

$$\{k \in \{0, \ldots, q\} : \ x_k \in \mathcal{B}(M_1)\} = \{0\}$$

and $y_0 = x_0$, $y_k = 0$ $(k = 1, \ldots, q)$, *then (3.6) is true.*

PROPOSITION 6.3.4 *Assume that* $M_1, M_2 > 0$ *and (3.5) holds. Then there exists a number* $M_3 > M_1 + M_2$ *such that for each natural number* q *and each sequence* $\{x_k\}_{k=0}^q \subset R^n$ *the following properties hold:*
  1. *If*

$$x_0, x_q \in \mathcal{B}(M_1), \ \max\{|x_k| : \ k = 0, \ldots, q\} > M_3, \qquad (3.7)$$

then there is a sequence $\{y_k\}_{k=0}^q \subset R^n$ *such that* $y_i = z_i$, $i = 0, q$, *and (3.6) holds.*
  2. *If*

$$x_0 \in \mathcal{B}(M_1), \ \max\{|x_k| : \ k = 0, \ldots, q\} > M_3, \qquad (3.8)$$

then there is a sequence $\{y_k\}_{k=0}^q \subset R^n$ *such that* $y_0 = z_0$ *and (3.6) holds.*

Proof. Let an integer $N > 6$ be as guaranteed in Proposition 6.3.3. Fix a large number $M_3 > M_1 + M_2$. We prove Assertion 1. Assume that $\{x_k\}_{k=0}^q \subset R^n$ satisfies (3.7). Then there is $j \in \{0, \ldots, q\}$ such that $|x_j| > M_3$. Set

$$i_1 = \max\{i \in \{0, \ldots, j\} : |x_i| \leq M_1\},$$

$$i_2 = \min\{i \in \{j, \ldots q\} : |x_i| \leq M_1\}.$$

If $i_2 - i_1 \geq N$ then the validity of Assertion 1 follows from the definition of N and Proposition 6.3.3. If $i_2 - i_1 < N$ we set

$$y_i = x_i, \ i \in \{0, \ldots, i_1\} \cup \{i_2, \ldots, q\}, \ y_i = 0, \ i = i_1 + 1, \ldots, i_2 - 1 \quad (3.9)$$

and it is easy to see that (3.6) holds if the constant M_3 is large enough.

We will prove Assertion 2. Assume that $\{x_i\}_{i=0}^q \subset R^n$ satisfies (3.8). Then there is $j \in \{1, \ldots, q\}$ such that $|x_j| > M_3$. Set $i_1 = \max\{i \in \{0, \ldots, j\} : |x_i| \leq M_1\}$.

If $|x_i| > M_1$ for $i = j, \ldots, q$ we set

$$y_i = x_i, \ i = 0, \ldots, i_1, \ y_i = 0, \ i = i_1 + 1, \ldots, q.$$

Otherwise we set

$$i_2 = \min\{k \in \{j, \ldots, q\} : |x_k| \leq M_1\}$$

and define $\{y_k\}_{k=0}^q$ by (3.9). It is easy to verify that in both cases (3.6) holds. This completes the proof of the proposition.

Propositions 3.5.4 and 6.3.4 imply the following result.

PROPOSITION 6.3.5 *There is $M > 0$ such that if $\{x_k\}_{k=0}^\infty \subset R^n$ and if the sequence*

$$\left\{\sum_{k=0}^N [w(x_k, x_{k+1}) - \mu(w)]\right\}_{N=0}^\infty$$

is bounded, then $\limsup_{k \to \infty} |x_k| \leq M$.

6.4.   Proof of Theorem 6.1.1

PROPOSITION 6.4.1 $\mu(v) = \mu\tau$ *(see (1.4), (2.2) and (3.2)).*

Proof. In view of (2.22), $\mu(v) \leq \mu\tau$. It follows from Propositions 6.3.1 and 3.5.2 that there exists a sequence $\{z_i\}_{i=0}^\infty \subset R^n$ such that

$$\left\{\left|\sum_{i=0}^N [v(z_i, z_{i+1}) - \mu(v)]\right| : N = 0, 1, \ldots\right\} < \infty.$$

By Proposition 6.2.5 and (2.22), for each natural number N,

$$\sum_{i=0}^{N-1} [v(z_i, z_{i+1}) - \mu(v)] = \sum_{i=0}^{N-1} \theta(z_i, z_{i+1}) + p'(z_0 - z_N) + N(\tau\mu - \mu(v)).$$

Proposition 3.5.4 implies that

$$\sup\{|z_i| : i = 0, 1, \ldots\} < \infty.$$

Therefore the sequence $\sup\{N(\mu\tau - \mu(v))\}_{N=1}^{\infty}$ is bounded from above. This completes the proof of the proposition.

Proof of Theorem 6.1.1 Assumption A implies that relation (i) holds if and only if

$$I(0, i\tau, x, u) - i\tau\mu \to \infty \text{ as } i \to \infty \tag{4.1}$$

for i integer, and relation (ii) holds if and only if

$$\sup\{|I(0, i\tau, x, u) - i\tau\mu| : i = 1, 2, \ldots\} < \infty. \tag{4.2}$$

It is easy to see that (4.2) holds if and only if

$$\sum_{i=0}^{\infty} [I(\tau i, \tau(i+1)), x, u) - v(x(i\tau), x((i+1)\tau))] < \infty \tag{4.3}$$

and

$$\sup\left\{\sum_{i=0}^{N-1} [v(x(i\tau), x((i+1)\tau)) - \mu\tau] : N = 1, 2, \ldots\right\} < \infty. \tag{4.4}$$

It is easy to see that (4.1) is not fulfilled if and only if relations (4.3) and (4.4) hold. Therefore either condition (i) or (ii) is valid.

Assume that relation (ii) holds so that (4.3) and (4.4) are satisfied. Propositions 3.5.4 and 6.4.1 imply that the sequence $\{x(i\tau)\}_{i=0}^{\infty}$ is bounded. Combined with (4.3), (4.4), (2.22) and Proposition 6.2.5 this implies that

$$\sum_{i=0}^{\infty} \theta(x(i\tau), x((i+1)\tau)) < \infty, \ x(i\tau) \to z^* \text{ as } i \to \infty. \tag{4.5}$$

The final assertion in the theorem now follows from (4.3), (4.5) and Proposition 6.2.6.

6.5.   Proof of Theorem 6.1.2

Let $x_0 \in R^n$. By Propositions 6.3.4 and 6.3.5 there exists a ball $D \subset R^n$ such that for every sequence $\{z_i\}_{i=0}^{\infty} \subset R^n$ not included in D with $z_0 = x_0$ there exists a sequence $\{x_i\}_{i=0}^{\infty} \subset D$ with $s_0 = z_0$ such that

$$\sum_{i=0}^{N} v(s_i, s_{i+1}) \leq \sum_{i=0}^{N} v(z_i, z_{i+1}) - 1 \text{ for all large } N.$$

Denote by $\mathcal{A}$ the set of all sequences $\{z_i\}_{i=0}^{\infty} \subset D$ such that

$$z_0 = x_0, \ \liminf_{N \to \infty} \sum_{i=0}^{N} [v(z_i, z_{i+1}) - \mu\tau] < \infty$$

(see (1.4), (2.2), (3.2) and Proposition 6.4.1). It is easy to see that

$$\pi^v(x_0) = \inf \left\{ \liminf_{N \to \infty} \sum_{i=0}^{N-1} [v(z_i, z_{i+1}) - \mu\tau] : \{z_i\}_{i=0}^{\infty} \in \mathcal{A} \right\}. \qquad (5.1)$$

It follows from (2.22) and Propositions 6.4.1 and 6.2.5 that for each $\{z_i\}_{i=0}^{\infty} \in \mathcal{A}$,

$$\sum_{i=0}^{\infty} \theta(z_i, z_{i+1}) < \infty, \ z_i \to z^* \text{ as } i \to \infty. \qquad (5.2)$$

Put

$$F(\{z_i\}_{i=0}^{\infty}) = \sum_{i=0}^{\infty} \theta(z_i, z_{i+1}) \text{ where } \{z_i\}_{i=0}^{\infty} \in \mathcal{A}.$$

By Proposition 6.2.5 the function θ is strictly convex. This implies that the function F has a unique minimizer which we denote by $\{y_i\}_{i=0}^{\infty} \in \mathcal{A}$. The proof of Theorem 6.1.2 will be based on the following auxiliary result.

LEMMA 6.5.1

$$\pi^v(x_0) = \liminf_{N \to \infty} \sum_{i=0}^{N} [v(y_i, y_{i+1}) - \mu\tau] = \lim_{N \to \infty} \sum_{i=0}^{N} [v(y_i, y_{i+1} - \mu\tau] \qquad (5.3)$$

and if a sequence $\{z_i\}_{i=0}^{\infty} \subset R^n$ *satisfies*

$$z_0 = x_0, \ \{z_i\}_{i=0}^{\infty} \neq \{y_i\}_{i=0}^{\infty}, \qquad (5.4)$$

then there is a natural number N_0 such that

$$\inf\left\{\sum_{i=0}^{N}[v(z_i, z_{i+1}) - v(y_i, y_{i+1})] : N \geq N_0\right\} > 0. \tag{5.5}$$

Proof. (5.2), (2.22) and Proposition 6.2.5 imply that for each $\{z_i\}_{i=0}^{\infty} \in \mathcal{A}$,

$$\lim_{N\to\infty} \sum_{i=0}^{N}[v(z_i, z_{i+1}) - \tau\mu] = \lim_{N\to\infty} \left[\sum_{i=0}^{N}\theta(z_i, z_{i+1}) + p'(x_0 - z_N)\right]$$

$$= \sum_{i=0}^{\infty}\theta(z_i, z_{i+1}) + p'(x_0 - z^*). \tag{5.6}$$

(5.1) and (5.6) imply (5.3).

Assume that a sequence $\{z_i\}_{i=0}^{\infty} \subset R^n$ satisfies (5.4). We show that there exists a natural number N_0 such that (5.5) holds. In view of our choice of D we may assume that $\{z_i\}_{i=0}^{\infty} \in \mathcal{A}$. Since $\{y_i\}_{i=0}^{\infty}$ is a unique minimizer of the function F it follows from (5.6) that

$$\lim_{N\to\infty} \sum_{i=0}^{N}[v(z_i, z_{i+1}) - v(y_i, y_{i+1})] = \sum_{i=0}^{\infty}\theta(z_i, z_{i+1})$$

$$- \sum_{i=0}^{\infty}\theta(y_i, y_{i+1}) > 0.$$

This completes the proof of the lemma.

By Corollary 6.2.1 there exists a trajectory-control pair $\tilde{x} : [0, \infty) \to R^n$, $\tilde{u} : [0, \infty) \to R^m$ such that for $i = 0, 1, \ldots,$

$$\tilde{x}(i\tau) = y_i, \ I(i\tau, (i+1)\tau, \tilde{x}, \tilde{u}) = v(\tilde{x}(i\tau), \tilde{x}((i+1)\tau)). \tag{5.7}$$

Assume that $x : [0, \infty) \to R^n$, $u : [0, \infty) \to R^m$ is a trajectory-control pair satisfying

$$x(0) = x_0, \ (x, u) \neq (\tilde{x}, \tilde{u}). \tag{5.8}$$

We show that there is a time T_0 such that

$$\inf\{I(0, T, x, u) - I(0, T, \tilde{x}, \tilde{u}) : T \in [T_0, \infty)\} > 0. \tag{5.9}$$

By the choice of $\tilde{x}, \tilde{u}$, Lemma 6.5.1 and Theorem 6.1.1,

$$\sup\{|I(0, T, \tilde{x}, \tilde{u}) - \mu T| : T \in [0, \infty)\} < \infty. \tag{5.10}$$

In view of Theorem 6.1.1 we may assume without loss of generality that

$$\sup\{|I(0, T, x, u) - \mu T| : T \in [0, \infty)\} < \infty, \qquad (5.11)$$

$$\sup\{|x(i\tau + t) - x^*(t)| : t \in [0, \tau]\} \to 0 \text{ as } i \to \infty \text{ over the integers.}$$
$$(5.12)$$

These relations imply that

$$I(i\tau, (i+1)\tau, x, u) - v(z^*, z^*) \to 0 \text{ as } i \to \infty. \qquad (5.13)$$

We show that there exist $\epsilon > 0$ and a natural number N_0 such that

$$I(0, N\tau, x, u) - I(0, N\tau, \tilde{x}, \tilde{u}) \geq 2\epsilon \qquad (5.14)$$

for all integers $N \geq N_0$.

If $x(i_0\tau) \neq \tilde{x}(i_0\tau)$ for some natural number i_0, then the existence of $\epsilon > 0$ and an integer $N_0 \geq 1$ such that (5.14) holds for all integers $N \geq N_0$ follows from Lemma 6.5.1 and the definition of $\tilde{x}$, $\tilde{u}$. If $x(i\tau) = \tilde{x}(i\tau)$ for all natural numbers i, then this existence follows from (5.7), (5.8), and Corollary 6.2.1.

The validity of the theorem now follows from (5.14), (5.7), (5.12), (5.13) and Proposition 6.2.7.

6.6.   Proof of Theorem 6.1.3

The following auxiliary result shows that if a sequence $\{x_i\}_{i=p}^q$ is an approximate solution of the related discrete-time optimal control problem and $q - p$ is large enough, then x_j and x_{j+1} are close to z^* for a certain $j \in \{p, \ldots, q-1\}$.

LEMMA 6.6.1 *Let M_1, M_2, ϵ be positive numbers. Then there exists a natural number $N \geq 4$ such that the following property holds:*
If a natural number $Q \geq N$, if a sequence $\{x_k\}_{k=0}^Q \subset R^n$ satisfies

$$x_k \in \mathcal{B}(M_1) \text{ for } k = 0, \ldots, Q,$$

$$\sum_{k=0}^{Q-1} v(x_k.x_{k+1})$$

$$\leq \inf\{\sum_{k=0}^{Q-1} v(y_k, y_{k+1}) : \{y_k\}_{k=0}^Q \subset R^n, y_0 = x_0, y_Q = x_Q\} + M_2, \quad (6.1)$$

and if integers $p, q \in \{0, \ldots, Q\}$ satisfy $q - p \geq N$, then there exists an integer $j \in \{p, \ldots, q-1\}$ such that

$$x_j, x_{j+1} \in \mathcal{B}(z^*, \epsilon). \tag{6.2}$$

Proof. Fix a large integer $N \geq 4$. Suppose that an integer $Q \geq N$, a sequence $\{x_k\}_{k=0}^{Q} \subset R^n$ satisfies (6.1), and integers $p, q \in \{0, \ldots, Q\}$ satisfy $q - p \geq N$. We show that there exists an integer $j \in \{p, \ldots, q-1\}$ satisfying (6.2).

Assume the converse and set

$$y_i = x_i \ (i \in \{0, \ldots, p\} \cup \{q, \ldots, Q\}), \ y_i = z^* \ (i \in \{p+1, \ldots, q-1\}).$$

By (6.1) and Proposition 6.2.5,

$$M_2 \geq \sum_{k=0}^{Q-1} [v(x_k, x_{k+1}) - v(y_k, y_{k+1})]$$

$$= \sum_{k=p}^{q-1} \theta(x_k, x_{k+1}) - \theta(x_p, z^*) - \theta(z^*, x_q)$$

$$\geq N \inf\{\theta(z_1, z_2) : z_1, z_2 \in \mathcal{B}(M_1), |z_1 - z^*| + |z_2 - z^*| \geq \epsilon\}$$

$$-2 \sup\{\theta(z_1, z_2) : z_1, z_2 \in \mathcal{B}(M_1 + |z^*|)\}.$$

When N is sufficiently large we obtain a contradiction which proves the lemma.

Theorem 6.1.1, Propositions 6.2.2 and 6.3.4 and Corollary 6.2.1 imply the following result.

LEMMA 6.6.2 *For each $M_0 > 0$ there exists a number $M_1 > 0$ such that the following property holds:*

If an overtaking optimal trajectory-control pair $x : [0, \infty) \to R^n$, $u : [0, \infty) \to R^m$ satisfies $|x(0)| \leq M_0$, then $|x(t)| \leq M_1$ for all $t \in [0, \infty)$.

Lemmas 6.6.1 and 6.6.2 and Corollary 6.2.1 imply the following result which shows that an overtaking optimal trajectory-control pair (x, u) reaches a neighborhood of z^* during a period of time which depends only on $|x(0)|$.

LEMMA 6.6.3 *For each $M, \delta > 0$ there exists an integer $N \geq 1$ such that the following property holds:*

If an overtaking optimal trajectory-control pair $x : [0, \infty) \to R^n$, $u : [0, \infty) \to R^m$ satisfies $|x(0)| \leq M$, then there exists an integer $i \in [0, N]$ such that

$$|x(i\tau) - z^*| \leq \delta.$$

The next lemma establishes the turnpike property for an overtaking optimal trajectory-control pair (x, u) such that $x(0)$ belongs to a sufficiently small neighborhood of z^*.

LEMMA 6.6.4 *For each $\epsilon > 0$ there exists a positive number δ such that the following property holds:*

If an overtaking optimal trajectory-control pair $x : [0, \infty) \to R^n$, $u : [0, \infty) \to R^m$ satisfies

$$|x(0) - z^*| \leq \delta, \tag{6.3}$$

then for all integers $i \geq 0$,

$$\sup\{|x(i\tau + t) - x^*(t)| : t \in [0, \tau]\} \leq \epsilon. \tag{6.4}$$

Proof. Let $\epsilon > 0$. It follows from Proposition 6.2.6 that there exists a number $\delta_0 \in (0, 1)$ such that the following property holds:

If a trajectory-control pair $x : [0, \tau] \to R^n$, $u : [0, \tau] \to R^m$ satisfies

$$|x(0) - z^*| \leq \delta_0, \ |x(\tau) - z^*| \leq \delta_0, \ I(0, \tau, x, u) - v(x(0), x(\tau)) \leq \delta_0, \tag{6.5}$$

then

$$|x(t) - x^*(t)| \leq \epsilon \ (t \in [0, \tau]). \tag{6.6}$$

There exists a number $M_1 > 0$ such that Lemma 6.6.2 holds with $M_0 = |z^*| + 1$. Fix a small constant $\delta > 0$.

Let $x : [0, \infty) \to R^n$, $u : [0, \infty) \to R^m$ be an overtaking optimal trajectory-control pair satisfying (6.3). By Corollary 6.2.1 there exists a trajectory-control pair $y : [0, \infty) \to R^n$, $w : [0, \infty) \to R^m$ such that

$$y(0) = x(0), \ y(i\tau) = z^* \text{ for } i = 1, 2, \ldots,$$

$$I(i\tau, (i + 1)\tau, y, w) = v(y(i\tau), y((i + 1)\tau)) \text{ for } i = 0, 1, \ldots.$$

By Theorem 6.1.1 $x(i\tau) \to z^*$ as $i \to \infty$. It is easy to see that

$$0 \geq \limsup_{k \to \infty} [I(0, k\tau, x, u) - I(0, k\tau, y, w)]$$

$$\geq \sum_{k=0}^{\infty} [I(k\tau, (k + 1)\tau, x, u) - v(x(k\tau), x((k + 1)\tau))]$$

$$+ \sum_{k=0}^{\infty} \theta(x(k\tau), x((k + 1)\tau)) - \sup\{\theta(z, z^*) : z \in R^N, |z - z^*| \leq \delta\}.$$

When the constant δ is sufficiently small it follows from this relation and the definition of δ_0 that (6.4) holds for all integers $i \geq 0$. The lemma is proved.

Theorem 6.1.3 now follows from Lemmas 6.6.3 and 6.6.4.

6.7. Proof of Theorem 6.1.4

Proposition 6.3.4, the boundedness below of f and Corollary 6.2.1 imply the following result.

LEMMA 6.7.1 *For each pair of positive numbers* M_0, M_1 *there exists* $M_2 > M_1$ *such that the following property holds:*

If N *is a natural number,* $T \in [N\tau, (N+1)\tau)$ *and if a trajectory-control pair* $x : [0, T] \to R^n$, $u : [0, T] \to R^m$ *satisfies*

$$|x(0)| \leq M_0, \ I(0, T, x, u) \leq \Delta(x(0), T) + M_1,$$

then

$$\sum_{k=0}^{N-1} v(x(k\tau), x((k+1)\tau)) \leq \sum_{k=0}^{N-1} v(y_k.y_{k+1}) + M_2$$

for each sequence $\{y_k\}_{k=0}^{N} \subset R^n$ *which satisfies* $y_0 = y(0)$.

Proposition 6.3.4 and Lemma 6.7.1 imply the following result.

LEMMA 6.7.2 *For each positive number* M_0 *there exists* $M > M_0$ *such that if a trajectory-control pair* $x : [0, T] \to R^n$, $u : [0, T] \to R^m$ *satisfies*

$$T \geq \tau, \ |x(0)| \leq M_0, \ I(0, T, x, u) \leq \Delta(x(0), T) + 1, \qquad (7.1)$$

then $|x(i\tau)| \leq M$ *for all integers* $i \in [0, \tau^{-1}T]$.

Lemmas 6.6.1, 6.7.1 and 6.7.2 imply the following result.

LEMMA 6.7.3 *For each* $M_0, \epsilon > 0$ *there exists a natural number* $N > 4$ *such that the following property holds:*

If $T \geq N\tau$, *a trajectory-control pair* $x : [0, T] \to R^n$, $u : [0, T] \to R^m$ *satisfies (7.1) and if a pair of integers* $p, q \in [0, \tau^{-1}T]$ *satisfy* $q - p \geq N$, *then there is an integer* $j \in \{p, \ldots, q-1\}$ *such that*

$$\max\{|x(j\tau) - z^*|, \ |x((j+1)\tau) - z^*|\} \leq \epsilon.$$

The next auxiliary result establishes the turnpike property for approximate optimal trajectory-control pairs (x, u) on finite intervals such that $x(0)$ belongs to a sufficiently small neighborhood of z^*.

LEMMA 6.7.4 *For each* $\epsilon > 0$ *there exist a natural number* N *and a positive number* $\delta > 0$ *such that the following property holds:*

If $T \geq N\tau$ *and if a trajectory-control pair* $x : [0, T] \to R^n$, $u : [0, T] \to R^m$ *satisfies*

$$I(0, T, x, u) \leq \Delta(x(0), T) + \delta, \ |x(0) - z^*| \leq \delta, \qquad (7.2)$$

then for all integers $i \in [0, \tau^{-1}T - N]$,

$$\sup\{|x(i\tau + t) - x^*(t)| : t \in [0, \tau]\} \le \epsilon. \tag{7.3}$$

Proof. Let $\epsilon > 0$. It follows from Proposition 6.2.6 that there exists $\delta_0 \in (0, 1)$ such that the following property holds:
If a trajectory-control pair $x : [0, \tau] \to R^n$, $u : [0, \tau] \to R^m$ satisfies

$$|x(0) - z^*|, \ |x(\tau) - z^*| \le \delta_0, \ I(0, \tau, x, u) - v(x(0), x(\tau)) \le \delta_0,$$

then for all $t \in [0, \tau]$,
$$|x(t) - x^*(t)| \le \epsilon.$$

There exists a number $M > |z^*| + 4$ such that Lemma 6.7.2 holds with $M_0 = |z^*| + 4$. By Proposition 6.2.5 there exists a positive number δ_1 such that

$$\delta_1 < \min\{1, \epsilon, \delta_0\}, \ \delta_1 < 4^{-1}\min\{\theta(z_1, z_2) : z_1, z_2 \in R^n, |z_1|, |z_2| \le M,$$

$$|z_1 - z^*| + |z_2 - z^*| \ge \delta_0\}. \tag{7.4}$$

Fix a positive number δ such that

$$\delta < 8^{-1}\delta_1, \ \sup\{\theta(z_1, z_2) : z_1, z_2 \in R^n, |z_1 - z^*| + |z_2 - z^*| \le \delta\} < 8^{-1}\delta_1. \tag{7.5}$$

There exists an integer $N_0 > 4$ such that Lemma 6.7.3 holds with $M_0 = |z^*| + 4$, $\epsilon = \delta$, and $N = N_0$. Fix an integer

$$N > 2N_0 + 4. \tag{7.6}$$

Let $T > N\tau$ and let $x : [0, T] \to R^n$, $u : [0, T] \to R^m$ be a trajectory-control pair satisfying (7.2). There exists an integer $q \ge N$ such that

$$q \le T\tau^{-1} < q + 1. \tag{7.7}$$

It follows from Lemma 6.7.3 and the definition of N_0 that there is an integer $j \in \{q - N_0, \ldots, q - 1\}$ such that

$$\max\{|x(j\tau) - z^*|, \ |x((j + 1)\tau) - z^*| \le \delta. \tag{7.8}$$

By Corollary 6.2.1 there exists a trajectory-control pair $y : [0, T] \to R^n$, $w : [0, T] \to R^m$ such that

$$y(t) = x(t), \ w(t) = u(t) \ (t \in [j\tau, T]), \ y(0) = x(0),$$

$$y(i\tau) = z^* \ (i = 1, \ldots, j - 1),$$

$$I(i\tau, (i + 1)\tau, y, w) = v(y(i\tau), y((i + 1)\tau)) \ (i = 0, \ldots, j - 1). \tag{7.9}$$

It follows from (7.2), (7.9) and Proposition 6.2.5 that

$$\delta \geq I(0, T, x, u) - I(0, T, y, w)$$

$$= \sum_{i=0}^{j-1} [I(i\tau, (i+1)\tau, x, u) - v(x(i\tau), x((i+1)\tau))]$$

$$+ \sum_{i=0}^{j-1} \theta(x(i\tau), x((i+1)\tau)) - \theta(x(0), z^*) - \theta(z^*, x(j\tau)).$$

Combined with (7.2), (7.8) and (7.5) this implies that for $i = 0, \ldots, j-1$,

$$\sup\{I(i\tau, (i+1)\tau, x, u) - v(x(i\tau), x((i+1)\tau)), \theta(x(i\tau), x((i+1)\tau))\}$$

$$\leq \delta + 2\sup\{\theta(z_1, z_2) : z_1, z_2 \in R^n, |z_1 - z^*| + |z_2 - z^*| \leq \delta\}$$

$$\leq 2^{-1}\delta. \tag{7.10}$$

It follows from Lemma 6.7.2, the choice of M, and (7.2) that

$$|x(i\tau)| \leq M \ (i = 0, \ldots, q). \tag{7.11}$$

Equations (7.4), (7.10) and (7.11) imply that

$$|x(i\tau) - z^*| \leq \delta_0 \text{ for } i = 0, \ldots, j. \tag{7.12}$$

It is not difficult to see that $j > T\tau^{-1} - 1 - N_0 \geq T\tau^{-1} + 3 - N$. It follows from (7.10), (7.12), and the definition of δ_0 that

$$|x(i\tau + t) - x^*(t)| \leq \epsilon \ (t \in [0, \tau], \ i = 0, \ldots, j - 1).$$

This completes the proof of the lemma.

Theorem 6.1.4 now follows from Lemmas 6.7.3 and 6.7.4.

Chapter 7

LINEAR SYSTEMS
WITH NONPERIODIC INTEGRANDS

In this chapter we analyze the existence and structure of optimal trajectories of linear control systems with nonperiodic convex integrands $f : [0, \infty) \times R^n \times R^m \to R^1$, and extend the results of Chapter 6 established for linear periodic control systems.

7.1.   Main results

We consider a linear control system defined by

$$x'(t) = Ax(t) + Bu(t), \ x(0) = x_0, \tag{1.1}$$

where A and B are given matrices of dimensions $n \times n$ and $n \times m$, $x(t) \in R^n$, $u(t) \in R^m$ and the admissible controls are measurable functions.

The performance of the above control system is measured on any finite interval $[T_1, T_2]$ by the integral functional

$$I(T_1, T_2, x, u) = \int_{T_1}^{T_2} f(t, x(t), u(t))dt \tag{1.2}$$

where $f : [0, \infty) \times R^n \times R^m \to R^1$ is a given integrand.

We assume that the linear system (1.1) is controllable and that the integrand f is a Borel measurable function.

In this chapter we extend the results of Chapter 6 to a cost function $f : [0, \infty) \times R^n \times R^m \to R^1$ which satisfies the following assumptions:

(A1)

(i) (uniform strict convexity; see [82]). For any $t \in [0, \infty)$ the function $f(t, \cdot, \cdot) : R^n \times R^m \to R^1$ is strictly convex and moreover, for any $\epsilon > 0$

there exists $\delta(\epsilon) > 0$ such that

$$f(t, 2^{-1}(x_1 + x_2), 2^{-1}(u_1 + u_2)) \leq 2^{-1}[f(t, x_1, u_1) + f(t, x_2, u_2)] - \delta(\epsilon)$$

for each $t \geq 0$, each $x_1, x_2 \in R^n$ and each $u_1, u_2 \in R^n$ satisfying

$$|x_1 - x_2| + |u_1 - u_2| \geq \epsilon;$$

(ii) $f(t, x, u)|u|^{-1} \to \infty$ as $|u| \to \infty$ uniformly in $(t, x) \in [0, \infty) \times R^n$;

(iii) the function f is bounded on any bounded subset of $[0, \infty) \times R^n \times R^m$;

(iv) $f(t, x, u) \to \infty$ as $|x| \to \infty$ uniformly in $(t, u) \in [0, \infty) \times R^m$;

(v) for any $(x, u) \in R^n \times R^m$ the function $f(\cdot, x, u) : [0, \infty) \to R^1$ is bounded;

(vi) the function f is bounded below on $[0, \infty) \times R^n \times R^m$;

(A2) For each $M, \epsilon > 0$ there exist $\Gamma, \delta > 0$ such that if $t \geq 0$, $x_1, x_2 \in R^n$ and if $u_1, u_2 \in R^m$ satisfy

$$|x_1| \leq M, \ |u_1| \geq \Gamma, \ |x_1 - x_2| + |u_1 - u_2| \leq \delta,$$

then

$$|f(t, x_1, u_1) - f(t, x_2, u_2)| \leq \epsilon(f(t, x_1, u_1) + 1);$$

(A3) For each $M, \epsilon > 0$ there exists $\delta > 0$ such that the following property holds:

If $t \geq 0$, $x_1, x_2 \in R^n$ and if $u_1, u_2 \in R^m$ satisfy

$$|x_i| |u_i| \leq M \ (i = 1, 2), \ |x_1 - x_2| + |u_1 - u_2| \leq \delta,$$

then

$$|f(t, x_1, u_1) - f(t, x_2, u_2)| \leq \epsilon.$$

We denote by $|\cdot|$ the Euclidean norm in R^n. For each $r > 0$ and each $x \in R^n$ set

$$\mathcal{B}(r) = \{y \in R^n : |y| \leq r\}, \ \mathcal{B}(x, r) = \{y \in R^n : |x - y| \leq r\}.$$

Remark 1. It is not difficult to see that if λ is a positive number and if Assumptions (A1)-(A3) hold with $f = f_i$, $i = 1, 2$, where $f_1, f_2 : [0, \infty) \times R^n \times R^m \to R^1$ are measurable functions, then Assumptions (A1)-(A3) hold with $f = \lambda f_1$ and $f = f_1 + f_2$.

Remark 2. It is not difficult to see that Assumptions (A1)-(A3) hold with a function $f : [0, \infty) \times R^n \times R^m \to R^1$ defined by

$$f(t, x, u) = g(t)(x - \Gamma(t))'Q(x - \Gamma(t))$$

$$+ h(t)u'Pu + H(t) \ (t \in [0, \infty), \ x \in R^n, \ u \in R^m),$$

where $\Gamma : [0, \infty) \to R^n$ is a measurable and bounded on $[0, \infty)$ function, P, Q are positive definite symmetric matrices, $H, h, g : [0, \infty) \to R^1$ are measurable bounded functions such that

$$\inf\{g(t) : t \in [0, \infty)\} > 0, \ \inf\{h(t) : t \in [0, \infty)\} > 0.$$

The optimal control problem with the autonomous plant (1.1) and this integrand f was studied in [3].

Remark 3. Assume that $h : [0, \infty) \to R^1$ is a bounded measurable function such that

$$\inf\{h(t) : t \in [0, \infty)\} > 0$$

and that a strictly convex function $g = g(x, u) \in C^1(R^{n+m})$ has the following properties:

$$g(x, u) \geq \max\{\psi(x), \ \psi(|u|)|u|\},$$

$$\max\{|\partial g/\partial x(x, u)|, \ |\partial g/\partial u(x, u)|\} \leq \psi_0(|x|)(1 + \psi|u|)|u|,$$

$x \in R^n$, $u \in R^m$, where $\psi : [0, \infty) \to (0, \infty)$, $\psi_0 : (0, \infty) \to [0, \infty)$ are monotone increasing functions, $\psi(t) \to \infty$ as $t \to \infty$; for any $\epsilon > 0$ there exists $\delta(\epsilon) > 0$ such that if $x_1, x_2 \in R^n$ and if $u_1, u_2 \in R^m$ satisfy $|x_1 - x_2| + |u_1 - u_2| \geq \epsilon$, then

$$g(2^{-1}(x_1 + x_2), 2^{-1}(u_1 + u_2)) \leq 2^{-1}[g(x_1, u_1) + g(x_2, u_2)] - \delta(\epsilon).$$

It is not difficult to see that the integrand $f(t, x, u) = h(t)g(x, u)$, $t \in [0, \infty)$, $x \in R^n$, $u \in R^m$ satisfies (A1), (A2) and (A3) and f is not taken from Remarks 1 and 2.

In this chapter we prove the following results which were obtained in [112].

THEOREM 7.1.1 *Assume that (A1) holds and $x_0 \in R^n$. Then there exists a trajectory-control pair $x^* : [0, \infty) \to R^n$, $u^* : [0, \infty) \to R^m$ such that $x^*(0) = x_0$, the function x^* is bounded on $[0, \infty)$ and for any trajectory-control pair $x : [0, \infty) \to R^n$, $u : [0, \infty) \to R^m$ either*
(i) $I(0, T, x, u) - I(0, T, x^*, u^*) \to \infty$ as $T \to \infty$
or
(ii) $\sup\{|I(0, T, x, u) - I(0, T, x^*, u^*)| : T > 0\} < \infty$.
Moreover, if the relation (ii) is true, then

$$\lim_{t \to \infty} (x(t) - x^*(t)) = 0.$$

We say that a trajectory-control pair $x : [0, \infty) \to R^n$, $u : [0, \infty) \to R^m$ is *good* if it satisfies relation (ii) of Theorem 7.1.1.

We say that a trajectory-control pair $\bar{x} : [0, \infty) \rightarrow R^n$, $\bar{u} : [0, \infty) \rightarrow R^m$ is *overtaking optimal* if

$$\limsup_{T \rightarrow \infty} [I(0, T, \bar{x}, \bar{u}) - I(0, T, x, u)] \leq 0$$

for any trajectory-control pair $x : [0, \infty) \rightarrow R^n$, $u : [0, \infty) \rightarrow R^m$ such that $x(0) = \bar{x}(0)$.

THEOREM 7.1.2 *Let (A1)-(A3) hold and let $x_0 \in R^n$. Then there exists an overtaking optimal trajectory-control pair $\bar{x} : [0, \infty) \rightarrow R^n$, $\bar{u} : [0, \infty) \rightarrow R^m$ such that $\bar{x}(0) = x_0$. Moreover if a trajectory-control pair $x : [0, \infty) \rightarrow R^n$, $u : [0, \infty) \rightarrow R^m$ satisfies $x(0) = x_0$ and $(x, u) \neq (\bar{x}, \bar{u})$, then there are a time T_0 and $\epsilon > 0$ such that for all $T \geq T_0$,*

$$I(0, T, x, u) \geq I(0, T, \bar{x}, \bar{u}) + \epsilon.$$

Theorem 7.1.3 describes the limit behavior of overtaking optimal trajectories.

THEOREM 7.1.3 *Let (A1)-(A3) hold, M, ϵ be positive numbers and let (x^*, u^*) be an overtaking optimal trajectory-control pair. Then there exists a number $T_0 > 0$ such that*

$$|x(t) - x^*(t)| \leq \epsilon, \ t \in [T_0, \infty)$$

for each overtaking optimal trajectory-control pair $x : [0, \infty) \rightarrow R^n$, $u : [0, \infty) \rightarrow R^m$ which satisfies $|x(0)| \leq M$. Moreover, there exists a positive number δ such that if $T \geq 0$ and if an overtaking optimal trajectory-control pair $x : [0, \infty) \rightarrow R^n$, $u : [0, \infty) \rightarrow R^m$ satisfy

$$|x(T) - x^*(T)| \leq \delta, \ |x(0)| \leq M,$$

then

$$|x(t) - x^*(t)| \leq \epsilon \ (t \in [T, \infty)).$$

For each $z \in R^n$ and $T > 0$ we set

$$\Delta(z, T) = \inf\{I(0, T, x, u) : \ x : [0, T] \rightarrow R^n, \ u : [0, T] \rightarrow R^m$$

is a trajectory-control pair satisfying $x(0) = z\}$.

We will see that $-\infty < \Delta(z, T) < \infty$.

Theorem 7.1.4 establishes the turnpike property for optimal trajectories with the turnpike $x^*(\cdot)$.

THEOREM 7.1.4 *Assume that (A1)-(A3) hold, M, ϵ are positive numbers and (x^*, u^*) is an overtaking optimal trajectory-control pair. Then there exist numbers $T_0, \delta > 0$ such that:*

(1) If $T > 2T_0$ and if a trajectory-control pair $x : [0, T] \to R^n$, $u : [0, T] \to R^m$ satisfies $|x(0)| \le M$,

$$I(0, T, x, u) \le \Delta(x(0), T) + \delta, \qquad (1.3)$$

then for all $t \in [T_0, T - T_0]$,

$$|x(t) - x^*(t)| \le \epsilon. \qquad (1.4)$$

(2) If $s \ge 0$, $T \ge s + T_0$ and if a trajectory-control pair $x : [0, T] \to R^n$, $u : [0, T] \to R^m$ satisfies (1.3) and

$$|x(0)| \le M, \ |x(s) - x^*(s)| \le \delta,$$

then (1.4) holds for all $t \in [s, T - T_0]$.

Chapter 7 is organized as follows. Section 7.2 contains auxiliary results. A discrete-time control system related to our optimal control problem is discussed in Section 7.3. Theorem 7.1.1 is proved in Section 7.4. Section 5 contains the proof of Theorem 7.1.2. Theorems 7.1.3 and 7.1.4 are proved in Section 7.6.

7.2. Preliminary results

Suppose that (A1) holds. We have the following result (see Proposition 2.3 of [41]).

PROPOSITION 7.2.1 *For every $\tilde{y}, \tilde{z} \in R^n$ and every $T > 0$ there exists a unique solution $x(\cdot), y(\cdot)$ of the following system:*

$$x' = Ax + BB^t y, \ y' = x - A^t y,$$

with the boundary conditions $x(0) = \tilde{y}$, $x(T) = \tilde{z}$ (where B^t denotes the transpose of B).

Let $s \ge 0$ and $\tau > 0$. Consider the function $v_\tau^s(y, z)$ defined on $R^n \times R^n$ by

$$v_\tau^s(y, z) = \min I(s, s + \tau, x, u) \qquad (2.1)$$

$$\text{subject to } x(s) = y, \ x(\tau + s) = z; \qquad (2.2)$$

here $x(\cdot)$ is the response to $u(\cdot)$.

It follows from Proposition 7.2.1 and Assumption (A1) that the function v_τ^s is convex and

$$-\infty < v_\tau^s(y, z) < \infty \text{ for each } y, z \in R^n. \tag{2.3}$$

The following Propositions 7.2.2, 7.2.3 and 7.2.4 can be established analogously to Propositions 6.2.2, 6.2.3 and 6.2.4 respectively.

PROPOSITION 7.2.2 *Let $M_1 > 0$ and $0 < \tau_0 < \tau_1$. Then there exists a number $M_2 > 0$ such that the following property holds:*

If numbers t_1, t_2 satisfy $0 \le t_1 < t_2$, $t_2 - t_1 \in [\tau_0, \tau_1]$ and if a trajectory-control pair $x : [t_1, t_2] \to R^n$, $u : [t_1, t_2] \to R^m$ satisfies

$$I(t_1, t_2, x, u) \le M_1, \tag{2.4}$$

then

$$|x(t)| \le M_2, \ t \in [t_1, t_2].$$

PROPOSITION 7.2.3 *Let $M_1, \epsilon, \tau_0, \tau_1 > 0$ and $0 < \tau_0 < \tau_1$. Then there exists a positive number δ such that the following property holds:*

If $0 \le T_1 < T_2$, $T_2 - T_1 \in [\tau_0, \tau_1]$ and if a trajectory-control pair $x : [T_1, T_2] \to R^n$, $u : [T_1, T_2] \to R^m$ satisfies

$$I(T_1, T_2, x, u) \le M_1,$$

then

$$|x(t_1) - x(t_2)| \le \epsilon$$

for each $t_1, t_2 \in [T_1, T_2]$ such that $|t_1 - t_2| \le \delta$.

PROPOSITION 7.2.4 *Let $M_1 > 0$, $0 \le t_1 < t_2$ and let $\mathcal{F}$ be the set of all trajectory-control pairs $x : [t_1, t_2] \to R^n$, $u : [t_1, t_2] \to R^m$ satisfying (2.4). Then for every sequence $\{(x_i, u_i)\}_{i=1}^\infty \subset \mathcal{F}$ there exist a subsequence $\{(x_{i_k}, u_{i_k})\}_{k=1}^\infty$ and $(x, u) \in \mathcal{F}$ such that $x_{i_k}(t) \to x(t)$ as $k \to \infty$ uniformly in $[t_1, t_2]$, $x'_{i_k} \to x'$ as $k \to \infty$ weakly in $L^1((t_1, t_2); R^n)$ and $u_{i_k} \to u$ as $k \to \infty$ weakly in $L^1((t_1, t_2); R^m)$.*

COROLLARY 7.2.1 *Let $t \ge 0$, $s > 0$ and let $x_1, x_2 \in R^n$. Then there exists a unique trajectory-control pair $x : [t, t + s] \to R^n$, $u : [t, t + s] \to R^m$ such that*

$$x(t) = x_1, \ x(t + s) = x_2$$

and $I(t, t + s, x, u) = v_s^t(x_1, x_2)$.

7.3. Discrete-time control systems

In this section we establish some useful properties of discrete-time control systems with cost function v_τ^s.

Assume that (A1) holds. Proposition 7.2.2 implies the following result.

PROPOSITION 7.3.1 *For each pair of positive numbers τ and M_1 there exists a positive number M_2 such that if $s \in [0, \infty)$ and if $x_1, x_2 \in R^n$ satisfy $|x_1| + |x_2| \geq M_2$, then*

$$v_\tau^s(x_1, x_2) \geq M_1.$$

PROPOSITION 7.3.2 *For each $\tau, M_1 > 0$ the set*

$$\{v_\tau^s(x_1, x_2) : \ s \in [0, \infty), \ x_1, x_2 \in B(M_1)\}$$

is bounded from above.

Proof. Let $\tau, M_1 > 0$. Define a function $f_0 : R^n \times R^m \to R^1$ by

$$f_0(x, u) = \sup\{f(t, x, u) : \ t \in [0, \infty)\}$$

for $x \in R^n$ and $u \in R^m$. It follows from Assumption (A1) that the function f_0 is well defined and convex. For $y, z \in R^n$ set

$$\tilde{v}(y, z) = \inf \int_0^\tau f_0(x(t), u(t)) dt \text{ subject to } x(0) = y, \ x(\tau) = z;$$

here $x(\cdot)$ is the response to $u(\cdot)$. By (A1), Proposition 7.2.1 and the convexity of f_0, the function $\tilde{v} : R^n \times R^n \to R^1$ is well defined and convex. The continuity of $\tilde{v}$ implies the validity of the proposition.

Analogously to Proposition 6.3.4 we can establish the following result.

PROPOSITION 7.3.3 *Let τ, M_1, M_2 be positive numbers such that*

$$\inf\{v_\tau^s(y, z) : \ s \in [0, \infty), \ y, z \in R^n, \ |y| + |z| \geq M_2\}$$

$$> 2 \sup\{|v_\tau^s(0, 0)| : \ ; s \in [0, \infty)\} + 1. \tag{3.1}$$

Then there exists a number $M_3 > M_2$ such that for every pair of integers q_1, q_2 satisfying $0 \leq q_1 < q_2$ and every sequence $\{z_k\}_{k=q_1}^{q_2} \subset R^n$ the following properties hold:
   If

$$z_{q_1}, z_{q_2} \in B(M_2), \max\{|z_k| : \ k = q_1, \ldots, q_2\} > M_3,$$

then there is a sequence $\{y_k\}_{k=q_1}^{q_2} \subset R^n$ *such that* $y_{q_i} = z_{q_i}$, $i = 1, 2$ *and*

$$\sum_{k=q_1}^{q_2-1} [v_\tau^{\tau k}(z_k, z_{k+1}) - v_\tau^{\tau k}(y_k, y_{k+1})] \leq M_1; \tag{3.2}$$

(2) *if* $z_{q_1} \in B(M_2)$ *and* $\sup\{|z_k| : k = q_1, \ldots, q_2\} > M_3$, *then there is a sequence* $\{y_k\}_{k=q_1}^{q_2} \subset R^n$ *such that* $y_{q_1} = z_{q_1}$ *and* (3.2) *holds.*

7.4.   Proof of Theorem 7.1.1

Let $x_0 \in R^n$. Fix a positive number τ. It follows from Proposition 7.2.4 that for any natural number k there exists a trajectory-control pair $x_k : [0, \tau k] \to R^n$, $u_k : [0, \tau_k] \to R^m$ such that $x_k(0) = x_0$ and

$$I(0, \tau k, x_k, u_k) = \Delta(x_0, \tau k).$$

By Corollary 7.2.1 and Proposition 7.3.3,

$$\sup\{|x_k(i\tau)| : k = 1, 2, \ldots, i = 0, \ldots, k\} < \infty. \tag{4.1}$$

There exists a subsequence of trajectory-control pairs $\{(x_{k_j}, u_{k_j})\}_{j=1}^\infty$ such that for any integer $i \geq 0$ there exists

$$x_i^* = \lim_{j \to \infty} x_{k_j}(i\tau). \tag{4.2}$$

In view of Corollary 7.2.1 there is a trajectory-control pair $x^* : [0, \infty) \to R^n$, $u^* : [0, \infty) \to R^m$ such that for each integer $i \geq 0$,

$$x^*(i\tau) = x_i^*, \quad I(i\tau, (i+1)\tau, x^*, u^*) = v_\tau^{\tau i}(x_i^*, x_{i+1}^*). \tag{4.3}$$

Put

$$r_0 = \sup\{|x_k(i\tau)| : k = 1, 2, \ldots, i = 0, \ldots, k\}. \tag{4.4}$$

It follows from Assumption (A1), (4.1), (4.2), (4.3), Proposition 7.2.2 and 7.3.2 that

$$\sup\{|I(\tau i, \tau(i + 1), x^*, u^*)| : i = 0, 1, \ldots\} < \infty,$$

$$\sup\{|x^*(t)| : t \in [0, \infty)\} < \infty. \tag{4.5}$$

Thus we have constructed the trajectory-control pair (x^*, u^*). We will show that (x^*, u^*) has the properties described in Theorem 7.1.1. In order to meet this goal we consider the analogous properties for the related discrete-time control systems.

LEMMA 7.4.1 *For each $S_0 > 0$ there exists $S_1 > S_0 + 8$ such that the following property holds:*

If $q_1 \geq 0$, $q_2 > q_1$ are integers and if a sequence $\{y_i\}_{i=q_1}^{q_2} \subset R^n$ satisfies $y_{q_1} \in B(S_0)$, then

$$\sum_{i=q_1}^{q_2-1} [v_\tau^{\tau i}(x_i^*, x_{i+1}^*) - v_\tau^{\tau i}(y_i, y_{i+1})] \leq S_1. \tag{4.6}$$

Proof. Let $S_0 > 0$. We may assume without loss of generality that (3.1) holds with $M_2 = S_0$. There exists a number $S_2 > S_0$ such that Proposition 7.3.3 holds with $M_1 = 1$, $M_2 = S_0$ and $M_3 = S_2$. Assumption (A1) and Proposition 7.3.2 imply that there exists a number S_1 such that

$$6|v_\tau^\sigma(h_1, h_2)| < S_1 - S_0 - 9 \text{ for each } \sigma \in [0, \infty),$$

$$\text{each } h_1, h_2 \in R^n \text{ satisfying } |h_1| + |h_2| \leq 2r_0 + 2S_2. \tag{4.7}$$

Let $0 \leq q_1 < q_2$ and let the sequence $\{y_i\}_{i=q_1}^{q_2} \subset R^n$ satisfy $y_{q_1} \in B(S_0)$. We will show that (4.6) holds.

Let us assume the converse. Then

$$\sum_{i=q_1}^{q_2-1} [v_\tau^{\tau i}(x_i^*, x_{i+1}^*) - v_\tau^{\tau i}(y_i, y_{i+1})] > S_1. \tag{4.8}$$

We may assume without loss of generality that

$$\sum_{i=q_1}^{q_2-1} v_\tau^{\tau i}(y_i, y_{i+1}) = \inf\{\sum_{i=q_1}^{q_2-1} v_\tau^{\tau i}(z_i, z_{i+1}) :$$

$$\{z_i\}_{i=q_1}^{q_2} \subset R^n \text{ and } z_{q_1} = y_{q_1}\}.$$

In view of Proposition 7.3.3 which holds with $M_1 = 1$, $M_2 = S_0$, $M_3 = S_2$,

$$y_i \in B(S_2) \text{ for all } i = q_1, \ldots, q_2. \tag{4.9}$$

There exists a natural number $k > q_2 + 1$ such that

$$\left| \sum_{i=q_1}^{q_2} [v_\tau^{i\tau}(x_i^*, x_{i+1}^*) - v_\tau^{i\tau}(x_k(i\tau), x_k((i+1)\tau))] \right| \leq 1. \tag{4.10}$$

It is easy to see that

$$\sum_{i=0}^{k-1} v_\tau^{i\tau}(x_k(i\tau), x_k((i+1)\tau))$$

$$= \sum_{i=0}^{k-1} \inf\{v_\tau^{i\tau}(z_i, z_{i+1}) : \{z_i\}_{i=0}^k \subset R^n \text{ and } z_0 = x_0\}. \tag{4.11}$$

Set

$$a_i = x_k(i\tau), \ i \in \{0, \ldots, q_1\} \cup \{q_2 + 1, \ldots, k\}, \ a_i = y_i, \ i = q_1 + 1, \ldots, q_2. \tag{4.12}$$

It follows from the definition of $\{a_i\}_{i=0}^k$, (4.11) with $z_i = a_i$ $(i = 0, \ldots, k)$, (4.10) and (4.8) that

$$0 \geq \sum_{i=0}^{k-1} [v_\tau^{\tau i}(x_k(i\tau), x_k((i+1)\tau)) - v_\tau^{\tau i}(a_i, a_{i+1})]$$

$$\geq -1 + S_1 + v_\tau^{q_2\tau}(x_{q_2}^*, x_{q_2+1}^*) + v_\tau^{q_1\tau}(y_{q_1}, y_{q_1+1})$$

$$- v_\tau^{q_1\tau}(a_{q_1}, a_{q_1+1}) - v_\tau^{q_2\tau}(a_{q_2}, a_{q_2+1}).$$

Using this relation, (4.12), (4.4), (4.2) and (4.9) we obtain an estimation for S_1 which is contradictory to (4.7). The obtained contradiction proves the lemma.

The next auxiliary result establishes a useful property for the discrete-time control system which implies the boundedness of any good trajectory.

LEMMA 7.4.2 *There exists a positive number S_0 such that if a sequence $\{y_i\}_{i=0}^\infty \subset R^n$ satisfies*

$$\limsup_{i\to\infty} |y_i| > S_0, \tag{4.13}$$

then

$$\sum_{j=0}^{N-1} [v_\tau^{i\tau}(y_i, y_{i+1}) - v_\tau^{\tau i}(x_i^*, x_{i+1}^*)] \to \infty \text{ as } N \to \infty. \tag{4.14}$$

Proof. It follows from Propositions 7.3.1 and 7.3.2 that there exists a positive number S_1 such that the inequality (3.1) holds with $M_2 = S_1$ and (4.14) is valid for each sequence $\{y_i\}_{i=0}^\infty$ satisfying

$$\liminf_{i\to\infty} |y_i| > S_1.$$

There exist numbers $S_0 > S_1 + 1$ and $Q > S_0 + 8$ such that Proposition 7.3.3 holds with $M_1 = 4$, $M_2 = S_1 + 1$, $M_3 = S_0$ and Lemma 7.4.1 holds with $S_1 = Q$.

Assume that a sequence $\{y_i\}_{i=0}^\infty \subset R^n$ satisfies (4.13). We will establish (4.14). In view of the choice of S_1 we may assume that

$$\liminf_{j\to\infty} |y_i| \leq S_1. \tag{4.15}$$

(4.13) and (4.15) imply that there exists a subsequence $\{y_{i_k}\}_{k=1}^{\infty}$ such that

$$0 < i_1, \ y_{i_k} \in B(S_1 + 1), \ \sup\{|y_i| : \ j = i_k, \ldots, i_{k+1}\} > S_0 \ (k = 1, 2, \ldots).$$
(4.16)

By Proposition 7.3.3 which holds with $M_1 = 4$, $M_2 = S_1 + 1$, $M_3 = S_0$, for any natural number k there exists a sequence $\{z_j\}_{j=i_k}^{i_k+1} \subset R^n$ such that

$$z_j = y_j \ (j \in \{i_k, i_{k+1}\}), \ \sum_{j=i_k}^{i_{k+1}-1} [v_T^{\tau j}(y_j, y_{j+1}) - v_T^{\tau j}(z_j, z_{j+1})] \geq 4. \ (4.17)$$

Relation (4.14) now follows from (4.17), (4.16) and Lemma 7.4.1 applied with $S_1 = Q$. The lemma is proved.

The next lemma follows from (4.3), Lemma 7.4.1, Assumption (A1), and the definition of (x^*, u^*).

LEMMA 7.4.3 *For each positive number S_0 there exists a positive number S_1 such that the following property holds:*

If an integer $q \geq 0$, $T \in (q\tau, \infty)$ and if a trajectory-control pair $x :$ $[q\tau, T] \to R^n$, $u : [q\tau, T] \to R^m$ satisfies $|x(q\tau)| \in B(S_0)$, then

$$I(q\tau, T, x^*, u^*) \leq I(q\tau, T, x, u) + S_1.$$

The next auxiliary result shows that a trajectory-control pair (x, u) is not good if the function $|x(t)|$ is not bounded at infinity by S_1. Its proof is based on Lemmas 7.4.2 and 7.4.3 and Proposition 7.2.2.

LEMMA 7.4.4 *There is a positive number S_1 such that if a trajectory-control pair $x : [0, \infty) \to R^n$, $u : [0, \infty) \to R^m$ satisfies*

$$\limsup_{t \to \infty} |x(t)| > S_1, \tag{4.18}$$

then

$$I(0, T, x, u) - I(0, T, x^*, u^*) \to \infty \ as \ T \to \infty. \tag{4.19}$$

Proof. It follows from Lemma 7.4.2 that there exists a positive number S_0 such that the following property holds:

If a sequence $\{y_i\}_{i=0}^{\infty} \subset R^n$ satisfies (4.13), then relation (4.14) holds.

Lemma 7.4.3, Proposition 7.2.2 and (4.5) imply that there exists a number $S_1 > S_0 + 1$ such that the following property holds:

If a trajectory-control pair $x : [0, \infty) \to R^n$, $u : [0, \infty) \to R^m$ satisfies (4.18) and

$$\limsup_{i \to \infty} |x(i\tau)| \leq S_0 \ \text{where} \ i \ \text{is an integer}, \tag{4.20}$$

then relation (4.19) holds.

Suppose that $x : [0, \infty) \to R^n$, $u : [0, \infty) \to R^m$ is a trajectory-control pair which satisfies (4.18). Then (4.19) follows if (4.20) holds. Otherwise (4.19) follows from Lemma 7.4.2, Assumption (A1), and the definition of (x^*, u^*). The lemma is proved.

The next lemma establishes the first part of Theorem 7.1.1.

LEMMA 7.4.5 *Let* $x : [0, \infty) \to R^n$, $u : [0, \infty) \to R^m$ *be a trajectory-control pair. Then either*

(A) $I(0, T, x, u) - I(0, T, x^*, u^*) \to \infty$ *as* $T \to \infty$

or

(B) $\sup\{|I(0, T, x, u) - I(0, T, x^*, u^*)| : T \in (0, \infty)\} < \infty$.

Proof. In view of Lemma 7.4.4 we may assume without loss of generality that the function $x : [0, \infty) \to R^n$ is bounded. Lemma 7.4.3 implies that there exists a positive number S_1 such that

$$I(q\tau, T, x^*, u^*) \le I(q\tau, T, x, u) + S_1 \qquad (4.21)$$

for each natural number q and each number $T > q\tau$.

Assume that (B) does not hold. By (4.5) and Assumption (A1) the sequence

$$\{|I(0, N\tau, x, u) - I(0, N\tau, x^*, u^*)|\}_{N=0}^{\infty}$$

is not bounded. Combined with (4.21) this implies (A). The lemma is proved.

The next auxiliary result proves the second part of Theorem 7.1.1.

LEMMA 7.4.6 *Assume that* $x : [0, \infty) \to R^n$, $u : [0, \infty) \to R^m$ *is a trajectory-control pair such that* $\lim_{t \to \infty} \sup |x(t) - x^*(t)| > 0$. *Then*

$$\lim_{T \to \infty} I(0, T, x, u) - I(0, T, x^*, u^*) = \infty. \qquad (4.22)$$

Proof. In view of Lemma 7.4.4 we may assume without loss of generality that the function $x : [0, \infty) \to R^n$ is bounded. Put

$$\epsilon = 2^{-1} \limsup_{t \to \infty} |x(t) - x^*(t)|, \qquad (4.23)$$

$$y(t) = 2^{-1}(x(t) + x^*(t)), \quad w(t) = 2^{-1}(u(t) + u^*(t)) \ (t \in [0, \infty)). \quad (4.24)$$

In view of the boundedness of x, the definition of (x^*, u^*), and Lemmas 7.4.3 and 7.4.5 we may assume that the sequence $\{I(i\tau, (i+1)\tau, x, u)\}_{i=0}^{\infty}$ is bounded from above. Then by (4.23) and Proposition 7.2.3 there exist a positive number $\delta < 1$ and a sequence of nonnegative numbers $\{T_i\}_{i=0}^{\infty}$ such that

$$T_{i+1} - T_i \geq 10, \ i = 0, 1, \ldots, \ |x(t) - x^*(t)| \geq 2^{-1}\epsilon$$

$$(t \in [T_i - \delta, T_i + \delta], \ i = 0, 1, 2, \ldots).$$

It follows from these relations, (4.24) and Assumption (A1) that

$$2^{-1}[I(0, T, x, u) + I(0, T, x^*, u^*)] - I(0, T, y, w) \to \infty \text{ as } T \to \infty. \quad (4.25)$$

By the boundedness of x, (4.24), (4.5) and Lemma 7.4.3, the function

$$T \to I(0, T, y, w) - I(0, T, x^*, u^*), \ T \in (0, \infty)$$

is bounded from below. Combined with (4.25) this implies (4.22). The lemma is proved.

Theorem 7.1.1 now follows from (4.5) and Lemmas 7.4.3, 7.4.5 and 7.4.6.

7.5.   Proof of Theorem 7.1.2

Assume that (A1)-(A3) hold. We begin with the following auxiliary result which establishes a useful continuity property of the function v_τ^s. It should be mentioned that the number δ in the statement of Lemma 7.5.1 depends only on τ, ϵ and M and does not depend on s.

LEMMA 7.5.1 *For each* $M, \tau, \epsilon > 0$ *there exists a positive number* δ *such that the following property holds:*
If $s \in [0, \infty)$ *and if* $y_1, y_2, z_1, z_2 \in R^n$ *satisfy*

$$|z_i|, |y_i| \in B(M), \ i = 1, 2, \ \max\{|y_1 - y_2|, |z_1 - z_2|\} \leq \delta, \quad (5.1)$$

then

$$|v_\tau^s(y_1, z_1) - v_\tau^s(y_2, z_2)| \leq \epsilon. \quad (5.2)$$

Proof. Let M, τ, ϵ be positive numbers. It follows from Proposition 7.2.1 that for each $\tilde{y}, \tilde{z} \in R^n$ there exists a unique solution $x(\cdot)$, $y(\cdot)$ of the following system:

$$(x', y')^t = C((x, y)^t), \quad (5.3)$$

with the boundary constraints $x(0) = \tilde{y}$, $x(\tau) = \tilde{z}$ and

$$C(x, y)^t = (Ax + BB^t y, x - A^t y)^t \quad (5.4)$$

(here $I : R^n \to R^n$ is the identity operator, $Iz = z$, $z \in R^n$).

For any initial value $(x_0, y_0) \in R^n \times R^n$ there exists a unique solution of (5.3) satisfying

$$(x(s), y(s))^t = e^{sC}(x_0, y_0)^s, \ s \in R^1.$$

Clearly for each $\tilde{y}, \tilde{z} \in R^n$ there exists a unique vector $D(\tilde{y}, \tilde{z}) \in R^n$ such that the function

$$(x(s), y(s)) = (e^{sC}(\tilde{y}, D(\tilde{y}, \tilde{z}))^t)^t, \ s \in R^1$$

satisfies (5.3) with the boundary constraints $x(0) = \tilde{y}$, $x(\tau) = \tilde{z}$. It is easy to see that $D : R^n \times R^n \to R^1$ is a linear operator. Set

$$M_0 = \sup\{|v_\tau^s(y, z)| : \ y, z \in B(M+1), \ s \in [0, \infty)\}. \qquad (5.5)$$

Assumption (A1) and Proposition 7.3.2 imply that M_0 is finite. It follows from Proposition 7.2.2 that there exists a positive number M_1 such that the following property holds:

If $s \geq 0$ and if a trajectory-control pair $x : [s, s + \tau] \to R^n$, $u : [s, s + \tau] \to R^m$ satisfies $I(s, s + \tau, x, u) \leq 4M_0 + 1$, then

$$x(t) \in B(M_1), \ t \in [s, s + \tau]. \qquad (5.6)$$

Choose numbers $c_1, \delta_1 > 0$ such that

$$f(t, x, u) \geq -c_1, \ ((t, x, u) \in R^{2n+1}), \ 4\delta(2\tau + c_1 + c_1\tau + M_0) \leq \epsilon. \ (5.7)$$

It follows from Assumption (A2) that there exist a positive number Γ_0 and $\delta_2 \in (0, 8^{-1})$ such that the following property holds:

If $t \geq 0$, if $x_1 \in B(M_1)$, $x_2 \in R^n$ and if $u_1, u_2 \in R^m$ satisfy

$$|u_1| \geq \Gamma_0, \ |x_1 - x_2| + |u_1 - u_2| \leq \delta_2, \qquad (5.8)$$

then

$$|f(t, x_1, u_1) - f(t, x_2, u_2)| \leq \delta_1(f(t, x_1, u_1) + 1). \qquad (5.9)$$

It follows from Assumption (A3) that there exists

$$\delta_3 \in (0, 4^{-1}\min\{\delta_1, \delta_2\}) \qquad (5.10)$$

such that the following property holds:

If $t \geq 0$, $x_1, x_2 \in R^n$ and if $u_1, u_2 \in R^m$ satisfy

$$|x_i|, |u_i| \leq \Gamma_0 + M_1 + 1, \ i = 1, 2, \ |x_1 - x_2| + |u_1 - u_2| \leq \delta_3, \quad (5.11)$$

then

$$|f(t, x_1, u_1) - f(t, x_2, u_2)| \leq \delta_1. \qquad (5.12)$$

There exists a number

$$\delta \in (0, 8^{-1}\delta_3) \tag{5.13}$$

such that for each $y, z \in B(\delta)$,

$$(1 + ||B||)|e^{sC}(y, D(y, z))^t| \le 2^{-1}\delta_3 \ (s \in [0, \tau]). \tag{5.14}$$

Assume that $s \in [0, \infty)$ and $y_1, y_2, z_1, z_2 \in R^n$ satisfy (5.1). By Corollary 7.2.1 there exists a trajectory-control pair $x_1 : [s, s + \tau] \to R^n$, $u_1 : [s, s + \tau] \to R^m$ such that

$$x_1(s) = y_1, \ x_1(s + \tau) = z_1, \ I(s, s + \tau, x_1, u_1) = v_\tau^s(y_1, z_1). \tag{5.15}$$

By (5.1), (5.15), (5.5) and the choice of M_1,

$$|x_1(t)| \in B(M_1), \ t \in [s, s + \tau]. \tag{5.16}$$

Define functions $h_1, h_2 : R^1 \to R^n$ by

$$(h_1(r), h_2(r))^t = e^{rC}(y_2 - y_1, D(y_2 - y_1, z_2 - z_1))^t, \ r \in R^1. \tag{5.17}$$

Put

$$x_2(t) = x_1(t) + h_1(t - s), \ u_2(t) = u_1(t) + B^t h_2(t - s) \ (t \in [s, s + \tau]). \tag{5.18}$$

By the definition of D_1, D_2, (5.17), (5.3), (5.4) and (5.15),

$$(x_2, u_2) \text{ is a trajectory-control pair, } x_2(s) = y_2, \ x_2(s + \tau) = z_2. \tag{5.19}$$

It follows from (5.1), (5.17), (5.19) and the choice of δ that

$$|x_2(t) - x_1(t)|, \ |u_2(t) - u_1(t)| \le 2^{-1}\delta_3 \ (t \in [s, s + \tau]). \tag{5.20}$$

By (5.1), (5.15), (5.7) and (5.5) for any measurable set $E \subset [s, s + \tau]$,

$$\int_E f(t, x_1(t), u_1(t))dt \le M_0 + c_1\tau. \tag{5.21}$$

Put

$$E_1 = \{t \in [s, s + \tau] : |u_1(t)| \ge \Gamma_0\}, \ E_2 = [s, s + \tau] \setminus E_1.$$

We will estimate separately

$$\int_{E_i} |f(t, x_1(t), u_1(t)) - f(t, x_2(t), u_2(t))|dt, \ i = 1, 2.$$

It follows from (5.16), (5.20), the definition of Γ_0, δ_2 and (5.21) that

$$\int_{E_1} |f(t, x_1(t), u_1(t)) - f(t, x_2(t), u_2(t))|dt$$

$$\leq \delta_1 \tau + \delta_1 \left(\int_{E_1} f(t, x_1(t), u_1(t)) dt \right) \qquad (5.22)$$

$$\leq \delta_1 \tau + c_1 \tau \delta_1 + \delta_1 M_0.$$

By (5.16), (5.20) and the definition of δ_3,

$$\int_{E_2} |f(t, x_1(t), u_1(t)) - f(t, x_2(t), u_2(t))| dt \leq \delta_1 \tau.$$

This relation, (5.22), (5.7), (5.15) and (5.19) imply that

$$v_\tau^s(y_2, z_2) \leq v_\tau^s(y_1, z_1) + \epsilon.$$

This completes the proof of the lemma.

Let $x_0 \in R^n$, $\tau > 0$. Consider the trajectory-control pairs x^* : $[0, \infty) \to R^n$, $u^* : [0, \infty) \to R^m$, $x_k : [0, \tau k] \to R^n$, $u_k : [0, \tau k] \to R^m$, $k = 1, 2, \ldots$ defined in Section 7.4 (see (4.1)-(4.3)).

We will show that (x^*, u^*) is a unique overtaking optimal trajectory-control pair which satisfies the initial condition $x^*(0) = x_0$. The following lemma is one of the important gradients in our proof.

LEMMA 7.5.2 *Let $0 \leq \tau_1 < \tau_2$ and let $x : [\tau_1, \tau_2] \to R^n$, $u : [\tau_1, \tau_2] \to R^m$ be a trajectory-control pair such that*

$$x(\tau_1) = x^*(\tau_i) \ (i = 1, 2), \ (x, u) \neq (x^*, u^*)|_{[\tau_1, \tau_2]} \qquad (5.23)$$

where $(x^*, u^*)|_{[\tau_1, \tau_2]}$ is the restriction of (x^*, u^*) to $[\tau_1, \tau_2]$. Then

$$I(\tau_1, \tau_2, x, u) > I(\tau_1, \tau_2, x^*, u^*).$$

Proof. By Assumption (A1) and (5.23) it is sufficient to show that

$$I(\tau_1, \tau_2, x, u) \geq I(\tau_1, \tau_2, x^*, u^*).$$

Let us assume the converse. Fix an integer q_0 and a number ϵ such that

$$\epsilon \in (0, 8^{-1}[I(\tau_1, \tau_2, x^*, u^*) - I(\tau_1, \tau_2, x, u)]), \ q_0 > \tau^{-1}\tau_2 + 5. \qquad (5.24)$$

By (4.1), (4.2), (4.3) and the continuity of $v_\tau^s : R^n \times R^n \to R^1$, there exists a natural number $k > 2q_0 + 4$ such that

$$|v_\tau^{\tau i}(x^*(i\tau), x^*((i+1)\tau)) - v_\tau^{\tau i}(x_k(i\tau), x_k((i+1)\tau))|$$

$$\leq (2q_0 + 1)^{-1}\epsilon \ (i = 0, \ldots, 2q_0 + 1), \qquad (5.25)$$

$$|v_\tau^{q_0\tau}(x^*(q_0\tau), x_k((q_0+1)\tau)) - v_\tau^{q_0\tau}(x_k(q_0\tau), x_k((q_0+1)\tau))| \leq (2q_0+1)^{-1}\epsilon.$$

It follows from Corollary 7.2.1, (5.23) and (5.24) that there exists a trajectory-control pair $y : [0, \tau k] \to R^n$, $w : [0, \tau k] \to R^m$ such that

$$y(t) = x^*(t), \; w(t) = u^*(t)$$

$$(t \in [0, \tau_1] \cup [\tau_2, q_0\tau]),$$

$$y(t) = x(t), \; w(t) = u(t) \; (t \in [\tau_1, \tau_2]),$$

$$y(t) = x_k(t), \; w(t) = u_k(t) \; (t \in [(q_0 + 1)\tau, k\tau]), \qquad (5.26)$$

$$I(q_0\tau, (q_0 + 1)\tau, y, w) = v_\tau^{q_0\tau}(y(q_0\tau), y((q_0 + 1)\tau)).$$

In view of the definition of x_k, u_k, y and w,

$$I(0, \tau k, y, w) \geq I(0, \tau k, x_k, u_k).$$

On the other hand by (5.26), (4.3), (5.25), (5.24) and the definition of (x_k, u_k),

$$I(0, \tau k, y, w) - I(0, \tau k, x_k, u_k) = I(0, \tau q_0, y, w) - I(0, \tau q_0, x^*, u^*)$$

$$+ v_\tau^{q_0\tau}(x^*(q_0\tau), x_k((q_0 + 1)\tau)) + I(0, \tau q_0, x^*, u^*) - I(0, \tau q_0, x_k, u_k)$$

$$- v_\tau^{q_0\tau}(x_k(q_0\tau), x_k((q_0 + 1)\tau)) = I(\tau_1, \tau_2, x, u) - I(\tau_1, \tau_2, x^*, u^*)$$

$$+ \sum_{i=0}^{q_0-1} [v_\tau^{\tau i}(x^*(i\tau), x^*((i + 1)\tau)) - v_\tau^{\tau i}(x_k(i\tau), x_k((i + 1)\tau))]$$

$$+ v_\tau^{q_0\tau}(x^*(q_0\tau), x_k((q_0 + 1)\tau)) - v_\tau^{q_0\tau}(x_k(q_0\tau), x_k((q_0 + 1)\tau))$$

$$\leq I(\tau_1, \tau_2, x, u) - I(\tau_1, \tau_2, x^*, u^*) + \epsilon < -\epsilon.$$

The obtained contradiction proves the lemma.

The following auxiliary result shows that (x^*, u^*) is an overtaking optimal trajectory-control pair.

LEMMA 7.5.3 *Let* $x : [0, \infty) \to R^n$, $u : [0, \infty) \to R^m$ *be a trajectory-control pair such that* $x(0) = x_0$. *Then*

$$\limsup_{T \to \infty}[I(0, T, x^*, u^*) - I(0, T, x, u)] \leq 0.$$

Proof. Let us assume the converse. Put

$$\epsilon = 2^{-1} \limsup_{T \to \infty}[I(0, T, x^*, u^*) - I(0, T, x, u)]. \qquad (5.27)$$

Lemma 7.4.6 implies that

$$\lim_{t \to \infty} (x(t) - x^*(t)) = 0. \qquad (5.28)$$

Put

$$M_0 = \sup\{|x(t)| + |x^*(t)| : t \in [0, \infty)\}. \tag{5.29}$$

By (5.28) and (4.5), M_0 is finite.

It follows from Lemma 7.5.1 that there exists a positive number δ_0 such that the following property holds:

If $s \geq 0$, $y_1, y_2, z_1, z_2 \in B(M_0 + 1)$ and if

$$\max\{|y_1 - y_2|, |z_1 - z_2|\} \leq \delta_0,$$

then

$$|v_\tau^s(y_1, z_1) - v_\tau^s(y_2, z_2)| \leq 16^{-1}\epsilon. \tag{5.30}$$

(5.28) implies that there exists a natural number N_0 such that

$$|x(t) - x^*(t)| \leq 8^{-1}\delta_0 \text{ for each } t \geq N_0. \tag{5.31}$$

It follows from (5.27) that there exists a number $T_0 > 2N_0 + 2$ such that

$$I(0, T_0, x^*, u^*) - I(0, T_0, x, u) \geq \epsilon. \tag{5.32}$$

Corollary 7.2.1 implies that there exists a trajectory-control pair $y : [0, \infty) \to R^n$, $w : [0, \infty) \to R^m$ such that

$$y(t) = x(t), \ w(t) = u(t) \quad (t \in [0, T_0]),$$

$$y(t) = x^*(t), \ w(t) = u^*(t) \quad (t \in [T_0 + \tau, \infty)),$$

$$I(T_0, T_0 + \tau, y, w) = v_\tau^{T_0}(x(T_0), x^*(T_0 + \tau)).$$

By Lemma 7.5.2,

$$I(T_0, T_0 + \tau, x^*, u^*) = v_\tau^{T_0}(x^*(T_0), x^*(T_0 + \tau)). \tag{5.33}$$

It follows from the definition of (y, w), (5.32), and (5.33) that

$$I(0, T_0 + \tau, y, w) - I(0, T_0, x^*, u^*) \leq I(0, T_0, x, u) - I(0, T_0, x^*, u^*)$$

$$+ v_\tau^{T_0}(x(T_0), x^*(T_0 + \tau)) - v_\tau^{T_0}(x^*(T_0), x^*(T_0 + \tau))$$

$$\leq -\epsilon + v_\tau^{T_0}(x(T_0), x^*(T_0 + \tau)) - v_\tau^{T_0}(x^*(T_0), x^*(T_0 + \tau)).$$

It follows from this relation, (5.29), (5.30), (5.31), and the definition of δ_0 that

$$I(0, T_0 + \tau, y, w) - I(0, T_0 + \tau, x^*, u^*) \leq -2^{-1}\epsilon.$$

This is contradictory to Lemma 7.5.2. The obtained contradiction proves the lemma.

Proof of Theorem 7.1.2. Let $x : [0, \infty) \to R^n$, $u : [0, \infty) \to R^m$ be a trajectory-control pair such that

$$x(0) = x_0, \ (x, u) \neq (x^*, u^*). \tag{5.34}$$

We set

$$y(t) = 2^{-1}(x(t) + x^*(t)), \ w(t) = 2^{-1}(u(t) + u^*(t)) \quad (t \in [0, \infty)). \tag{5.35}$$

Clearly, (y, w) is a trajectory-control pair satisfying $y(0) = x_0$. By Assumption (A1) (i) and (5.34) there exists a number $\gamma > 0$ such that for all T,

$$I(0, T, y, w) \leq 2^{-1} I(0, T, x, u) + 2^{-1} I(0, T, x^*, u^*) - \gamma. \tag{5.36}$$

By Lemma 7.5.3,

$$\limsup_{T \to \infty} [I(0, T, x^*, u^*) - I(0, T, y, w)] \leq 0.$$

Together with (5.36) this implies the validity of the theorem.

7.6. Proofs of Theorems 7.1.3 and 7.1.4

In this section we establish the turnpike property of optimal trajectory-control pairs.

Assume that (A1)-(A3) hold, $x_0 \in R^n$, $\tau > 0$ and consider the trajectory-control pairs $x^* : [0, \infty) \to R^n$, $u^* : [0, \infty) \to R^m$, $x_k : [0, \tau k] \to R^n$, $u_k : [0, \tau k] \to R^m$ ($k = 1, 2, \ldots$) defined in Section 7.4 (see (4.1)-(4.3)). It was shown in Section 7.5 that (x^*, u^*) is an overtaking optimal trajectory-control pair.

We begin with the following auxiliary result which shows that all overtaking optimal trajectory-control pairs starting from a given bounded set stay in a certain ball.

LEMMA 7.6.1 *For each $M_0 > 0$ there exists a positive number M_1 such that the following property holds:*

If an overtaking optimal trajectory-control pair $x : [0, \infty) \to R^n$, $u : [0, \infty) \to R^m$ satisfies $x(0) \in B(M_0)$, then $x(t) \in B(M_1)$ for all $t \in [0, \infty)$.

Proof. Let M_0 be a positive number. It follows from Lemma 7.4.4 that there exists a positive number S_1 such that for any overtaking optimal trajectory-control pair (x, u),

$$\limsup_{t \to \infty} |x(t)| \leq S_1.$$

Combined with Proposition 7.3.3 this inequality implies that there exists a number $S_2 > 0$ such that the following property holds:

If an overtaking optimal trajectory-control pair (x, u) satisfies $x(0) \in B(M_0)$, then $x(\tau i) \in B(S_2)$ for all integers $i \geq 0$. The validity of the lemma now follows from Propositions 7.2.2 and 7.3.2.

Propositions 7.3.2, 7.3.3 and Assumption (A1) imply the following result.

LEMMA 7.6.2 *For each pair of positive numbers M_0, M_1 there exists $M_2 > M_1$ such that the following property holds:*

If N is a natural number, $T \in [N\tau, (N+1)\tau)$, if a trajectory-control pair $x : [0, T] \to R^n$, $u : [0, T] \to R^m$ satisfies

$$x(0) \in B(M_0), \; I(0, T, x, u) \leq \Delta(x(0), T) + M_1 \tag{6.1}$$

and if a sequence $\{y_k\}_{k=0}^N \subset R^n$ satisfies $y_0 = x(0)$, then

$$\sum_{k=0}^{N-1} v_\tau^{k\tau}(x(k\tau), x((k+1)\tau)) \leq \sum_{k=0}^{N-1} v_\tau^{k\tau}(y_k, y_{k+1}) + M_2.$$

Propositions 7.3.3, 7.3.2, 7.2.2 and Lemma 7.6.2 imply the following result which shows that for an approximate optimal trajectory-control pair (x, u) satisfying (6.1) the function $|x(t)|$ is bounded by a constant which depends only on M_0, M_1 and does not depend on T.

LEMMA 7.6.3 *For each pair of positive numbers M_0, M_1 there exists $M_2 > 0$ such that the following property holds:*

If N is a natural number, $T \in [N\tau, (N+1)\tau)$ and if a trajectory-control pair $x : [0, T] \to R^n$, $u : [0, T] \to R^m$ satisfies (6.1), then $x(t) \in B(M_2)$ for all $t \in [0, N\tau]$.

The next lemma is an important tool in our proof. It shows that if an approximate optimal trajectory-control pair (x, u) is defined on an interval $[0, T]$ with a sufficiently large T, then $x(t)$ is close to $x^*(t)$ at some point $t \in [0, T]$.

LEMMA 7.6.4 *Let $\epsilon, M_0, M_1 > 0$. Then there is a natural number N_0 such that the following property holds:*

If $T \geq N_0\tau$ and if a trajectory-control pair $x : [0, T] \to R^n$, $u : [0, T] \to R^m$ satisfies

$$x(t) \in B(M_0), \; t \in [0, T], \; I(0, T, x, u) \leq v_T^0(x(0), x(T)) + M_1, \tag{6.2}$$

then

$$\inf\{|x(t) - x^*(t)| \, : \, t \in [s, s + N_0\tau]\} \leq \epsilon \tag{6.3}$$

for each number s satisfying $0 \leq s \leq T - \tau N_0$.

Proof. We may assume that

$$x^*(t) \in B(M_0), \ t \in [0, \infty). \tag{6.4}$$

Assumption (A1) and Proposition 7.3.2 imply that there exists a number

$$M_2 > \sup\{|v_\tau^s(z_1, z_2)| : \ z_1, z_2 \in B(2M_0), \ s \in [0, \infty)\} + 1. \tag{6.5}$$

It follows from Lemma 7.4.3 that there exists a positive number S_1 such that the following property holds:

If an integer $q \geq 0$, $T > q\tau$ and if a trajectory-control pair $x : [q\tau, T] \to R^n$, $u : [q\tau, T] \to R^m$ satisfies $x(q\tau) \in B(M_0)$, then

$$I(q\tau, T, x^*, u^*) \leq I(q\tau, T, x, u) + S_1. \tag{6.6}$$

It follows from Assumption (A1) that there exists a positive number δ_0 such that the following property holds:

If $t \in [0, \infty)$ and if $x_1, x_2 \in R^n$, $u_1, u_2 \in R^m$ satisfy

$$|x_1 - x_2| + |u_1 - u_2| \geq 8^{-1}\epsilon,$$

then

$$2^{-1}f(t, x_1, u_1) + 2^{-1}f(t, x_2, u_2) - f(t, 2^{-1}(x_1 + x_2), 2^{-1}(u_1 + u_2)) \geq \delta_0. \tag{6.7}$$

Choose a natural number

$$N_0 \geq 4 + (\delta_0 \tau)^{-1}(4M_2 + 2M_1 + S_1 + 1). \tag{6.8}$$

Let $T \geq N_0\tau$, $0 \leq s \leq T - \tau N_0$ and let $x : [0, T] \to R^n$, $u : [0, T] \to R^m$ be a trajectory-control pair satisfying (6.2). We will show that (6.3) holds.

Let us assume the converse. Then

$$|x(t) - x^*(t)| > \epsilon \text{ for all } t \in [s, s + N_0\tau]. \tag{6.9}$$

By Corollary 7.2.1 there exists a trajectory-control pair $x_1 : [0, T] \to R^n$, $u_1 : [0, T] \to R^m$ such that

$$x_1(0) = x(0), \ x_1(t) = x^*(t),$$

$$u_1(t) = u^*(t) \ (t \in [\tau, T - \tau]), \ x_1(T) = x(T),$$

$$I(j\tau, (j+1)\tau, x_1, u_1) = v_\tau^{j\tau}(x_1(j\tau), x_1((j+1)\tau)), \ j = 0, T - \tau. \tag{6.10}$$

Since (x^*, u^*) is an overtaking optimal trajectory-control pair it follows from (6.10), (6.5), (6.2), and (6.4) that

$$|I(0, T, x^*, u^*) - I(0, T, x_1, u_1)| \leq 4M_2. \tag{6.11}$$

In view of (6.2) and the choice of S_1,

$$I(0, T, x^*, u^*) \leq I(0, T, x, u) + S_1. \tag{6.12}$$

Define a trajectory-control pair $x_2 : [0, T] \to R^n$, $u_2 : [0, T] \to R^m$ as follows:

$$x_2(t) = 2^{-1}(x_1(t) + x(t)), \ u_2(t) = 2^{-1}(u_1(t) + u(t)) \text{ for } t \in [0, T]. \tag{6.13}$$

It follows from (6.2), (6.13) and (6.10) that

$$I(0, T, x, u) \leq I(0, T, x_2, u_2) + M_1. \tag{6.14}$$

By (6.13), (6.10), (6.9), Assumption (A1) and the choice of δ_0,

$$I(0, T, x_2, u_2) \leq 2^{-1}I(0, T, x, u) + 2^{-1}I(0, T, x_1, u_1) - \delta_0(N_0 - 2)\tau. \tag{6.15}$$

(6.11) and (6.12) imply that $I(0, T, x_1, u_1) \leq I(0, T, x, u) + 4M_2 + S_1$. Combining this relation and (6.14) and (6.15) we obtain a relation which is contradictory to (6.8). The obtained contradiction proves the lemma.

The following auxiliary result is another important ingredient in our proof. It establishes that if $x(T_i)$ is close enough to $x^*(T_i)$, $i = 1, 2$, then an approximate optimal trajectory-control pair (x, u) defined on an interval $[T_1, T_2]$ is close to x^* at any point of $[T_1, T_2]$.

LEMMA 7.6.5 *For each $\epsilon, M_0 > 0$ there exists a positive number δ such that the following property holds:*

If $T_1 \geq 0$, $T_2 \geq T_1 + 3\tau$ and if a trajectory-control pair $x : [T_1, T_2] \to R^n$, $u : [T_1, T_2] \to R^m$ satisfies

$$x(t) \in B(M_0), \ t \in [T_1, T_2], \ |x(T_i) - x^*(T_1)| \leq \delta, \ i = 1, 2,$$

$$I(T_1, T_2, x, u) \leq v_{T_2-T_1}^{T_1}(x(T_1), x(T_2)) + \delta, \tag{6.16}$$

then

$$|x(t) - x^*(t)| \leq \epsilon \text{ for all } t \in [T_1, T_2]. \tag{6.17}$$

Proof. Let ϵ, M_0 be positive numbers. We may assume that

$$x^*(t) \in B(M_0) \text{ for all } t \in [0, \infty). \tag{6.18}$$

Assumption (A1) and Proposition 7.3.2 imply that there exists a number

$$M_1 > \sup\{|v_{2\tau}^s(z_1, z_2)| : s \in [0, \infty), z_1, z_2 \in R^n, |z_1| + |z_2| \leq 2M_0 + 1\}. \tag{6.19}$$

It follows from Proposition 7.2.3 that there exists a number $\Delta \in (0, 4^{-1}\tau)$ such that the following property holds:

If $s \geq 0$ and if a trajectory-control pair $x : [s, s + 2\tau] \rightarrow R^n$, $u : [s, s + 2\tau] \rightarrow R^m$ satisfies

$$I(s, s + 2\tau, x, u) \leq 2M_1 + 2, \tag{6.20}$$

then

$$|x(t_1) - x(t_2)| \leq 8^{-1}\epsilon$$

for each $t_1, t_2 \in [s, s + 2\tau]$ such that $|t_1 - t_2| \leq \Delta$.

By Assumption (A1) (i) there is a positive number δ_1 such that the following property holds:

If $t \geq 0$ and if $x_1, x_2 \in R^n$ and $u_1, u_2 \in R^m$ satisfy

$$|x_1 - x_2| + |u_1 - u_2| \geq 8^{-1}\epsilon,$$

then

$$2^{-1}f(t, x_1, u_1) + 2^{-1}f(t, x_2, u_2) - f(t, 2^{-1}(x_1 + x_2), 2^{-1}(u_1 + u_2)) \geq \delta_1. \tag{6.21}$$

It follows from Lemma 7.5.1 that there exists a positive number

$$\delta < \min\{16^{-1}, 16^{-1}\epsilon, 16^{-1}\delta_1\Delta\} \tag{6.22}$$

such that the following property holds:

If $s \geq 0$ and if $y_1, y_2, z_1, z_2 \in R^n$ satisfy

$$z_i, y_i \in B(M_0 + 1) \ (i = 1, 2), \ \max\{|y_1 - y_2|, |z_1 - z_2|\} \leq \delta, \tag{6.23}$$

then

$$|v_\Delta^s(y_1, z_1) - v_\Delta^s(y_2, z_2)| \leq 16^{-1}\delta_1\Delta. \tag{6.24}$$

Let $T_1 \geq 0$, $T_2 \geq T_1 + 3\tau$ and let $x : [T_1, T_2] \rightarrow R^n$, $u : [T_1, T_2] \rightarrow R^m$ be a trajectory-control pair which satisfies (6.16). We will show that (6.17) holds.

Let us assume the converse. Then there exists a number t_0 such that

$$t_0 \in [T_1, T_2], \ |x(t_0) - x^*(t_0)| > \epsilon. \tag{6.25}$$

Since (x^*, u^*) is an overtaking optimal trajectory-control pair it follows from (6.18) and (6.19) that for any $s \in [0, \infty)$,

$$|I(s, s + 2\tau, x^*, u^*)| = |v_{2\tau}^s(x^*(s), x^*(s + 2\tau))| \leq M_1. \tag{6.26}$$

By (6.16), (6.22) and (6.19),

$$I(s, s + 2\tau, x, u) \leq v_{2\tau}^s(x(s), x(s + 2\tau)) + 1 \leq M_1 + 1 \qquad (6.27)$$

for any $s \in [T_1, T_2 - 2\tau]$. It follows from (6.25)-(6.27) and the choice of Δ that the following property holds:
   If $t \in [T_1, T_2]$ satisfies $|t_0 - t_1| \leq \Delta$, then

$$|x(t_0) - x(t)| \leq 8^{-1}\epsilon, \ |x^*(t_0) - x^*(t)| \leq 8^{-1}\epsilon, \ |x(t) - x^*(t)| \geq 3 \cdot 4^{-1}\epsilon.$$

Combined with (6.16) and (6.22) this property implies that

$$T_1 < t_0 - \Delta, \ t_0 + \Delta < T_2, \ |x(t) - x^*(t)| \geq 3 \cdot 4^{-1}\epsilon, \qquad (6.28)$$

$$t \in [t_0 - \Delta, t_0 + \Delta].$$

It follows from Corollary 7.2.1 that there exist trajectory-control pairs $x_1 : [0, \infty) \to R^n$, $u_1 : [0, \infty) \to R^m$ and $x_2 : [T_1, T_2] \to R^n$, $u_2 : [T_1, T_2] \to R^m$ which satisfy

$$x_1(t) = x^*(t), \ u_1(t) = u^*(t) \ (t \in [0, T_1] \cup [T_2, \infty)),$$

$$x_1(t) = x(t), \ u_1(t) = u(t) \ (t \in [T_1 + \Delta, T_2 - \Delta]),$$

$$I(a, a + \Delta, x_1, u_1) \leq v_\Delta^a(x_1(a), x_1(a + \Delta)) \ (a \in \{T_1, T_2 - \Delta\}), \quad (6.29)$$

$$x_2(T_i) = x(T_i), \ i = 1, 2, \ x_2(t) = x^*(t),$$

$$u_2(t) = u^*(t) \ (t \in [T_1 + \Delta, T_2 - \Delta]),$$

$$I(a, a + \Delta, x_2, u_2) \leq v_\Delta^a(x_2(a), x_2(a + \Delta)) \ (a \in \{T_1, T_2 - \Delta\}). \quad (6.30)$$

It follows from (6.30), (6.29), (6.16) and Lemma 7.5.2 that

$$I(T_1, T_2, x_1, u_1) + I(T_1, T_2, x_2, u_2) - I(T_1, T_2, x^*, u^*) - I(T_1, T_2, x, u)$$

$$\leq [v_\Delta^{T_1}(x^*(T_1), x(T_1 + \Delta)) - v_\Delta^{T_1}(x(T_1), x(T_1 + \Delta))]$$

$$+ [v_\Delta^{T_1}(x(T_1), x^*(T_1 + \Delta)) - v_\Delta^{T_1}(x^*(T_1), x^*(T_1 + \Delta))]$$

$$+ [v_\Delta^{T_2 - \Delta}(x(T_2 - \Delta), x^*(T_2)) - v_\Delta^{T_2 - \Delta}(x(T_2 - \Delta), x(T_2))]$$

$$+ [v_\Delta^{T_2 - \Delta}(x^*(T_2 - \Delta), x(T_2)) - v_\Delta^{T_2 - \Delta}(x^*(T_2 - \Delta), x^*(T_2))]. \quad (6.31)$$

By (6.31), (6.16), (6.18) and the definition of δ (see (6.22)-(6.24)),

$$I(T_1, T_2, x_1, u_1) + I(T_1, T_2, x_2, u_2)$$

$$- I(T_1, T_2, x^*, u^*) - I(T_1, T_2, x, u) \leq 4^{-1}\delta_1\Delta. \qquad (6.32)$$

Define trajectory-control pairs $y_i : [T_1, T_2] \to R^n$, $w_i : [T_1, T_2] \to R^m$ for $i = 1, 2$ as follows:

$$y_1(t) = 2^{-1}(x_1(t) + x^*(t)), \ w_1(t) = 2^{-1}(u_1(t) + u^*(t)), \ (t \in [T_1, T_2]),$$

$$y_2(t) = 2^{-1}(x_2(t) + x(t)), \ w_2(t) = 2^{-1}(u_2(t) + u(t)), \ t \in [T_1, T_2]. \quad (6.33)$$

By (6.28) there exists a number d such that

$$t_0 \in [d, d + \Delta] \subset [T_1 + \Delta, T_2 - \Delta]. \quad (6.34)$$

It follows from (6.33), (6.29), (6.30), (6.28), (6.34) and the definition of δ_1 that for $i = 1, 2$,

$$I(d, d + \Delta, y_i, w_i) \le 2^{-1} I(d, d + \Delta, x^*, u^*) + 2^{-1} I(d, d + \Delta, x, u) - \delta_1 \Delta. \quad (6.35)$$

(6.33), (6.29), (6.30), (6.16), Assumption (A1) and (6.32) imply that

$$-\delta \le I(T_1, T_2, y_1, w_1) + I(T_1, T_2, y_2, w_2)$$

$$-I(T_1, T_2, x^*, w^*) - I(T_1, T_2, x, w)$$

$$\le -2\delta_1 \Delta + 2^{-1}[I(T_1, T_2, x_1, u_1) + I(T_1, T_2, x_2, u_2)$$

$$-I(T_1, T_2, x^*, w^*) - I(T_1, T_2, x, w)]$$

$$\le -2\delta_1 \Delta + 8^{-1}\delta_1 \Delta \le -\delta_1 \Delta.$$

This is contradictory to (6.22). The obtained contradiction proves the lemma.

Theorem 7.1.3 now follows from Lemmas 7.6.1, 7.6.4 and 7.6.5. Theorem 7.1.4 now follows from Lemmas 7.6.3, 7.6.4, 7.6.5.

Chapter 8

DISCRETE-TIME CONTROL SYSTEMS

In this chapter we study the structure of "approximate" solutions for a discrete-time control system determined by a sequence of continuous functions $v_i : X \times X \to R^1$, $i = 0, \pm 1, \pm 2, \ldots$ where X is a complete metric space. We show that for a generic sequence of functions $\{v_i\}_{i=-\infty}^{\infty}$ there exists a sequence $\{y_i\}_{i=-\infty}^{\infty} \subset X$ (the "turnpike") such that the following properties hold: (i) $\{y_i\}_{i=k_1}^{k_2}$ is an optimal solution for any finite interval $[k_1, k_2]$; (ii) given $\epsilon > 0$, each "approximate" solution on an interval $[k_1, k_2]$ with sufficiently large $k_2 - k_1$ is within ϵ of the turnpike for all $i \in \{L + k_1, \ldots, k_2 - L\}$ where L is a constant which depends only on ϵ.

8.1. Convex infinite dimensional control systems

Let X be a Banach space, $||\cdot||$ be the norm on X, and let $K \subset X$ be a closed convex bounded set. Denote by $\mathcal{A}$ the set of all bounded convex functions $v : K \times K \to R^1$ which satisfy the following assumption:

(A) For each positive number ϵ there is a positive number δ such that if $x_1, x_2, y_1, y_2 \in K$ satisfy

$$||x_i - y_i|| \leq \delta, \ i = 1, 2,$$

then

$$|v(x_1, x_2) - v(y_1, y_2)| \leq \epsilon.$$

We equip the space $\mathcal{A}$ with the metric ρ defined by

$$\rho(u, v) = \sup\{|v(x, y) - u(x, y)| : x, y \in K\}, \quad u, v \in \mathcal{A}.$$

It is easy to see that the metric space $\mathcal{A}$ is complete.

We study the structure of "approximate" solutions of optimization problems

$$\sum_{i=0}^{n-1} v(x_i, x_{i+1}) \to \min, \qquad (P)$$

$$\{x_i\}_{i=0}^n \subset K, \quad x_0 = y, \ x_n = z$$

where $v \in \mathcal{A}$, $y, z \in K$ and n is a natural number.

The interest in these discrete-time optimal problems stems from the study of various optimization problems which can be reduced to this framework, e.g., continuous-time control systems which are represented by ordinary differential equations whose cost integrand contains a discounting factor [39], the infinite-horizon control problem of minimizing $\int_0^T L(z, z') dt$ as $T \to \infty$ [42] and the analysis of a long slender bar of a polymeric material under tension [20, 44, 90-92]. Similar optimization problems are also considered in mathematical economics [47, 48, 56-61, 69, 76, 77].

In [99] we show that for a generic function $v \in \mathcal{A}$ there exists $y_v \in K$ such that the following turnpike property holds:

For all large enough n and each $y, z \in K$ an "approximate" solution $\{x_i\}_{i=0}^n$ of problem (P) is contained in a small neighborhood of y_v for all $i \in \{N, \ldots, n - N\}$ where N is a constant which depends on the neighborhood and does not depend on n.

In almost all studies of discrete time control systems the turnpike property was considered for a single cost function v and a space of states K which was a compact convex set in a finite dimensional space. In these studies the compactness of K plays an important role. Specifically for the optimization problems considered in [99] if a function v has the turnpike property then as we will see later, its turnpike y_v is a unique solution of the optimization problem

$$v(x, x) \to \min, \quad x \in K.$$

The existence of a solution of this problem is guaranteed only if K satisfies some compactness assumptions. To obtain the uniqueness of the solution we need additional assumptions on v such as its strict convexity.

In [99], instead of considering the turnpike property for a single cost function v, we investigate it for the space of all such functions equipped with some natural metric, and show that this property holds for most of these functions. This allows us to establish the turnpike property without compactness assumptions on the space of states and assumptions on functions themselves.

For each $v \in \mathcal{A}$, each pair of integers $m_1, m_2 > m_1$ and each $y_1, y_2 \in K$ we define

$$\sigma(v, m_1, m_2) = \inf\{ \sum_{i=m_1}^{m_2-1} v(z_i, z_{i+1}) : \{z_i\}_{i=m_1}^{m_2} \subset K\},$$

$$\sigma(v, m_1, m_2, y_1, y_2) = \inf\{ \sum_{i=m_1}^{m_2-1} v(z_i, z_{i+1}) :$$

$$\{z_i\}_{i=m_1}^{m_2} \subset K, \; z_{m_1} = y_1, z_{m_2} = y_2\},$$

and the minimal growth rate

$$\mu(v) = \inf\{\liminf_{N \to \infty} N^{-1} \sum_{i=0}^{N-1} v(z_i, z_{i+1}) : \{z_i\}_{i=0}^{\infty} \subset K\}.$$

For each $x \in K$ and each positive number r set

$$B(x, r) = \{y \in K : ||x - y|| \le r\}.$$

We establish the existence of a set $\mathcal{F} \subset \mathcal{A}$ which is a countable intersection of open everywhere dense subsets of $\mathcal{A}$ and such that the following two theorems are valid. The first theorem shows that for $v \in \mathcal{F}$ the minimization problem $v(z, z) \to \min$, $z \in K$ has a unique solution y_v while the second theorem establishes the turnpike property for v with the turnpike y_v.

THEOREM 8.1.1 *For each $v \in \mathcal{F}$ there exists a unique $y_v \in K$ such that $v(y_v, y_v) = \mu(v)$ and the following assertion holds:*

For each positive number ϵ there exist a neighborhood $\mathcal{U}$ of v in $\mathcal{A}$ and a positive number δ such that if $u \in \mathcal{U}$ and if $y \in K$ satisfies

$$u(y, y) \le \mu(u) + \delta,$$

then $||y - y_v|| \le \epsilon$.

THEOREM 8.1.2 *Let $w \in \mathcal{F}$ and ϵ be a positive number. Then there exist a neighborhood $\mathcal{U}$ of w in $\mathcal{A}$, $\delta \in (0, \epsilon)$ and a natural number N such that the following property holds:*

If $u \in \mathcal{U}$, a natural number $n \ge 2N$ and if a sequence $\{x_i\}_{i=0}^{n} \subset K$ satisfies

$$\sum_{i=0}^{n-1} u(x_i, x_{i+1}) \le \sigma(u, 0, n, x_0, x_n) + \delta,$$

then there exist $\tau_1 \in \{0, \ldots, N\}$ *and* $\tau_2 \in \{n - N, \ldots, n\}$ *such that*

$$||x_t - y_w|| \le \epsilon, \quad t = \tau_1, \ldots, \tau_2.$$

Moreover, if $||x_0 - y_w|| \le \delta$, *then* $\tau_1 = 0$, *and if* $||y_w - x_n|| \le \delta$, *then* $\tau_2 = n$.

Theorems 8.1.1 and 8.1.2 were obtained in [99].

In Section 8.2 we prove three auxiliary lemmas. Theorems 8.1.1 and 8.1.2 are proved in Section 8.3.

8.2.   Preliminary results

Set

$$D_0 = \sup\{||x|| : x \in K\}. \tag{2.1}$$

For each bounded function $u : K \times K \to R^1$ we set

$$||u|| = \sup\{|u(x, y)| : x, y \in K\}. \tag{2.2}$$

PROPOSITION 8.2.1 *For each* $v \in \mathcal{A}$

$$\mu(v) = \inf\{v(z, z) : z \in K\}.$$

Proof. Let $v \in \mathcal{A}$. Evidently

$$\inf\{v(x, x) : x \in K\} \ge \mu(v). \tag{2.3}$$

It is not difficult to see that for each natural number $m \ge 1$,

$$\sigma(v, 0, m) \le m\mu(v). \tag{2.4}$$

Let $\epsilon > 0$. Since the function v is uniformly continuous there exists $\delta \in (0, \epsilon)$ such that the following property holds:
If $x_1, x_2, y_1, y_2 \in K$ satisfy $||x_i - y_i|| \le \delta$, $i = 1, 2$, then

$$|v(x_1, x_2) - v(y_1, y_2)| \le \epsilon. \tag{2.5}$$

Fix a natural number m such that

$$8m^{-1}(D_0 + 1) \le \delta. \tag{2.6}$$

Consider a sequence $\{y_i\}_{i=0}^m \subset K$ such that

$$\sum_{i=0}^{m-1} v(y_i, y_{i+1}) \le \sigma(v, 0, m) + \delta.$$

Put

$$z_0 = m^{-1} \sum_{i=0}^{m-1} y_i, \quad z_1 = m^{-1} \sum_{i=1}^{m} y_i.$$

It is not difficult to see that

$$\|z_0 - z_1\| \leq 2m^{-1} D_0 < \delta \tag{2.7}$$

and

$$v(z_0, z_1) \leq m^{-1} \sum_{i=0}^{m-1} v(y_i, y_{i+1}) \leq m^{-1}[\sigma(v, 0, m) + \delta]. \tag{2.8}$$

It follows from (2.7) and the choice of δ (see (2.5)) that

$$|v(z_0, z_0) - v(z_0, z_1)| \leq \epsilon.$$

Combined with (2.8) and (2.4) this inequality implies that

$$v(z_0, z_0) \leq 2\epsilon + m^{-1}\sigma(v, 0, m) \leq \mu(v) + 2\epsilon.$$

Since ϵ is an arbitrary positive number we conclude that $\inf\{v(z, z) : z \in K\} \leq \mu(v)$. This completes the proof of the proposition.

The next auxiliary result shows that for $v \in \mathcal{A}$ and $\epsilon > 0$ we can find $u \in \mathcal{A}$ which is close to v in $\mathcal{A}$ such that all approximate solutions of the minimization problem $u(y, y,) \to \min$, $y \in K$ are contained in a small ball with radius ϵ.

PROPOSITION 8.2.2 *For each $v \in \mathcal{A}$ and each $\epsilon \in (0, 1)$ there exist $\delta \in (0, \epsilon)$, $u \in \mathcal{A}$ and $z_0 \in K$ such that*

$$0 \leq u(x, y) - v(x, y) \leq \epsilon, \quad x, y \in K, \quad \mu(v) + \delta \geq v(z_0, z_0) \tag{2.9}$$

and that the following property holds:
  If $y \in K$ satisfies
$$u(y, y) \leq \mu(u) + \delta,$$
then $\|y - z_0\| \leq \epsilon$.

Proof. Let $v \in \mathcal{A}$ and $\epsilon \in (0, 1)$. Choose numbers δ, γ such that

$$\gamma \in (0, (8D_0 + 4)^{-1}\epsilon), \quad \delta \in (0, 8^{-1}\gamma\epsilon). \tag{2.10}$$

Proposition 8.2.1 implies that there is $z_0 \in K$ such that

$$v(z_0, z_0) < \mu(v) + \delta.$$

Define a function $u : K \times K \to R^1$ by

$$u(x, y) = v(x, y) + \gamma(||x - z_0|| + ||y - z_0||), \quad x, y \in K. \qquad (2.11)$$

It is not difficult to see that $u \in \mathcal{A}$ and (2.9) is valid.
Assume that $y \in K$ and

$$u(y, y) \leq \mu(u) + \delta. \qquad (2.12)$$

By Proposition 8.2.1, (2.11), (2.12), (2.10) and the choice of z_0,

$$\mu(v) \leq \mu(u) \leq u(z_0, z_0) = v(z_0, z_0) \leq \mu(v) + \delta,$$

$$2\gamma||y - z_0|| + v(y, y) = u(y, y) \leq \mu(u) + \delta$$

$$\leq \mu(v) + 2\delta \leq v(y, y) + 2\delta, \quad ||y - z_0|| \leq \delta\gamma^{-1} < \epsilon.$$

This completes the proof of the proposition.

The next result shows that for a generic $v \in \mathcal{A}$ the minimization problem $v(z, z) \to \min$, $z \in K$ has a unique solution.

PROPOSITION 8.2.3 *There exists a set $\mathcal{F}_0$ which is a countable inter-section of open everywhere dense subsets of $\mathcal{A}$ and such that for each $v \in \mathcal{F}_0$ there exists a unique $y_v \in K$ which satisfies $v(y_v, y_v) = \mu(v)$ and that the following property holds:*

For each positive number ϵ there exist a neighborhood $\mathcal{U}$ of v in $\mathcal{A}$ and a positive number δ such that if $u \in \mathcal{U}$ and if $y \in K$ satisfies

$$u(y, y) \leq \mu(u) + \delta,$$

then $||y - y_v|| \leq \epsilon$.

Proof. Let $w \in \mathcal{A}$ and i be a natural number. It follows from Proposition 8.2.2 that there exist a positive number $\delta(w, i) < 4^{-i}$, a function $u^{(w,i)} \in \mathcal{A}$ and $z(w, i) \in K$ such that

$$0 \leq u^{(w,i)}(x, y) - w(x, y) \leq 4^{-i}, \quad x, y \in K, \qquad (2.13)$$

$$w(z(w, i), z(w, i)) \leq \mu(w) + \delta(w, i)$$

and the following property holds:
If $z \in K$ satisfies

$$u^{(w,i)}(z, z) \leq \mu(u^{(w,i)}) + \delta(w, i),$$

then $z \in B(z(w, i), 4^{-i})$.

Put
$$\mathcal{U}(w, i) = \{u \in \mathcal{A} : \rho(u, u^{(w,i)}) < 8^{-1}\delta(w, i)\}. \qquad (2.14)$$

Assume that $u \in \mathcal{U}(w, i)$ and $z \in K$ satisfies
$$u(z, z) \leq \mu(u) + 8^{-1}\delta(w, i).$$

Then it follows from (2.14) that
$$u^{(w,i)}(z, z) \leq u(z, z) + 8^{-1}\delta(w, i) \leq \mu(u) + 4^{-1}\delta(w, i)$$
$$\leq \mu(u^{(w,i)}) + 2^{-1}\delta(w, i)$$

and in view of the choice of $\delta(w, i)$, $u^{(w,i)}$ and $z(w, i)$,
$$z \in B(z(w, i), 4^{-i}).$$

Thus we have shown that the following property holds:
(a) If $u \in \mathcal{U}(w, i)$ and if $z \in K$ satisfies
$$u(z, z) \leq \mu(u) + 8^{-1}\delta(w, i),$$

then $z \in B(z(w, i), 4^{-i})$.
Set
$$\mathcal{F}_0 = \cap_{q=1}^{\infty} \cup \{\mathcal{U}(w, i) : w \in \mathcal{A}, \quad i = q, q+1, \ldots\}.$$

It is easy to see that $\mathcal{F}_0$ is a countable intersection of open everywhere dense subsets of $\mathcal{A}$.

Assume that $v \in \mathcal{F}_0$. By Proposition 8.2.1 there exists a sequence $\{x_j\}_{j=1}^{\infty} \subset K$ such that
$$\lim_{j \to \infty} v(x_j, x_j) = \mu(v). \qquad (2.15)$$

By property (a) and the definition of $\mathcal{F}_0$, $\{x_j\}_{j=1}^{\infty}$ is a Cauchy sequence. Since K is a closed subset of the Banach space X and the function v is continuous we obtain that there exists $\lim_{j \to \infty} x_j$ and
$$v(\lim_{j \to \infty} x_j, \lim_{j \to \infty} x_j) = \mu(v).$$

Since any sequence $\{x_j\}_{j=1}^{\infty} \subset K$ satisfying (2.15), converges in K, we conclude that there exists a unique $y_v \in K$ such that $v(y_v, y_v) = \mu(v)$.

Let $\epsilon > 0$. Choose a natural number q such that
$$2^{-q} < 8^{-1}\epsilon. \qquad (2.16)$$

In view of the definition of $\mathcal{F}_0$ there are $w \in \mathcal{A}$ and a natural number $i \geq q$ such that $v \in \mathcal{U}(w, i)$. By property (a),
$$z(w, i) \in B(y_v, 4^{-i}). \qquad (2.17)$$

(2.17), property (a) and (2.16) imply that if $u \in \mathcal{U}(w, i)$ and if $y \in K$ satisfies

$$u(y, y) \leq \mu(u) + 8^{-1}\delta(w, i),$$

then $y \in B(y_v, \epsilon)$ holds. This completes the proof of the proposition.

8.3.   Proofs of Theorems 8.1.1 and 8.1.2

Let the set $\mathcal{F}_0$ be as guaranteed in Proposition 8.2.3. For each $w \in \mathcal{F}_0$ there exists a unique $y_w \in K$ such that

$$w(y_w, y_w) = \mu(w). \tag{3.1}$$

Now for any $w \in \mathcal{F}_0$ we construct $\tilde{w} \in \mathcal{A}$ which is close to w in $\mathcal{A}$ and has the turnpike property.

Let $v \in \mathcal{F}_0$, $\gamma \in (0, 1)$. Define

$$v_\gamma(x, y) = v(x, y) + \gamma(||x - y_v|| + ||y - y_v||), \quad x, y \in K. \tag{3.2}$$

Clearly $v_\gamma \in \mathcal{A}$. We show that v_γ has the turnpike property. We begin with the following lemma which establishes that for any approximate solution $\{x_i\}_{i=0}^n$ with large enough n there is $j \in \{0, \ldots, n-1\}$ such that x_j and x_{j+1} are close enough to y_v.

LEMMA 8.3.1 *For each $\epsilon \in (0, 1)$ there exists a natural number n such that the following property holds:*

If a sequence $\{x_i\}_{i=0}^n \subset K$ satisfies

$$\sum_{i=0}^{n-1} v_\gamma(x_i, x_{i+1}) \leq \sigma(v_\gamma, 0, n, x_0, x_n) + 4, \tag{3.3}$$

then there exists $j \in \{0, \ldots, n-1\}$ such that

$$x_j, x_{j+1} \in B(y_v, \epsilon). \tag{3.4}$$

Proof. Let $\epsilon \in (0, 1)$. Choose a natural number

$$n > (\epsilon\gamma)^{-1}(5 + 4(||v_\gamma|| + ||v||)) + 4. \tag{3.5}$$

We show that

$$n\mu(v) \leq \sigma(v, 0, n) + 2||v||. \tag{3.6}$$

Let Δ be a positive number. There exists a sequence $\{z_i\}_{i=0}^n$ such that

$$\sum_{i=0}^{n-1} v(z_i, z_{i+1}) \leq \sigma(v, 0, n) + \Delta.$$

Then it follows from Proposition 8.2.1 and the convexity of v that

$$n\mu(v) \leq nv(n^{-1}\sum_{i=0}^{n-1} z_i, n^{-1}\sum_{i=0}^{n-1} z_i)$$

$$= nv(n^{-1}[\sum_{i=0}^{n-2}(z_i, z_{i+1}) + (z_{n-1}, z_0)])$$

$$\leq \sum_{i=0}^{n-2} v(z_i, z_{i+1}) + v(z_{n-1}, z_0) \leq \sum_{i=0}^{n-1} v(z_i, z_{i+1}) + 2||v||$$

$$\leq \sigma(v, 0, n) + 2|v|| + \Delta.$$

Since Δ is an arbitrary positive number we obtain the inequality (3.6).

Assume that $\{x_i\}_{i=0}^n \subset K$ satisfies (3.3). Put

$$y_i = x_i, \quad i = 0, n, \quad y_i = y_v, \quad i = 1, \ldots, n-1. \tag{3.7}$$

By (3.2), (3.3), (3.7), (3.1), (3.6) and (3.5),

$$\sigma(v, 0, n) + \gamma \sum_{i=0}^{n-1}(||x_i - y_v|| + ||x_{i+1} - y_v||) \leq \sum_{i=0}^{n-1} v_\gamma(x_i, x_{i+1})$$

$$\leq \sum_{i=0}^{n-1} v_\gamma(y_i, y_{i+1}) + 4 \leq 4||v_\gamma|| + 4 + nv(y_v, y_v)$$

$$= 4 + 4||v_\gamma|| + n\mu(v) \leq 4 + 4||v_\gamma|| + \sigma(v, 0, n) + 2||v||,$$

$$\min\{||x_i - y_v|| + ||x_{i+1} - y_v|| : i = 0, \ldots, n-1\}$$

$$\leq (n\gamma)^{-1}(4 + 4||v_\gamma|| + 2||v||) < \epsilon.$$

This completes the proof of the lemma.

Lemma 8.3.1 implies the following auxiliary result which shows that the convergence property established in Lemma 8.3.1 for v_γ also holds for approximate solutions with the cost function u belonging to a small neighborhood of v_γ in $\mathcal{A}$.

LEMMA 8.3.2 *For each $\epsilon \in (0,1)$ there exist a neighborhood $\mathcal{U}$ of v_γ in $\mathcal{A}$ and a natural number n such that the following property holds:*

If $u \in \mathcal{U}$ and if a sequence $\{x_i\}_{i=0}^n \subset K$ satisfies

$$\sum_{i=0}^{n-1} u(x_i, x_{i+1}) \leq \sigma(u, 0, n, x_0, x_n) + 3,$$

there is $j \in \{0, \ldots, n-1\}$ for which

$$x_j, x_{j+1} \in B(y_v, \epsilon).$$

The next auxiliary result is an important tool in our proof. It shows that if x_0, x_n are close to y_v, then the approximate solution $\{x_i\}_{i=0}^n$ is close to y_v for all $i = 0, \ldots, n$.

LEMMA 8.3.3 *For each $\epsilon \in (0,1)$ there exists $\delta \in (0, \epsilon)$ such that the following property holds:*
    If n is a natural number and if a sequence $\{x_i\}_{i=0}^n \subset K$ satisfies

$$x_0, x_n \in B(y_v, \delta), \quad \sum_{i=0}^{n-1} v_\gamma(x_i, x_{i+1}) \leq \sigma(v_\gamma, 0, n, x_0, x_n) + \delta, \qquad (3.8)$$

then

$$x_i \in B(y_v, \epsilon), \quad i = 0, \ldots, n. \qquad (3.9)$$

Proof. We show that for each natural number m,

$$\sigma(v, 0, m, y_v, y_v) = m\mu(v). \qquad (3.10)$$

Let m be a natural number. Evidently

$$\sigma(v, 0, m, y_v, y_v) \leq \mu(v).$$

Assume that $\{z_i\}_{i=0}^m \subset K$ and $z_0, z_m = y_v$. Then it follows from the convexity of v and Proposition 8.2.1 that

$$m^{-1} \sum_{i=0}^{m-1} v(z_i, z_{i+1}) \geq v(m^{-1} \sum_{i=0}^{m-1} (z_i, z_{i+1}))$$

$$= v(m^{-1} \sum_{i=0}^{m-1} (z_i, z_i)) \geq \mu(v).$$

Therefore the equality (3.10) is true.
    Let $\epsilon \in (0,1)$. Fix

$$\delta_0 \in (0, 8^{-1}\gamma\epsilon). \qquad (3.11)$$

Since the function v is uniformly continuous there exists a positive number $\delta < 2^{-1}\delta_0$ such that the following property holds:
    If $x_1, x_2, y_1, y_2 \in K$ satisfy $\|x_i - y_i\| \leq \delta$, $i = 1, 2$, then

$$|v(x_1, x_2) - v(y_1, y_2)| \leq 64^{-1}\delta_0. \qquad (3.12)$$

Assume that n is a natural number and a sequence $\{x_i\}_{i=0}^n \subset K$ satisfies (3.8). We will show that (3.9) is valid.

Let us assume the converse. Then $n \geq 2$ and there exists $j \in \{1, \ldots, n-1\}$ such that

$$||x_j - y_v|| > \epsilon. \tag{3.13}$$

Set

$$z_i = x_i, \quad i = 0, n, \quad z_i = y_v, \quad i = 1, \ldots, n-1, \tag{3.14}$$

$$h_i = y_v, i = 0, n, \ h_i = x_i, \ i = 1, \ldots, n-1.$$

By (3.2), (3.13), (3.8), (3.14) and (3.1),

$$\gamma\epsilon + \sum_{i=0}^{n-1} v(x_i, x_{i+1}) \leq \sum_{i=0}^{n-1} v_\gamma(x_i, x_{i+1}) \leq \sum_{i=0}^{n-1} v_\gamma(z_i, z_{i+1}) + \delta \tag{3.15}$$

$$= \delta + v_\gamma(x_0, y_v) + v_\gamma(y_v, x_n) + (n-2)\mu(v).$$

It follows from (3.8), (3.1) and the choice of δ (see (3.12)) that

$$|v(x_i, y_v) - \mu(v)|, \quad |v(y_v, x_i) - \mu(v)| \leq 64^{-1}\delta_0, \quad i = 0, n, \tag{3.16}$$

$$|v(y_v, x_1) - v(x_0, x_1)| \leq 64^{-1}\delta_0, \quad |v(x_{n-1}, x_n) - v(x_{n-1}, y_v)| \leq 64^{-1}\delta_0.$$

By (3.16), (3.8), (3.15), (3.10), (3.14) and (3.2),

$$\gamma\epsilon + \sum_{i=0}^{n-1} v(h_i, h_{i+1}) \leq \gamma\epsilon + \sum_{i=0}^{n-1} v(x_i, x_{i+1}) + 32^{-1}\delta_0$$

$$\leq 32^{-1}\delta_0 + \delta + (n-2)\mu(v) + v_\gamma(x_0, y_v) + v_\gamma(y_v, x_n)$$

$$\leq 32^{-1}\delta_0 + \delta + \mu(v)n + 32^{-1}\delta_0 + 2\gamma\delta \leq \mu(v)n + 16^{-1}\delta_0$$

$$+3\delta = \sigma(v, 0, n, y_v, y_v) + 16^{-1}\delta_0 + 3\delta.$$

Combined with (3.14) this implies that $\gamma\epsilon \leq 4\delta_0$. This is contradictory to (3.11). The obtained contradiction proves the lemma.

The next lemma shows that the convergence property established in Lemma 8.3.3 for v_γ also holds for all $u \in \mathcal{A}$ which are close to v_γ.

LEMMA 8.3.4 *For each $\epsilon \in (0, 1)$ there exist $\delta \in (0, \epsilon)$, a neighborhood $\mathcal{U}$ of v_γ in $\mathcal{A}$ and a natural number N such that the following property holds:*

If $u \in \mathcal{U}$, a natural number $n \geq 2N$ and if a sequence $\{x_i\}_{i=0}^n \subset K$ satisfies

$$\sum_{i=0}^{n-1} u(x_i, x_{i+1}) \leq \sigma(u, 0, n, x_0, x_n) + \delta, \tag{3.17}$$

then there exist $\tau_1 \in \{0, \ldots, N\}$, $\tau_2 \in \{-N + n, n\}$ such that

$$x_i \in B(y_v, \epsilon), \quad t = \tau_1, \ldots, \tau_2, \tag{3.18}$$

and moreover if $x_0 \in B(y_v, \delta)$, then $\tau_1 = 0$, and if $x_n \in B(y_v, \delta)$ then $\tau_2 = n$.

Proof. Let $\epsilon \in (0, 1)$. It follows from Lemma 8.3.3 that there exists $\delta \in (0, 4^{-1}\epsilon)$ such that the following property holds:

If n is a natural number and if a sequence $\{x_i\}_{i=0}^n \subset K$ satisfies

$$x_i \in B(y_v, 4\delta), \ i = 0, n, \quad \sum_{i=0}^{n-1} v_\gamma(x_i, x_{i+1}) \le \sigma(v_\gamma, 0, n, x_0, x_n) + 4\delta,$$

$$\tag{3.19}$$

then

$$x_i \in B(y_v, \epsilon), \quad i = 0, \ldots, n. \tag{3.20}$$

It follows from Lemma 8.3.2 that there exist a natural number N and a neighborhood $\mathcal{U}_1$ of v_γ in $\mathcal{A}$ such that the following property holds:

If $u \in \mathcal{U}_1$ and if a sequence $\{x_i\}_{i=0}^N \subset K$ satisfies

$$\sum_{i=0}^{N-1} u(x_i, x_{i+1}) \le \sigma(u, 0, N, x_0, x_N) + 3, \tag{3.21}$$

then there is $j \in \{0, \ldots N - 1\}$ such that

$$x_j, x_{j+1} \in B(y_v, \delta). \tag{3.22}$$

Define

$$\mathcal{U} = \{u \in \mathcal{U}_1 : \rho(u, v_\gamma) \le (16N)^{-1}\delta\}. \tag{3.23}$$

Assume that $u \in \mathcal{U}$, a natural number $n \ge 2N$ and a sequence $\{x_i\}_{i=0}^n \subset K$ satisfies (3.17). It follows from (3.17) and the definition of $\mathcal{U}_1$, N (see (3.21), (3.22)) that there exist integers τ_1, τ_2 such that:

$$\tau_1 \in \{0, \ldots, N\}, \quad \tau_2 \in \{n - N, \ldots, n\}, \quad x_{\tau_i} \in B(y_v, \delta), \ i = 1, 2;$$
$$\tag{3.24}$$

if $x_0 \in B(y_v, \delta)$, then $\tau_1 = 0$; if $x_n \in B(y_v, \delta)$, then $\tau_2 = n$.

We will show that (3.18) is valid. Let us assume the converse. Then there is an integer $s \in (\tau_1, \tau_2)$ for which

$$\|x_s - y_v\| > \epsilon. \tag{3.25}$$

It follows from (3.17), (3.24) and the definition of $\mathcal{U}_1$, N (see (3.21), (3.22)) that there exist integers t_1, t_2 such that

$$\sup\{\tau_1, s - N\} \le t_1 < s, \quad s < t_2 \le \inf\{\tau_2, s + N\},$$

$$x_{t_i} \in B(y_v, \delta), \quad i = 1, 2. \tag{3.26}$$

In view of (3.23), (3.26) and (3.17),

$$\sum_{i=t_1}^{t_2-1} v_\gamma(x_i, x_{i+1}) \le \sigma(v_\gamma, t_1, t_2, x_{t_1}, x_{t_2}) + 2\delta. \tag{3.27}$$

By (3.26), (3.27) and the choice of δ (see (3.19), (3.20)),

$$\|x_t - y_t\| \le \epsilon, \quad t = t_1, \dots, t_2.$$

This is contradictory to (3.25). The obtained contradiction proves that (3.18) is valid. This completes the proof of the lemma.

Clearly the set $\{v_\gamma : v \in \mathcal{F}_0, \gamma \in (0,1)\}$ is everywhere dense in $\mathcal{A}$. Now we are ready to construct the set $\mathcal{F}$.

Let $v \in \mathcal{F}_0$, $\gamma \in (0,1)$ and let j be a natural number. There exist a natural number $N(v, \gamma, j)$, an open neighborhood $\mathcal{U}_0(v, \gamma, j)$ of v_γ in $\mathcal{A}$ and a number $\delta(v, \gamma, j) \in (0, 2^{-j})$ such that Lemma 8.3.4 holds with $v, \gamma, \epsilon = 2^{-j}$, $\delta = \delta(v, \gamma, j)$, $\mathcal{U} = \mathcal{U}_0(v, \gamma, j)$, $N = N(v, \gamma, j)$.

There are an open neighborhood $\mathcal{U}(v, \gamma, j)$ of v_γ in $\mathcal{A}$ and a natural number $N_1(v, \gamma, j)$ such that $\mathcal{U}(v, \gamma, j) \subset \mathcal{U}_0(v, \gamma, j)$ and Lemma 8.3.2 holds with $v, \gamma, \mathcal{U} = \mathcal{U}(v, \gamma, j)$, $n = N_1(v, \gamma, j)$, $\epsilon = 4^{-j}\delta(v, \gamma, j)$.

Define

$$\mathcal{F} = [\cap_{q=1}^\infty \cup \{\mathcal{U}(v, \gamma, j) : v \in \mathcal{F}_0, \gamma \in (0,1), j = q, q+1, \dots\}] \cap \mathcal{F}_0.$$

Clearly $\mathcal{F}$ is a countable intersection of open everywhere dense subsets of $\mathcal{A}$.

It is easy to see that Theorem 8.1.1 follows from Proposition 8.2.3 and the definition of $\mathcal{F}$.

Proof of Theorem 8.1.2. Let $w \in \mathcal{F}$, $\epsilon > 0$. We may assume that $\epsilon < 1$. Choose a natural number q such that

$$64 \cdot 2^{-q} < \epsilon. \tag{3.28}$$

There exist $v \in \mathcal{F}_0$, $\gamma \in (0,1)$ and a natural number $j \ge q$ such that

$$w \in \mathcal{U}(v, \gamma, j). \tag{3.29}$$

In view of (3.29), Lemma 8.3.2 which holds with $\mathcal{U} = \mathcal{U}(v, \gamma, j)$, $n = N_1(v, \gamma, j)$, $\epsilon = 4^{-j}\delta(v, \gamma, j)$, v, γ, and the equality

$$\sigma(w, 0, N_1(v, \gamma, j), y_w, y_w) = N_1(v, \gamma, j)\mu(w)$$

we have

$$y_v \in B(y_w, 4^{-j}\delta(v, \gamma, j)). \tag{3.30}$$

Put

$$\mathcal{U} = \mathcal{U}(v, \gamma, j), \quad N = N(v, \gamma, j), \quad \delta = 4^{-j}\delta(v, \gamma, j). \tag{3.31}$$

Assume that $u \in \mathcal{U}$, a natural number $n \geq 2N$ and a sequence $\{x_i\}_{i=0}^n \subset K$ satisfies

$$\sum_{i=0}^{n-1} u(x_i, x_{i+1}) \leq \sigma(u, 0, n, x_0, x_n) + \delta. \tag{3.32}$$

By (3.32), (3.31), the choice of

$$\mathcal{U}_0(v, \gamma, j), \; N(v, \gamma, j), \; \delta(v, \gamma, j)$$

and Lemma 8.3.4, there exist $\tau_1 \in \{0, \ldots, N\}$, $\tau_2 \in \{n - N, \ldots, n\}$ such that

$$x_i \in B(y_v, 2^{-j}), \quad t = \tau_1, \ldots, \tau_2.$$

Moreover if

$$x_0 \in B(y_v, \delta(v, \gamma, j)),$$

then $\tau_1 = 0$, and if $x_n \in B(y_v, \delta(v, \gamma, j))$, then $\tau_2 = n$. Combined with (3.30), (3.28), (3.31) this implies that:

$$x_i \in B(y_w, \epsilon), \quad i = \tau_1 \ldots, \tau_2;$$

if $x_0 \in B(y_w, \delta)$, then $x_0 \in B(y_v, \delta(v, \gamma, j))$ and $\tau_1 = 0$; if $x_n \in B(y_w, \delta)$, then $x_n \in B(y_v, \delta(v, \gamma, j))$ and $\tau_2 = n$. This completes the proof of the theorem.

8.4. Nonautonomous control systems in metric spaces

Let X be a complete metric space and let $d(\cdot, \cdot)$ be the metric on X. We equip the set $X \times X$ with a metric $d_1(\cdot, \cdot)$ defined by

$$d_1((x_1, x_2), (y_1, y_2)) = d(x_1, y_1) + d(x_2, y_2), \quad x_1, x_2, y_1, y_2 \in X.$$

Denote by $\mathcal{M}$ the set of all sequences of functions $\mathbf{v} = \{v_i\}_{i=-\infty}^{\infty}$ which satisfy the following assumptions:
A(i) $v_j : X \times X \to R^1$ is a continuous function for each integer j;
A(ii) (uniform boundedness)

$$\sup\{|v_j(x, y)| : \quad x, y \in X, \; j = 0, \pm 1, \pm 2, \ldots\} < \infty;$$

A(iii) (uniform continuity) for each positive number ϵ there exists a positive number δ such that if $x_1, x_2, y_1, y_2 \in X$ satisfy

$$d(x_1, x_2), \quad d(y_1, y_2) \leq \delta,$$

then

$$|v_j(x_1, y_1) - v_j(x_2, y_2)| \leq \epsilon$$

for each integer j.

Such a sequence of functions $\{v_i\}_{i=-\infty}^{\infty} \in \mathcal{M}$ will occasionally be denoted by a boldface $\mathbf{v}$ (similarly $\{u_i\}_{i=-\infty}^{\infty}$ will be denoted by $\mathbf{u}$, etc.)

We endow the set $\mathcal{M}$ with the metric

$$\rho(\mathbf{v}, \mathbf{w}) = \sup\{|v_i(x, y) - w_i(x, y)| :$$

$$x, y \in X, \ i = 0, \pm 1, \pm 2, \ldots\}, \quad \mathbf{v}, \mathbf{w} \in \mathcal{M}. \tag{4.1}$$

Clearly the metric space $(\mathcal{M}, \rho)$ is complete.

In this chapter we investigate the structure of "approximate" solutions of the optimization problem

$$\sum_{i=k_1}^{k_2-1} v_i(x_i, x_{i+1}) \to \min, \quad \{x_i\}_{i=k_1}^{k_2} \subset X, \quad x_{k_1} = y, \ x_{k_2} = z \tag{P}$$

where $\mathbf{v} = \{v_i\}_{i=-\infty}^{\infty} \in \mathcal{M}$, $y, z \in X$ and $k_2 > k_1$ are integers.

For each $\mathbf{v} \in \mathcal{M}$, each pair of integers $m_2 > m_1$ and each $y_1, y_2 \in X$ we define

$$\sigma(\mathbf{v}, m_1, m_2) = \inf\{\sum_{i=m_1}^{m_2-1} v_i(z_i, z_{i+1}) : \quad \{z_i\}_{i=m_1}^{m_2} \subset X\}, \tag{4.2}$$

$$\sigma(\mathbf{v}, m_1, m_2, y_1, y_2) = \inf\{\sum_{i=m_1}^{m_2-1} v_i(z_i, z_{i+1}) :$$

$$\{z_i\}_{i=m_1}^{m_2} \subset X, \ z_{m_1} = y_1, z_{m_2} = y_2\}.$$

If the space of states X is compact, then the problem (P) has a solution for each $\mathbf{v} \in \mathcal{M}$, $y, z \in X$ and each pair of integers $k_2 > k_1$. For the noncompact space X the existence of solutions of the problem (P) is not guaranteed and in this situation we consider δ-*approximate solutions*.

Let $\mathbf{v} \in \mathcal{M}$, $y, z \in X$, $k_2 > k_1$ be integers and let δ be a positive number. A sequence $\{x_i\}_{i=k_1}^{k_2} \subset X$ which satisfies $x_{k_1} = y$, $x_{k_2} = z$ is called a δ-approximate solution of the problem (P) if

$$\sum_{i=k_1}^{k_2-1} v_i(x_i, x_{i+1}) \leq \sigma(\mathbf{v}, k_1, k_2, y, z) + \delta.$$

The following optimality criterion for infinite horizon problems was introduced by Aubry and Le Daeron [6].

Let $\mathbf{v} = \{v_i\}_{i=-\infty}^{\infty} \in \mathcal{M}$. We say that a sequence $\{x_i\}_{i=-\infty}^{\infty}$ is $(\mathbf{v})$-minimal if

$$\sum_{i=m_1}^{m_2-1} v_i(x_i, x_{i+1}) = \sigma(\mathbf{v}, m_1, m_2, x_{m_1}, x_{m_2}) \tag{4.3}$$

for each pair of integers $m_1 < m_2$.

If the space of states X is compact, then a $(\mathbf{v})$-minimal sequence can be constructed as a limit of a sequence of optimal solutions on finite intervals. For the noncompact space X the problem is more difficult and less understood.

We say that a sequence $\mathbf{v} = \{v_i\}_{i=-\infty}^{\infty} \in \mathcal{M}$ has the *turnpike property* if there is a $(\mathbf{v})$-minimal sequence $\{x_i^{\mathbf{v}}\}_{i=-\infty}^{\infty} \subset X$ which satisfies the following condition:

For each $\epsilon > 0$ there exist a positive number δ and a natural number L such that each δ-approximate solution $\{x_i\}_{i=k_1}^{k_2}$ of the problem (P) with $y, z \in X$, and $k_2 \geq k_1 + 2L$ satisfies

$$d(x_i, x_i^{\mathbf{v}}) \leq \epsilon, \quad i = k_1 + L, \ldots, k_2 - L. \tag{4.4}$$

In this chapter we prove the following result.

THEOREM 8.4.1 *There exists a set $\mathcal{F} \subset \mathcal{M}$ which is a countable intersection of open everywhere dense sets in $\mathcal{M}$ such that each $\mathbf{v} \in \mathcal{F}$ satisfies the following conditions:*

(a) there is a unique $(\mathbf{v})$-minimal sequence $\{x_i^{\mathbf{v}}\}_{i=-\infty}^{\infty} \subset X$;

(b) for each positive number ϵ there exist $\delta \in (0, \epsilon)$, a natural number N and a neighborhood U of $\mathbf{v}$ in $\mathcal{M}$ such that the following property holds:

If $\mathbf{w} \in U$, $k_1, k_2 > k_1 + 2N$ are integers and if a sequence $\{y_i\}_{i=k_1}^{k_2} \subset X$ satisfies

$$\sum_{i=k_1}^{k_2} w_i(y_i, y_{i+1}) \leq \sigma(\mathbf{w}, k_1, k_2, y_{k_1}, y_{k_2}) + \delta,$$

then

$$d(y_i, x_i^{\mathbf{v}}) \leq \epsilon, \quad i = k_1 + L_1, \ldots, k_2 - L_2,$$

where the integers $L_1, L_2 \in [0, N]$, and moreover, if $d(y_{k_1}, x_{k_1}^{\mathbf{v}}) \leq \delta$, then $L_1 = 0$, and if $d(y_{k_2}, x_{k_2}^{\mathbf{v}}) \leq \delta$, then $L_2 = 0$.

Theorem 8.4.1 was obtained in [104].

In Section 8.5 we construct a set $\mathcal{F}_0 \subset \mathcal{M}$ which is a countable intersection of open everywhere dense subsets of $\mathcal{M}$ such that for each $\mathbf{v} \in \mathcal{F}_0$

there is a **v**-minimal sequence $\{x_i^{\mathbf{v}}\}_{i=-\infty}^{\infty}$. In Section 8.6 we construct a set $\mathcal{F} \subset \mathcal{F}_0$ which is a countable intersection of open everywhere dense subsets of $\mathcal{M}$ such that for each $\mathbf{v} \in \mathcal{F}$ the turnpike property holds with the turnpike $\{x_i^{\mathbf{v}}\}_{i=-\infty}^{\infty}$.

8.5.   An auxiliary result

For each $\mathbf{v} \in \mathcal{M}$ we define

$$||\mathbf{v}|| = \sup\{|v_i(x,y)| : x, y \in X, \ i = 0, \pm 1, \pm 2, \ldots\}.$$

Denote by $\mathrm{Card}(A)$ the cardinality of a set A, by $\mathbf{Z}$ the set of all integers and by $\mathbf{N}$ the set of all natural numbers.

In this section we will establish the following result.

PROPOSITION 8.5.1 *There exists a set $\mathcal{F}_0 \subset \mathcal{M}$ which is a countable intersection of open everywhere dense sets in $\mathcal{M}$ such that for each $\mathbf{v} \in \mathcal{F}_0$ there is a $(\mathbf{v})$-minimal sequence $\{x_i^{\mathbf{v}}\}_{i=-\infty}^{\infty} \subset X$.*

Construction of the set $\mathcal{F}_0$.

For each $\mathbf{v} \in \mathcal{M}$ and each natural number q we define a number $\epsilon(\mathbf{v}, q) > 0$, an integer $n(\mathbf{v}, q) \geq 1$, a sequence $\{z_i(\mathbf{v}, q)\}_{i=-n(\mathbf{v},q)}^{n(\mathbf{v},q)} \subset X$, $\mathbf{u}^{(\mathbf{v},q)} \in \mathcal{M}$, a positive number $\gamma(\mathbf{v}, q)$ and an open neighborhood $U(\mathbf{v}, q)$ of $\mathbf{u}^{(\mathbf{v},q)}$ in $\mathcal{M}$.

Let $\mathbf{v} = \{v_i\}_{i=-\infty}^{\infty} \in \mathcal{M}$ and let $q \in \mathbf{N}$. It follows from Assumption A(iii) that there exists a number

$$\epsilon(\mathbf{v}, q) \in (0, 2^{-20}q^{-2}) \cap \{i^{-1} : i \in \mathbf{N}\} \tag{5.1}$$

such that the following property holds:

If $x_1, x_2, y_1, y_2 \in X$ satisfy

$$d(x_1, x_2), d(y_1, y_2) \leq 4\epsilon(\mathbf{v}, q),$$

then for each $j \in \mathbf{Z}$,

$$|v_j(x_1, y_1) - v_j(x_2, y_2)| \leq 2^{-8}q^{-2}. \tag{5.2}$$

Choose a natural number

$$n(\mathbf{v}, q) > 4 \cdot 64^2 (8||\mathbf{v}|| + 8)q\epsilon(\mathbf{v}, q)^{-1}. \tag{5.3}$$

Consider a sequence $\{z_i(\mathbf{v}, q)\}_{i=-n(\mathbf{v},q)}^{n(\mathbf{v},q)} \subset X$ such that

$$\sum_{i=-n(\mathbf{v},q)}^{n(\mathbf{v},q)-1} v_i(z_i(\mathbf{v}, q), z_{i+1}(\mathbf{v}, q)) \leq \sigma(\mathbf{v}, -n(\mathbf{v}, q), n(\mathbf{v}, q)) + 2^{-10}q^{-2}.$$

$$\tag{5.4}$$

Define $\mathbf{u}^{(\mathbf{v},q)} = \{u_i^{(\mathbf{v},q)}\}_{i=-\infty}^{\infty}$ by

$$u_i^{(\mathbf{v},q)} = v_i, \quad i \in \mathbf{Z} \setminus [-n(\mathbf{v},q), n(\mathbf{v},q) - 1], \tag{5.5}$$

$$u_i^{(\mathbf{v},q)}(x,y) = v_i(x,y) + (4q)^{-1} \min\{d(x, z_i(\mathbf{v},q)) + d(y, z_{i+1}(\mathbf{v},q)), 1\},$$
$$x, y \in X, \ i \in \mathbf{Z} \cap [-n(\mathbf{v},q), n(\mathbf{v},q) - 1].$$

Evidently $\mathbf{u}^{(\mathbf{v},q)} \in \mathcal{M}$. Choose a number

$$\gamma(\mathbf{v},q) \in (0, 2^{-1}(64n(\mathbf{v},q))^{-1}\epsilon(\mathbf{v},q)) \tag{5.6}$$

and put

$$U(\mathbf{v},q) = \{\mathbf{w} \in \mathcal{M} : \ \rho(\mathbf{w}, \mathbf{u}^{(\mathbf{v},q)}) < \gamma(\mathbf{v},q)\}. \tag{5.7}$$

Define

$$\mathcal{F}_0 = \cap_{m=1}^{\infty} \cup \{U(\mathbf{v},q) : \ \mathbf{v} \in \mathcal{M} \text{ and } q \in \mathbf{N} \cap [m,\infty)\}. \tag{5.8}$$

It is easy to see that $\mathcal{F}_0$ is a countable intersection of open everywhere dense sets in $\mathcal{M}$.

We preface the proof of Proposition 8.5.1 with the following auxiliary lemma which shows that any approximate solution $\{z_i\}_{i=-n(\mathbf{v},q)}^{n(\mathbf{v},q)}$ with respect to $\{u_i^{(\mathbf{v},q)}\}_{i=-\infty}^{\infty}$ is close enough to $\{z_i(\mathbf{v},q)\}_{i=-n(\mathbf{v},q)}^{n(\mathbf{v},q)}$.

LEMMA 8.5.1 *Let $\mathbf{v} \in \mathcal{M}$, $q \in \mathbf{N}$ and let a sequence $\{z_i\}_{i=-n(\mathbf{v},q)}^{n(\mathbf{v},q)} \subset X$ satisfy*

$$\sum_{i=-n(\mathbf{v},q)}^{n(\mathbf{v},q)-1} u_i^{(\mathbf{v},q)}(z_i, z_{i+1}) \le \sigma(\mathbf{u}^{(\mathbf{v},q)}, -n(\mathbf{v},q), n(\mathbf{v},q)) + 32^{-1}q^{-2}. \tag{5.9}$$

Then

$$d(z_i, z_i(\mathbf{v},q)) \le (4q)^{-1}, \quad i = -n(\mathbf{v},q), \dots, n(\mathbf{v},q). \tag{5.10}$$

Proof. It follows from (5.4), (5.5) and (5.9) that

$$\sum_{i=-n(\mathbf{v},q)}^{n(\mathbf{v},q)-1} v_i(z_i, z_{i+1}) + (4q)^{-1} \sum_{i=-n(\mathbf{v},q)}^{n(\mathbf{v},q)-1} \inf\{1, d(z_i, z_i(\mathbf{v},q))$$

$$+ d(z_{i+1}, z_{i+1}(\mathbf{v},q))\}$$

$$= \sum_{i=-n(\mathbf{v},q)}^{n(\mathbf{v},q)-1} u_i^{(\mathbf{v},q)}(z_i, z_{i+1}) \le 32^{-1}q^{-2} + \sigma(\mathbf{u}^{(\mathbf{v},q)} - n(\mathbf{v},q), n(\mathbf{v},q))$$

$$\leq 32^{-1}q^{-2} + \sum_{i=-n(\mathbf{v},q)}^{n(\mathbf{v},q)-1} u_i^{(\mathbf{v},q)}(z_i(\mathbf{v},q), z_{i+1}(\mathbf{v},q))$$

$$= 32^{-1}q^{-2} + \sum_{i=-n(\mathbf{v},q)}^{n(\mathbf{v},q)-1} v_i(z_i(\mathbf{v},q), z_{i+1}(\mathbf{v},q))$$

$$\leq 32^{-1}q^{-2} + \sum_{i=-n(\mathbf{v},q)}^{n(\mathbf{v},q)-1} v_i(z_i, z_{i+1}) + 2^{-10}q^{-2}.$$

This imples (5.10). The lemma is proved.

The next auxiliary result gives an important estimation of the cardinality of the set of all integers i such that z_i is not close enough to $z_i(\mathbf{v}, q)$ where $\{z_i\}_{i=-n(\mathbf{v},q)}^{n(\mathbf{v},q)}$ is an approximate solution with respect to $\{u_i^{(\mathbf{v},q)}\}_{i=-\infty}^{\infty}$.

LEMMA 8.5.2 *Let* $\mathbf{v} \in \mathcal{M}$, $q \in \mathbf{N}$ *and let* $\{z_i\}_{i=-n(\mathbf{v},q)}^{n(\mathbf{v},q)} \subset X$ *satisfy*

$$\sum_{i=-n(\mathbf{v},q)}^{n(\mathbf{v},q)-1} u_i^{(\mathbf{v},q)}(z_i, z_{i+1}) \leq \sigma(\mathbf{u}^{(\mathbf{v},q)}, -n(\mathbf{v},q), n(\mathbf{v},q)) + 8(\|\mathbf{v}\| + 2).$$
$$(5.11)$$

Then

$$Card\{i \in \{-n(\mathbf{v},q), \ldots, n(\mathbf{v},q)-1\}: \quad d(z_i, z_i(\mathbf{v},q)) + d(z_{i+1}, z_{i+1}(\mathbf{v},q)) \tag{5.12}$$

$$\geq \epsilon(\mathbf{v},q)\} \leq (8\|\mathbf{v}\| + 3)4q\epsilon(\mathbf{v},q)^{-1}.$$

Proof. It follows from (5.5), (5.11) and (5.4) that

$$\sum_{i=-n(\mathbf{v},q)}^{n(\mathbf{v},q)-1} v_i(z_i, z_{i+1}) + (4q)^{-1} \sum_{i=-n(\mathbf{v},q)}^{n(\mathbf{v},q)-1} \inf\{1, d(z_i, z_i(\mathbf{v},q))$$

$$+d(z_{i+1}, z_{i+1}(\mathbf{v},q))\} = \sum_{i=-n(\mathbf{v},q)}^{n(\mathbf{v},q)-1} u_i^{(\mathbf{v},q)}(z_i, z_{i+1})$$

$$\leq \sigma(\mathbf{u}^{(\mathbf{v},q)}, -n(\mathbf{v},q), n(\mathbf{v},q)) + 8(\|\mathbf{v}\| + 2)$$

$$\leq \sum_{i=-n(\mathbf{v},q)}^{n(\mathbf{v},q)-1} u_i^{(\mathbf{v},q)}(z_i(\mathbf{v},q), z_{i+1}(\mathbf{v},q)) + 8(\|\mathbf{v}\| + 2)$$

$$\leq \sum_{i=-n(\mathbf{v},q)}^{n(\mathbf{v},q)-1} v_i(z_i(\mathbf{v},q), z_{i+1}(\mathbf{v},q)) + 8(\|\mathbf{v}\| + 2)$$

$$\leq \sum_{i=-n(\mathbf{v},q)}^{n(\mathbf{v},q)-1} v_i(z_i, z_{i+1}) + 2^{-10}q^{-2} + 8(\|\mathbf{v}\| + 2).$$

This implies (5.12). The lemma is proved.

Lemma 8.5.1, (5.1), (5.6) and (5.7) imply the following result.

LEMMA 8.5.3 *For each* $\mathbf{v} \in \mathcal{M}$, *each* $q \in \mathbf{N}$, *each* $\mathbf{w} \in U(\mathbf{v}, q)$ *and each* $\{z_i\}_{i=-n(\mathbf{v},q)}^{n(\mathbf{v},q)} \subset X$ *which satisfy*

$$\sum_{i=-n(\mathbf{v},q)}^{n(\mathbf{v},q)} w_i(z_i, z_{i+1}) \leq \sigma(\mathbf{w}, -n(\mathbf{v},q), n(\mathbf{v},q)) + 32^{-1}q^{-2} - 16^{-1}\epsilon(\mathbf{v},q),$$

the inequality (5.10) holds.

Lemma 8.5.2, (5.1), (5.6), (5.7) and (5.5) imply the following result which shows that the estimation obtained in Lemma 8.5.2 for

$$\{u_i^{(\mathbf{v},q)}\}_{i=-\infty}^{\infty}$$

also holds for any $\mathbf{w}$ which is close enough to $\mathbf{u}^{(\mathbf{v},q)}$ in $\mathcal{M}$.

LEMMA 8.5.4 *For each* $\mathbf{v} \in \mathcal{M}$, *each* $q \in \mathbf{N}$, *each* $\mathbf{w} \in U(\mathbf{v}, q)$ *and each* $\{z_i\}_{i=-n(\mathbf{v},q)}^{n(\mathbf{v},q)} \subset X$ *which satisfy*

$$\sum_{i=-n(\mathbf{v},q)}^{n(\mathbf{v},q)-1} w_i(z_i, z_{i+1}) \leq \sigma(\mathbf{w}, -n(\mathbf{v},q), n(\mathbf{v},q)) + 8\|\mathbf{w}\| + 13,$$

the inequality (5.12) holds.

The next lemma plays a crucial role in our proof. It shows that two approximate solutions with respect to $\mathbf{w} \in U(\mathbf{v}, q)$ are close for all i belonging to a certain subinterval of $[-n(\mathbf{v},q), n(\mathbf{v},q)]$ with a large length.

LEMMA 8.5.5 *Let* $\mathbf{v} \in \mathcal{M}$, $q \geq 1$ *be an integer and let* $\mathbf{w} \in U(\mathbf{v}, q)$. *Assume that a sequence* $\{z_i\}_{i=-n(\mathbf{v},q)}^{n(\mathbf{v},q)} \subset X$ *satisfies*

$$\sum_{i=-n(\mathbf{v},q)}^{n(\mathbf{v},q)-1} w_i(z_i, z_{i+1}) \leq \sigma(\mathbf{w}, -n(\mathbf{v},q), n(\mathbf{v},q)) + 64^{-2}q^{-2}, \qquad (5.13)$$

a natural number $m > n(\mathbf{v}, q)$ and that a sequence $\{y_i\}_{i=-m}^{m} \subset X$ satisfies

$$\sum_{i=-m}^{m-1} w_i(y_i, y_{i+1}) \le \sigma(\mathbf{w}, -m, m) + 64^{-2}q^{-2}.$$

Then

$$d(z_i, y_i) \le (2q)^{-1}, \quad i \in \{-n(\mathbf{v}, q) + 2 + 8q(8\|\mathbf{v}\| + 3)\epsilon(\mathbf{v}, q)^{-1}, \dots,$$
$$n(\mathbf{v}, q) - 2 - 8q(8\|\mathbf{v}\| + 3)\epsilon(\mathbf{v}, q)^{-1}\}. \tag{5.14}$$

Proof. We show that

$$\sum_{i=-n(\mathbf{v},q)}^{n(\mathbf{v},q)-1} w_i(y_i, y_{i+1}) \le \sigma(\mathbf{w}, -n(\mathbf{v}, q), n(\mathbf{v}, q)) + 4\|\mathbf{w}\| + 1. \tag{5.15}$$

Set

$$x_i = y_i, \quad i \in \{-m, \dots, -n(\mathbf{v}, q) - 1\} \cup \{n(\mathbf{v}, q) + 1, \dots, m\}, \tag{5.16}$$
$$x_i = z_i, \quad i \in \{-n(\mathbf{v}, q), \dots, n(\mathbf{v}, q)\}.$$

In view of (5.16) and (5.13),

$$64^{-2}q^{-2} \ge \sum_{i=-m}^{m-1} w_i(y_i, y_{i+1}) - \sum_{i=-m}^{m-1} w_i(x_i, x_{i+1})$$

$$\ge \sum_{i=-n(\mathbf{v},q)}^{n(\mathbf{v},q)-1} w_i(y_i, y_{i+1}) - \sum_{i=-n(\mathbf{v},q)}^{n(\mathbf{v},q)-1} w_i(z_i, z_{i+1}) - 4\|\mathbf{w}\|.$$

Combined with (5.13) this inequality implies (5.15). It follows from (5.13), (5.15) and Lemma 8.5.4 that

$$\text{Card}\{i \in \{-n(\mathbf{v}, q), \dots, n(\mathbf{v}, q) - 1\} : \quad d(z_i, z_i(\mathbf{v}, q)) \tag{5.17}$$

$$+d(z_{i+1}, z_{i+1}(\mathbf{v}, q)) \ge \epsilon(\mathbf{v}, q)\} \le (8\|\mathbf{v}\| + 3)4q\epsilon(\mathbf{v}, q))^{-1},$$
$$\text{Card}\{i \in \{-n(\mathbf{v}, q), \dots, n(\mathbf{v}, q) - 1\} : \quad d(y_i, z_i(\mathbf{v}, q))$$
$$+d(y_{i+1}, z_{i+1}(\mathbf{v}, q)) \ge \epsilon(\mathbf{v}, q)\} \le (8\|\mathbf{v}\| + 3)4q\epsilon(\mathbf{v}, q))^{-1}.$$

By (5.17) and (5.3) there exist integers

$$j_1, j_2 \in [-n(\mathbf{v}, q) + 1, n(\mathbf{v}, q) - 1] \tag{5.18}$$

for which

$$j_1 \le -n(\mathbf{v}, q) + 2 + 8q(8\|\mathbf{v}\| + 3)\epsilon(\mathbf{v}, q)^{-1}, \tag{5.19}$$

$$j_2 \geq n(\mathbf{v}, q) - 2 - 8q(8||\mathbf{v}|| + 3)\epsilon(\mathbf{v}, q)^{-1},$$

$$d(y_{j_p}, z_{j_p}(\mathbf{v}, q)), \ d(z_{j_p}, z_{j_p}(\mathbf{v}, q)) \leq \epsilon(\mathbf{v}, q), \quad p = 1, 2.$$

Define sequences $\{h_i\}_{i=-n(\mathbf{v},q)}^{n(\mathbf{v},q)} \subset X$ and $\{s_i\}_{i=-m}^{m} \subset X$ by

$$h_i = z_i, \quad i \in \{-n(\mathbf{v}, q), \ldots, j_1 - 1\} \cup \{j_2 + 1, \ldots, n(\mathbf{v}, q)\}, \quad (5.20)$$

$$h_i = y_i, \ i \in \{j_1, \ldots, j_2\}, \quad s_i = y_i, \ i \in \{-m, \ldots, j_1 - 1\} \cup \{j_2 + 1, \ldots, m\},$$

$$s_i = z_i, \quad i \in \{j_1, \ldots, j_2\}.$$

We estimate

$$\sum_{i=-n(\mathbf{v},q)}^{n(\mathbf{v},q)-1} w_i(h_i, h_{i+1}) - \sum_{i=-n(\mathbf{v},q)}^{n(\mathbf{v},q)-1} w_i(z_i, z_{i+1}).$$

In view of the choice of $\epsilon(\mathbf{v}, q)$ (see (5.1), (5.2)), (5.20), (5.19) and (5.18),

$$|v_p(h_p, h_{p+1}) - v_p(z_p, z_{p+1})|, \ |v_p(s_p, s_{p+1}) - v_p(y_p, y_{p+1})| \qquad (5.21)$$

$$\leq 2^{-8}q^{-2}, \ p = j_1 - 1, j_2.$$

(5.18), (5.20) and (5.19) imply that for $p = j_1 - 1, j_2,$

$$|\inf\{1, d(h_p, z_p(\mathbf{v}, q)) + d(h_{p+1}, z_{p+1}(\mathbf{v}, q))\}$$

$$- \inf\{1, d(z_p, z_p(\mathbf{v}, q)) + d(z_{p+1}, z_{p+1}(\mathbf{v}, q))\}|$$

$$\leq |d(h_p, z_p(\mathbf{v}, q)) + d(h_{p+1}, z_{p+1}(\mathbf{v}, q))$$

$$- d(z_p, z_p(\mathbf{v}, q)) - d(z_{p+1}, z_{p+1}(\mathbf{v}, q))|$$

$$\leq d(z_p, h_p) + d(h_{p+1}, z_{p+1}) < 2\epsilon(\mathbf{v}, q),$$

$$|\inf\{1, d(s_p, z_p(\mathbf{v}, q)) + d(s_{p+1}, z_{p+1}(\mathbf{v}, q))\}$$

$$- \inf\{1, d(y_p, z_p(\mathbf{v}, q)) + d(y_{p+1}, z_{p+1}(\mathbf{v}, q))\}|$$

$$\leq |d(s_p, z_p(\mathbf{v}, q)) + d(s_{p+1}, z_{p+1}(\mathbf{v}, q))$$

$$- d(y_p, z_p(\mathbf{v}, q)) - d(y_{p+1}, z_{p+1}(\mathbf{v}, q))|$$

$$\leq d(s_p, y_p) + d(s_{p+1}, y_{p+1}) < 2\epsilon(\mathbf{v}, q).$$

Combined with (5.5) and (5.21) these inequalities imply that for $p = j_1 - 1, j_2,$

$$|u_p^{(\mathbf{v},q)}(h_p, h_{p+1}) - u_p^{(\mathbf{v},q)}(z_p, z_{p+1})|, \quad |u_p^{(\mathbf{v},q)}(s_p, s_{p+1})$$

$$- u_p^{(\mathbf{v},q)}(y_p, y_{p+1})| \leq 2^{-8}q^{-2} + (4q)^{-1}2\epsilon(\mathbf{v}, q).$$

It follows from this inequality, (5.3), (5.6) and (5.7) that for $p = j_1 - 1, j_2$,

$$|w_p(h_p, h_{p+1}) - w_p(z_p, z_{p+1})|, \quad |w_p(s_p, s_{p+1}) - w_p(y_p, y_{p+1})| \quad (5.22)$$

$$\leq 2^{-8}q^{-2} + \epsilon(\mathbf{v}, q)[(2q)^{-1} + (32q)^{-1}].$$

(5.20), (5.13) and (5.22) imply that

$$64^{-2}q^{-2} \geq \sum_{i=-m}^{m-1} w_i(y_i, y_{i+1}) - \sum_{i=-m}^{m-1} w_i(s_i, s_{i+1})$$

$$= \sum_{i=j_1-1}^{j_2} w_i(y_i, y_{i+1}) - \sum_{i=j_1-1}^{j_2} w_i(s_i, s_{i+1})$$

$$= \sum_{i=j_1}^{j_2-1} w_i(y_i, y_{i+1}) - \sum_{i=j_1}^{j_2-1} w_i(z_i, z_{i+1}) + w_{j_1-1}(y_{j_1-1}, y_{j_1})$$

$$+ w_{j_2}(y_{j_2}, y_{j_2+1}) - w_{j_1-1}(s_{j_1-1}, s_{j_1}) - w_{j_2}(s_{j_2}, s_{j_2+1})$$

$$\geq \sum_{i-j_1}^{j_2-1} w_i(y_i, y_{i+1}) - \sum_{i-j_1}^{j_2-1} w_i(z_i, z_{i+1})$$

$$- 2[2^{-8}q^{-2} + \epsilon(\mathbf{v}, q)((2q)^{-1} + (32q)^{-1})],$$

$$\sum_{i=j_1}^{j_2-1} w_i(y_i, y_{i+1}) - \sum_{i=j_1}^{j_2-1} w_i(z_i, z_{i+1}) \quad (5.23)$$

$$\leq 64^{-2}q^{-2} + 2^{-7}q^{-2} + \epsilon(\mathbf{v}, q)(q^{-1} + (16q)^{-1}).$$

By (5.23), (5.22) and (5.20),

$$\sum_{i=-n(\mathbf{v},q)}^{n(\mathbf{v},q)-1} w_i(h_i, h_{i+1}) - \sum_{i=-n(\mathbf{v},q)}^{n(\mathbf{v},q)-1} w_i(z_i, z_{i+1})$$

$$= \sum_{i=j_1-1}^{j_2} w_i(h_i, h_{i+1}) - \sum_{i=j_1-1}^{j_2} w_i(z_i, z_{i+1})$$

$$= \sum_{i=j_1}^{j_2-1} w_i(y_i, y_{i+1}) - \sum_{i=j_1}^{j_2-1} w_i(z_i, z_{i+1}) + w_{j_1-1}(h_{j_1-1}, h_{j_1})$$

$$+ w_{j_2}(h_{j_2}, h_{j_2+1}) - w_{j_1-1}(z_{j_1-1}, z_{j_1}) - w_{j_2}(z_{j_2}, z_{j_2+1})$$

$$\leq 64^{-2}q^{-2} + 2^{-7}q^{-2} + \epsilon(\mathbf{v}, q)(q^{-1} + (16q)^{-1})$$

$$+ 2[2^{-8}q^{-2} + \epsilon(\mathbf{v}, q)((2q)^{-1} + (32q)^{-1})]$$

$$\leq 64^{-2}q^{-2} + 2^{-6}q^{-2} + \epsilon(\mathbf{v}, q)(2q^{-1} + (8q)^{-1})$$

and

$$\sum_{i=-n(\mathbf{v},q)}^{n(\mathbf{v},q)-1} w_i(h_i, h_{i+1}) - \sum_{i=-n(\mathbf{v},q)}^{n(\mathbf{v},q)-1} w_i(z_i, z_{i+1})$$

$$\leq 64^{-2}q^{-2} + 2^{-6}q^{-2} + \epsilon(\mathbf{v}, q)(2q^{-1} + (8q)^{-1}).$$

It follows from this inequality, (5.13) and (5.1) that

$$\sum_{i=-n(\mathbf{v},q)}^{n(\mathbf{v},q)-1} w_i(h_i, h_{i+1}) \leq \sigma(\mathbf{w}, -n(\mathbf{v}, q), n(\mathbf{v}, q))$$

$$+2 \cdot 64^{-2}q^{-2} + 2^{-6}q^{-2} + \epsilon(\mathbf{v}, q)(2q^{-1} + (8q)^{-1})$$

$$\leq \sigma(\mathbf{w}, -n(\mathbf{v}, q), n(\mathbf{v}, q)) + 32^{-1}q^{-2} - 16^{-1}\epsilon(\mathbf{v}, q).$$

In view of this inequality and Lemma 8.5.3,

$$d(h_i, z_i(\mathbf{v}, q)) \leq (4q)^{-1}, \quad i = -n(\mathbf{v}, q), \ldots, n(\mathbf{v}, q).$$

Combined with (5.19) and (5.20) this inequality implies that

$$d(y_i, z_i(\mathbf{v}, q)) \leq (4q)^{-1}, \quad i = -n(\mathbf{v}, q) + 2 \tag{5.24}$$

$$+8q(8||\mathbf{v}|| + 3)\epsilon(\mathbf{v}, q)^{-1}, \ldots, n(\mathbf{v}, q) - 2 - 8q(8||\mathbf{v}|| + 3)\epsilon(\mathbf{v}, q)^{-1}.$$

By (5.13) and Lemma 8.5.3

$$d(z_i, z_i(\mathbf{v}, q)) \leq (4q)^{-1}, \quad i = -n(\mathbf{v}, q), \ldots, n(\mathbf{v}, q).$$

Combined with (5.24) this implies (5.14). The lemma is proved.

The next auxiliary result follows from Lemma 8.5.5.

LEMMA 8.5.6 *For each* $\mathbf{w} \in \mathcal{F}_0$, *each* $\epsilon \in (0, 1)$ *and each* $k \in \mathbf{N}$ *there exist* $\epsilon_0 \in (0, \epsilon)$ *and a natural number* $k_0 > k$ *such that the following property holds:*
If $m_1, m_2 \geq k_0$ *are integers and if* $\{z_i^p\}_{i=-m_p}^{m_p} \subset X$, $p = 1, 2$ *satisfies*

$$\sum_{i=-m_p}^{m_p-1} w_i(z_i^p, z_{i+1}^p) \leq \sigma(\mathbf{w}, -m_p, m_p) + \epsilon_0, \ p = 1, 2, \tag{5.25}$$

then

$$d(z_i^1, z_i^2) \leq \epsilon, \quad i = -k, \ldots, k. \tag{5.26}$$

Proof. Let $\mathbf{w} \in \mathcal{F}_0$, $\epsilon \in (0,1)$ and let $k \in \mathbf{N}$. Choose $q_0 \in \mathbf{N}$ such that

$$q_0 > 8k + 8 + 8\epsilon^{-1}. \tag{5.27}$$

In view of (5.8) there exist $q \in \mathbf{N}$ and $\mathbf{v} \in \mathcal{M}$ such that $q \geq q_0$ and

$$\mathbf{w} \in U(\mathbf{v}, q). \tag{5.28}$$

Put

$$k_0 = n(\mathbf{v}, q) + 1 + k, \quad \epsilon_0 = 64^{-2}q^{-2}. \tag{5.29}$$

Consider a sequence $\{x_i\}_{i=-n(\mathbf{v},q)}^{n(\mathbf{v},q)} \subset X$ such that

$$\sum_{i=-n(\mathbf{v},q)}^{n(\mathbf{v},q)-1} w_i(x_i, x_{i+1}) \leq \sigma(\mathbf{w}, -n(\mathbf{v}, q), n(\mathbf{v}, q)) + 64^{-2}q^{-2}. \tag{5.30}$$

Let $m_1, m_2 \geq k_0$ be integers and let sequences $\{z_i^p\}_{i=-m_p}^{m_p} \subset X$, $p = 1, 2$ satisfy (5.25). By Lemma 8.5.5, (5.28), (5.30), (5.29) and (5.25),

$$d(x_i, z_i^p) \leq (2q)^{-1},$$

$$i \in \{-n(\mathbf{v}, q) + 2 + 8q(8\|\mathbf{v}\| + 3)\epsilon(\mathbf{v}, q)^{-1}, \ldots,$$

$$n(\mathbf{v}, q) - 2 - 8q(8\|\mathbf{v}\| + 3)\epsilon(\mathbf{v}, q)^{-1}\}, \quad p = 1, 2.$$

Combined with (5.27) and (5.3) this inequality implies that $d(z_i^1, z_i^2) \leq q^{-1} \leq \epsilon$ for any integer $i \in [-2^{-1}n(\mathbf{v}, q), 2^{-1}n(\mathbf{v}, q)] \supset [-k, k]$. This completes the proof of the lemma.

Proof of Proposition 8.5.1. Let $\mathbf{w} \in \mathcal{F}_0$. By Lemma 8.5.6 there exist a strictly increasing sequence of natural numbers $\{s_k\}_{k=1}^{\infty}$ and a strictly decreasing sequence of positive numbers $\{\delta_k\}_{k=1}^{\infty}$ such that

$$\delta_k < k^{-1}, \quad s_k > k, \quad k = 1, 2, \ldots \tag{5.31}$$

and that for each $k \in \mathbf{N}$ the following property holds:
If integers $m_1, m_2 \geq s_k$ and if sequences $\{z_i^p\}_{i=-m_p}^{m_p} \subset X$, $p = 1, 2$ satisfy

$$\sum_{i=-m_p}^{m_p-1} w_i(z_i^p, z_{i+1}^p) \leq \sigma(\mathbf{w}, -m_p, m_p) + \delta_k, \quad p = 1, 2,$$

then $d(z_i^1, z_i^2) \leq (2k)^{-1}$, $i = -k, \ldots, k$.

For any $k \in \mathbf{N}$ we fix a sequence $\{y_i^k\}_{i=-s_k}^{s_k} \subset X$ such that

$$\sum_{i=-s_k}^{s_k-1} w_i(y_i^k, y_{i+1}^k) \leq \sigma(\mathbf{w}, -s_k, s_k) + \delta_k. \tag{5.32}$$

It is not difficult to see that for each $i \in \mathbf{N}$ there exists $x_i = \lim_{k\to\infty} y_i^k$. Assume that $k_1, k_2 > k_1$ are integers. It follows from (5.32), (5.31) that

$$\sum_{i=k_1}^{k_2-1} w_i(x_i, x_{i+1}) = \lim_{k\to\infty} \sum_{i=k_1}^{k_2-1} w_i(y_i^k, y_{i+1}^k)$$

$$\leq \limsup_{k\to\infty} \sigma(\mathbf{w}, k_1, k_2, y_{k_1}^k, y_{k_2}^k) = \sigma(\mathbf{w}, k_1, k_2, x_{k_1}, x_{k_2}).$$

This completes the proof of the proposition.

8.6.   Proof of Theorem 8.4.1

It follows from Proposition 8.5.1 there exists a set $\mathcal{F}_0 \subset \mathcal{M}$ which is a countable intersection of open everywhere dense sets in $\mathcal{M}$ such that for each $\mathbf{v} \in \mathcal{F}_0$ there is a $(\mathbf{v})$-minimal sequence $\{x_i^{\mathbf{v}}\}_{i=-\infty}^{\infty} \subset X$ satisfying

$$\sum_{i=k_1}^{k_2-1} v_i(x_i^{\mathbf{v}}, x_{i+1}^{\mathbf{v}}) = \sigma(\mathbf{v}, k_1, k_2, x_{k_1}^{\mathbf{v}}, x_{k_2}^{\mathbf{v}}) \tag{6.1}$$

for each pair of integers $k_1 < k_2$.

For each $\mathbf{v} \in \mathcal{F}_0$ and each $r \in (0, 1]$ define $\mathbf{v}^r = \{v_i^r\}_{i=-\infty}^{\infty}$ as

$$v_i^r(x, y) = v_i(x, y) + r \inf\{d(x, x_i^{\mathbf{v}}), 1\}, \tag{6.2}$$

$$x, y \in X, \ i = 0, \pm 1, \ldots.$$

Clearly $\mathbf{v}^r \in \mathcal{M}$. We will see that $\mathbf{v}^r$ has the turnpike property with the turnpike $\{x_i^{\mathbf{v}}\}_{i=-\infty}^{\infty}$. We will preface the proof of Theorem 8.4.1 with the following auxiliary lemmas.

LEMMA 8.6.1 *For each $\mathbf{v} \in \mathcal{F}_0$ and each pair of integers $k_2 > k_1$,*

$$\sum_{i=k_1}^{k_2-1} v_i(x_i^{\mathbf{v}}, x_{i+1}^{\mathbf{v}}) \leq \sigma(\mathbf{v}, k_1, k_2) + 1 + 4||\mathbf{v}||.$$

Proof. Let $\mathbf{v} \in \mathcal{F}_0$ and $k_2 > k_1$ be integers. We may assume that $k_2 - k_1 \geq 3$. Consider a sequence $\{y_i\}_{i=k_1}^{k_2} \subset X$ such that

$$\sum_{i=k_1}^{k_2-1} v_i(y_i, y_{i+1}) \leq \sigma(\mathbf{v}, k_1, k_2) + 8^{-1}.$$

Set

$$z_i = x_i^{\mathbf{v}}, \; i = k_1, k_2, \quad z_i = y_i, \; i = k_1 + 1, \ldots, k_2 - 1.$$

It is not difficult to see that

$$\sum_{i=k_1}^{k_2-1} v_i(x_i^{\mathbf{v}}, x_{i+1}^{\mathbf{v}}) \leq \sum_{i=k_1}^{k_2-1} v_i(z_i, z_{i+1}) \leq \sum_{i=k_1}^{k_2-1} v_i(y_i, y_{i+1})$$

$$+ 4||\mathbf{v}|| \leq \sigma(\mathbf{v}, k_1, k_2) + 4||\mathbf{v}|| + 8^{-1}.$$

The lemma is proved.

The following auxiliary result shows that any approximate solution $\{y_i\}_{i=p}^{p+N}$ with respect to $\mathbf{v}^r$ and with large enough N is close to the turnpike for some $i \in [p, p + N]$.

LEMMA 8.6.2 *Assume that* $\mathbf{v} \in \mathcal{F}_0$, $r \in (0, 1]$, $\delta \in (0, 1)$ *and* $M > 0$. *Then there exists an integer* $N \geq 4$ *such that for each integer* p *and each sequence* $\{y_i\}_{i=p}^{p+N} \subset X$ *satisfying*

$$\sum_{i=p}^{p+N-1} v_i^r(y_i, y_{i+1}) \leq \sigma(\mathbf{v}^r, p, p + N, y_p, y_{p+N}) + M \qquad (6.3)$$

the following relation holds:

$$\inf\{d(y_i, x_i^{\mathbf{v}}) : \quad i \in \{p, \ldots, p + N\}\} < \delta. \qquad (6.4)$$

Proof. Choose a natural number

$$N \geq 6 + (M + 8||\mathbf{v}|| + 8)(\delta r)^{-1}. \qquad (6.5)$$

Assume that $p \in \mathbf{Z}$, $\{y_i\}_{i=p}^{p+N} \subset X$ and (6.3) holds. To prove the lemma it is sufficient to show that (6.4) holds.

Assume the contrary. Then

$$d(y_i, x_i^{\mathbf{v}}) \geq \delta, \quad i = p, \ldots, p + N. \qquad (6.6)$$

Define $\{z_i\}_{i=p}^{p+N} \subset X$ by

$$z_i = y_i, \; i = p, p + N, \quad z_i = x_i^{\mathbf{v}}, \; i = p + 1, \ldots, p + N - 1. \qquad (6.7)$$

By (6.2), (6.3), (6.6) and (6.7),

$$\sum_{i=p}^{p+N-1} v_i^r(z_i, z_{i+1}) + M \geq \sigma(\mathbf{v}^r, p, p+N, y_p, y_{p+N}) + M$$

$$\geq \sum_{i=p}^{p+N-1} v_i^r(y_i, y_{i+1}) \geq \sum_{i=p}^{p+N-1} v_i(y_i, y_{i+1}) + r\delta N.$$

It follows from this inequality, (6.7), Lemma 8.6.1 and (6.2) that

$$r\delta N + \sigma(\mathbf{v}, p, p+N) \leq M + \sum_{i=p}^{p+N-1} v_i^r(z_i, z_{i+1}) \leq M$$

$$+ \sum_{i=p}^{p+N-1} v_i^r(x_i^{\mathbf{v}}, x_{i+1}^{\mathbf{v}}) + 4(||\mathbf{v}|| + 1) \leq M + 4||\mathbf{v}|| + 4$$

$$+\sigma(\mathbf{v}, p, p+N) + 1 + 4||\mathbf{v}||.$$

This is contradictrory to (6.5). The obtained contradiction proves the lemma.

Lemma 8.6.2 implies the following result which shows that the property established in Lemma 8.6.2 for $\mathbf{v}^r$ also holds for all $\mathbf{w}$ belonging to a small neighborhood of $\mathbf{v}^r$ in $\mathcal{M}$.

LEMMA 8.6.3 *For each* $\mathbf{v} \in \mathcal{F}_0$, *each* $r \in (0,1]$, *each* $\delta \in (0,1)$ *and each positive number* M *there exist a natural number* $N \geq 4$ *and a neighborhood* U *of* $\mathbf{v}^r$ *in* $\mathcal{M}$ *such that the following property holds:*
If $\mathbf{w} \in U$, $p \in \mathbf{Z}$ *and if a sequence* $\{y_i\}_{i=p}^{p+N} \subset X$ *satisfies*

$$\sum_{i=p}^{p+N-1} w_i(y_i, y_{i+1}) \leq \sigma(\mathbf{w}, p, p+N, y_p, y_{p+N}) + M,$$

then the inequality (6.4) holds.

The next lemma shows that if an approximate solution $\{y_i\}_{i=k_1}^{k_2}$ with respect to $\mathbf{v}^r$ is close to the turnpike for $i = k_1, k_2$, then it is close to the turnpike for all $i = k_1, \ldots, k_2$.

LEMMA 8.6.4 *Let* $\mathbf{v} \in \mathcal{F}_0$, $r \in (0,1]$ *and* $\epsilon \in (0,1)$. *Then there exists* $\delta \in (0, \epsilon)$ *such that the following property holds:*
If $k_1, k_2 \geq k_1 + 2$ *are integers and if a sequence* $\{y_i\}_{i=k_1}^{k_2} \subset X$ *satisfies*

$$d(y_i, x_i^{\mathbf{v}}) \leq \delta, \quad i = k_1, k_2, \tag{6.8}$$

$$\sum_{i=k_1}^{k_2-1} v_i^r(y_i, y_{i+1}) \leq \sigma(\mathbf{v}^r, k_1, k_2, y_{k_1}, y_{k_2}) + \delta,$$

then

$$d(y_i, x_i^{\mathbf{v}}) \leq \epsilon, \quad i = k_1, \ldots, k_2. \tag{6.9}$$

Proof. Choose a positive number

$$\epsilon_0 < 8^{-1} r\epsilon. \tag{6.10}$$

It follows from Assumption A(iii) that there exists a positive number $\delta < 8^{-1}\epsilon_0$ such that the following property holds:
  If i is an integer and if $x_1, x_2, y_1, y_2 \in X$ satisfy

$$d(x_1, y_1) + d(x_2, y_2) \leq 8\delta, \tag{6.11}$$

then

$$|v_i(x_1, x_2) - v_i(y_1, y_2)|, \ |v_i^r(x_1, x_2) - v_i^r(y_1, y_2)| \leq 64^{-1}\epsilon_0. \tag{6.12}$$

Assume that k_1 and $k_2 \geq k_1 + 2$ are integers and $\{y_i\}_{i=k_1}^{k_2} \subset X$ satisfies (6.8). To prove the lemma it is sufficient to show that (6.9) holds.
  Let us assume the converse. Then

$$\sup\{d(y_i, x_i^{\mathbf{v}}) : \ i = k_1 + 1, \ldots, k_2 - 1\} > \epsilon. \tag{6.13}$$

Put

$$z_i = x_i^{\mathbf{v}}, \ i = k_1, k_2, \quad z_i = y_i, \ i = k_1 + 1, \ldots, k_2 - 1. \tag{6.14}$$

By (6.8) and the choice of δ (see (6.11), (6.12)),

$$|\sigma(\mathbf{v}^r, k_1, k_2, y_{k_1}, y_{k_2}) - \sigma(\mathbf{v}^r, k_1, k_2, x_{k_1}^{\mathbf{v}}, x_{k_2}^{\mathbf{v}})| \leq 16^{-1}\epsilon_0. \tag{6.15}$$

It follows from (6.14), (6.8) and the choice of δ (see (6.11), (6.12)) that

$$|\sum_{i=k_1}^{k_2-1} v_i^r(y_i, y_{i+1}) - \sum_{i=k_1}^{k_2-1} v_i^r(z_i, z_{i+1})| \leq 16^{-1}\epsilon_0.$$

In view of this inequality, (6.15), (6.8), (6.1) and (6.2),

$$\sum_{i=k_1}^{k_2-1} v_i^r(z_i, z_{i+1}) \leq 8^{-1}\epsilon_0 + \delta + \sigma(\mathbf{v}^r, k_1, k_2, x_{k_1}^{\mathbf{v}}, x_{k_2}^{\mathbf{v}})$$

$$= 8^{-1}\epsilon_0 + \delta + \sigma(\mathbf{v}, k_1, k_2, x_{k_1}^{\mathbf{v}}, x_{k_2}^{\mathbf{v}}).$$

Combined with (6.2), (6.13) and (6.14) this relation implies that

$$\sigma(\mathbf{v}, k_1, k_2, x_{k_1}^{\mathbf{v}}, x_{k_2}^{\mathbf{v}}) + \delta + 8^{-1}\epsilon_0 \geq \sum_{i=k_1}^{k_2-1} v_i^r(z_i, z_{i+1})$$

$$\geq \sum_{i=k_1}^{k_2-1} v_i(z_i, z_{i+1}) + r\epsilon, \quad r\epsilon \leq \delta + 8^{-1}\epsilon_0 \leq 4^{-1}\epsilon_0.$$

This is contradictory to (6.10). The obtained contradiction proves the lemma.

The following lemma shows that the property established in Lemma 8.6.4 for $\mathbf{v}^r$ also holds for all $\mathbf{w}$ belonging to a small neighborhood of $\mathbf{v}^r$ in $\mathcal{M}$.

LEMMA 8.6.5 *For each* $\mathbf{v} \in \mathcal{F}_0$, *each* $r \in (0, 1]$ *and each* $\epsilon \in (0, 1)$ *there exist* $\delta \in (0, \epsilon)$ *and a neighborhood* U *of* $\mathbf{v}^r$ *in* $\mathcal{M}$ *such that the following property holds:*
If $\mathbf{w} \in U$, $k_1, k_2 \geq k_1 + 2$ *are integers and if a sequence* $\{y_i\}_{i=k_1}^{k_2} \subset X$ *satisfies*

$$d(y_i, x_i^{\mathbf{v}}) \leq \delta, \quad i = k_1, k_2, \tag{6.16}$$

$$\sum_{i=k_1}^{k_2-1} w_i(y_i, y_{i+1}) \leq \sigma(\mathbf{w}, k_1, k_2, y_{k_1}, y_{k_2}) + \delta,$$

then

$$d(y_i, x_i^{\mathbf{v}}) \leq \epsilon, \quad i = k_1, \ldots, k_2. \tag{6.17}$$

Proof. Let $\mathbf{v} \in \mathcal{F}_0$, $r \in (0, 1]$ and $\epsilon \in (0, 1)$. It follows from Lemma 8.6.4 that there exists

$$\delta \in (0, 4^{-1}\epsilon) \tag{6.18}$$

such that the following property holds:
If $k_1, k_2 \geq k_1 + 2$ are integers and if a sequence $\{y_i\}_{i=k_1}^{k_2} \subset X$ satisfies

$$d(y_i, x_i^{\mathbf{v}}) \leq 4\delta, \quad i = k_1, k_2, \tag{6.19}$$

$$\sum_{i=k_1}^{k_2-1} v_i^r(y_i, y_{i+1}) \leq \sigma(\mathbf{v}_r, k_1, k_2, y_{k_1}, y_{k_2}) + 4\delta,$$

then the inequality (6.17) holds.
It follows from Lemma 8.6.3 that there exist a neighborhood U_1 of $\mathbf{v}^r$ in $\mathcal{M}$ and a natural number $N \geq 4$ such that the following property holds:

If $\mathbf{w} \in U_1$, p is an integer and if a sequence $\{y_i\}_{i=p}^{p+N} \subset X$ satisfies

$$\sum_{i=p}^{p+N-1} w_i(y_i, y_{i+1}) \leq \sigma(\mathbf{w}, p, p+N, y_p, y_{p+N}) + 8, \qquad (6.20)$$

then

$$\inf\{d(y_i, x_i^{\mathbf{v}}) : \quad i = p, \ldots, p+N\} < \delta. \qquad (6.21)$$

Define

$$U = \{\mathbf{w} \in U_1 : \quad \rho(\mathbf{w}, \mathbf{v}^r) < \delta(16N)^{-1}\}. \qquad (6.22)$$

Assume that $\mathbf{w} \in U$, $k_1, k_2 \geq k_1 + 2$ are integers and $\{y_i\}_{i=k_1}^{k_2} \subset X$ satisfies (6.16). To prove the lemma it is sufficient to show that (6.17) holds.

Assume the contrary. Then there is an integer

$$i_0 \in [k_1 + 1, k_2 - 1] \qquad (6.23)$$

such that

$$d(y_{i_0}, x_{i_0}^{\mathbf{v}}) > \epsilon. \qquad (6.24)$$

It follows from (6.16) and the definition of U_1, N (see (6.20), (6.21)) that there exist integers $i_1, i_2 \in [k_1, k_2]$ for which

$$i_0 - N \leq i_1 < i_0 < i_2 \leq i_0 + N, \quad d(y_{i_p}, x_{i_p}^{\mathbf{v}}) \leq \delta, \; p = 1, 2. \qquad (6.25)$$

In view of (6.22), (6.16) and (6.25),

$$\sum_{i=i_1}^{i_2-1} v_i^r(y_i, y_{i+1}) \leq \sigma(\mathbf{v}^r, i_1, i_2, y_{i_1}, y_{i_2}) + 2\delta.$$

It follows from this inequality, (6.25) and the choice of δ (see (6.18), (6.19)) that

$$d(y_i, x_i^{\mathbf{v}}) \leq \epsilon, \quad i = i_1, \ldots, i_2.$$

This is contradictory to (6.24). The obtained contradiction proves the lemma.

Lemma 8.6.3 and 8.6.5 imply the following result.

LEMMA 8.6.6 *Let* $\mathbf{v} \in \mathcal{F}_0$, $r \in (0,1]$ *and* $\epsilon \in (0,1)$. *Then there exist a neighborhood* U *of* $\mathbf{v}^r$ *in* $\mathcal{M}$, *a positive number* $\delta < \epsilon$ *and a natural number* N *such that the following property holds:*

If $\mathbf{w} \in U$, $k_1, k_2 \geq k_1 + 2N$ *are integers and if a sequence* $\{y_i\}_{i=k_1}^{k_2} \subset X$ *satisfies*

$$\sum_{i=k_1}^{k_2-1} w_i(y_i, y_{i+1}) \leq \sigma(\mathbf{w}, k_1, k_2, y_{k_1}, y_{k_2}) + \delta,$$

then

$$d(y_i, x_i^{\mathbf{v}}) \leq \epsilon, \quad i = L_1 + k_1, \ldots, k_2 - L_2$$

where integers $L_1, L_2 \in [0, N]$, *and moreover, if* $d(y_{k_1}, x_{k_1}^{\mathbf{v}}) \leq \delta$, *then* $L_1 = 0$, *and if* $d(y_{k_2}, x_{k_2}^{\mathbf{v}}) \leq \delta$, *then* $L_2 = 0$.

Completion of the proof of Theorem 8.4.1. For each $\mathbf{v} \in \mathcal{F}_0$, each $r \in (0, 1]$ and each natural number i there exist an open neighborhood $U(\mathbf{v}, r, i)$ of $\mathbf{v}^r$ in $\mathcal{M}$, numbers

$$\delta(\mathbf{v}, r, i) \in (0, (2i)^{-1}), \quad \delta_0(\mathbf{v}, r, i) \in (0, 2^{-1}\delta(\mathbf{v}, r, i)) \qquad (6.26)$$

and natural numbers $N(\mathbf{v}, r, i), N_0(\mathbf{v}, r, i)$ such that Lemma 8.6.6 holds with $U = U(\mathbf{v}, r, i)$, $\epsilon = (2i)^{-1}$, $N = N(\mathbf{v}, r, i)$, $\delta = \delta(\mathbf{v}, r, i)$ and also holds with $\epsilon = 2^{-1}\delta(\mathbf{v}, r, i)$, $\delta = \delta_0(\mathbf{v}, r, i)$, $N = N_0(\mathbf{v}, r, i)$, $U = U(\mathbf{v}, r, i)$. Define

$$\mathcal{F} = \mathcal{F}_0 \cap \cap_{i=1}^{\infty} \cup \{U(\mathbf{v}, r, i) : v \in \mathcal{F}_0, \ r \in (0, 1]\}. \qquad (6.27)$$

Clearly $\mathcal{F}$ is a countable intersection of open everywhere dense sets in $\mathcal{M}$.

Let $\mathbf{w} \in \mathcal{F}$. We show that properties (a) and (b) hold. Let $\epsilon \in (0, 1)$. Choose a natural number

$$i > 8\epsilon^{-1} + 8. \qquad (6.28)$$

It follows from (6.27) that there exists $\mathbf{v} \in \mathcal{F}_0$ and $r \in (0, 1]$ such that

$$\mathbf{w} \in U(\mathbf{v}, r, i). \qquad (6.29)$$

Assume that $\{y_i\}_{i=-\infty}^{\infty} \subset X$ is a $(\mathbf{w})$-minimal sequence. Then for each pair of integers $k_2 > k_1$,

$$\sum_{i=k_1}^{k_2-1} w_i(y_i, y_{i+1}) = \sigma(\mathbf{w}, k_1, k_2, y_{k_1}, y_{k_2}). \qquad (6.30)$$

By (6.30), the choice of $\{x_i^{\mathbf{w}}\}_{i=-\infty}^{\infty} \subset X$ (see (6.1)), Lemma 8.6.6 and the choice of $U(\mathbf{v}, r, i)$, $\delta_0(\mathbf{v}, r, i)$, $\delta(\mathbf{v}, r, i)$ and $N_0(\mathbf{v}, r, i)$,

$$d(y_j, x_j^{\mathbf{v}}), \quad d(x_j^{\mathbf{w}}, x_j^{\mathbf{v}}) \leq 2^{-1}\delta(\mathbf{v}, r, i), \quad j = 0, \pm 1, \ldots. \qquad (6.31)$$

(6.31), (6.26) and (6.28) imply that

$$d(y_j, x_j^{\mathbf{w}}) \leq \epsilon, \quad j = 0, \pm 1, \pm 2, \ldots.$$

Since ϵ is any positive number from the interval $(0, 1)$, we conclude that $y_j = x_j^{\mathbf{w}}$ for any integer j. Therefore property (a) holds.

Assume that $\mathbf{u} \in U(\mathbf{v}, r, i)$, $k_1, k_2 > k_1 + 2N(\mathbf{v}, r, i)$ are integers, $\{y_i\}_{k_1}^{k_2} \subset X$ and

$$\sum_{i=k_1}^{k_2-1} u_i(y_i, y_{i+1}) \leq \sigma(\mathbf{u}, k_1, k_2, u_{k_1}, u_{k_2}) + \delta(\mathbf{v}, r, i). \tag{6.32}$$

It follows from (6.32), the definition of $U(\mathbf{v}, r, i)$, $N(\mathbf{v}, r, i)$, $\delta(\mathbf{v}, r, i)$ and Lemma 8.6.6 that

$$d(y_j, x_j^{\mathbf{v}}) \leq (2i)^{-1}, \quad j = k_1 + L_1, \ldots, k_2 - L_2,$$

where integers $L_1, L_2 \in [0, N(\mathbf{v}, r, i)]$, and if $d(y_{k_1}, x_{k_1}^{\mathbf{v}}) \leq \delta(\mathbf{v}, r, i)$, then $L_1 = 0$, and if $d(y_{k_2}, x_{k_2}^{\mathbf{v}}) \leq \delta(\mathbf{v}, r, i)$, then $L_2 = 0$. Together with (6.31), (6.26) and (6.28) this implies that

$$d(y_j, x_j^{\mathbf{w}}) \leq \epsilon, \quad j = k_1 + L_1, \ldots, k_2 - L_2,$$

and if $d(y_{k_1}, x_{k_1}^{\mathbf{w}}) \leq \delta_0(\mathbf{v}, r, i)$, then $L_1 = 0$, and if $d(y_{k_2}, x_{k_2}^{\mathbf{w}}) \leq \delta_0(\mathbf{v}, r, i)$, then $L_2 = 0$. Therefore property (b) holds. This completes the proof of the theorem.

Chapter 9

CONTROL PROBLEMS
IN HILBERT SPACES

In this chapter we analyze the structure of optimal solutions for infinite-dimensional optimal continuous-time control problems. We show that an optimal trajectory defined on an interval $[0, \tau]$ is contained in a small neighborhood of the optimal steady-state in the weak topology for all $t \in [0, \tau] \setminus E$, where $E \subset [0, \tau]$ is a measurable set such that the Lebesgue measure of E does not exceed a constant which depends only on the neighborhood of the optimal steady-state and does not depend on τ. Moreover, we show that the set E is a finite union of intervals and their number does not exceed a constant which depends only on the neighborhood.

9.1. Main results

We consider a system described by the following input-output relationship:

$$x(t) = S(t)x_0 + \int_0^T S(t-s)Bu(s)ds, \ t \in I, \tag{1.1}$$

where I is either $[0, \infty)$ or $[0, T]$ $(0 \le T < \infty)$, E and F are separable Hilbert spaces, $x_0 \in E$, $\{S(t) : t \ge 0\}$ is a strongly continuous semigroup on E with generator A, $u(\cdot) \in L^2_{loc}(I; F)$, the space of all strongly measurable functions $u(\cdot) : I \to F$, which are square-integrable on every finite interval $\Delta \subset I$, and $B : F \to E$ is a bounded linear operator.

Thus $x(\cdot)$ is the mild solution of the state equation

$$x'(t) = Ax(t) + Bu(t), \ t \in I, \tag{1.2}$$

$$x(0) = x_0, \tag{1.3}$$

where A is a possibly unbounded, closed, and densely defined operator in E.

In addition we know (see [7]) that although a mild solution need not be absolutely continuous it does satisfy the following mild differential equation for any $y \in D(A^*)$:

$$(d/dt) < x(t), y > = < x(t), A^*y > + < Bu(t), y > \quad \text{a.e. } t \in I, \tag{1.4}$$

$$\lim_{t \to 0^+} < x(t), y > = < x_0, y >, \tag{1.5}$$

where A^* is the adjoint operator associated with A, with domain $\mathcal{D}(A^*)$.

We assume that

$$x(t) \in X, \ t \in I \text{ where } X \text{ is a convex and closed subset of } E \tag{1.6}$$

and

$$u(t) \in U(x(t)) \subset F, \ t \in I \text{ where } U(\cdot) : X \to 2^F$$

is a point to set mapping which is convex valued and satisfies

$$\alpha U(x_1) + (1 - \alpha)U(x_2) \subset U(\alpha x_1 + (1 - \alpha)x_2), \ x_1, x_2 \in X, \ \alpha \in [0, 1],$$

if $u_n \to u, x_n \to x$ as $n \to \infty$ in the weak topology,

$$u_n \in U(x_n), \ n = 1, 2, \ldots, \text{ then } u \in U(x). \tag{1.7}$$

The performance of the control system is measured on any finite interval $[T_1, T_2]$ by the integral functional

$$I(T_1, T_2, x, u) = \int_{T_1}^{T_2} f(x(t), u(t))dt, \tag{1.8}$$

where $f : E \times F \to R^1$ is a convex function. We assume that f is lower semicontinuous on $E \times F$ and that there exist positive numbers K_1 and K such that

$$f(x, u) \geq K(||x||^2 + ||u||^2)$$

for each $x \in E, \ u \in F$ satisfying $||x||^2 + ||u||^2 > K_1$. \tag{1.9}

A function $x : I \to E$ where I is either $[0, \infty)$ or $[0, T]$ $(T > 0)$ is called a trajectory if there exists $u(\cdot) \in L^2_{loc}(I; F)$ (referred to as a control) such that the pair (x, u) satisfies (1.1) and

$$x(t) \in X, \ u(t) \in U(x(t)), \ t \in I. \tag{1.10}$$

A trajectory-control pair $\widehat{x} : [0, \infty) \to E$, $\widehat{u} : [0, \infty) \to F$ is overtaking (resp. weakly overtaking) optimal if for any other trajectory-control pair $x : [0, \infty) \to E$, $u : [0, \infty) \to F$ satisfying $x(0) = \widehat{x}(0)$,

$$\limsup_{t \to \infty}[I(0, t, \widehat{x}, \widehat{u}) - I(0, T, x, u)] \leq 0$$

$$(\text{resp. } \liminf_{t \to \infty}[I(0, t, \widehat{x}, \widehat{u}) - I(0, T, x, u)] \leq 0). \tag{1.11}$$

Assume the following.

Assumption 1. The optimal steady state problem (OSSP)

$$\text{Min } f(x, u) \text{ over all } (x, u) \in E \times F \text{ satisfying}$$

$$0 = <x, A^*y> + <Bu, y> \text{ for any } y \in \mathcal{D}(A^*) \; x \in X, \; u \in U(x),$$

has a solution $(\bar{x}, \bar{u})$ with $\bar{x}$ uniquely defined.

It is not difficult to see that the OSSP is a convex programming problem in a Hilbert space. Therefore there exists $\bar{p} \in \mathcal{D}(A^*)$ such that (see [30, 73])

$$f(\bar{x}, \bar{u}) \leq f(x, u) - <x, A^*\bar{p}> - <Bu, \bar{p}> \tag{1.12}$$

for every $x \in X$ and $u \in U(x)$. Define a function $L : E \times F \to [0, \infty)$ by

$$L(x, u) = f(x, u) - f(\bar{x}, \bar{u}) - <x, A^*\bar{p}> - <Bu, \bar{p}>$$

$$\text{if } x \in X \text{ and } u \in U(x), L(x, u) = \infty \text{ otherwise.} \tag{1.13}$$

Then we have $L(\bar{x}, \bar{u}) = 0$. Since L differs from f through an affine function of x and u, it still satisfies the growth property (1.9) with f replaced by L.

Let I be either $[0, \infty)$ or $[0, T]$ $(T > 0)$, $x : I \to E$, $u : I \to F$ be a trajectory-control pair, and $T_1, .T_2 \in I$, $T_1 < T_2$. We define

$$I_L(T_1, T_2, x, u) = \int_{T_1}^{T_2} L(x(t), u(t))dt. \tag{1.14}$$

For a trajectory-control pair $x : [0, \infty) \to E$, $u : [0, \infty) \to F$ we define

$$I_L(0, \infty, x, u) = \int_0^\infty L(x(t), u(t))dt.$$

For each $T > 0$ and each $z \in E$ we define

$$\sigma(z, T) = \inf\{I(0, T, x, u) : x : [0, T] \to E, \; u : [0, T] \to F \text{ is a}$$

$$\text{trajectory-control pair, } x(0) = z\}. \tag{1.15}$$

In this chapter we prove the following results.

THEOREM 9.1.1 *Suppose that Assumption 1 holds and* $x : [0, \infty) \to E$, $u : [0, \infty) \to F$ *is a trajectory-control pair. Then either*
(i) $\sup\{|I(0, T, x, u) - Tf(\bar{x}, \bar{u})| : T \in (0, \infty)\} < \infty$
or
(ii) $I(0, T, x, u) - Tf(\bar{x}, \bar{u}) \to \infty$ *as* $T \to \infty$.
Moreover (i) holds if and only if $I_L(0, \infty, x, u) < \infty$.

The next theorem establishes that $||x(t)||$ is bounded by some constant $\Delta > 0$ for all $t \in [0, T]$ if $(x, u) : [0, T] \to E \times F$ is an approximate optimal trajectory-control pair and T is large enough. Note that Δ does not depend on T.

THEOREM 9.1.2 *Suppose that Assumption 1 holds and* $r_1, r_2, r_3 > 0$. *Then there exist positive numbers* Δ, r *such that if* $T > 0$ *and if a trajectory-control pair* $x : [0, T] \to E$, $u : [0, T] \to F$ *satisfies the following conditions:*
(a) $||x(0)|| \le r_2$, $I(0, T, x, u) \le \sigma(x(0), T) + r_3$;
(b) *there is a trajectory-control pair* $y : [0, \infty) \to E$, $v : [0, \infty) \to F$ *satisfying*

$$y(0) = x(0), \quad I_L(0, \infty, y, v) \le r_1,$$

then

$$||x(t)|| \le \Delta, \ t \in [0, T], \ I_L(0, T, x, u) \le r.$$

The following result is an extension of Theorem 1 of [18] to optimal trajectories defined on finite intervals. It shows that if $(x, u) : [0, T] \to E \times F$ is an optimal trajectory-control pair, then the average of x on any subinterval of $[0, T]$ with a sufficiently large length belongs to a small neighborhood of $\bar{x}$ in the weak topology.

THEOREM 9.1.3 *Suppose that Assumption 1 holds,* $r_1, r_2, r_3 > 0$, *and* V *is a neighborhood of* $\bar{x}$ *in the weak topology. Then there exists a number* $l > 0$ *such that the following property holds:*
If $T \ge l$ *and if a trajectory-control pair* $x : [0, T] \to E$, $u : [0, T] \to F$ *satisfies conditions (a) and (b) from Theorem 9.1.2, then*

$$(T_2 - T_1)^{-1} \int_{T_1}^{T_2} x(t)dt \in V$$

for each $T_1, T_2 \in [0, T]$ *satisfying* $T_2 - T_1 \ge l$.

Denote by $\mathcal{F}$ the set of all trajectory-control pairs $x : [0, \infty) \to E$, $u : [0, \infty) \to F$ such that

$$L(x(t), u(t)) = 0 \text{ a.e. on } [0, \infty). \tag{1.16}$$

We say that $\mathcal{F}$ has property $\mathcal{G}$ if for any neighborhood V of $\bar{x}$ in the weak topology there exists a number $t_v > 0$ such that

$$x(t) \in V \text{ for each } t \geq t_v$$

and each trajectory-control pair $(x, u) \in \mathcal{F}$.

This property appears in [18] and corresponds to property (S) in [40].

We establish the following result which describes the structure of optimal solutions on finite intervals.

THEOREM 9.1.4 *Suppose that Assumptiom 1 holds and $\mathcal{F}$ has property $\mathcal{G}$. Let r_1, r_2, r_3 be positive numbers and let V be a neighborhood of $\bar{x}$ in the weak topology. Then there exist a natural number Q and a positive number l such that the following property holds:*

If $T > 0$ and if a trajectory-control pair $x : [0, T] \to E$, $u : [0, T] \to F$ satisfies conditions (a) and (b) from Theorem 9.1.2, then there exists a sequence of intervals $[b_j, c_j]$, $j = 1, \ldots, q$ such that

$$1 \leq q \leq Q,\ 0 < c_j - b_j \leq l,\ j = 1, \ldots, q,\ and$$

$$x(t) \in V \text{ for each } t \in [0, T] \setminus \cup_{j=1}^{q} [b_j, c_j].$$

The following result is a generalization of Theorem 4 in [18] which establishes the existence of an overtaking optimal solution in the subclass of bounded trajectories.

THEOREM 9.1.5 *Suppose that Assumption 1 holds and $\mathcal{F}$ has property $\mathcal{G}$. Let $\tilde{x} : [0, \infty) \to E$, $\tilde{u} : [0, \infty) \to F$ be a trajectory-control pair satisfying $I_L(0, \infty, \tilde{x}, \tilde{u}) < \infty$. Then there exists an overtaking optimal trajectory-control pair $x^* : [0, \infty) \to E$, $u^* : [0, \infty) \to F$ such that $x^*(0) = \tilde{x}(0)$.*

The results of this chapter were obtained in [105].

Chapter 9 is organized as follows. Section 9.2 contains four simple auxiliary results. Theorems 9.1.1-9.1.3 are proved in Section 9.3. Section 9.4 contains the proofs of Theorems 9.1.4 and 9.1.5. In Section 9.5 we discuss systems with distributed parameters and boundary controls.

9.2. Preliminary results

Denote by $\mathbf{N}$ the set of all natural numbers. For each positive number r set

$$B_E(r) = \{x \in E : ||x|| \leq r\},$$

$$B_F(r) = \{x \in F : ||x|| \le r\}.$$

In the sequel we use the following lemmas which can be established in a straightforward manner.

LEMMA 9.2.1 *The set* $\{||S(t)|| : t \in [0,1]\}$ *is bounded.*

LEMMA 9.2.2 *Let* $-\infty < T_1 < T_2 < \infty$, $u \in L^1([T_1, T_2]; F)$. *Then*

$$|| \int_{T_1}^{T_2} u(t)dt|| \le \int_{T_1}^{T_2} ||u(t)dt.$$

LEMMA 9.2.3 *Let* $-\infty < T_1 < T_2 < \infty$, $D \subset E \times F$ *be a closed convex subset,* $x \in L^1([T_1, T_2]; E)$, $y \in L^1([T_1, T_2]; F)$, *and*

$$(x(t), y(t)) \in D, \ t \in [T_1, T_2].$$

Then $((T_2 - T_1)^{-1} \int_{T_1}^{T_2} x(t)dt, \ (T_2 - T_1)^{-1} \int_{T_1}^{T_2} y(t)dy) \in D$.

LEMMA 9.2.4 *Suppose that Assumption 1 holds. Then for any* $T > 0$ *the function*

$$\phi : (x(\cdot), u(\cdot)) \rightarrow \int_0^T L(x(t), u(t))dt$$

is convex and weakly lower semicontinuous on

$$L^2([0, T]; E) \times L^2([0, T]; F).$$

9.3.   Proof of Theorems 9.1.1-9.1.3

In this section we assume that Assumption 1 holds.

LEMMA 9.3.1 *There exist positive numbers* K_2, K_3 *such that*

$$f(x, u), \ L(x, u) \ge K_2(||x||^2 + ||u||^2) - K_3 \text{ for all } (x, u) \in E \times F. \ (3.1)$$

Lemma 9.3.1 follows from equations (1.9), (1.12), (1.13), convexity and lower semicontinuity of f.
   (1.13) and (1.2) imply the following auxiliary result.

LEMMA 9.3.2 *For each* $T_1 \ge 0$, $T_2 > T_1$, $T \ge T_2$ *and each trajectory-control pair* $x : [0, T] \rightarrow E$, $u : [0, T] \rightarrow F$,

$$I(T_1, T_2, x, u) = (T_2 - T_1)f(\bar{x}, \bar{u}) + I_L(T_1, T_2, x, u) + < \bar{p}, x(T_2) - x(T_1) > .$$

The following auxiliary result shows that for any approximate optimal with respect to the functional $I_L(0, T, \cdot, \cdot)$ trajectory-control pair (x, u) : $[0, T] \to E \times F$ with large enough T the function $||x(t)||$ is bounded by some constant which does not depend on T.

LEMMA 9.3.3 *For each pair of positive numbers r, l there exists a positive number N such that the following property holds:*

If a trajectory-control pair $x : [0, T] \to E$, $u : [0, T] \to F$ satisfies

$$T \geq l, \quad x(0) \in B_E(r), \quad I_L(0, T, x, u) \leq r, \tag{3.2}$$

then $x(t) \in B_E(N)$ for all $t \in [0, T]$.

Proof. Let $r, l > 0$. Choose numbers

$$\delta \in (0, 8^{-1} \min\{1, l\}), \quad h > 1 + r, \tag{3.3}$$

such that

$$\delta(K_2 h - K_3) \geq 20r \tag{3.4}$$

(see (3.1)) and choose a number

$$N > (1 + \sup\{||S(t)|| : t \in [0, 1]\})(h + 1)(h + 1 + ||B||)(1 + K_2^{-1}(r + K_3)). \tag{3.5}$$

Assume that $T \geq l$ and $x : [0, T] \to E$, $u : [0, T] \to F$ is a trajectory-control pair which satisfies (3.2). We show that $x(t) \in B_E(N)$ for all $t \in [0, T]$.

Assume that there exists $t_0 \in [0, T]$ for which

$$||x(t_0)|| > N. \tag{3.6}$$

It follows from (3.1)-(3.6) that there is a number t_1 such that

$$t_1 \in [\max\{0, t_0 - \delta\}, t_0], \quad x(t_1) \in B_E(h). \tag{3.7}$$

By (3.3) and (3.7),

$$||x(t_0)|| = \left\| S(t_0)x(0) + \int_0^{t_0} S(t_0 - s)Bu(s)ds \right\|$$

$$\leq \left\| S(t_0 - t_1) \left[S(t_1)x(0) + \int_0^{t_1} S(t_1 - s)Bu(s)ds \right] \right\|$$

$$+ \left\| \int_{t_1}^{t_0} S(t_0 - s)Bu(s)ds \right\|$$

$$\leq \sup\{||S(\tau)|| : \tau \in [0, 1]\} \left(h + ||B|| \left(1 + \int_{t_1}^{t_0} ||u(s)||^2 ds \right) \right).$$

Combined with (3.6), (3.1) and (3.2) this implies that

$$N < \sup\{||S(\tau)|| : \tau \in [0,1]\}(h + ||B||(1 + (r + K_3)K_2^{-1})).$$

This is contradictory to (3.5). This completes the proof of the lemma.

Proof of Theorem 9.1.1. Assume that $x : [0,\infty) \to E$, $u : [0,\infty) \to F$ is a trajectory-control pair such that $I_L(0,\infty,x,u) < \infty$. Lemma 9.3.3 implies that the function $x(t)$ is bounded. Combined with Lemma 9.3.2 this implies relation (i).

Assume that $I_L(0,\infty,x,u) = \infty$. We show that relation (ii) holds. There exists $\Delta > ||x(0)||$ such that

$$f(w,p) \geq 8 + |f(\bar{x},\bar{u})| \text{ for each } w \in B_E(\Delta)$$

$$\text{and each } p \in F. \tag{3.8}$$

For any $T > 0$ put

$$\tau(T) = \sup\{t \in [0,T] : x(t) \in B_E(\Delta)\}. \tag{3.9}$$

We may assume without loss of generality that $\tau(T) \to \infty$ as $T \to \infty$. It follows from Lemma 9.3.2, (3.8), and (3.9) that for any $T > 0$,

$$I(0,T,x,u) - Tf(\bar{x},\bar{u}) \geq I(0,\tau(T),x,u) - \tau(T)f(\bar{x},\bar{u})$$

$$\geq I_L(0,\tau(T),x,u) + < \bar{p}, x(\tau(T)) - x(0) >$$

$$\geq I_L(0,\tau(T),x,u) - 2||\bar{p}||\Delta \to \infty \text{ as } T \to \infty.$$

This completes the proof of the theorem.

In order to prove Theorem 9.1.2 we need the following lemma which is its weakened version. Instead of the inequality $||x(t)|| \leq \Delta$ for all $t \in [0,T]$ in Theorem 9.1.2 the lemma guarantees that $||x(t)|| \leq \Delta$ for some $t \in [0,T]$.

LEMMA 9.3.4 *Let r_1, r_2, r_3 be positive numbers and let*

$$\Delta > r_2 \text{ and } f(w,p) \geq 4 + |f(\bar{x},\bar{u})| \text{ for each } w \in E$$

$$\text{satisfying } ||w|| \geq \Delta \text{ and each } p \in F. \tag{3.10}$$

Then there exists $T_\Delta > 1$ such that the following property holds:
If $T \geq T_\Delta$ and if a trajectory-control pair $x : [0,T] \to E$, $u : [0,T] \to F$ satisfies conditions (a) and (b) of Theorem 9.1.2, then

$$\inf\{||x(t)|| : t \in [T - T_\Delta, T]\} \leq \Delta. \tag{3.11}$$

Proof. It follows from Lemma 9.3.3 that there exists a positive number N_1 such that the following property holds:

If $T \geq 1$ and if a trajectory-control pair $y : [0, T] \to E$, $v : [0, T] \to F$ satisfies

$$y(0) \in B_E(r_1 + r_2), \quad I_L(0, T, y, v) \leq r_1 + r_2, \tag{3.12}$$

then

$$y(t) \in B_E(N_1), \quad t \in [0, T]. \tag{3.13}$$

Choose a number

$$T_\Delta > r_3 + r_1 + 1 + 2||\bar{p}||\Delta + 2||\bar{p}||(N_1 + r_2). \tag{3.14}$$

Assume that $T \geq T_\Delta$ and a trajectory-control pair $x : [0, T] \to E$, $u : [0, T] \to F$ satisfies conditions (a) and (b) of Theorem 9.1.2. There exists a trajectory-control pair $y : [0, \infty) \to E$, $v : [0, \infty) \to F$ such that

$$y(0) = x(0), \quad I_L(0, \infty, y, v) \leq r_1. \tag{3.15}$$

It follows from the choice of N_1, condition (a) of Theorem 9.1.2, and (3.15) that

$$y(t) \in B_E(N_1), \quad t \in [0, \infty). \tag{3.16}$$

(3.15), (3.16) and Lemma 9.3.2 imply that

$$I(0, T, y, v) \leq Tf(\bar{x}, \bar{u}) + r_1 + 2||\bar{p}||N_1. \tag{3.17}$$

Put

$$\tau = \sup\{t \in [0, T] : x(t) \in B_E(\Delta)\}. \tag{3.18}$$

In view of (3.18), (3.10), and Lemma 9.3.2,

$$I(0, T, x, u) \geq (T - \tau)[4 + |f(\tilde{x}, \tilde{u})|] + \tau f(\bar{x}, \bar{u}) - 2||\bar{p}||\Delta. \tag{3.19}$$

Since (x, u) satisfies condition (a) of Theorem 9.1.2 it follows from (3.15) and (3.17) that

$$r_3 \geq I(0, T, x, u) - I(0, T, y, v) \geq 4(T - \tau) - 2||\bar{p}||\Delta - r_1 - 2||\bar{p}||N_1.$$

It follows from this inequality and (3.14) that $T - \tau < T_\Delta$. This completes the proof of the lemma.

The next lemma plays a crucial role in the proof of Theorem 9.1.2. Its proof is based on Lemmas 9.3.2-9.3.4. This lemma can be considered as a weakened version of Theorem 9.1.2. Instead of the inequalities of Theorem 9.1.2 the lemma establishes analogous inequalities but with larger bounds.

LEMMA 9.3.5 *Assume that $r_1, r_2, r_3 > 0$, a positive number Δ satisfies (3.10) and a number $T_\Delta > 1$ is as guaranteed in Lemma 9.3.4. Then there exist $Q > \Delta$, $h > 0$ such that the following property holds:*

If $T \geq T_\Delta + 1$ and if a trajectory-control pair $x : [0,T] \to E$, $u : [0,T] \to F$ satisfies conditions (a) and (b) of Theorem 9.1.2, then

$$x(t) \in B_E(Q) \text{ for all } t \in [0,T] \text{ and } I_L(0,T,x,u) \leq h. \qquad (3.20)$$

Proof. It follows from Lemma 9.3.3 that there exists $\Delta_0 > \Delta$ such that the following property holds:

If $T \geq 1$ and if a trajectory-control pair $y : [0,T] \to E$, $v : [0,T] \to F$ satisfies

$$y(0) \in B_E(r_2 + 1), \quad I_L(0,T,y,v) \leq r_1 + r_3 + 2 + 2||\bar{p}||(\Delta + r_2), \quad (3.21)$$

then

$$y(t) \in B_E(\Delta_0) \text{ for all } t \in [0,T].$$

It follows from Lemma 9.3.3 that there exists $\Delta_1 < \Delta_0$ such that the following property holds:

If $T \geq 1$ and if a trajectory-control pair $y : [0,T] \to E$, $v : [0,T] \to F$ satisfies

$$y(0) \in B_E(r_2+1), \quad I_L(0,T,y,v) \leq r_1+r_3+4+2||\bar{p}||(2\Delta+\Delta_0+r_2), \quad (3.22)$$

then $y(t) \in B_E(\Delta_1)$ for all $t \in [0,T]$.

Choose numbers

$$Q > \Delta_1 + 2 + K_2^{-1}(||B|| + 1)(\sup\{||S(t)|| : t \in [0,T_\Delta\})$$

$$\times [2 + ||\bar{p}||(2\Delta + \Delta_0 + r_2) + (K_2 + K_3)T_\Delta + K_2\Delta_1 + r_3 + r_1$$

$$+ T_\Delta|f(\bar{x},\bar{u})], \quad h > 1 + r_3 + r_1 + ||\bar{p}||(\Delta_0 + r_2 + 2Q). \qquad (3.23)$$

Assume that $T \geq T_\Delta + 1$ and that $x : [0,T] \to E$, $u : [0,T] \to F$ is a trajectory-control pair which satisfies conditions (a) and (b) of Theorem 9.1.2. Then there exists a trajectory-control pair $y : [0,\infty) \to E$, $v : [0,\infty) \to F$ such that

$$y(0) = x(0), \quad I_L(0,\infty,y,v) \leq r_1. \qquad (3.24)$$

It follows from (3.24) and the choice of Δ_0 that

$$y(t) \in B_E(\Delta_0), \quad t \in [0,\infty). \qquad (3.25)$$

(3.24), condition (a) of Theorem 9.1.2, (3.25), and Lemma 9.3.2 imply that

$$I(0,T,y,v) \leq Tf(\bar{x},\bar{u}) + r_1 + ||\bar{p}||(\Delta_0 + r_2). \qquad (3.26)$$

Since the trajectory-control pair (x, u) satisfies condition (a) of Theorem 9.1.2 it follows from (3.26) and (3.24) that

$$I(0, T, x, u) \leq I(0, T, y, v) + r_3 \leq r_3 + T f(\bar{x}, \bar{u}) + r_1 + ||\bar{p}||(\Delta_0 + r_2). \quad (3.27)$$

In view of Lemma 9.3.4 and the choice of T_Δ there exists a number t_1 such that

$$t_1 \in [T - T_\Delta, T], \ x(t_1) \in B_E(\Delta), \ ||x(t)|| \geq \Delta, \ t \in [t_1, T]. \quad (3.28)$$

By (3.27), (3.28), and (3.10),

$$I(0, t_1, x, u) = I(0, T, x, u) - I(t_1, T, x, u)$$

$$\leq r_3 + r_1 + ||\bar{p}||(\Delta_0 + r_2) + t_1 f(\bar{x}, \bar{u}). \quad (3.29)$$

It follows from Lemma 9.3.2, (3.28) and (3.10) that

$$I(0, t_1, x, u) \geq t_1 f(\bar{x}, \bar{u}) + J_L(0, t_1, x, u) - 2||\bar{p}||\Delta. \quad (3.30)$$

(3.29) and (3.30) imply that

$$I_L(0, t_1, x, u) \leq r_3 + r_1 + ||\bar{p}||(2\Delta + \Delta_0 + r_2). \quad (3.31)$$

By the choice of Δ_1, (3.31) and (3.28),

$$x(t) \in B_E(\Delta_1) \text{ for all } t \in [0, t_1]. \quad (3.32)$$

We show that $x(t) \in B_E(Q)$ for all $t \in [t_1, T]$. Assume the contrary. Then there exists a number t_2 such that

$$t_2 \in (t_1, T] \text{ and } ||x(t_2)|| > Q. \quad (3.33)$$

(3.33) and (3.28) imply that

$$t_2 - t_1 \in (0, T_\Delta], \quad (3.34)$$

$$x(t_2) = S(t_2)x(0) + \int_0^{t_2} S(t_2 - s)Bu(s)ds$$

$$= S(t_2 - t_1)\left[S(t_1)x(0) + \int_0^{t_1} S(t_1 - s)Bu(s)ds\right] + \int_{t_1}^{t_2} S(t_2 - s)Bu(s)ds$$

$$= S(t_2 - t_1)x(t_1) + \int_{t_1}^{t_2} S(t_2 - s)Bu(s)ds. \quad (3.35)$$

By (3.1),

$$\int_{t_1}^{t_2} ||u(s)||ds \leq \int_{t_1}^{t_2} [1 + K_2^{-1}(f(x(s), u(s)) + K_3)]ds. \quad (3.36)$$

It follows from (3.34), (3.33), (3.32), and (3.35) that

$$\sup\{||S(t)|| : t \in [0, T_\Delta]\}||B|| \int_{t_1}^{t_2} ||u(s)||ds$$

$$\geq Q - \Delta_1 \sup\{||S(t)|| : t \in [0, T_\Delta]\}.$$

Combined with (3.36) and (3.34) this inequality implies that

$$\int_{t_1}^{t_2} f(x(s), u(s))ds \geq \left(\int_{t_1}^{t_2} ||u(s)||ds - (1 + K_2^{-1}K_3)T_\Delta \right) K_2$$

$$\geq -(K_2 + K_3)T_\Delta + K_2 Q(||B|| + 1)^{-1}$$

$$\times (\sup\{||S(t)|| : t \in [0, T_\Delta]\})^{-1} - K_2\Delta_1(||B|| + 1)^{-1}. \qquad (3.37)$$

By (3.33), (3.28), (3.10), (3.30) and (3.37),

$$I(0, T, x, u) \geq I(0, t_1, x, u) + I(t_1, t_2, x, u)$$

$$\geq t_1 f(\bar{x}, \bar{u}) - 2||\bar{p}||\Delta - (K_2 + K_3)T_\Delta - K_2\Delta_1$$

$$+ K_2 Q(||B|| + 1)^{-1}(\sup\{||S(t)|| : t \in [0, T_\Delta]\})^{-1}.$$

It follows from this inequality, (3.27) and (3.28) that

$$K_2 Q(||B|| + 1)^{-1}(\sup\{||S(t)|| : t \in [0, T_\Delta]\})^{-1}$$

$$\leq 2||\bar{p}||\Delta + (K_2 + K_3)T_\Delta + K_2\Delta_1 + r_3 + T_\Delta|f(\bar{x}, \bar{u})| + r_1 + ||\bar{p}||(\Delta_0 + r_2).$$

This is contradictory to (3.23). The obtained contradiction proves that

$$x(t) \in B_E(Q) \text{ for all } t \in [t_1, T].$$

Combined with (3.32) and (3.23) this implies that

$$x(t) \in B_E(Q) \text{ for all } t \in [0, T].$$

It follows from this inequality, Lemma 9.3.2, (3.27) and (3.23) that

$$I_L(0, T, x, u) + T f(\bar{x}, \bar{u}) - 2||\bar{p}||Q \leq I(0, T, x, u)$$

$$\leq r_3 + T f(\bar{x}, \bar{u}) + r_1 + ||\bar{p}||(\Delta_0 + r_2),$$

$$I_L(0, T, x, u) \leq h.$$

This completes the proof of the lemma.

Proof of Theorem 9.1.2. We may assume that $\bar{x} \in B_E(r_2)$. It follows from Lemma 9.3.5 that there exist positive numbers T_0, Q, h such that the following property holds:

If $T \geq T_0$ and if a trajectory-control pair $x : [0, T] \to E$, $u : [0, T] \to F$ satisfies conditions (a) and (b) of Theorem 9.1.2, then

$$x(t) \in B_E(Q) \text{ for all } t \in [0, T], \ I_L(0, T, x, u) \leq h. \tag{3.38}$$

Lemma 9.3.3 implies that there exists a positive number N_0 such that the following property holds:

If $T \geq 1$ and if a trajectory-control pair $y : [0, T] \to E$, $v : [0, T] \to F$ satisfies

$$y(0) \in B_E(r_2), \ I_L(0, T, y, v) \leq r_1, \tag{3.39}$$

then $y(t) \in B_E(N_0)$ for all $t \in [0, T]$.

Choose numbers

$$\Delta > Q + 1 + 2r_2 + [(r_3 + r_1 + 2\|\bar{p}\| N_0 + T_0(|f(\bar{x}, \bar{u})| + K_3 + K_2))K_2^{-1}$$

$$+r_2(1 + \|B\|)^{-1}](1 + \|B\|)\sup\{\|S(t)\| : t \in [0, T_0]\}, \tag{3.40}$$

$$r > h + 1 + r_3 + r_1 + 2\|\bar{p}\|(\Delta + N_0) + 2T_0|f(\bar{x}, \bar{u})|.$$

Assume that $T > 0$ and $x : [0, T] \to E$, $u : [0, T] \to F$ is a trajectory-control pair which satisfies conditions (a) and (b) of Theorem 9.1.2. In view of (3.40) and the choice of T_0, Q, h we may assume that

$$T < T_0. \tag{3.41}$$

Assume that there exists $t_0 \in [0, T]$ such that

$$\|x(t_0)\| > \Delta.$$

By this inequality, (3.41), and condition (a) of Theorem 9.1.2,

$$\Delta - r_2 < \|x(t_0) - x(0)\| = \left\| S(t_0)x(0) + \int_0^{t_0} S(t_0 - s)Bu(s)ds - x(0) \right\|$$

$$\leq r_2(1 + \sup\{\|S(t)\| : t \in [0, T_0]\})$$

$$+\|B\|\sup\{\|S(t)\| : t \in [0, T_0]\} \int_0^t \|u(s)\|ds.$$

It follows from this inequality, (3.1) and (3.41) that

$$\int_0^T f(x(t), u(t))dt \geq -K_3 T_0 + K_2 \int_0^{t_0} \|u(t)\|^2 dt$$

$$-K_3 T_0 - K_2 T_0 + K_2 \int_0^{t_0} \|u(t)\|dt$$

$$\geq -T_0(K_3 + K_2) + K_2[(\Delta - 2r_2)(1 + \|B\|)^{-1}$$

$$\times (\sup\{||S(t)|| : t \in [0, T_0]\})^{-1} - r_2(1 + ||B||)^{-1}]. \qquad (3.42)$$

Since the trajectory-control pair $x : [0, T] \to E$, $u : [0, T] \to F$ satisfies condition (b) of Theorem 9.1.2 there exists a trajectory-control pair $y : [0, \infty) \to E$, $v : [0, \infty) \to F$ such that

$$y(0) = x(0), \quad I_L(0, \infty, y, v) \leq r_1. \qquad (3.43)$$

By the choice of N_0 (see (3.39)) and (3.43),

$$y(t) \in B_E(N_0), \quad t \in [0, \infty).$$

It follows from this inclusion, (3.43), (3.41) and Lemma 9.3.2 that

$$I(0, T, y, v) \leq T_0|f(\bar{x}, \bar{u})| + r_1 + 2||\bar{p}||N_0.$$

Combined with (3.43) and condition (a) of Theorem 9.1.2 this inequality implies that

$$I(0, T, x, u) \leq I(0, T, y, v) + r_3 \leq T_0|f(\bar{x}, \bar{u})| + r_3 + r_1 + 2||\bar{p}||N_0. \quad (3.44)$$

(3.44) and (3.42) imply that

$$\Delta \leq [(r_3 + r_1 + 2||\bar{p}||N_0 + T_0(|f(\bar{x}, \bar{u})| + K_2 + K_3))K_2^{-1} + r_2(1 + ||B||)^{-1}]$$

$$\times (1 + ||B||)\sup\{||S(t)|| : t \in [0, T_0]\} + 2r_2.$$

This is contradictory to (3.40). The obtained contradiction proves that

$$x(t) \in B_E(\Delta) \text{ for all } t \in [0, T].$$

By this inequality, Lemma 9.3.2, (3.41), (3.44) and (3.40),

$$I_L(0, T, x, u) \leq I(0, T, x, u) + T_0|f(\bar{x}, \bar{u})| + 2||\bar{p}||\Delta$$

$$\leq 2T_0|f(\bar{x}, \bar{u})| + r_1 + r_3 + 2||\bar{p}||(\Delta + N_0) \leq r.$$

This completes the proof of the theorem.

We preface the proof of Theorem 9.1.3 with the following auxiliary result.

LEMMA 9.3.6 *For each pair of positive numbers r_1, r_2 and each neighborhood V of $\bar{x}$ in the weak topology there exists a number $l > 0$ such that the following property holds:*

If $T > l$ and if a trajectory-control pair $x : [0, T] \to E$, $u : [0, T] \to F$ satisfies

$$x(0) \in B_E(r_2), \quad I_L(0, T, x, u) \leq r_1, \qquad (3.45)$$

then

$$T^{-1} \int_0^T x(t)dt \in V. \tag{3.46}$$

Proof. Let $r_1, r_2 > 0$ and let V be a neighborhood of $\bar{x}$ in the weak topology. Assume that the lemma is not true. Then there exists a sequence of trajectory-control pairs $x_j : [0, T_j] \to E$, $u_j : [0, T_j] \to F$, $j \in \mathbf{N}$ such that

$$T_j < T_{j+1}, \; j \in \mathbf{N}, \; T_j \to \infty \text{ as } j \to \infty, \; x_j(0) \leq B_E(r_2),$$

$$I_L(0, T_j, x_j, u_j) \leq r_1, \; T_j^{-1} \int_0^{T_j} x_j(t)dt \notin V, \; j \in \mathbf{N}. \tag{3.47}$$

Lemma 9.3.3 and (3.47) imply that there exists a positive number Δ_0 such that

$$x_j(t) \in B_E(\Delta_0) \text{ for all } t \in [0, T_j], \; j \in \mathbf{N}. \tag{3.48}$$

Assume that $j \in \mathbf{N}$ and $y \in \mathcal{D}(A^*)$. Integrating the state equation in the mild form (1.4) with $x = x_j$, $u = u_j$, we have

$$T_j^{-1} < x_j(T_j) - x_j(0), y > = \left\langle T_j^{-1} \int_0^{T_j} x_j(t)dt, A^* y \right\rangle$$

$$+ \left\langle T_j^{-1} B \left(\int_0^{T_j} u_i(t)dt \right), y \right\rangle. \tag{3.49}$$

By (3.48), (3.47) and (3.1) the set

$$\Omega = \left\{ \left(T_j^{-1} \int_0^{T_j} x_j(t)dt, T_j^{-1} \int_0^{T_j} u_j(t)dt \right) : j \in \mathbf{N} \right\}$$

is bounded. In view of Lemma 9.2.3, (1.6) and (1.7),

$$\bar{\Omega} \subset \{(x, u) : x \in X, \; u \in U(x)\}, \tag{3.50}$$

where $\bar{\Omega}$ is the closure of Ω in the weak topology.

Let (x^*, u^*) be a cluster point of the set Ω. By (3.48), (3.49), and (3.50), (x^*, u^*) is an admissible point for the OSSP. It follows from Lemma 9.3.2, (3.47) and (3.48) that for $j \in \mathbf{N}$,

$$I(0, T_j, x_j, u_j) \leq T_j f(\bar{x}, \bar{u}) + r_1 + 2\|\bar{p}\|\Delta_0.$$

Combined with the lower semicontinuity of f and and Jensen's inequality this inequality implies that

$$f(x^*, u^*) \leq f(\bar{x}, \bar{u}).$$

Since (x^*, u^*) is an admissible point for the OSSP it follows from Assumption 1 that $x^* = \bar{x}$. This is contradictory to (3.47), The contradiction we have reached proves the lemma.

Proof of Theorem 9.1.3. It follows from Theorem 9.1.2 that there exist positive numbers Δ_1, Δ_2 such that the following property holds:

If $T > 0$ and if a trajectory-control pair $x : [0, T] \rightarrow E$, $u : [0, T] \rightarrow F$ satisfies conditions (a) and (b) of Theorem 9.1.2, then

$$x(t) \in B_E(\Delta_1) \text{ for all } t \in [0, T] \text{ and } I_L(0, T, x, u) \leq \Delta_2. \qquad (3.51)$$

It follows from Lemma 9.3.6 that there exists a positive number l such that the following property holds:

If $T \geq l$ and if a trajectory-control pair $x : [0, T] \rightarrow E$, $u : [0, T] \rightarrow F$ satisfies

$$x(0) \in B_E(\Delta_2 + \Delta_1 + r_1 + r_2), \ I_L(0, T, x, u) \leq \Delta_2 + \Delta_1 + r_1 + r_2, \quad (3.52)$$

then

$$T^{-1} \int_0^T x(t) dt \in V.$$

Let $T \geq l$, $T_1, T_2 \in [0, T]$, $T_2 - T_1 \geq l$ and let $x : [0, T] \rightarrow E$, $u : [0, T] \rightarrow F$ be a trajectory-control pair which satisfies conditions (a) and (b) of Theorem 9.1.2. Therefore (3.51) holds. Define

$$\tilde{x}(t) = x(t + T_1), \ \tilde{u}(t) = u(t + T_1), \ t \in [0, T_2 - T_1]. \qquad (3.53)$$

For $t \in [0, T_2 - T_1]$ we have

$$\tilde{x}(t) = x(t + T_1) = S(t + T_1)x(0) + \int_0^{t+T} S(t + T_1 - s)Bu(s)ds$$

$$= S(t) \left[S(T_1)x(0) + \int_0^{T_1} S(T_1 - s)Bu(s)ds \right]$$

$$+ \int_{T_1}^{t+T_1} S(t + T_1 - s)Bu(s)ds$$

$$= S(t)\tilde{x}(0) + \int_0^t S(t - s)B\tilde{u}(s)ds.$$

By this relation and (3.53), $\tilde{x} : [0, T_2 - T_1] \rightarrow E$, $\tilde{u} : [0, T_2 - T_1] \rightarrow F$ is a trajectory-control pair. It follows from (3.51), (3.53), (3.52) and the definition of l that

$$(T_2 - T_1)^{-1} \int_{T_1}^{T_2} x(t)dt = (T_2 - T_1)^{-1} \int_0^{T_2 - T_1} \tilde{x}(t)dt \in V.$$

This completes the proof of the theorem.

9.4. Proof of Theorems 9.1.4 and 9.1.5

In this section we assume that Assumption 1 holds and the set $\mathcal{F}$ has property $\mathcal{G}$.

We begin with the following auxiliary result which shows that if $(x, u) : [0, T] \to E \times F$ is an approximate optimal trajectory-control pair with respect to the functional $I_L(0, T, \cdot, \cdot)$ and if T is large enough, then $x(t)$ belongs to a small neighborhood of $\bar{x}$ in the weak topology for most $t \in [0, T]$.

PROPOSITION 9.4.1 *For each neighborhood V of $\bar{x}$ in the weak topolgy there exist positive numbers δ, l such that the following property holds:*

If $T \geq 2l$ and if a trajectory-control pair $x : [0, T] \to E$, $u : [0, T] \to F$ satisfies $I_L(0, T, x, u) \leq \delta$, then

$$x(t) \in V, \ t \in [l, T - l].$$

Proof. Let V be a neighborhood of $\bar{x}$ in the weak topology. Assume that the proposition is not true. Then there exist a sequence of numbers $\{t_j\}_{j=1}^{\infty}$ and a sequence of trajectory-control pairs $x_j : [0, T_j] \to E$, $u_j : [0, T_j] \to F$, $j \in \mathbf{N}$ such that

$$T_j \geq 8j, \ I_L(0, T_j, x_j, u_j) \leq j^{-1},$$

$$t_j \in [4j, T_j - 4j], \ x_j(t_j) \notin V, \ j \in \mathbf{N}. \tag{4.1}$$

Lemma 9.3.1 implies that there exists a positive number r_0 such that

$$L(x, u) \geq 8 \ \text{for each} \ (x, u) \in E \times F \ \text{satisfying} \ ||x|| \geq r_0. \tag{4.2}$$

In view of Lemma 9.3.3 there exists a positive number r_1 such that the following property holds:

If $T \geq 1$ and if a trajectory-control pair $x : [0, T] \to E$, $u : [0, T] \to F$ satisfies

$$x(0) \in B_E(r_0), \ I_L(0, T, x, u) \leq 1, \tag{4.3}$$

then

$$x(t) \in B_E(r_1), \ t \in [0, T]. \tag{4.4}$$

There exists a neighborhood V_0 of 0 in the weak topology such that

$$\bar{x} + 3V_0 \subset V. \tag{4.5}$$

It follows from property $\mathcal{G}$ that there exists $\tau_v > 0$ such that for each trajectory-control pair $(x, u) \in \mathcal{F}$,

$$x(t) \in V + \bar{x} \ \text{for each} \ t \geq \tau_v. \tag{4.6}$$

(4.1) and (4.2) imply that for any $j \in \mathbf{N}$ there exists a number h_j which satisfies

$$h_j \in [0,1], \ x(h_j) \in B_E(r_0). \tag{4.7}$$

For $j \in \mathbf{N}$ we define

$$\tilde{x}_j(t) = x_j(t + h_j), \ \tilde{u}_j(t) = u_j(t + h_j), \ t \in [0, T_j - h_j]. \tag{4.8}$$

It is easy to verify that

$$\tilde{x}_j : [0, T_j - h_j] \to E, \ \tilde{u}_j : [0, T_j - h_j] \to F$$

is a trajectory-control pair for any $j \in \mathbf{N}$. Therefore by (4.1), (4.7), (4.8) and the definition of r_1 (see (4.3), (4.4)),

$$x_j(t) \in B_E(r_1), \ t \in [1, T_j], \ j \in \mathbf{N}. \tag{4.9}$$

For any integer $j \geq \tau_v$ we define

$$\tilde{x}_j(t) = x_j(t + t_j - \tau_v), \ \tilde{u}_j(t) = u_j(t + t_j - \tau_v), \ t \in [0, T_j - t_j + \tau_v]. \tag{4.10}$$

It is easy to verify that $\tilde{x}_j : [0, T_j - t_j + \tau_v] \to E, \ \tilde{u}_j : [0, T_j - t_j + \tau_v] \to F$ is a trajectory-control pair for any $j \geq \tau_v$. By (4.1) and (4.10),

$$\tilde{x}_j(\tau_v) \notin V \ \text{for any integer } j \geq \tau_v. \tag{4.11}$$

By (4.10), (4.9) and (4.1),

$$\tilde{x}_j(t) \in B_E(r_1) \ \text{for each } t \in [0, T_j - t_j + \tau_v] \ \text{and each integer } j \geq \tau_v, \tag{4.12}$$

$$I_L(0, T_j - t_j + \tau_v, \tilde{x}_j, \tilde{u}_j) \leq j^{-1} \ \text{for each integer } j \geq \tau_v. \tag{4.13}$$

In view of (4.13) and (3.1) for any positive T the sequence $\{\tilde{u}_j(\cdot)\}$ is bounded in $L^2([0,T]; F)$. Thus, extracting a subsequence if necessary, we may suppose that there exist $x_0^* \in E$ and $u^* \in L^2_{loc}([0,\infty); F)$ such that

$$\tilde{x}_j(0) \to x_0^* \ \text{as } j \to \infty \ \text{weakly in } E, \tag{4.14}$$

$$\tilde{u}_j \to u^* \ \text{as } j \to \infty \ \text{weakly in } L^2([0,T]; F) \ \text{for any } T > 0. \tag{4.15}$$

(4.12), (4.14), and (4.15) imply that there is a function $x^* : [0,\infty) \to E$ such that (x^*, u^*) is a trajectory-control pair defined on $[0,\infty)$,

$$x^*(0) = x_0^*,$$

$$\tilde{x}_j(t) \to x^*(t) \ \text{as } j \to \infty \ \text{weakly in } E \ \text{for any } t \in [0,\infty),$$

$$\tilde{x}_j \to x^* \ \text{as } j \to \infty \ \text{weakly in } L^2([0,T]; E) \ \text{for any } T > 0. \tag{4.16}$$

By (4.16), (4.15), (4.13) and Lemma 9.2.4,

$$I_L(0, T, x^*, u^*) = 0 \text{ for any } T > 0.$$

Since the function L is nonnegative this equality implies that

$$L(x^*(t), u^*(t)) = 0$$

a.e. on $[0, \infty)$. Hence $(x^*, u^*) \in \mathcal{F}$, and by the definition of τ_v (see (4.6)),

$$x^*(\tau_v) \in V_0 + \bar{x}.$$

It follows from this relation, (4.10), (4.16) and (4.5) that for j sufficiently large we have

$$x_j(t_j) = \tilde{x}_j(\tau_v) \in \bar{x} + 2V_0 \subset V,$$

which contradicts (4.1). The obtained contradiction proves the proposition.

COROLLARY 9.4.1 *Assume that* $x : [0, \infty) \to E$, $u : [0, \infty) \to F$ *is a trajectory-control pair such that* $I_L(0, \infty, x, u) < \infty$. *Then* $x(t)$ *converges weakly to* $\bar{x}$ *as* $t \to \infty$.

Proof of Theorem 9.1.4. It follows from Theorem 9.1.2 that there exist positive numbers Δ, r such that the following properties hold:

If $T > 0$ and if a trajectory-control pair $x : [0, T] \to E$, $u : [0, T] \to F$ satisfies conditions (a) and (b) (see the statement of Theorem 9.1.2), then

$$x(t) \in B_E(\Delta) \text{ for all } t \in [0, T] \text{ and } I_L(0, T, x, u) \leq r. \quad (4.17)$$

Proposition 9.4.1 implies that there exist positive numbers δ, l_0 such that the following property holds:

If $T \geq 2l_0$ and if a trajectory-control pair $x : [0, T] \to E$, $u : [0, T] \to F$ satisfies $I_L(0, T, x, u) \leq \delta$, then

$$x(t) \in V, \ t \in [l_0, T - l_0]. \quad (4.18)$$

Choose $Q \in \mathbf{N}$ and a number $l > 0$ such that

$$Q > 4r\delta^{-1} + 8, \ l > 1 + 2l_0. \quad (4.19)$$

Assume that $T > 0$ and $x : [0.T] \to E$, $u : [0, T] \to F$ is a trajectory-control pair which satisfies conditions (a) and (b) of Theorem 9.1.2. Then (4.17) follows from the definition of Δ, r. There exists a sequence of numbers $\{t_j\}_{j=0}^N$ such that

$$t_0 = 0, \ t_j < t_{j+1}, \ j = 0, \ldots, N-1, \ t_N = T,$$

$$I_L(t_j, t_{j+1}, x, u) = \delta \text{ if } 0 \le j < N - 1,$$

$$I_L(t_{N-1}, T, x, u) \le \delta. \tag{4.20}$$

(4.17) and (4.20) imply that

$$N \le r\delta^{-1} + 1. \tag{4.21}$$

Put

$$\mathcal{A} = \{j \in \{0, \dots, N-1\} : t_{j+1} - t_j \ge l\}. \tag{4.22}$$

Let $j \in \mathcal{A}$ and define

$$x_j(t) = x(t + j), \ u_j(t) = u(t + t_j), \ t \in [0, t_{j+1} - t_j]. \tag{4.23}$$

Clearly $x_j : [0, t_{j+1} - t_j] \to E$, $u_j : [0, t_{j+1} - t_j] \to F$ is a trajectory-control pair and

$$I_L(0, t_{j+1} - t_j, x_j, u_j) \le \delta.$$

In view of this inequality, (4.22), (4.19) and the choice of δ, l_0 (see (4.18)),

$$x_j(t) \in V, \ t \in [l_0, t_{j+1} - t_j - l_0].$$

Combined with (4.23) this relation implies that

$$x(t) \in V, \ t \in [t_j + l_0, t_{j+1} - l_0], \ j \in \mathcal{A}.$$

This completes the proof of the theorem.

We have the following result (see Lemma 2 of [18]).

PROPOSITION 9.4.2 *Assume that $z \in E$ and there exists a trajectory-control pair $\tilde{x} : [0, \infty) \to E$, $\tilde{u} : [0, \infty) \to F$ satisfying $I_L(0, \infty, \tilde{x}, \tilde{u}) < \infty$, $\tilde{x}(0) = z$. Then there exists a trajectory-control pair $x^* : [0, \infty) \to E$, $u^* : [0, \infty) \to F$ such that $x^*(0) = z$, $I_L(0, \infty, x^*, u^*) \le I_L(0, \infty, x, u)$ for each trajectory-control pair $x : [0, \infty) \to E$, $u : [0, \infty) \to F$ satisfying $x(0) = z$.*

Proof of Theorem 9.1.5. It follows from Proposition 9.4.2 that there exists a trajectory-control pair $x^* : [0, \infty) \to E$, $u^* : [0, \infty) \to F$ such that

$$x^*(0) = \tilde{x}(0)$$

and

$$I_L(0, \infty, x^*, u^*) \le I_L(0, \infty, x, u) \tag{4.24}$$

for each trajectory-control pair $x : [0, \infty) \to E$, $u : [0, \infty) \to F$ satisfying $x(0) = \tilde{x}(0)$. Evidently

$$I_L(0, \infty, x^*, u^*) < \infty. \tag{4.25}$$

Let $x : [0, \infty) \to E$, $u : [0, \infty) \to F$ be a trajectory-control pair such that

$$x(0) = x^*(0).$$

We show that

$$\limsup_{T \to \infty}[I(0, T, x^*, u^*) - I(0, T, x, u)] \leq 0.$$

In view of (4.25) and Theorem 9.1.1 we may assume that

$$I_L(0, \infty, x, u) < \infty. \tag{4.26}$$

(4.25), (4.26) and Corollary 9.4.1 imply that

$$x^*(t) \to \bar{x}, \ x(t) \to \bar{x} \text{ as } t \to \infty \text{ in the weak topology.}$$

It follows from this relation, (4.24) and Lemma 9.3.2 that

$$\limsup_{T \to \infty}[I(0, T, x^*, u^*) - I(0, T, x, u)]$$

$$= \limsup_{T \to \infty}[I_L(0, T, x^*, u^*) - I_L(0, T, x, u) + <\bar{p}, x^*(t) - x(T)>] \leq 0.$$

This completes the proof of the theorem.

9.5. Systems with distributed and boundary controls

We extend Theorems 9.1.1-9.1.5 to the cases of infinite-dimensional control problems with distributed and boundary controls considered in Section 4 of [18].

We consider the following control system:

$$x'(t) = \sigma x(t) + B_1 u(t), \tag{5.1}$$

$$\gamma x(t) = B_2 u_2(t), \tag{5.2}$$

with initial condition

$$x(0) = x^0, \tag{5.3}$$

and the following additional constraints:

$$x(t) \in X \subset E_1, \ u_j(t) \in U_j(x(t)) \subset F_j, \ j = 1, 2, \ t \in I, \tag{5.4}$$

where I is either $[0, \infty)$ or $[0, T]$ $(T > 0)$, E_1, E_2, F_1 and F_2 are separable Hilbert spaces, $x^0 \in E_1$, σ is a closed linear and densely defined operator on E_1, γ is a linear operator (the boundary operator) with domain in

E_1 and range in E_2, and $B_i : F_i \to E_i$, $i = 1, 2$, are linear continuous operators.

The sets X and $U_j(x)$, $j = 1, 2$, $x \in X$ satisfy the assumptions (1.6) and (1.7).

A solution to the system (5.1)-(5.3) satisfies the following input-output relation:

$$x(t) = S(t)x_0 + \int_0^t S(t - s)(B_1 u_1(s) + \sigma B u_2(s))ds$$

$$-A \int_0^t S(t - s) B u_2(s) ds, \tag{5.5}$$

where the operators A, B and $S(\cdot)$ are defined in the assumptions below.

Assumption 2. We assume that $\mathcal{D}(\sigma) \subset \mathcal{D}(\gamma)$ (here $\mathcal{D}(\cdot)$ denotes the domain) and that the restriction of γ to $\mathcal{D}(\sigma)$ is continuous with respect to the graph norm of $\mathcal{D}(\sigma)$.

Assumption 3. The operator $A : \mathcal{D}(A) \subset E_1 \to E_1$ defined by

$$Ay = \sigma y, \ y \in \mathcal{D}(A) = \{y \in \mathcal{D}(\sigma) : \ \gamma y = 0\}, \tag{5.6}$$

is the infinitesimal generator of a strongly continuous semigroup $\{S(t) : t \geq 0\}$ on E_1.

Assumption 4. There exists a linear continuous operator $B : F_2 \to E_1$ such that

$$\sigma B \in \mathcal{L}(F_2, E_1), \ \gamma(Bv) = B_2 v \ \text{ for all } v \in F_2,$$

$$\|Bv\|_{E_1} \leq c\|B_2\|_{E_2} \ \text{ for all } v \in F_2. \tag{5.7}$$

Assumption 5. For each $t \geq 0$ and $v \in L^2_{loc}([0, \infty); F_2)$,

$$\int_0^t S(t - s) Bv(s) ds \in \mathcal{D}(A) \tag{5.8}$$

and there exists $\delta \in L^2_{loc}(0, \infty)$ such that

$$\|AS(t)B\| \leq \delta(t) \ \text{a.e.}, \tag{5.9}$$

for each $t \geq 0$, each $h \in [0, 1]$ we have

$$\int_t^{t+h} \delta(s) ds \leq K, \tag{5.10}$$

where K is a constant.

Under the above hypotheses, the expression (5.5) is well defined for $u_j(\cdot) \in L^2_{loc}([0, \infty); F_j)$, $j = 1, 2$, and agrees with the formulation presented in [8].

The performance of the system is evaluated by the cost functional

$$J(T_1, T_2, x, u_1, u_2) = \int_{T_1}^{T_2} f(x(t), u_1(t), u_2(t))dt, \qquad (5.11)$$

where $f : E_1 \times F_1 \times F_2 \to R^1$ is a convex lower semicontinuous functional which satisfies the following coercivity assumption:

there exist $K_1 > 0$ and $K > 0$ such that

$$f(x, u_1, u_2) \geq K(||x||^2 + ||u_1||^2 + ||u_2||^2) \text{ for each } x \in E_1, \ u_1 \in F_1,$$

$$u_2 \in F_2 \text{ satisfying } ||x||^2 + ||u_1||^2 + ||u_2||^2 > K_1. \qquad (5.12)$$

Assume the following.

Assumption 6. The optimal steady state problem (OSSP) consisting of

$$\text{Min } f(x, u_1, u_2) \text{ subject to}$$

$$0 = <x - Bu_2, A^*z> + <B_1u_1 + \sigma Bu_2, z>$$

$$\text{for all } z \in \mathcal{D}(A^*) \ x \in X, \ u_1 \in U_1(x), \ u_2 \in U_2(x) \qquad (5.13)$$

has a unique solution $(\bar{x}, \bar{u}_1, \bar{u}_2)$.

Once again the convexity assumptions we have imposed allow us to define the nonnegative convex lower semicontinuous functional $L : E_1 \times F_1 \times F_2 \to R^1$ by

$$L(x, u_1, u_2) = f(x, u_1, u_2) - f(\bar{x}, \bar{u}_1, \bar{u}_2)$$

$$- <x - Bu_2, A^*\bar{p}> - <B_1u_1 + \sigma Bu_2, \bar{p}>$$

$$\text{if } x \in X \text{ and } u_j \in U_j(x), \ j = 1, 2,$$

$$L(x, u) = \infty \text{ otherwise}, \qquad (5.14)$$

where $\bar{p} \in \mathcal{D}(A^*)$.

A function $x : I \to E_1$ where I is either $[0, \infty)$ or $[0, T]$ $(T > 0)$ is called a trajectory if there exists $u_j(\cdot) \in L^2_{loc}(I; F_j), j = 1, 2$ (referred to as a control) such that the pair (x, u_1, u_2) satisfies (5.5) for all $t \in I$ and (5.4).

Let I be either $[0, \infty)$ or $[0, T]$ $(T > 0)$, $x : I \to E_1$, $u_j : I \to F_j$, $j = 1, 2$ be a trajectory-control pair, and $T_1, T_2 \in I, T_1 < T_2$. We define

$$J_L(T_1, T_2, x, u_1, u_2) = \int_{T_1}^{T_2} L(x(t), u_1(t), u_2(t))dt. \qquad (5.15)$$

For a trajectory-control pair $x : [0, \infty) \rightarrow E_1$, $u_j : [0, \infty) \rightarrow F_j$, $j = 1, 2$ we define

$$J_L(0, \infty, x, u_1, u_2) = \int_0^\infty L(x(t), u_1(t), u_2(t))dt.$$

For each $T > 0$ and each $z \in E_1$ we define

$$\sigma(z, T) = \inf\{J(0, T, x, u_1, u_2) :$$

$$x : [0, T] \rightarrow E_1, \ u_j : [0, T] \rightarrow F_j, \ j = 1, 2 \text{ is a}$$

$$\text{trajectory-control pair, } x(0) = z\}. \tag{5.16}$$

The proof of the following theorems is identical to the proof of Theorems 9.1.1-9.1.5.

THEOREM 9.5.1 *Suppose that Assumptions 2-6 hold and $x : [0, \infty) \rightarrow E_1$, $u_j : [0, \infty) \rightarrow F_j$, $j = 1, 2$ is a trajectory-control pair. Then one of the following relations holds:*
   (i) $\sup\{|J(0, T, x, u_1, u_2) - Tf(\bar{x}, \bar{u}_1, \bar{u}_2)| : T \in (0, \infty)\} < \infty$.
   (ii) $J(0, T, x, u_1, u_2) - Tf(\bar{x}, \bar{u}_1, \bar{u}_2) \rightarrow \infty$ as $T \rightarrow \infty$.
Moreover (i) holds if and only if $J_L(0, \infty, x, u_1, u_2) < \infty$.

THEOREM 9.5.2 *Suppose that Assumptions 2-6 hold and r_1, r_2, r_3 are positive numbers. Then there exist $\Delta, r > 0$ such that*

$$\|x(t)\| \leq \Delta, \ t \in [0, T], \ J_L(0, T, x, u_1, u_2) \leq r$$

for each $T > 0$ and each trajectory-control pair $x : [0, T] \rightarrow E_1$, $u_j : [0, T] \rightarrow F_j$, $j = 1, 2$ which has the following properties:
   (a) $\|x(0)\| \leq r_2$, $J(0, T, x, u_1, u_2) \leq \sigma(x(0), T) + r_3$;
   (b) there is a trajectory-control pair $y : [0, \infty) \rightarrow E_1$, $v_j : [0, \infty) \rightarrow F_j$, $j = 1, 2$, satisfying

$$y(0) = x(0), \ J_L(0, \infty, y, v_1, v_2) \leq r_1.$$

THEOREM 9.5.3 *Suppose that Assumptions 2-6 hold, r_1, r_2, r_3 are positive numbers, and V is a neighborhood of $\bar{x}$ in the weak topology. Then there exists a number $l > 0$ such that*

$$(T_2 - T_1)^{-1} \int_{T_1}^{T_2} x(t)dt \in V$$

for each $T \geq l$, each trajectory-control pair $x : [0, T] \rightarrow E_1$, $u_j : [0, T] \rightarrow F_j$, $j = 1, 2$, which has properties (a) and (b) from Theorem 9.5.2 and each $T_1, T_2 \in [0, T]$ satisfying $T_2 - T_1 \geq l$.

Denote by $\mathcal{F}$ the set of all trajectory-control pairs $x : [0, \infty) \to E_1$, $u_j : [0, \infty) \to F_j$, $j = 1, 2$ such that

$$L(x(t), u_1(t), u_2(t)) = 0 \text{ a.e. on } [0, \infty).$$

We say that $\mathcal{F}$ has property $\mathcal{G}$ if for any neighborhood V of $\bar{x}$ in the weak topology there exists a number $t_v > 0$ such that

$$x(t) \in V \text{ for each } t \geq t_v$$

and each trajectory-control pair $(x, u_1, u_2) \in \mathcal{F}$.

THEOREM 9.5.4 *Suppose that Assumptions 2-6 hold and $\mathcal{F}$ has property $\mathcal{G}$. Let r_1, r_2, r_3 be positive numbers and let V be a neighborhood of $\bar{x}$ in the weak topology. Then there exist an integer $Q \geq 1$ and a number $l > 0$ such that for each $T > 0$ and each trajectory-control pair $x : [0, T] \to E_1$, $u_j : [0, T] \to F_j$, $j = 1, 2$, which has properties (a) and (b) from Theorem 9.5.2, there exists a sequence of intervals $[b_j, c_j]$, $j = 1, \ldots, q$ such that*

$$1 \leq q \leq Q, \ 0 < c_j - b_j \leq l, \ j = 1, \ldots, q, \ and$$

$$x(t) \in V \text{ for each } t \in [0, T] \setminus \cup_{j=1}^{q} [b_j, c_j].$$

THEOREM 9.5.5 *Suppose that Assumptions 2-6 hold and $\mathcal{F}$ has property $\mathcal{G}$. Let $\tilde{x} : [0, \infty) \to E_1$, $\tilde{u}_j : [0, \infty) \to F_j$, $j = 1, 2$, be a trajectory-control pair satisfying $J_L(0, \infty, \tilde{x}, \tilde{u}_1, \tilde{u}_2) < \infty$. Then there exists an overtaking optimal trajectory-control pair $x^* : [0, \infty) \to E_1$, $u_j^* : [0, \infty) \to F_j$, $j = 1, 2$ such that $x^*(0) = \tilde{x}(0)$.*

Chapter 10

A CLASS OF
DIFFERENTIAL INCLUSIONS

In this chapter we study the turnpike property for a class of differential inclusions of the form $x'(t) \in b(x(t)) - x(t)$ arising in economic dynamics where b belongs to a space of convex processes (superlinear set-valued mappings). This class of set-valued mappings was introduced by Rubinov in [76] and studied in [83, 84, 86, 96]. We establish the existence of an open everywhere dense subset E of the space of set-valued mappings such that each mapping from E has the turnpike property.

10.1. Main result

Let R_+^n be the cone of the elements of the Euclidean space

$$R^n = \{x = (x^1, \ldots, x^n) : x^i \in R^1, i = 1, \ldots, n\}$$

that have nonnegative coordinates and let K_n be the interior of R_+^n. We suppose that the space R^n is ordered by the cone R_+^n. Let $I_n : R^n \to R^n$ be the identity operator and let $||x|| = \sup\{|x^i| : i = 1, \ldots, n\}$ for each $x \in R^n$. Denote by $< \cdot, \cdot >$ the scalar product in R^n. For each matrix $V = (v^{ij})$ of a dimension $m \times n$ we set

$$||V|| = \sup\{|v^{ij}| : i = 1, \ldots, m, j = 1, \ldots, n\}.$$

We will identify a linear operator $A : R^m \to R^n$ with its matrix in the standard basis.

For each metric space Q denote by $\Pi(Q)$ the collection of nonempty compact subsets of Q endowed with the Hausdorff metric $dist(\cdot, \cdot)$.

A mapping $b : R_+^n \to \Pi(R_+^n)$ is called a convex process if for each $x, y \in R_+^n$ and each $\lambda \in (0, \infty)$,

$$b(x + y) \supset b(x) + b(y) \text{ and } b(\lambda x) = \lambda b(x).$$

For each mapping $b : R_+^n \to \Pi(R_+^n)$ denote by $Gr(b)$ the graph of b. A mapping $b : R_+^n \to \Pi(R_+^n)$ is called normal if

$$b(x) = (b(x) - R_+^n) \cap R_+^n \text{ for each } x \in R_+^n.$$

Note that convex processes are set-valued analogs of linear mappings [48, 71, 75, 76].

Consider a normal convex process $b : R_+^n \to \Pi(R_+^n)$ with a closed graph such that $b(0) = \{0\}$. The von Neumann growth rate of the mapping b is a number $\alpha(b)$ defined by

$$\alpha(b) = \sup\{\beta \in R_+^1 : \text{ there is } y \in R_+^n \setminus \{0\} \text{ for which } \beta y \in b(y)\} \quad (1.1)$$

(see [48, 75]).

It is not difficult to see that $\alpha(b) \in [0, \infty)$ and there exists $X \in R_+^n \setminus \{0\}$ such that $\alpha(b)X \in b(X)$.

Assume that $b(1, 1, \ldots, 1) \cap K_n \neq \emptyset$. Denote by b' the dual mapping $b' : R_+^n \to \Pi(R_+^n)$ defined as

$$b'(f) = \{g \in R_+^n :< f, x > \geq < g, y > \text{ for each } x \in R_+^n, y \in b(x)\}. \quad (1.2)$$

There exists a generalized equilibrium state $(\alpha(b), (X, \alpha(b)X), p)$ where

$$X, p \in R_+^n, ||X||, ||p|| = 1, \alpha(b)X \in b(X), \ p \in b'(\alpha(b)p). \quad (1.3)$$

Let $b : R_+^n \to \Pi(R_+^n)$ be a normal convex process with a closed graph such that $b(0) = \{0\}$.

A function $u : I \to R_+^n$ where I is either $[T, \infty)(T \geq 0)$ or $[T_1, T_2](0 \leq T_1 < T_2)$ will be called a trajectory of the model $\mathbf{Z}(b)$ generated by the mapping b, if u is absolutely continuous (a.c.) and satisfies the relation

$$u'(t) \in b(u(t)) - u(t), \text{ a.e. } t \in I. \quad (1.4)$$

Assume that $b(1, 1, \ldots, 1) \cap K_n \neq \emptyset$. Let $0 \leq T_1 < T_2, u : [T_1, T_2] \to R_+^n$ be a trajectory of the model $\mathbf{Z}(b)$ and let $f : [T_1, T_2] \to R_+^n$ be an a.c. function which satisfies

$$f'(t) \in f(t) - b^*(f(t)), \quad (1.5)$$

a.e. $t \in [T_1, T_2]$ where

$$b^*(g) = \{h \in R_+^n : g \in b'(h)\}, \ g \in R_+^n. \quad (1.6)$$

It is easy to verify that the function $t \to\, < f(t), u(t) >, t \in [T_1, T_2]$ is monotone decreasing.

Assume that there exists a generalized equilibrium state

$$(\alpha(b), (X, \alpha(b)X), p)$$

with $X, p \in K_n$ such that (1.3) holds. It is easy to see that the function $t \to exp((\alpha(b) - 1)t)X$, $t \in [0, \infty)$ is a trajectory of $\mathbf{Z}(b)$, the function $f(t) = exp((1 - \alpha(b))t)p$ satisfies (1.5) for a.e. $t \in [0, \infty)$, and for each trajectory $u : [0, \infty) \to R_+^n$ of the model $\mathbf{Z}(b)$ the function $exp((1 - \alpha(b))t) < u(t), p >$, $t \in [0, \infty)$ is monotone decreasing.

We say that a trajectory $u : [0, \infty) \to R^n$ of the model $\mathbf{Z}(b)$ has the von Neumann growth rate if

$$\lim_{t \to \infty} exp((1 - \alpha(b))t) < p, u(t) >> 0. \qquad (1.7)$$

It follows from Propositions 0.1 and 1.1 of [86] that for each $x \in K_n$ there exists a trajectory $u : [0, \infty) \to R^n$ which has the von Neumann growth rate and satisfies $u(0) = x$.

We say that the mapping b has the turnpike property if for each trajectory $u : [0, \infty) \to R_+^n$ of $\mathbf{Z}(b)$ which has the von Neumann growth rate, there exists a positive number λ such that

$$\lim_{t \to \infty} exp((1 - \alpha(b))t)u(t) = \lambda X.$$

Most studies which are concerned with the turnpike property for convex processes assume the strict convexity of their graphs. In this chapter we establish the turnpike property without such assumptions.

Let m, n be natural numbers such that $m \leq n$. Denote by $\mathcal{L}_0$ the set of all matrices $V = (v^{ij})$ of a dimension $m \times n$ such that $v^{ij} \in [0, 1]$ for each $i = 1, \ldots, m, j = 1, \ldots, n$. For each $V \in \mathcal{L}_0$ and each $i \in \{1, \ldots, m\}$ denote by V^i a diagonal matrix of a dimension $n \times n$ with diagonal elements $v^{i1}, \ldots, v^{in}$.

Assume that $V \in \mathcal{L}_0$ and $a = (a^1, \ldots, a^m)$, $a^i : R_+^n \to \Pi(R_+^n)$, $i = 1, \ldots, m$ are normal convex processes such that $Gr(a_i)$ is a closed set and $a^i(0) = \{0\}, i = 1, \ldots, m$. Consider the mapping

$$Q(a, V) : (R_+^n)^m \to \Pi((R_+^n)^m)$$

defined as

$$Q(a, V)(x) = \{y = (y^1, \ldots, y^m) \in (R_+^n)^m : y^i \in R_+^n, \qquad (1.8)$$

$$y^i \leq V^i x^i + d^i, d^i \in R_+^n, i = 1, \ldots, m, \sum_{i=1}^m d^i \in \sum_{i=1}^m a^i(x^i)\}$$

(here $x = (x^1, \ldots, x^m) \in (R_+^n)^m$, $x^i = (x^{i1}, \ldots, x^{in}) \in R_+^n$, $i = 1, \ldots, m$).

It is easy to see that $Q(a, V)$ is a normal convex process, the graph of $Q(a, V)$ is closed and $Q(a, V)(0) = \{0\}$. Set

$$\alpha(a, V) = \alpha(Q(a, V)). \tag{1.9}$$

Assume that

$$\mathcal{I}_s \subset \{1, \ldots, n\}, \mathcal{I}_s \neq \emptyset, s = 1, \ldots, m, \cup\{\mathcal{I}_s : s = 1, \ldots, m\} = \{1, \ldots, n\}, \tag{1.10}$$

$$\mathcal{I}_k \setminus \cup\{\mathcal{I}_s : s \in \{1, \ldots, m\} \setminus \{k\}\} \neq \emptyset \text{ for each } k \in \{1, \ldots, m\},$$

$$\mathcal{P} \subset \{(i, j) : i = 1, \ldots, m, j = 1, \ldots, n\}. \tag{1.11}$$

Let $\mathcal{L}$ be the set of all matrices $V \in \mathcal{L}_0$ such that $v^{ij} = 0$ for each $(i, j) \in \mathcal{P}$.

Denote by $\mathcal{M}$ the set of all mappings

$$a = (a^1, \ldots, a^m) : (R_+^n)^m \to (\Pi(R_+^n))^m$$

such that for each $s \in \{1, \ldots, m\}$ the normal convex process $a^s : R_+^n \to \Pi(R_+^n)$ has the following properties:

(i) the graph of a^s is closed, $a^s(0) = \{0\}$ and

$$a^s(R_+^n) \subset \{x = (x^1, \ldots, x^n) \in R^n : x^i = 0 \text{ for each } i \in \{1, \ldots, n\} \setminus \mathcal{I}_s\};$$

(ii) for each $x, g \in K_n$ the optimization problem

$$< g, z > \to \max, z \in a^s(x)$$

has a unique solution;

(iii) for each $g, x_1 \in K_n$ and each $x_2 \in R_+^n \setminus \{\lambda x_1 : \lambda \in [0, \infty)\}$,

$$\sup\{< g, z > : z \in a^s(x_1) + a^s(x_2)\} < \sup\{< g, z > : z \in a^s(x_1 + x_2)\}.$$

Our interest in a set-valued mapping $Q(a, V)$ with $Q \in \mathcal{M}$ and $V \in \mathcal{L}$ stems from the work by Rubinov [76] who studied an analogous class of set-valued mappings describing multisector models of economic dynamics.

Let $a = (a^1, \ldots, a^m) \in \mathcal{M}, V \in \mathcal{L}$. The mapping $Q(a, V)$ describes the model of economic dynamics with m sectors and n products such that for each $i \in \{1, \ldots, m\}$, the production process of the ith sector is described by the pair (a^i, V^i). Rubinov [76] (see also [86]) studied a special case of this model given in the example below, such that $m = n$ and each sector produces one product. For this case the production process of the ith sector is described by the pair (ϕ^i, V^i), where ϕ^i : $R^n_+ \to R^1$ is a superlinear function. Rubinov [76] studied the discrete-time model. In this chapter we consider a generalization of his model. For the discrete-time generalization of the model by Rubinov a state of the economy at time $t = 0, 1, \ldots$ is a vector $x_t = (x_t^1, \ldots, x_t^m) \in (R^n_+)^m$ where $x_t^i = (x_t^{i1}, \ldots, x_t^{in}) \in R^n_+$ is the state of the ith sector of the economy, $i = 1, \ldots, m$. During the time interval $[t, t+1]$ the ith sector produces a vector of products $y_i \in a_i(x_i)$. It also has a vector of the old products in the amount $V^i x_i$. At time $t+1$ the newly produced vector of products $\sum_{i=1}^m y_i$ is distributed between m sectors. We assume that the sth sector produces only products with indices $i \in \mathcal{I}_s$ (see (1.10) and (i)) which is nonempty. Relations (1.10) mean that each product is produced by some sector and that for each sector there is a product which is produced only by this sector. In this chapter we study the continuous-time version of the model.

For each pair of normal convex processes $b_1, b_2 : R^l_+ \to \Pi(R^l_+)$ which have closed graphs and satisfy $b_i(0) = \{0\}, i = 1, 2$ we set

$$d(b_1, b_2) = dist(\{(x, y) \in Gr(b_1), ||x|| \leq 1\}, \qquad (1.12)$$

$$\{(x, y) \in Gr(b_2), ||x|| \leq 1\}).$$

For the set $\mathcal{M}$ we consider the product topology generated by the metric $d(\cdot, \cdot)$. It is easy to verify that the mapping $(a, V) \to Q(a, V), a \in \mathcal{M}, V \in \mathcal{L}$ is continuous.

In this chapter we will establish the following result.

THEOREM 10.1.1 *There exists an open everywhere dense set $E \subset \mathcal{M} \times \mathcal{L}$ such that each $(a, V) \in E$ satisfies the following conditions:*

There is a generalized equilibrium state $(\alpha(a, V), (X, \alpha(a, V)X), p)$ such that $\alpha(a, V) > ||V||, X, p \in (K_n)^m$; the mapping $Q(a, V)$ has the turnpike property.

Example. Assume that $m = n, \mathcal{P} = \emptyset, \mathcal{I}_s = \{s\}, s = 1, \ldots, n$ and consider the spaces $\mathcal{M}, \mathcal{L}$ defined above. Let $e_1, \ldots, e_n$ be the standard basis in R^n. It is easy to see that $(a^1, \ldots, a^n) \in \mathcal{M}$ if and only if there exist continuous functions $\phi^i : R^n_+ \to R^1_+, i = 1, \ldots, n$ such that for each $i \in \{1, \ldots, n\}$ the following properties hold:

$$a^i(x) = [0, \phi^i(x)]e_i, x \in R^n_+;$$

$$\phi^i(\lambda x) = \lambda \phi^i(x), \phi^i(x + y) \geq \phi^i(x) + \phi^i(y), x, y \in R^n_+, \lambda \in [0.\infty);$$

for each $x_1 \in K_n$ and each $x_2 \in R^n_+ \setminus \{\lambda x_1 : \lambda \in [0, \infty)\}$ the inequality $\phi^i(x_1) + \phi^i(x_2) < \phi^i(x_1 + x_2)$ holds.

In [86] we studied the turnpike property for the space $\mathcal{M} \times \mathcal{L}$ described in this example. Theorem 10.1.1 which is a generalization of Theorem 3.1 of [86] was obtained in [96].

Chapter 10 is organized as follows. In Section 10.2 we study a generalized equilibrium state of the model $Q(a, V)$ and obtain useful auxiliary results. In Section 10.3 we obtain a sufficient condition for the turnpike property. This condition is some property of a generalized equilibrium state of the model. Section 10.4 contains a number of auxiliary results. Theorem 10.1.1 is proved in Section 10.5.

10.2.   Preliminary results

We have the following result (see Proposition 1.2 of [86]).

PROPOSITION 10.2.1 *Let* $b : R^n_+ \to \Pi(R^n_+)$ *be a normal convex process such that* $Gr(b)$ *is closed,* $b(0) = \{0\}$, $b(1, 1, \ldots, 1) \cap K_n \neq \emptyset$ *and let* $(\alpha(b), (X, \alpha(b)X), p)$ *be a generalized equilibrium state such that* $X, p \in K_n$. *Assume that* $x : [0, \infty) \to R^n_+$ *is a trajectory of the model* $\mathbf{Z}(b)$ *which has the von Neumann growth rate and put*

$$\Omega = \{y \in R^n_+ : \ \text{there exists a sequence } \{t_i\}^\infty_{i=1} \subset (0, \infty) \ \text{such that}$$

$$t_i \to \infty, \ exp((1 - \alpha(b))t_i)x(t_i) \to y \ as \ i \to \infty\}.$$

Then for each $y_0 \in \Omega$ *there exists an a.c. function* $y : R^1 \to R^n_+$ *such that* $y(0) = y_0$,

$$exp((1 - \alpha(b))t)y(t) \in \Omega, \ t \in R^1,$$

$$y'(t) \in b(y(t)) - y(t) \ a.e. \ t \in R^1.$$

Denote by $Card(B)$ the cardinality of a set B.

Consider the spaces $\mathcal{M}$ and $\mathcal{L}$ defined in Section 10.1. For each $g = (g^1, \ldots, g^m) \in (R^n_+)^m$ where $g^i = (g^{i1}, \ldots, g^{in}) \in R^n, i = 1, \ldots, m$ we define

$$r(g, j) = \sup\{g^{sj} : s = 1, \ldots, m\}, j = 1, \ldots, n, \qquad (2.1)$$

$$r(g) = (r(g, 1), \ldots, r(g, n)).$$

Let $V \in \mathcal{L}$, $a = (a^1, \ldots, a^m) \in \mathcal{M}$. Consider the dual mapping $Q(a, V)' : (R^n_+)^m \to 2^{(R^n_+)^m} \setminus \{\emptyset\}$ which is defined as

$$Q(a, V)'(f) = \{g \in (R^n_+)^m :< f, x >\geq< g, y > \ \text{for each} \qquad (2.2)$$

$$x \in (R^n_+)^m, y \in Q(a, V)(x)\}, f = (f^1, \ldots, f^m) \in (R^n_+)^m.$$

Analogously to Propositions 1 and 2 in [76] we can establish the following results which describe some useful relations between $Q(a, V)$ and its dual mapping.

LEMMA 10.2.1 *Let $f, g \in (R_+^n)^m$. Then $g \in Q(a, V)'(f)$ if and only if*

$$< f^i - V^i g^i, u >\geq \sup\{< r(g), z >: z \in a^i(u)\}$$

for each $u \in R_+^n$ and each $i \in \{1, \ldots, m\}$.

LEMMA 10.2.2 *Let $f, g, x, y \in (R_+^n)^m$, $g \in Q(a, V)'(f)$, $y \in Q(a, V)(x)$,*

$$y^i \leq V^i x^i + d^i, d^i \geq 0, \sum_{i=1}^m d^i = \sum_{i=1}^m z^i$$

where $z^i \in a^i(x^i), i = 1, \ldots, m$. Then $< f, x >= < g, y >$ if and only if for each $i \in \{1, \ldots m\}$ the following relations hold:
(a) $< g^i, d^i + V^i x^i - y^i >= 0$;
(b) $< r(g) - g^i, d^i >= 0$;
(c) $< r(g), z^i >= \sup\{< r(g), u >: u \in a^i(x^i)\}$;
(d) $< f^i - V^i g^i, x^i >= \sup\{< r(g), u >: u \in a^i(x^i)\}$.

Lemmas 10.2.1 and 10.2.2 imply the following auxiliary result which gives an important property of a generalized equilibrium state.

LEMMA 10.2.3 *Let $\alpha = \alpha(a, V) > \|V\|$, $X \in (K_n)^m$, $\alpha(a, V)X \in Q(a, V)X, p \in (R_+^n)^m \setminus \{0\}$. Then $p \in Q(a, V)'(\alpha p)$ if and only if the following relations hold:*
(a) $p = (r, \ldots, r)$ where $r \in R_+^n \setminus \{0\}$;
(b) $< (\alpha I_n - V^i)r, u >\geq \sup\{< r, d >: d \in a^i(u)\}$ for each $u \in R_+^n$ and each $i \in \{1, \ldots, m\}$;
(c) $< (\alpha I_n - V^i)r, X^i >= \sup\{< r, d >: d \in a^i(X^i)\}, i = 1, \ldots, m$.

Let $e_1, \ldots, e_n$ be the standard basis in R^n. A function $\psi : R_+^n \to R_+^1$ is called a CES-function if

$$\psi(x) = \gamma(\sum_{i=1}^n \beta_i(x^i)^\rho)^{1/\rho}, x \in R_+^n \tag{2.3}$$

where $\gamma > 0, \rho \in (-\infty, 1) \setminus \{0\}, \beta = (\beta_1, \ldots, \beta_n) \in K_n, \sum_{i=1}^n \beta_i = 1$. The vector $(\gamma, \beta, \rho)(\psi) = (\gamma, \beta, \rho)$ is called the coordinates of the CES-function ψ. In this chapter we consider only CES-functions ψ for which $\rho(\psi) > 0$.

Let $(a, V) \in \mathcal{M} \times \mathcal{L}$ and let

$$(\alpha(a, V), (X, \alpha(a, V)X), p)$$

be its generalized equilibrium state. In view of Lemma 10.2.3 it is important that this generalized equilibrium state satisfies $\alpha(a, V) > ||V||$ and that all coordinates of X and p are positive. In general these conditions do not hold. In the sequel for any $(a, V) \in \mathcal{M} \times \mathcal{L}$ we construct $(\tilde{a}, \tilde{V}) \in \mathcal{M} \times \mathcal{L}$ which is close to (a, V) in $\mathcal{M} \times \mathcal{L}$ and for which these properties hold (see Lemmas 10.2.4-10.2.6).

Let $(a, V) \in \mathcal{M} \times \mathcal{L}$ and let $\phi^i : R_+^n \to R_+^1, i = 1, \ldots, n$ be CES-functions. We define $a \oplus \phi = (a^s \oplus \phi)_{s=1}^n$ by

$$(a^s \oplus \phi)(x) = a^s(x) + \sum_{i \in \mathcal{I}_s} [0, \phi^i(x)] e_i, \qquad (2.4)$$

$$x \in R_+^n, \ s = 1, \ldots, m.$$

It is easy to see that $a \oplus \phi \in \mathcal{M}$.

LEMMA 10.2.4 *Let $(a, V) \in \mathcal{M} \times \mathcal{L}$ and let $\phi^i : R_+^n \to R_+^1, i = 1, \ldots, n$ be CES functions. Then $\alpha(a \oplus \phi, V) > ||V||$.*

Proof. There exist $p \in \{1, \ldots, m\}, q \in \{1, \ldots, n\}$ such that $v^{pq} = ||V||$. It is easy to see that if $q \in \mathcal{I}_p$ then $\alpha(a \oplus \phi, V) > ||V||$. Assume that $q \notin \mathcal{I}_p$. Then there exists $s \in \{1, \ldots, m\} \setminus \{p\}$ for which $q \in \mathcal{I}_s$.

Fix any $r \in \mathcal{I}_p$. Clearly $r \neq q$. We define $X \in (R_+^n)^m$ as

$$X^j = 0, j \in \{1, \ldots, m\} \setminus \{s, p\}, X^p = e_p, X^s = 2^{-1}\phi^r(X^p)e_r.$$

It is easy to verify that there exists a number $\alpha > ||V||$ for which $\alpha X \in Q(a \oplus \phi, V)$. The lemma is proved.

LEMMA 10.2.5 *Let $(a, V) \in \mathcal{M} \times \mathcal{L}$ and let $\phi^i : R_+^n \to R_+^1, i = 1, \ldots, n$ be CES-functions. Assume that*

$$X \in (R_+^n)^m \setminus \{0\}, \quad \alpha(a \oplus \phi, V)(X) \in Q(a \oplus \phi, V)(X). \qquad (2.5)$$

Then $X^i > 0, i = 1, \ldots, m$.

Proof. Assume the contrary. Then there exists $i \in \{1, \ldots, m\}$ for which $X^i = 0$. By Lemma 10.2.4,

$$\alpha(a \oplus \phi, V) > ||V||. \qquad (2.6)$$

Set

$$E_1 = \{i \in \{1, \ldots, m\} : X^i > 0\}, E_2 = \{1, \ldots, m\} \setminus E_1. \qquad (2.7)$$

We have that $E_1 \neq \emptyset, E_2 \neq \emptyset$. There exist $d_i \in R_+^n, i = 1, \ldots, m$ such that

$$\alpha(a \oplus \phi, V)X^i \leq V^i X^i + d^i, i = 1, \ldots, m, \qquad (2.8)$$

$$\sum_{i=1}^{m} d^i \in \sum_{i=1}^{m} (a^i \oplus \phi)(X^i).$$

We may assume that

$$d^i = 0, i \in E_2. \tag{2.9}$$

By (2.8), (2.9) and (2.4) for each $j \in E_1$ there exist $y^j \in a^j(X^j)$ and $z^j \in R^n_+$ such that

$$z^j \leq \sum_{i \in \mathcal{I}_j} \phi^i(X^j)e_i, \tag{2.10}$$

$$\sum_{i \in E_1} d^i = \sum_{j \in E_1} (y^j + z^j). \tag{2.11}$$

Fix $k \in E_2$. There exists

$$p \in \mathcal{I}_k \setminus \cup \{\mathcal{I}_j : j \in \{1, \ldots, m\} \setminus \{k\}\}. \tag{2.12}$$

It follows from (2.6) and (2.12) that

$$X^{ip} = 0, i \in E_1. \tag{2.13}$$

For all small enough $h > 0$ we set

$$X_h^i = 0, i \in E_2 \setminus \{k\}, X_h^i = X^i + he_p, \ i \in E_1. \tag{2.14}$$

By (2.13) and (2.14) there exist constants $\rho, c_0 \in (0, 1)$ such that for all small enough $h > 0$, each $j \in E_1$ and each $r \in \mathcal{I}_j$,

$$\phi^r(X_h^j) \geq \phi^r(X^j) + c_0 h^\rho. \tag{2.15}$$

We set

$$c_1 = \min\{\phi^p(e_j) : j = 1, \ldots, n\}. \tag{2.16}$$

For all small enough $h > 0$ we define

$$X_h^{kr} = [4(\alpha(a \oplus \phi, V) + 1)]^{-1} c_0 h^\rho \text{ if} \tag{2.17}$$

$$r \in \cup\{\mathcal{I}_j : j \in E_1\}, \quad \text{otherwise } X_h^{kr} = 0.$$

(2.6), (2.8), (2.9), (2.10) and (2.11) imply that

$$\sum_{i \in E_1} [\alpha(a \oplus \phi, V)X^i - V^i X^i] \leq \sum_{i \in E_1} d^i \leq \sum_{i \in E_1} (y^i + z^i) \tag{2.18}$$

$$\subset \{x \in R^n_+ : x^i = 0 \text{ for each } i \in \{1, \ldots, n\} \setminus \cup\{\mathcal{I}_j : j \in E_1\}\}.$$

It is easy to see that for all small enough $h > 0$,

$$[\alpha(a \oplus \phi, V) + 1]X_h^k \leq 4^{-1} \sum_{j \in E_1} [\sum_{r \in \mathcal{I}_j} (\phi^r(X_h^j) - \phi^r(X^j))e_r] \tag{2.19}$$

$$\leq [4\alpha(a \oplus \phi, V) + 4][m \sum_{i \in E_1} (X_h^i - X^i)] \leq \phi^p(X_h^k)e_p.$$

For all small enough $h > 0$ we define

$$y_h^j = 0, j \in E_2, z_h^j = 0, j \in E_2 \setminus \{k\}, \tag{2.20}$$

$$z_h^k = \phi^p(X_h^k)e_p, \ y_h^j = y^j, i \in E_1,$$

$$z_h^j = z^j + \sum_{i \in \mathcal{I}_j} (\phi^i(X_h^j) - \phi^i(X^j))e_i, j \in E_1,$$

$$d_h^j = 0, j \in E_2 \setminus \{k\}, d_h^k = 4^{-1} \sum_{j \in E_1} (\sum_{r \in \mathcal{I}_j} (\phi^r(X_h^j) - \phi^r(X^j))e_r), \quad (2.21)$$

$$d_h^j = d^j + 4^{-1} \sum_{s \in E_1} m^{-1}[\sum_{r \in \mathcal{I}_s} (\phi^r(X_h^s) - \phi^r(X^s))e_r] + m^{-1}\phi^p(X_h^k)e_p, j \in E_1.$$

By (2.20), (2.21), (2.14), (2.8), (2.10), (2.11) and (2.4) for all small enough $h > 0$,

$$y_h^j, z_h^j, d_h^j \in R_+^n, \ j = 1, \ldots, m, \quad \sum_{j=1}^m d_h^j \leq \sum_{j=1}^m (y_h^j + z_h^j), \tag{2.22}$$

$$y_h^j + z_h^j \in (a^j \oplus \phi)(X_h^j), \ j = 1, \ldots, m.$$

By (2.14), (2.17), (2.21) and (2.19) for each small enough $h > 0$,

$$d_h^i + V^i X_h^i \geq [\alpha(a \oplus \phi, V) + 1)]X_h^i, \ i \in E_2. \tag{2.23}$$

It follows from (2.21), (2.14), (2.8) and (2.19) that for each small enough $h > 0$ and each $j \in E_1$,

$$d_h^j + V^j X_h^j \geq d^j + V^j X^j + m^{-1}\phi^p(X_h^k)e_p \tag{2.24}$$

$$+(4m)^{-1} \sum_{s \in E_1} [\sum_{r \in \mathcal{I}_s} (\phi^r(X_h^s) - \phi^r(X^s))e_r] \geq \alpha(a \oplus \phi, V)X^j + m^{-1}\phi^p(X_h^k)e_p$$

$$+(4m)^{-1} \sum_{s \in E_1} [\sum_{r \in \mathcal{I}_s} (\phi^r(X_h^s) - \phi^r(X^s))e_r] \geq \alpha(a \oplus \phi, V)X^j$$

$$+4(\alpha(a \oplus \phi, V) + 1)(X_h^j - X^j)$$

$$+(4m)^{-1} \sum_{s \in E_1} [\sum_{r \in \mathcal{I}_s} (\phi^r(X_h^s) - \phi^r(X^s))e_r].$$

(2.6), (2.18) and (2.14) imply that for each small enough $h > 0$ there exists a number $\lambda_h > 0$ which satisfies

$$\sum_{s \in E_1} [\sum_{r \in \mathcal{I}_s} (\phi^r(X_h^s) - \phi^r(X^s))e_r] \geq \lambda_h X^i, \ i \in E_1.$$

Together with (2.23), (2.24), (2.22) this implies that for each small enough $h > 0$ there exists a number $\alpha_h > \alpha(a \oplus \phi, V)$ such that

$$d_h^j + V^j X_h^j \geq \alpha_h X_h^j, \ j \in \{1, \ldots, m\}, \ \alpha_h X_h \in Q(a \oplus \phi, V)(X_h).$$

This is contradictory to the definition of $\alpha(a \oplus \phi, V)$. The obtained contradiction proves the lemma.

LEMMA 10.2.6 *Let* $(a, V) \in \mathcal{M} \times \mathcal{L}$, $\phi^i : R_+^n \rightarrow R_+^1, i = 1, \ldots, n$ *be CES-functions and let* $X \in (R_+^n)^m \setminus \{0\}$,

$$\alpha(a \oplus \phi, V)X \in Q(a \oplus \phi, V)(X). \tag{2.25}$$

Then $\alpha(a \oplus \phi, V) > ||V||, X \in (K_n)^m$. *Moreover for each* $p \in (R_+^n)^m \setminus \{0\}$ *satisfying* $p \in Q(a \oplus \phi, V)'(\alpha(a \oplus \phi, V)p)$ *there exists* $r \in K_n$ *such that* $p = (r, \ldots, r)$.

Proof. By Lemmas 10.2.4 and 10.2.5,

$$\alpha(a \oplus \phi, V) > ||V||, \ X^i > 0, i = 1, \ldots m. \tag{2.26}$$

Assume that $X \notin (K_n)^m$. Then there exist $k \in \{1, \ldots, m\}, j, j_1 \in \{1, \ldots, n\}$ such that $X^{kj} = 0, X^{kj_1} > 0$.

For all small enough $h > 0$ we define $\hat{X}_h \in (R_+^n)^m$. For all small enough $h > 0$ we set $X_h^k = X^k + he_j$. There exist constants $c_1 > 0, \rho \in (0, 1)$ such that for each small enough $h > 0$ and each $i \in \{1, \ldots, m\}$, $\phi^i(X_h^k) \geq \phi^i(X^k) + c_1 h^\rho$. There are two cases: (i) $j \in \mathcal{I}_k$; (ii) $j \notin \mathcal{I}_k$. In the case (i) we set $X_h^s = X^s, s \in \{1, \ldots, m\} \setminus \{k\}$.

Consider the case (ii). There exists $q \in \{1, \ldots, m\} \setminus \{k\}$ for which $j \in \mathcal{I}_q$. We set

$$X_h^s = X^s, s \in \{1, \ldots, m\} \setminus \{q, k\}, X_h^{qi} = X^{qi}, i \in \{1, \ldots, m\} \setminus \mathcal{I}_k,$$

$$X_h^{qi} = X^{qi} + (\alpha(a \oplus \phi, V))^{-1} c_1 h^\rho, i \in \mathcal{I}_k.$$

In the case (ii) there exists a constant $c_2 \in (0, c_1)$ such that $\phi^j(X_h^q) \geq \phi^j(X^q) + c_2 h^\rho$. In both cases we set

$$\hat{X}_h^s = X_h^s, s \in \{1, \ldots, m\} \setminus \{k\}, \hat{X}_h^k = X_h^k + (2\alpha(a \oplus \phi, V))^{-1} c_2 h^\rho e_j.$$

It is easy to verify that for all small enough $h > 0$,

$$\alpha(a \oplus \phi, V)\hat{X}_h \in Q(a \oplus \phi, V)(X_h). \tag{2.27}$$

Since for all small enough $h > 0$,

$$\text{Card}\{(i, j) \in \{1, \ldots, m\} \times \{1, \ldots, n\} : X_h^{ij} > 0\} \tag{2.28}$$

$$> \text{Card}\{(i,j) \in \{1,\ldots,m\} \times \{1,\ldots,n\} : X^{ij} > 0\}$$

and X is any vector from $(R_+^n)^m \setminus \{0\}$ satisfying (2.25) we conclude that there exists $X_0 \in (K_n)^m$ for which $\alpha(a \oplus \phi, V)X_0 \in Q(a \oplus \phi, V)(X_0)$. There exists $p \in (R_+^n)^m \setminus \{0\}$ satisfying

$$p \in Q(a \oplus \phi, V)'(\alpha(a \oplus \phi, V)p). \tag{2.29}$$

By Lemma 10.2.3 there exists $r \in R_+^n \setminus \{0\}$ such that

$$p = (r,\ldots,r), \quad < (\alpha(a \oplus \phi, V)I_n - V^s)r, u > \tag{2.30}$$

$$\geq \sup\{< r, d >: d \in a^s(u)\} + \sum_{i \in \mathcal{I}_s} r^i \phi^i(u), s = 1,\ldots,m, u \in R_+^n,$$

$$< (\alpha(a \oplus \phi, V)I_n - V^s)r, X_0^s >= \sup\{< r, d >: d \in a^s(X_0^s)\}$$

$$+ \sum_{i \in \mathcal{I}_s} r^i \phi^i(X_0^s), s = 1,\ldots,m.$$

Since the functions $\phi^i, i = 1,\ldots n$ are strictly monotone increasing these relations imply that $r \in K_n$. It follows from this relation, (2.29), (2.27), (2.30) and the definition of $\hat{X}_h$ that for all small enough $h > 0$,

$$< \alpha(a \oplus \phi, V)p, X_h >\geq< p, \alpha(a \oplus \phi, V)\hat{X}_h >$$

$$>< p, \alpha(a \oplus \phi, V)X_h > .$$

The obtained contradiction proves the lemma.

10.3.   Sufficient condition for the turnpike property

We need the following result established by Leizarowitz [38].

PROPOSITION 10.3.1 *Let $G \subset R^n \times R^n$ be a convex compact set,*

$$domG = \{x \in R^n : \text{ there exists } y \in R^n$$

$$\text{such that } (x, y) \in G\},$$

$$G(x) = \{y \in R^n : (x, y) \in G\}, x \in domG.$$

Assume that the following properties hold:
(i) There exists a unique $Y \in domG$ such that $0 \in G(Y)$;
(ii) if $a, b \in R^n$, a scalar $\alpha \neq 0$ and the function $z(t) = Y + \cos(\alpha t)a + \sin(\alpha t)b, t \in [0, \infty)$ satisfies

$$z'(t) \in G(z(t)) \tag{3.1}$$

almost everywhere in $[0, \infty)$ *then* $a, b = 0$.

Denote by $\mathcal{E}$ the set of all a.c. functions $z : [0, \infty) \to domG$ which satisfy (3.1) a.e. in $[0, \infty)$. Then for each $\epsilon > 0$ there exists $T_\epsilon > 0$ such that $|z(t) - Y| \le \epsilon$ for each $z(\cdot) \in \mathcal{E}$ and each $t \ge T_\epsilon$.

Denote by $\mathbf{C}$ the set of complex numbers, for each linear operator $M : \mathbf{C}^q \to \mathbf{C}^k$ denote by M' the dual operator and denote by $\sigma(M)$ the spectrum of M. For each $x = (x^1, \ldots, x^s) \in R^s, y = (y^1, \ldots, y^q) \in R^q$ we set

$$[x, y] = (x^1, \ldots, x^s, y^1, \ldots, y^q) \in R^{s+q}. \tag{3.2}$$

For each matrix $M = (m^{ij}), N = (n^{ij})$ of a dimension $s \times q$ we define

$$\sigma(M, N) = \{\theta \in \mathbf{C} : \text{ there exists } \lambda \in \mathbf{C}^s \setminus \{0\} \tag{3.3}$$

$$\text{for which } M\theta\lambda = N\lambda\}$$

and denote by $M*N$ the matrix $B = (b^{ij})$ of the dimension $s \times q$ such that $b^{ij} = m^{ij}n^{ij}, i = 1, \ldots, s, j = 1, \ldots, q$. For each $Y = (Y^1, \ldots, Y^m) \in (R^n)^m$ denote by $\hat{Y} = (\hat{y}^{ij})$ the matrix of the dimension $m \times n$ for which $\hat{y}^{ij} = Y^{ij}, i = 1, \ldots, m, j = 1, \ldots, n$.

In this section we assume that

$$V \in \mathcal{L}, a = (a^1, \ldots, a^m) \in \mathcal{M}, Q = Q(a, V), \alpha = \alpha(a, V) > ||V||, \tag{3.4}$$

$$X \in (K_n)^m, r \in K_n, p = (r, \ldots, r), \alpha X \in Q(X),$$

$$p \in Q'(\alpha p), <p, X> = 1, Z^i \in a^i(X^i),$$

$$i = 1, \ldots, m, \sum_{i=1}^m ((\alpha I_n - V^i)X^i - Z^i) = 0.$$

The next simple but important lemma describes the set of all pairs $x, y \in (R_+^n)^m$ such that $y \in Q(x)$ and $<\alpha px> = (p, y>$.

LEMMA 10.3.1 *Assume that*

$$x \in (R_+^n)^m, \ y \in Q(x), \ d^i \in R_+^n, \ i = 1, \ldots, m,$$

$$z^i \in a^i(x^i), i = 1, \ldots, m, \ y^i \le V^i x^i + d^i, \ i = 1, \ldots, m,$$

$$\sum_{i=1}^m (d^i - z^i) = 0, \ < \alpha p, x > = < p, y > .$$

Then there exist numbers $\lambda^i \ge 0$, $i = 1, \ldots, m$ *such that* $[x^i, z^i] = \lambda^i[X^i, Z^i]$, $i = 1, \ldots, m$.

Proof. It follows from Lemma 10.2.2 that

$$y^i = V^i x^i + d^i, < (\alpha I_n - V^i) r, x^i >=< r, z^i >$$

$$= \sup\{< r, u >: u \in a^i(x^i)\}, \ i = 1, \ldots, m.$$

Combined with Lemma 10.2.3 and property (ii) (see Section 10.1) this implies the validity of the lemma.

The following auxiliary result establishes a sufficient condition for the turnpike property. It is important that this condition depends only on α, X, V, Z. The proof of this lemma is based on Lemma 10.3.1 and Proposiion 10.3.1.

LEMMA 10.3.2 *Assume that*

$$\{\lambda \in \mathbf{C}^n : (\alpha \hat{X} - V * \hat{X} - \hat{Z})'\lambda = 0\} = \{\gamma(1, 1, \ldots, 1) : \gamma \in \mathbf{C}\}, \quad (3.5)$$

$$\{\theta \in \sigma(\hat{X}', (V * \hat{X} + \hat{Z})') : Re\theta = \alpha\} = \{\alpha\}.$$

Then the mapping Q has the turnpike property.

Proof. Let $x : [0, \infty) \rightarrow (R^n_+)^m$ be a trajectory of the model $\mathbf{Z}(Q)$ which has the von Neumann growth rate. We will show that

$$\lim_{t \to \infty} \exp((1 - \alpha)t)x(t) = \lambda X \text{ with some } \lambda > 0.$$

We may assume without loss of generality that

$$\lim_{t \to \infty} \exp((1 - \alpha)t) < x(t), p >= 1. \quad (3.6)$$

Put

$$\Omega = \{y \in (R^n_+)^m :< y, p >= 1\},$$

$$G = \{(y, z) \in \Omega \times R^n : z \in Q(y) - \alpha y\}, \quad (3.7)$$

$$\Omega_0 = \{y \in (R^n_+)^m : \text{ there exists a sequence } \{t_j\}_{j=1}^\infty \subset (0, \infty)$$

such that $t_j \to \infty, \exp((1 - \alpha)t_j)x(t_j) \to y$ as $j \to \infty\}.$

Clearly G is a compact convex set. It follows from (3.4), (3.7), (3.5) and Lemma 10.3.1 that $(Y, 0) \in G$ if and only if $Y = X$.

Let $a = (a^1, \ldots, a^m), b = (b^1, \ldots, b^m) \in (R^n)^m, \beta \neq 0$ and let

$$z(t) = X + \cos(\beta t)a + \sin(\beta t)b, \ t \in [0, \infty). \quad (3.8)$$

Assume that

$$(z(t), z'(t)) \in G \text{ for each } t \in [0, \infty). \quad (3.9)$$

We will show that $a, b = 0$. It follows from (3.8), (3.9) and (3.7) that

$$X + \cos(\beta t)a + \sin(\beta t)b \in \Omega, t \in [0, \infty), \tag{3.10}$$

$$\alpha(X + \cos(\beta t)a + \sin(\beta t)b) + \sin(\beta t)(-\beta)a + \cos(\beta t)\beta b$$
$$\in Q(X + \cos(\beta t)a + \sin(\beta t)b), t \in [0, \infty),$$
$$< p, -\sin(\beta t)\beta a + \cos(\beta t)b > = < p, z'(t) > = 0, \tag{3.11}$$
$$< p, \alpha X + (\cos \beta t)(\alpha a + \beta b) + \sin(\beta t)(\alpha b - \beta a) > = \alpha.$$

By (3.10), (3.11) and Lemma 10.3.1 there exist numbers $\gamma_a^i, \gamma_b^i, i = 1, \ldots, m$ such that

$$b^i = \gamma_b^i X^i, a^i = \gamma_a^i X^i, i = 1, \ldots, m, \tag{3.12}$$

$$\sum_{i=1}^{m} [\alpha X^i + (\cos \beta t)(\alpha \gamma_a^i + \beta \gamma_b^i)X^i + (\sin \beta t)(\alpha \gamma_b^i X^i - \beta \gamma_a^i X^i) \tag{3.13}$$

$$-V^i(X^i + \cos(\beta t)\gamma_a^i X^i + \sin(\beta t)\gamma_b^i X^i)]$$
$$= \sum_{i=1}^{m} (1 + \cos(\beta t)\gamma_a^i + (\sin \beta t)\gamma_b^i)Z^i$$

for each $t \in [0, \infty)$.

(3.4) and (3.13) which holds for each $t \in [0, \infty)$ imply that

$$\sum_{i=1}^{m} [(\cos(\beta t)(\alpha \gamma_a^i + \beta \gamma_b^i) + \sin(\beta t)(\alpha \gamma_b^i - \beta \gamma_a^i))X^i \tag{3.14}$$

$$-V^i(\cos(\beta t)\gamma_a^i + \sin(\beta t)\gamma_b^i)X^i] = \sum_{i=1}^{m} (\cos(\beta t)\gamma_a^i + \sin(\beta t)\gamma_b^i)Z^i, t \in [0, \infty).$$

This implies that

$$\sum_{i=1}^{m} (\alpha \gamma_a^i + \beta \gamma_b^i)X^i = \sum_{i=1}^{m} \gamma_a^i(V^i X^i + Z^i), \tag{3.15}$$

$$\sum_{i=1}^{m} (\alpha \gamma_b^i - \beta \gamma_a^i)X^i = \sum_{i=1}^{m} \gamma_b^i(V^i X^i + Z^i).$$

By (3.15),

$$\hat{X}'((\alpha + i\beta)\gamma) = (V * \hat{X} + \hat{Z})'\gamma,$$

where

$$\gamma = (\gamma_b^1, \ldots, \gamma_b^m) + i(\gamma_a^1, \ldots, \gamma_a^m).$$

Combined with (3.5), (3.3) and (3.12) this implies that $a, b = 0$. Therefore the set G has properties (i) and (ii) stated in Proposition 10.3.1. It follows from Proposition 10.3.1 that for each a.c. function $z : R^1 \to \Omega$ which satisfies

$$(z(t), z'(t)) \in G \text{ a.e. in } R^1 \tag{3.16}$$

the following relation holds:

$$z(t) = X, \ t \in R^1. \tag{3.17}$$

Consider any $y_0 \in \Omega_0$. In view of Proposition 10.2.1 there exists an a.c. function $y : R^1 \to R_+^n$ such that

$$y(0) = y_0, \ \exp((1 - \alpha)t)y(t) \in \Omega_0, \ t \in R^1$$

and

$$y'(t) \in Q(y(t)) - y(t) \text{ a.e. } t \in R^1.$$

It is easy to see that (3.16) holds a.e. in R^1 for $z(t) = \exp((1-\alpha)t)y(t)$. This implies (3.17) and the equality $y_0 = X$. Therefore $\Omega_0 = \{X\}$. The lemma is proved.

10.4.   Preliminary lemmas

The main result of this section is Lemma 10.4.6 which is an important tool in the proof of Theorem 10.1.1. With any $(a, V) \in \mathcal{M} \times \mathcal{L}$ we associate a generalized equilibrium state $(\alpha, (X, \alpha X), p)$ and a vector Z as in Section 10.3. We consider the set $\mathcal{K}$ which consists of all vectors (α, X, V, Z) which are associated with all $(a, V) \in \mathcal{M} \times \mathcal{L}$. Let $(a, V) \in \mathcal{M} \times \mathcal{L}$ and let a vector $(\alpha, X, V, Z) \in \mathcal{K}$ be associated with (a, V). Lemma 10.4.6 shows that there is $(\widehat{\alpha}, \widehat{X}, \widehat{V}, \widehat{Z}) \in \mathcal{K}$ which is close to (α, X, V, Z) and which satisfies the condition (3.5) in Lemma 10.3.2. In Section 10.5 we construct a pair $(\widehat{a}, \widehat{V})$ such that an element of $\mathcal{K}$ associated with $(\widehat{a}, \widehat{V})$ is $(\widehat{\alpha}, \widehat{X}, \widehat{V}, \widehat{Z})$. Then by Lemma 10.3.2 $(\widehat{a}, \widehat{V})$ has the turnpike property.

Let G be a metric space with a metric $d(\cdot, \cdot)$. For each $x \in G$ and each $h > 0$ we set

$$\mathbf{B}(x, h) = \{y \in G : d(x, y) < h\}.$$

Let $A = (a^{ij}), B = (b^{ij})$ be matrices of a dimension $m \times n, a^{ij}, b^{ij} \in \mathbf{C}, i = 1, \ldots, m, j = 1, \ldots, n$. Consider the function $\lambda \to \lambda A' - B', \lambda \in \mathbf{C}$. For each subset $F \subset \{1, \ldots, n\}$ satisfying $\mathrm{Card}(F) = m$ denote by $P_F(A, B)$ the polynomial such that for each $\lambda \in \mathbf{C}$ the number $P_F(A, B)(\lambda)$ is the determinant of the matrix of the dimension $m \times m$ with rows $(\lambda a^{1j} - b^{1j}, \ldots, \lambda a^{mj} - b^{mj}), j \in F$. Denote by $P(A, B)$ the greatest common divizor of all polynomials $P_F(A, B)$ where $F \subset \{1, \ldots, n\}, \mathrm{Card}(F) = m$. It is easy to see that $\lambda \in \sigma(A', B')$ if and only if $P(A, B)(\lambda) = 0$.

Remark. If polynomials $P_1, \ldots, P_s = 0$, then their greatest common divizor is 0. Otherwise their greatest common divizor is a polynomial whose leading coefficient is equal to 1.

Let $P(A, B)(\lambda) = \prod_{i=1}^{s}(\lambda - \lambda_i)^{m_i}, \lambda \in \mathbf{C}$ where $s \geq 1$, $\lambda_i \neq \lambda_j$ for each $i, j \in \{1, \ldots, s\}$ satisfying $i \neq j$. Set

$$m(A, B) = \sum_{i=1}^{s} m_i \leq m.$$

If $P(A, B) = 1$ we set $m(A, B) = 0$.

Let g be a polynomial. By $\deg(g)$ we denote the degree of g.

LEMMA 10.4.1 *Let A_0, B_0 be matrices of the dimension $m \times n$,*

$$m(A_0, B_0) \geq 1,$$

$$P(A_0, B_0)(\lambda) = \prod_{i=1}^{m(A_0, B_0)} (\lambda - \lambda_i^0), \lambda \in \mathbf{C}$$

where $\lambda_i^0 \in \mathbf{C}, i = 1, \ldots, m(A_0, B_0)$. Assume that $\epsilon, M > 0$,

$$|\lambda_i^0| < M, \ i = 1, \ldots, m(A_0, B_0),$$

$$\mathbf{B}(\lambda_i^0, \epsilon) \cap \mathbf{B}(\lambda_j^0, \epsilon) = \emptyset$$

for each $i, j \in \{1, \ldots, m(A_0, B_0)\}$ satisfying $\lambda_i^0 \neq \lambda_j^0$.

Then there exists $\delta > 0$ such that for each $(A, B) \in \mathbf{B}((A_0, B_0), \delta)$ the polynomial $P(A, B) \neq 0$ and one of the following conditions holds:

(a) $P(A, B)(\lambda) = 1, \lambda \in \mathbf{C}$;

(b) $1 \leq m(A, B) \leq m$ and there exist numbers

$$\lambda_i \in \mathbf{C}, \ i = 1, \ldots, m(A, B),$$

an integer $s \in \{0, \ldots, m(A, B)\}$ and a bijection $\tau : \{1, \ldots, m(A_0, B_0)\} \to \{1, \ldots, m(A_0, B_0)\}$ such that

$$P(A, B)(\lambda) = \prod_{i=1}^{m(A,B)} (\lambda - \lambda_i), \lambda \in \mathbf{C},$$

$$|\lambda_i - \lambda^0_{\tau(i)}| < \epsilon, 1 \le i \le s, |\lambda_i| \ge M \ \ if \ s + 1 \le i \le m(A, B).$$

Proof. We define

$$\mathcal{E} = \{F \subset \{1, \ldots, n\} : \mathrm{Card}(F) = m, P_F(A_0, B_0) \ne 0\}.$$

We will prove the following

Assertion. Assume that $m_1 \ge 1$ is an integer, $A_t, B_t, t = 1, 2 \ldots$ are matrices of the dimension $m \times n$, $A_t \to A_0, B_t \to B_0$ as $t \to \infty$, P_t is the greatest common divizor of the polynomials $P_F(A_t, B_t), F \in \mathcal{E}, t = 1, 2, \ldots$, $P_t \ne 0, P_t = Q_t Q^1_t$ where Q_t, Q^1_t are polynomials, $deg(Q_t) = m_1, deg(Q^1_t) \ge 0, t = 1, 2 \ldots$, $Q_t = \prod^{m_1}_{i=1}(\lambda - \lambda^t_i), |\lambda^t_i| \le M, i = 1, \ldots, m_1, t = 1, 2, \ldots, \lambda^t_i \to \Lambda_i$ as $t \to \infty, i = 1, \ldots, m_1$.
Then $m_1 \le m(A_0, B_0)$ and there exists a bijection

$$\tau : \{1, \ldots, m(A_0, B_0)\} \to \{1, \ldots, m(A_0, B_0)\}$$

such that $\Lambda_i = \lambda^0_{\tau(i)}, i = 1, \ldots, m_1$.

Proof of the assertion. Consider any $F \in \mathcal{E}$. For each $t \in \{1, 2, \ldots\}$ there exists a polynomial Q^F_t such that

$$\deg(Q^F_t) \ge 0, P_F(A_t, B_t) = Q_t Q^1_t Q^F_t. \tag{4.1}$$

It is easy to see that

$$P_F(A_t, B_t) \to P_F(A_0, B_0), Q_t(\lambda) \to \prod^{m_1}_{i=1}(\lambda - \Lambda_i), \lambda \in \mathbf{C} \ as \ t \to \infty. \tag{4.2}$$

This implies that the set of all coefficients of the polynomials $Q^1_t Q^F_t, t = 1, 2, \ldots$ is bounded. Therefore we may assume that there exists a polynomial $\hat{Q}$ for which

$$Q^1_t Q^F_t \to \hat{Q} \ as \ t \to \infty. \tag{4.3}$$

Then by (4.1)-(4.3) $P_F(A_0, B_0)(\lambda) = \prod^{m_1}_{i=1}(\lambda - \Lambda_i)\hat{Q}, \lambda \in \mathbf{C}$. Since F is any element of $\mathcal{E}$ we conclude that $\prod^{m_1}_{i=1}(\lambda - \Lambda_i)$ is a divizor of $P(A_0, B_0)$. This completes the proof of the assertion.

Assume that the lemma is wrong. Then there exist sequences of matrices $\{A_t\}^\infty_{t=1}, \{B_t\}^\infty_{t=1}$ of the dimension $m \times n$ such that:
(i) $A_t \to A_0, B_t \to B_0$ as $t \to \infty$ and for each integer $t \ge 1$ the following relations hold: $P(A_t, B_t) \ne 0, m(A_t, B_t) \ge 1$,

$$P(A_t, B_t)(\lambda) = \prod^{m(A_t, B_t)}_{i=1}(\lambda - \lambda^t_i), \lambda \in \mathbf{C};$$

(ii) $|\lambda_i^t| \geq M$ if and only if $s_t + 1 \leq i \leq m(A_t, B_t)$ with some $s_t \in \{0, \ldots, m(A_t, B_t)\}$;

(iii) for each bijection $\tau : \{1, \ldots, m(A_0, B_0)\} \to \{1, \ldots, m(A_0, B_0)\}$ the relation $s_t \leq m(A_0, B_0)$ implies that $|\lambda_i^t - \lambda_{\tau(i)}^0| \geq \epsilon$ for some natural number $i \leq s_t$.

We may assume without loss of generality that there exist an integer $m_1 \geq 1$ and numbers $\Lambda_i \in \mathbf{C}$, $i = 1, \ldots, m_1$ for which

$$s_t = m_1, \ t = 1, 2, \ldots, \lambda_i^t \to \Lambda_i \text{ as } t \to \infty, \ i = 1, \ldots, m_1. \quad (4.4)$$

For each integer $t \geq 1$ we set

$$Q_t(\lambda) = \prod_{i=1}^{m_1} (\lambda - \lambda_i^t), \ \lambda \in \mathbf{C} \quad (4.5)$$

and denote by P_t the greatest common divizor of polynomials

$$P_F(A_t, B_t), \ F \in \mathcal{E}.$$

For each integer $t \geq 1$ there exists a polynomial Q_t^1 for which

$$\deg(Q_t^1) \geq 0, P_t = Q_t Q_t^1. \quad (4.6)$$

We may assume without loss of generality that $P_t \neq 0, t = 1, 2 \ldots$. It follows from the Assertion, conditions (i),(ii), the definition of $P_t, t = 1, 2, \ldots$ and (4.4)-(4.6) that $m_1 \leq m(A_0, B_0)$ and there exists a bijection $\tau : \{1, \ldots, m(A_0, B_0)\} \to \{1, \ldots, m(A_0, B_0)\}$ for which $\Lambda_i = \lambda_{\tau(i)}^0, i = 1, \ldots, m_1$.

Together with (4.4) this implies that for all large enough t,

$$|\lambda_i^t - \lambda_{\tau(i)}^0| \leq 4^{-1}\epsilon, i = 1, \ldots, m_1.$$

This is contradictory to condition (iii). The obtained contradiction proves the lemma.

We define

$$\mathcal{K} = \{z \in (R_+^n)^m : z^{ij} = 0 \ (i \in \{1, \ldots, m\}, j \notin \mathcal{I}_i)\}, \quad (4.7)$$

$$\mathcal{A} = \{(\alpha, X, V, Z) \in K_1 \times (K_n)^m \times \mathcal{L} \times \mathcal{K} :$$

$$(\alpha \hat{X} - V * \hat{X} - \hat{Z})'(1, 1, \ldots, 1) = 0\}. \quad (4.8)$$

We consider the topological subspace $\mathcal{A} \subset R^1 \times (R^n)^m \times \mathcal{L} \times (R^n)^m$ with the relative topology.

LEMMA 10.4.2 *Let $\epsilon > 0, (\alpha_0, X_0, V_0, Z_0) \in \mathcal{A}$,*

$$\{\theta \in \mathbf{C}^m : (\alpha_0 \hat{X}_0 - V_0 * \hat{X}_0 - \hat{Z}_0)'\theta = 0\} = \{\lambda(1, 1, \ldots, 1) : \lambda \in \mathbf{C}\}.$$

Assume that there exists $r_0 \in K_n$ such that

$$||r_0|| = 1, (\alpha_0 \hat{X}_0 - V_0 * \hat{X}_0 - \hat{Z}_0)r_0 = 0. \tag{4.9}$$

Then there exists a neighborhood U of $(\alpha_0, X_0, V_0, Z_0)$ in $\mathcal{A}$ such that for each $(\alpha, X, V, Z) \in U$ there exists $r \in K_n$ which has the following properties:

$$||r|| = 1, ||r - r_\epsilon|| < \epsilon, (\alpha \hat{X} - V * \hat{X} - \hat{Z})r = 0, \tag{4.10}$$

$$\{\theta \in \mathbf{C}^m : (\alpha \hat{X} - V * \hat{X} - \hat{Z})'\theta = 0\} = \{\lambda(1, 1, \ldots, 1) : \lambda \in \mathbf{C}\}. \tag{4.11}$$

Proof. Since the rank of the matrix $\alpha_0 \hat{X}_0 - V_0 * \hat{X}_0 - \hat{Z}_0$ is $m - 1$ there exists a submatrix B of the matrix $\alpha_0 \hat{X}_0 - V_0 * \hat{X}_0 - \hat{Z}_0$ such that the dimension of B is $(m - 1) \times (m - 1)$ and the determinant of B is not zero.

Set

$$I(st) = \{i \in \{1, \ldots, m\} : i\text{th row of } \alpha_0 \hat{X}_0 - V_0 * \hat{X}_0 - \hat{Z}_0 \text{ belongs to } B\},$$

$$I(sl) = \{i \in \{1, \ldots, n\} :$$

$$i\text{th column of } \alpha_0 \hat{X}_0 - V_0 * \hat{X}_0 - \hat{Z}_0 \text{ belongs to } B\},$$

It is easy to see that there exists a neighborhood U_0 of the

$$(\alpha_0, X_0, V_0, Z_0)$$

in $\mathcal{A}$ such that for each $(\alpha, X, V, Z) \in U_0$ the determinant of the submatrix of $(\alpha \hat{X} - V * \hat{X} - \hat{Z})$ which has a dimension $(m - 1) \times (m - 1)$ and is generated by the sets $I(st), I(sl)$ is not zero and (4.11) holds. We define

$$E_1 = \{x \in R^m : x^s = 0 \text{ for } s \notin \{1, \ldots, m\} \setminus I(st)\}, \tag{4.12}$$

$$E_2 = \{x \in R^m : x^s = 0 \text{ for } s \in I(st)\},$$

$$E_3 = \{x \in R^n : x^s = 0 \text{ for } s \in \{1, \ldots, m\} \setminus I(sl)\},$$

$$E_4 = \{x \in R^n : x^s = 0 \text{ for } s \in I(sl)\}.$$

Let $P_i : R^m \to E_i, i = 1, 2, P_i : R^n \to E_i, i = 3, 4$ be linear projectors. We have by (4.9) that

$$P_1[\alpha_0 \hat{X}_0 - V_0 * \hat{X}_0 - \hat{Z}_0]P_3 r_0 \tag{4.13}$$

$$= -P_1[\alpha_0 \hat{X}_0 - V_0 * \hat{X}_0 - \hat{Z}_0]P_4 r_0.$$

For each $(\alpha, X, V, Z) \in \mathcal{A}$ we denote by $\pi(\alpha, X, V, Z)$ the restriction of $P_1(\alpha \hat{X} - V * \hat{X} - \hat{Z})$ to E_3. It is easy to see that

$$\pi(\alpha, X, V, Z)(E_3) \subset E_1, \text{ the operator}$$

$$\pi(\alpha_0, X_0, V_0, Z_0) : E_3 \to E_1 \text{ is invertible.} \qquad (4.14)$$

There exists a neighborhood $U_1 \subset U_0$ of $(\alpha_0, X_0, V_0, Z_0)$ in $\mathcal{A}$ such that for each $(\alpha, X, V, Z) \in U_1$ the operator $\pi(\alpha, X, V, Z)$ is invertible. For $(\alpha, X, V, Z) \in U_1$ we set

$$r(\alpha, X, V, Z) = -\pi(\alpha, X, V, Z)^{-1}[P_1(\alpha \hat{X} - V * \hat{X} - \hat{Z})P_4 r_0] + P_4 r_0 \in R^n. \qquad (4.15)$$

By (4.12)-(4.15),

$$r(\alpha_0, X_0, V_0, Z_0) = r_0, P_1(\alpha \hat{X} - V * \hat{X} - \hat{Z})r(\alpha, X, V, Z) = 0. \quad (4.16)$$

Since each row of the matrix $\alpha \hat{X} - V * \hat{X} - \hat{Z}$ is a linear combination of rows $((\alpha \hat{X} - V * \hat{X} - \hat{Z})^{ij}, j = 1, \ldots, n), i \in I(st)$ we conclude that $(\alpha \hat{X} - V * \hat{X} - \hat{Z})r(\alpha, X, V, Z) = 0$.

The assertion of the lemma now follows from the continuity of the function $(\alpha, V, X, Z) \to r(\alpha, V, X, Z), (\alpha, V, X, Z) \in U_1$.

Let $(\alpha, X, V, Z) \in \mathcal{A}$, $P(\hat{X}, V * \hat{X} + \hat{Z}) \neq 0$. Clearly $P(\hat{X}, V * \hat{X} + \hat{Z})(\alpha) = 0$. There exist $\lambda_1, \ldots, \lambda_k \in \mathbf{C}$ such that $k = m(\hat{X}, V * \hat{X} + \hat{Z})$,

$$P(\hat{X}, V * \hat{X} + \hat{Z})(\lambda) = \prod_{i=1}^{k}(\lambda - \lambda_i), \lambda \in \mathbf{C}.$$

Set

$$m(\alpha, X, V, Z) = \text{Card}\{i \in \{1, \ldots, k\} : Re\lambda_i = \alpha\}. \qquad (4.17)$$

LEMMA 10.4.3 Let $(\alpha_0, X_0, V_0, Z_0) \in \mathcal{A}$, $\alpha_0 > ||V_0||$ and U be a neighborhood of $(\alpha_0, X_0, V_0, Z_0)$ in $\mathcal{A}$. Then there exists $(\alpha_0, X, V, Z) \in U \cap \mathcal{A}$ such that the vectors $(X^{i1}, \ldots, X^{in}), i = 1, \ldots, m$ are linearly independent.

Proof. It is easy to see that for each $\gamma > 0$ there exists $X(\gamma) \in (K_n)^m$ such that $||X(\gamma) - X_0|| \leq \gamma$, $(X^{i1}(\gamma), \ldots, X^{in}(\gamma)), i = 1, \ldots, m$ are linearly independent,

$$\hat{Z}'_0(1, 1, \ldots, 1) = (\alpha_0 \hat{X}_0 - V_0 * \hat{X}_0)'(1, 1, \ldots, 1) \in K_n,$$

$$(\alpha_0 \hat{X}(\gamma) - V_0 * \hat{X}(\gamma))'(1, 1, \ldots, 1) \to$$

$$(\alpha_0 \hat{X}_0 - V_0 * \hat{X}_0)'(1, 1, \ldots, 1) \text{ as } \gamma \to 0.$$

There exists $\gamma_0 > 0$ such that for each $\gamma \in (0, \gamma_0)$,

$$2^{-1}(\alpha_0 \hat{X}_0 - V_0 * \hat{X}_0)'(1, 1, \ldots, 1) \le (\alpha_0 \hat{X}(\gamma) - V_0 * \hat{X}(\gamma))'(1, 1, \ldots, 1)$$

$$\le 2(\alpha_0 \hat{X}_0 - V_0 * \hat{X}_0)'(1, 1, \ldots, 1).$$

It is easy to see that for each $\gamma \in (0, \gamma_0)$ there exists a diagonal matrix $D(\gamma) = (d^{ij}(\gamma)), i, j = 1, \ldots, n$ such that

$$D(\gamma)[(\alpha_0 \hat{X}_0 - V_0 * \hat{X}_0)'(1, 1, \ldots, 1)] = (\alpha_0 \hat{X}(\gamma) - V_0 * \hat{X}(\gamma))'(1, 1, \ldots, 1),$$

$$d^{ii}(\gamma) \in [2^{-1}, 2], i = 1, \ldots, n.$$

For each $\gamma \in (0, \gamma_0)$ we define $Z(\gamma) \in (R_+^n)^m$ as

$$Z^i(\gamma) = D(\gamma) Z_0^i, \ i = 1, \ldots, m, \ Z(\gamma) = (Z^1(\gamma), \ldots, Z^m(\gamma)).$$

Clearly $Z(\gamma) \in \mathcal{K}, \gamma \in (0, \gamma_0)$. It is easy to see that for each $\gamma \in (0, \gamma_0)$ the relation $(\alpha_0, X(\gamma), V_0, Z(\gamma)) \in \mathcal{A}$ holds and $(X(\gamma), Z(\gamma)) \to (X_0, Z_0)$ as $\gamma \to 0$. The lemma is proved.

LEMMA 10.4.4 *Assume that* $(\alpha_0, X_0, V_0, Z_0) \in \mathcal{A}$, $X_0^i, i = 1, \ldots, m$ *are linearly independent. Then there are* $M > 0$ *and a neighborhood* U *of* $(\alpha_0, X_0, V_0, Z_0)$ *in* $\mathcal{A}$ *such that for each* $(\alpha, X, V, Z) \in U$ *and each* $\lambda \in \sigma(\hat{X}', (V * \hat{X} + \hat{Z})')$ *the relation* $|\lambda| \le M$ *holds.*

Proof. Let us assume the converse. Then there exist sequences $\{(\alpha_t, X_t, V_t, Z_t)\}_{t=1}^\infty \in \mathcal{A}, \{\lambda_t\}_{t=1}^\infty \subset \mathbf{C}$ such that

$$(\alpha_t, X_t, V_t, Z_t) \to (\alpha_0, X_0, V_0, Z_0), |\lambda_t| \to \infty \text{ as } t \to \infty,$$

$$\lambda_t \in \sigma(\hat{X}_t', (V_t * \hat{X}_t + \hat{Z}_t)'), t = 1, 2, \ldots.$$

For each integer $t \ge 1$ there exists $\theta_t \in \mathbf{C}^n \setminus \{0\}$ such that $\|\theta_t\| = 1$, $(\lambda_t \hat{X}_t' - (V_t * \hat{X}_t + \hat{Z}_t)')\theta_t = 0$.

We may assume without loss of generality that there exists $\theta_0 \in \mathbf{C}^n \setminus \{0\}$ for which $\theta_t \to \theta_0$ as $t \to \infty$. This implies that

$$\hat{X}_0' \theta_0 = \lim_{t \to \infty} \hat{X}_t' \theta_t = \lim_{t \to \infty} \lambda_t^{-1} (V_t * \hat{X}_t + \hat{Z}^t)' \theta_t = 0.$$

This is contradictory to the linear independence of $X_0^i, i = 1, \ldots, m$. The obtained contradiction proves the lemma.

LEMMA 10.4.5 *Assume that* $(\alpha_0, X_0, V_0, Z_0) \in \mathcal{A}$, $X_0^i, i = 1, \ldots, m$ *are linearly independent,* $\epsilon > 0, \|V_0\| < 1, \alpha_0, P(\hat{X}_0, V_0 * \hat{X}_0 + \hat{Z}_0) \ne 0,$

$\lambda_1^0 \in \sigma(\hat{X}_0', (V_0 * \hat{X}_0 + \hat{Z}_0)'), \lambda_1^0 \neq \alpha_0, Re(\lambda_1^0) = \alpha_0$. *Then there exists* $(\alpha_0, X, V, Z) \in \mathcal{A} \cap \mathbf{B}((\alpha_0, X_0, V_0, Z_0), \epsilon)$ *for which* $m(\alpha_0, X, V, Z) < m(\alpha_0, X_0, V_0, Z_0), X^i, i = 1, \ldots, m$ *are linearly independent.*

Proof. We may assume without loss of generality that $P(\hat{X}, V * \hat{X} + \hat{Z}) \neq 0$ and $X^i, i = 1, \ldots, m$ are linearly independent for each $(\alpha, X, V, Z) \in \mathcal{A} \cap \mathbf{B}((\alpha_0, X_0, V_0, Z_0), \epsilon)$. There are integers $k \geq 1, m_i \geq 1, i = 1, \ldots, k$, numbers $\lambda_i \in \mathbf{C}, i = 1, \ldots, k$ such that

$$P(\hat{X}_0, V_0 * \hat{X}_0 + \hat{Z}_0)(\lambda) = \prod_{i=0}^{k} (\lambda - \lambda_i^0)^{m_i}, \lambda \in \mathbf{C}, \qquad (4.18)$$

$\lambda_0^0 = \alpha_0, \lambda_i^0 \neq \lambda_j^0$ for each $i, j \in \{0, \ldots, k\}$ satisfying $i \neq j$.

We may assume that there exists an integer $k_0 \in \{1, \ldots, k\}$ such that

$$Re(\lambda_i^0) = \alpha_i \text{ if and only if } 0 \leq i \leq k_0. \qquad (4.19)$$

There exists $\delta \in (0, \epsilon)$ such that

$$\mathbf{B}(\lambda_i^0, 8\delta) \cap \mathbf{B}(\lambda_j^0, 8\delta) = \emptyset \text{ for each } i, j \in \{0, \ldots, k\} \qquad (4.20)$$

$$\text{satisfying } i \neq j,$$

$$\mathbf{B}(\lambda_i^0, 8\delta) \cap \{y \in \mathbf{C} : Rey = \alpha_0\} = \emptyset \qquad (4.21)$$

for each $i \in \{1, \ldots, k\}$ satisfying $i > k_0$.

It follows from Lemma 10.4.4 that there exist a neighborhood U_1 of $(\alpha_0, X_0, V_0, Z_0)$ in $\mathcal{A}$ and a number

$$M \geq 8(\alpha_0 + \epsilon + 1 + \sum_{i=0}^{k} |\lambda_i^0|) \qquad (4.22)$$

such that for each $(\alpha, X, V, Z) \in U_1$, each $\lambda \in \sigma(\hat{X}', (V * \hat{X} + \hat{Z})')$ the relation $|\lambda| \leq 2^{-1}M$ holds. By Lemma 10.4.1 there exists $\epsilon_0 \in (0, \epsilon)$ such that for each $(\alpha, X, V, Z) \in \mathcal{A} \cap \mathbf{B}((\alpha_0, X_0, V_0, Z_0), \epsilon_0)$ the following properties hold:

(i) $\deg(P(\hat{X}, V * \hat{X} + \hat{Z})) \geq 1$,
$P(\hat{X}, V * \hat{X} + \hat{Z})(\lambda) = \prod_{i=0}^{s} (\prod_{j=q(i)}^{q(i+1)-1} (\lambda - \lambda_j))Q(\lambda)$, where $0 \leq s \leq k, q(0) = 0, q(i+1) > q(i), i = 0, \ldots, s$, Q is a polynomial;

(ii) if $\deg(Q) \geq 1$ then for each root z of Q the relation $|z| \geq M$ holds;

(iii) there exists a bijection $\tau : \{0, \ldots, k\} \to \{0, \ldots, k\}$ such that

$$q(i+1) - q(i) \leq m(\tau(i)), |\lambda_{\tau(i)}^0 - \lambda_{q(i)+j}| \leq \delta, \qquad (4.23)$$

$$0 \leq j < q(i+1) - q(i) - 1, i = 0, \ldots, s.$$

It follows from the definition of U_1, M that for each $(\alpha, X, V, Z) \in U_1 \cap$ $\mathbf{B}((\alpha_0, X_0, V_0, Z_0), \epsilon)$, properties (i), (ii), (iii) hold with $Q(\lambda) = 1, \lambda \in \mathbf{C}$. For each $u = f + ig \in \mathbf{C}^m$ with $f, g \in R^m$ we set

$$Reu = f, Imu = g, \bar{u} = f - ig. \tag{4.24}$$

There exists $\theta \in \mathbf{C}^m \setminus \{0\}$ for which

$$\hat{X}_0' \lambda_1^0 \theta = (V_0 * \hat{X}_0 + \hat{Z}_0)' \theta. \tag{4.25}$$

Since vectors $X_0^i, i = 1, \ldots, m$ are linearly independent we have $\hat{X}_0' \theta \neq 0$. There are two cases: 1) $\lambda_1^0 \in R^1$; 2) $\lambda_1^0 \notin R^1$. In the first case we can assume that $\theta \in R^m \setminus \{0\}$ and vectors $\theta, (1, 1, \ldots, 1)$ are linearly independent.

Consider the case 2) and set $f = Re\theta, g = Im\theta$. We will show that vectors $f, g, (1, 1 \ldots, 1)$ are linearly independent. Assume the contrary. Then there exist $c_1, c_2, c_3 \in \mathbf{C}$ such that $|c_1| + |c_2| + |c_3| \neq 0$,

$$c_1 f + c_2 g + c_3 (1, 1, \ldots, 1) = 0. \tag{4.26}$$

It is easy to see that

$$\hat{X}_0' \bar{\lambda}_1^0 (f - ig) = (\hat{X}_0 + V_0 * \hat{Z}_0)' (f - ig) \tag{4.27}$$

and vectors $f + ig, f - ig$ are linearly independent over $\mathbf{C}$. Together with (4.26) this implies that there exist $c_4, c_5 \in \mathbf{C}$ for which

$$(1, 1, \ldots, 1) = c_4 \theta + c_5 \bar{\theta}. \tag{4.28}$$

Set

$$E = \{\gamma_1 \theta + \gamma_2 \bar{\theta} : \gamma_1, \gamma_2 \in \mathbf{C}\} \tag{4.29}$$

and denote by A, B the restriction of operators $\hat{X}_0', (V_0 * \hat{X}_0 + \hat{Z}_0)'$ to E. By (4.8), (4.25), (4.27) and (4.29), $\alpha_0, \lambda_1^0, \bar{\lambda}_1^0 \in \sigma(A, B), P(A', B') = 0$, $\sigma(A, B) = \mathbf{C}$.

Then

$$\sigma(\hat{X}_0', (V_0 * \hat{X}_0 + \hat{Z}_0)') = \mathbf{C}, P(\hat{X}_0, V_0 * \hat{X}_0 + \hat{Z}_0) = 0.$$

This is contradictory to (4.18). The obtained contradiction proves that vectors $f, g, (1, 1, \ldots, 1)$ are linearly independent.

Let

$$\epsilon_1 \in (0, \epsilon_0), \gamma \in (0, 1), 2|1 - \gamma|\alpha_0 < \delta. \tag{4.30}$$

We will show that in both cases if $1 - \gamma$ is small enough then there exists a matrix $\Delta = (\Delta^{ij}), i = 1, \ldots, m, j = 1, \ldots, n$ for which

$$\|\Delta\| < \epsilon_1, \tag{4.31}$$

$$(\alpha_0(\hat{X}_0 + \Delta) - V_0 * \hat{X}_0 - \hat{Z}_0)'(1, 1, \ldots, 1) = 0, \tag{4.32}$$
$$(\gamma\lambda_1^0(\hat{X}_0 + \Delta)' - (V_0 * \hat{X}_0 + \hat{Z}_0)')(\theta) = 0.$$

It follows from (4.8), (4.25) that (4.32) holds if and only if

$$\Delta'(1, 1, \ldots, 1) = 0, (\gamma\Delta')\theta = (1 - \gamma)\hat{X}_0'\theta.$$

It is easy to see that (4.32) holds if and only if

$$< \Delta^i, (1, 1, \ldots, 1) >= 0, < \Delta^i, Re\theta >= \gamma^{-1}(1 - \gamma) < \hat{X}_0^i, Re\theta >, \tag{4.33}$$

$$< \Delta^i, Im\theta >= \gamma^{-1}(1 - \gamma) < \hat{X}_0^i, Im\theta >, \; i = 1, \ldots, n$$

(here $\Delta^i = (\Delta^{1i}, \ldots, \Delta^{mi})$, $\hat{X}_0^i = (X_0^{1i}, \ldots, X_0^{mi})$, $i = 1, \ldots, n$).

Since in the case (1) $\theta, (1, 1, \ldots, 1)$ are linearly independent and in the case (2) $f, g, (1, 1, \ldots, 1)$ are linearly independent we conclude that there exists a neighborhood W of 1 in R^1 such that for each $\gamma \in W$ there exist $\Delta^i(\gamma) \in R^m, i = 1, \ldots, n$ such that (4.33) holds with $\Delta^i = \Delta^i(\gamma)$ for $i = 1, \ldots, m$ and moreover

$$\Delta^i(\gamma) \to 0 \text{ as } \gamma \to 1 \text{ for } i = 1, \ldots, n. \tag{4.34}$$

For each $\gamma \in W$ we define a matrix $\Delta(\gamma)$ of the dimension $m \times n$ which has columns $\Delta^1(\gamma), \ldots, \Delta^n(\gamma)$. Clearly for each $\gamma \in W$ (4.32) holds with $\Delta = \Delta(\gamma)$.

For each $\gamma \in W$ there exists $X(\gamma) \in (R^n)^m$ such that

$$\hat{X}(\gamma) = \hat{X}_0 + \Delta(\gamma). \tag{4.35}$$

By (4.34) we may assume without loss of generality that for each $\gamma \in W$,

$$||\Delta(\gamma)|| < \epsilon_1, X(\gamma) \in (K_n)^m. \tag{4.36}$$

For $\gamma \in W, i = 1, \ldots, m, j = 1, \ldots, n$ we set

$$v^{ij}(\gamma) = v_0^{ij} X_0^{ij}(X^{ij}(\gamma))^{-1}. \tag{4.37}$$

By (4.36), (4.34) and (4.35) we may assume without loss of generality that

$$V(\gamma) \in \mathcal{L}, ||V(\gamma)|| < \alpha_0 \text{ for each } \gamma \in W. \tag{4.38}$$

Since for each $\gamma \in W$ (4.32) holds with $\Delta = \Delta(\gamma)$ it follows from (4.32), (4.35), (4.37), (4.8), (4.36), (4.38) that for each $\gamma \in W$ the following relations hold:

$$(\alpha_0\hat{X}(\gamma) - V(\gamma) * \hat{X}(\gamma) - \hat{Z}_0)'(1, 1, \ldots, 1) = 0, \tag{4.39}$$

$$(\gamma \lambda_1^0 \hat{X}(\gamma)' - (V(\gamma) * \hat{X}(\gamma) + \hat{Z}_0)')\theta = 0,$$

$$(\alpha_0, X(\gamma), V(\gamma), Z_0) \in \mathcal{A}.$$

By (4.34), (4.35), (4.37), (4.39) there exists a neighborhood $W_1 \subset W$ of 1 in R^1 such that

$$W_1 \subset (1 - \delta M^{-1}, 1 + \delta M^{-1}), \qquad (4.40)$$

$(\alpha_0, X(\gamma), V(\gamma), Z_0) \in U_1 \cap \mathbf{B}((\alpha_0, X_0, V_0, Z_0), \epsilon_0)$ for each $\gamma \in W_1$.

It follows from the definition of ϵ_0, properties (i), (ii), (iii) which holds for each $(\alpha, X, V, Z) \in U_1 \cap \mathbf{B}((\alpha_0, X_0, V_0, Z_0), \epsilon)$ with $Q \equiv 1$, (4.40) and (4.22) that for each $\gamma \in W_1$,

$$m(\alpha_0, X(\gamma), V(\gamma), Z_0) < m(\alpha_0, X_0, V_0, Z_0).$$

This completes the proof of the lemma.

The next auxiliary result plays a crucial role in the proof of Theorem 10.1.1.

LEMMA 10.4.6 *Assume that*

$$(\alpha_0, X_0, V_0, Z_0) \in \mathcal{A}, \ \epsilon > 0, \ ||V_0|| < 1, \alpha_0,$$

$$P(\hat{X}_0, V_0 * \hat{X}_0 + \hat{Z}_0) \neq 0,$$

U_0 *is an open neighborhood of* $(\alpha_0, X_0, V_0, Z_0)$ *in* $\mathcal{A}$,

$$r_0 \in K_n, ||r_0|| = 1, (\alpha \hat{X}_0 - V_0 * \hat{X}_0 - \hat{Z}_0)r_0 = 0,$$

$$\{\theta \in \mathbf{C}^m : (\alpha_0 \hat{X}_0 - V_0 * \hat{X}_0 - \hat{Z}_0)'\theta = 0\} = \{\lambda(1, 1, \dots, 1) : \lambda \in \mathbf{C}^1\}.$$

Then there exists an open nonempty set $U_1 \subset U_0$ *such that for each* $(\alpha, X, V, Z) \in U_1$ *the following properties hold:*
   (a) $||V|| < \alpha, 1, P(\hat{X}, V * \hat{X} + \hat{Z}) \neq 0$, *there exists* $r \in K_n$ *for which*

$$||r|| = 1, ||r - r_0|| \le \epsilon, (\alpha \hat{X} - V * \hat{X} - \hat{Z})r = 0;$$

   (b) $\{\theta \in \mathbf{C}^m : (\alpha \hat{X} - V * \hat{X} - \hat{Z})'\theta = 0\}$

$$= \{\lambda(1, 1, \dots, 1) : \lambda \in \mathbf{C}^1\};$$

   (c) $\sigma(\hat{X}', (V * \hat{X} + \hat{Z})') \cap \{z \in \mathbf{C} : Re z = \alpha\} = \{\alpha\}.$

Proof. By Lemma 10.4.2 we may assume that for each $(\alpha, X, V, Z) \in U_0$ properties (a),(b) hold. By Lemma 10.4.3 there exists an open set

$U_2 \subset U_0$ such that for each $(\alpha, X, V, Z) \in U_2$ the vectors $X^i, i = 1, \ldots, m$ are linearly independent. There exists $(\alpha_1, X_1, V_1, Z_1) \in U_2$ such that

$$m(\alpha_1, X_1, V_1, Z_1) = \inf\{m(\alpha, X, V, Z) : (\alpha, X, V, Z) \in U_2\}. \quad (4.41)$$

By Lemma 10.4.4 there is an open neighborhood $U_3 \subset U_2$ of

$$(\alpha_1, X_1, V_1, Z_1)$$

in $\mathcal{A}$ and a number $M > 0$ such that for each $(\alpha, X, V, Z) \in U_3$ and each $\lambda \in \sigma(\hat{X}', (V * \hat{X} + \hat{Z})')$ the relation $|\lambda| \leq 2^{-1}M$ holds.

There exist integers $k \geq 1, m_i \geq 1, i = 1, \ldots, k$ and numbers $\lambda_i^1 \in \mathbf{C}$, $i = 1, \ldots, k$ such that

$$P(\hat{X}_1, V_1 * \hat{X}_1 + \hat{Z}_1)(\lambda) = \prod_{i=1}^{k} (\lambda - \lambda_i^1)^{m_i}, \lambda \in \mathbf{C},$$

$\lambda_1^1 = \alpha_1, \lambda_i^1 \neq \lambda_j^1$ for each $i, j \in \{1, \ldots, k\}$ satisfying $i \neq j$. It follows from Lemma 10.4.5 and (4.41) that

$$Re\lambda_i^1 \neq \alpha_1, i \in \{1, \ldots, k\} \setminus \{1\}. \quad (4.42)$$

There exists $\epsilon_1 \in (0, 2^{-1}\epsilon)$ such that

$$\mathbf{B}(\lambda_i^1, 4\epsilon_1) \cap \mathbf{B}(\lambda_j^1, 4\epsilon_1) = \emptyset \quad (4.43)$$

for each $i, j \in \{1, \ldots, k\}$ satisfying $i \neq j$,

$$|Re\lambda_i^1 - \alpha_1| \geq 8\epsilon_1, i \in \{1, \ldots, k\} \setminus \{1\}. \quad (4.44)$$

By Lemma 10.4.1, the definition of U_3, M, ϵ_1 (see (4.43),(4.44)) there exists an open neighborhood U_1 of $(\alpha_1, X_1, V_1, Z_1)$ in $\mathcal{A}$ such that

$$U_1 \subset U_3 \cap \mathbf{B}((\alpha_1, X_1, V_1, Z_1), 4^{-1}\epsilon_1) \quad (4.45)$$

and for each $(\alpha, X, V, Z) \in U_1$ the following properties hold:

(d) $P(\hat{X}, V * \hat{X} + \hat{Z})(\lambda) = \prod_{i=1}^{s} \prod_{j=q(i)}^{q(i+1)-1} (\lambda - \lambda_j), \lambda \in \mathbf{C}$ where $1 \leq s \leq k, q(1) = 1, q(i+1) > q(i), i = 1, \ldots, s;$

(e) there exists a bijection $\tau : \{1, \ldots, k\} \to \{1, \ldots, k\}$ such that $|\lambda_{\tau(i)}^1 - \lambda_{q(i)+j}| \leq \epsilon_1, 0 \leq j < q(i+1) - q(i) - 1, i = 1, \ldots, s, q(i+1) - q(i) \leq m_{\tau(i)}, i = 1, \ldots, s.$

Let $(\alpha, X, V, Z) \in U_1$. We will show that

$$m(\alpha, X, V, Z) = m(\alpha_1, X_1, V_1, Z_1).$$

It is sufficient to show that $m(\alpha, X, V, Z) \leq m(\alpha_1, X_1, V_1, Z_1)$. Let numbers $s, q(i), i = 1, \ldots, s$ and a bijection τ be as guaranteed in properties

(d), (e). By (4.8) $P(\hat{X}, V * \hat{X} + \hat{Z})(\alpha) = 0$. It follows from properties
(d),(e) that there exists $i_0 \in \{1, \ldots, k\}$ for which

$$|\lambda^1_{\tau(i_0)} - \alpha| \leq \epsilon_1. \tag{4.46}$$

(4.43), (4.45), (4.46) and (4.44) imply that

$$|\alpha - \alpha_1| \leq 4^{-1}\epsilon_1, \lambda^1_{\tau(i_0)} = \alpha_1, \tau(i_0) = 1. \tag{4.47}$$

Assume that $i \in \{1, \ldots, s\}$, an integer $t \in \{q(i), \ldots, q(i+1)-1\}$ and
$Re\lambda_t = \alpha$. By this relation and property (e), (4.47) $|Re\lambda^1_{\tau(i)} - \alpha| \leq$
$|\lambda^1_{\tau(i)} - \lambda_t| \leq \epsilon_1$, $|Re\lambda^1_{\tau(i)} - \alpha_1| \leq 2\epsilon_1$. Together with (4.44) and (4.47)
this implies that $\tau(i) = 1, i = i_0, \{t \in \{1, \ldots, q(s+1)-1\} : Re\lambda_t = \alpha\} \subset$
$[q(i_0), q(i_0+1)-1]$. It follows from these relations, (4.17), property (e),
(4.47) and (4.42) that $m(\alpha, X, V, Z) \leq q(i_0+1) - q(i_0) \leq m_{\tau(i_0)} = m_1 =$
$m(\alpha_1, X_1, V_1, Z_1)$. Therefore

$$m(\alpha, X, V, Z) = m(\alpha_1, X_1, V_1, Z_1) \text{ for each } (\alpha, X, V, Z) \in U_1. \tag{4.48}$$

Assume that there exists $(\alpha_2, X_2, V_2, Z_2) \in U_1$ which does not have
property (c). Then there exists $\lambda \in \sigma(\hat{X}'_2, (V_2 * \hat{X}_2 + \hat{Z}_2)')$ for which
$\lambda \neq \alpha_2, Re\lambda = \alpha_2$. By Lemma 10.4.5 there exists $(\alpha_3, X_3, V_3, Z_3) \in U_1$
such that $m(\alpha_3, X_3, V_3, Z_3) < m(\alpha_2, X_2, V_2, Z_2) = m(\alpha_1, X_1, V_1, Z_1)$ (see
(4.48)). This is contradictory to (4.41). The obtained contradiction
proves that each $(\alpha, X, V, Z) \in U_1$ has property (c). This completes the
proof of the lemma.

10.5.   Proof of Theorem 10.1.1

The next lemma plays an important role in the proof of Theorem
10.1.1. It allows us to construct an open everywhere dense subset of
$\mathcal{M} \times \mathcal{L}$ such that for each pair (a, V) belonging to this subset the sufficient
condition for the turnpike property holds.

LEMMA 10.5.1 *Assume that* $a_0 \in \mathcal{M}, V_0 \in \mathcal{L}, \alpha(a_0, V_0) > \|V_0\|,$

$$\{x \in (R^n_+)^m \setminus \{0\} : \alpha(a_0, V_0)x \in Q(a_0, V_0)(x)\} \subset (K_n)^m,$$

$$\{p \in (R^n_+)^m \setminus \{0\} : p \in Q(a_0, V_0)'(\alpha(a_0, V_0)p)\} \subset (K_n)^m.$$
Then there exists an open neighborhood U *of* (a_0, V_0) *in* $\mathcal{M} \times \mathcal{L}$ *such
that for each* $(a, V) \in U$ *the following properties hold:*
    (a) $\alpha(a, V) > \|V\|, \{p \in (R^n_+)^m \setminus \{0\} : p \in Q(a, V)'(\alpha(a, V)p)\} \subset$
$(K_n)^m;$

(b) there exists $X(a, V) \in (K_n)^m$ such that $||X(a, V)|| = 1$,

$$\{x \in (R_+^n)^m : \alpha(a, V)x \in Q(a, V)(x)\} = \{\lambda X(a, V) : \lambda \in [0, \infty)\};$$

(c) there exists a unique vector $Z(a, V) = (Z^1(a, V), \ldots, Z^m(a, V)) \in (R_+^n)^m$ such that $Z^i(a, V) \in a^i(X^i(a, V)), i = 1, \ldots, m,$
$\sum_{i=1}^m [(\alpha(a, V)I_n - V^i)X^i(a, V) - Z^i(a, V)] = 0;$
*(d) $\{\theta \in R^m : (\alpha(a, V)\hat{X}(a, V) - V * \hat{X}(a, V) - \hat{Z}(a, V))'\theta = 0\}$
$= \{\lambda(1, 1, \ldots, 1) : \lambda \in R^1\};$*
(e) there exists $r(a, V) \in K_n$ such that $||r(a, V)|| = 1$,

$$(\alpha(a, V)\hat{X}(a, V) - V * \hat{X}(a, V) - \hat{Z}(a, V))r(a, V) = 0,$$

$(r(a, V), \ldots, r(a, V)) \in Q(a, V)'(\alpha(a, V)(r(a, V), \ldots, r(a, V))$.
Moreover the functions $(a, V) \rightarrow \alpha(a, V), X(a, V), Z(a, V), ((a, V) \in U)$ are continuous.

Proof. Clearly there exists an open neighborhood U_0 of (a_0, V_0) in $\mathcal{M} \times \mathcal{L}$ such that for each $(a, V) \in U_0$ property (a) holds,

$$\{x \in (R_+^n)^m \setminus \{0\} : \alpha(a, V)(x) \in Q(a, V)(x)\} \subset (K_n)^m \qquad (5.1)$$

and moreover a function $(a, V) \rightarrow \alpha(a, V), (a, V) \in U_0$ is continuous.
 Let $(a, V) \in U_0, X \in (K_n)^m$,

$$||X|| = 1, \alpha = \alpha(a, V), \alpha X \in Q(a, V)(X). \qquad (5.2)$$

There exist $Z^i \in a^i(X^i), i = 1, \ldots, m$ such that

$$\sum_{i=1}^m ((\alpha I_n - V^i)X^i - Z^i)) = 0. \qquad (5.3)$$

Therefore

$$(\alpha\hat{X} - V * \hat{X} - \hat{Z})'\theta = 0 \qquad (5.4)$$

with $\theta = (1, 1, \ldots, 1)$.
 Assume that equation (5.4) has a solution which belongs to $R^m \setminus \{\lambda(1, 1, \ldots, 1) : \lambda \in R^1\}$. Then equation (5.4) has a solution $\theta \in R_+^m \setminus (\{0\} \cup K_m)$. This implies that

$$\alpha(\theta^1 X^1, \ldots, \theta^i X^i, \ldots, \theta^m X^m) \in Q(a, V)(\theta^1 X^1, \ldots, \theta^i X^i, \ldots, \theta^m X^m)$$

and $(\theta^i X^i)_{i=1}^m \notin (K_n)^m$. This contradicts (5.1). The contradiction we have reached proves that $\{\lambda(1, 1, \ldots, 1) : \lambda \in R^1\}$ is the set of all $\theta \in R^m$ which satisfy (5.4).

Let $p, Y \in (R_+^n)^m$ satisfy

$$||p|| = 1, \alpha Y \in Q(a, V)(Y), p \in Q(a, V)'(\alpha p). \qquad (5.5)$$

By Lemma 10.2.3, (5.5) and property (a), there exists $r \in K_n$ for which $p = (r, \ldots, r)$. It follows from Lemma 10.3.1 and (3.4) that there exists $\theta = (\theta^i)_{i=1}^m \in R_+^m$ which satisfies (5.4) and such that $Y^i = \theta^i X^i$, $i = 1, \ldots, m$. This implies that $\theta = \beta(1, 1, \ldots, 1)$ with some $\beta \geq 0$. Therefore $Y = \beta X$. We set

$$X(a, V) = X, \ Z(a, V) = Z, \ r(a, V) = r.$$

Clearly we have already established properties (a),(b),(d) for (a, V). Property (c) for (a, V) follows from relation (c) from Lemma 10.2.2 and property (ii) (see Section 10.1). Relations (c), (d) from Lemma 10.2.2 imply that property (e) holds for (a, V). It is easy now to see that the functions $(a, V) \to X(a, V), Z(a, V)$, $(a, V) \in U_0$ are continuous. This completes the proof of the lemma.

A function $\phi : R_+^n \to R_+^1$ is called superlinear if it is superadditive and positively homogeneous. A function $\phi : R_+^n \to R_+^1$ is called strictly superlinear if it is superlinear and relations $x \in K_n, y \in R_+^n, \phi(x + y) = \phi(x) + \phi(y)$ imply that $y \in \{\lambda x : \lambda \in [0, \infty)\}$.

A function $\phi : R_+^n \to R_+^1$ is called a CD-function if

$$\phi(x) = \gamma(x^1)^{\beta^1} \ldots (x^n)^{\beta^n}, \ x \in R_+^n$$

where $\gamma > 0, \beta \in K_n, \sum_{i=1}^n \beta^i = 1$.

A CD-function $\phi : R_+^n \to R_+^1$ is strictly superlinear [75] and for each $p, \theta \in K_n$ there exists a unique CD-function $\phi : R_+^n \to R_+^1$ such that $\phi'(\theta) = p$ (here $\phi'(\theta) = (\partial \phi / \partial x_i(\theta))_{i=1}^n)$. (see [86]).

For each number $\gamma > 0$ and each $\beta \in K_s$ satisfying $\sum_{i=1}^s \beta^i = 1$ we define a function $\phi(\gamma, \beta) : R^s \to R^1$ as

$$\phi(\gamma, \beta)(x) = \gamma(\sum_{i=1}^s \beta^i (x^i)^2)^{1/2}, x \in R^s. \qquad (5.6)$$

The function $\phi(\gamma, \beta)$ is positively homogeneous and subadditive.

Let $p, \theta \in K_s$. It is easy to see that there exist $\gamma > 0, \beta \in K_s$ such that $\sum_{i=1}^s \beta^i = 1$, $\phi(\gamma, \beta)'(\theta) = p$. For $s \in \{1, \ldots, m\}$ we set $L_s = \{x \in R^n : x^i = 0, i \in \{1, \ldots, n\} \setminus \mathcal{I}_s\}$.
Clearly L_s is isomorfic to $R^{q(s)}$ where $q(s) = \text{Card}(\mathcal{I}_s)$.

Proof of Theorem 10.1.1. Let $(a_0.V_0) \in \mathcal{M} \times \mathcal{L}$ and let W_0 be an open neighborhood of (a_0, V_0) in $\mathcal{M} \times \mathcal{L}$. In order to prove the theorem it is sufficient to show that there exists a nonempty open set $W \subset W_0$ such that each $(a, V) \in W$ satisfies the following conditions:

There exists a generalized equilibrium state

$$(\alpha(a, V), (X, \alpha(a, V)X), p)$$

with $\alpha(a, V) > ||V||, X, p \in (K_n)^m$;

the mapping $Q(a, V)$ has the turnpike property.

Lemma 10.2.6 implies that there exists $(a_1, V_1) \in W_0$ such that

$$1, \alpha(a_1, V_1) > ||V_1|| \tag{5.7}$$

and $\{x \in (R_+^n)^m \setminus \{0\} : \alpha(a_1, V_1)x \in Q(a_1, V_1)(x)\} \subset (K_n)^m$, and there exist $Y \in (K_n)^m, Z \in (R_+^n)^m$ such that $Z^i \in a^i(Y^i), i = 1, \ldots, m$, $Z^{ij} > 0$ for each $i \in \{1, \ldots, m\}, j \in \mathcal{I}_i$ and

$$\sum_{i=1}^{m} (\alpha(a_1, V_1)Y^i - V^i X^i - Z^i) = 0,$$

$$\{p \in (R_+^n) \setminus \{0\} : p \in Q(a_1, V_1)'(\alpha(a_1, V_1)p)\} \subset (K_n)^m.$$

By these relations and Lemma 10.5.1 there is an open neighborhood W_1 of (a_1, V_1) in $\mathcal{M} \times \mathcal{L}$ such that $W_1 \subset W_0$ and that for each $(a, V) \in W_1$ properties (a)-(e) from Lemma 10.5.1 hold.

For each $(a, V) \in W_1$ there exist

$$X(a, V) \in (K_n)^m, \ Z(a, V) \in (R_+^n)^m, \ r(a, V) \in K_n$$

guaranteed by properties (b),(c),(e) from Lemma 10.5.1. In view of Lemma 10.5.1 we may assume that the functions

$$(a, V) \to \alpha(a, V), \ X(a, V), \ Z(a, V), \ (a, V) \in W_1$$

are continuous. By (5.7) and properties (b), (c), (e) from Lemma 10.5.1, we may assume without loss of generality that for each $(a, V) \in W_1$,

$$Z^{sj}(a, V) > 0, s \in \{1, \ldots, m\}, \ j \in \mathcal{I}_s, ||V|| < 1. \tag{5.8}$$

Put

$$X_1 = X(a_1, V_1), \ Z_1 = Z(a_1, V_1), \ \alpha_1 = \alpha(a_1, V_1), \ r_1 = r(a_1, V_1). \tag{5.9}$$

Clearly $(\alpha_1, X_1, V_1, Z_1) \in \mathcal{A}$. We will show that $P(\hat{X}_1, V_1 * \hat{X}_1 + \hat{Z}_1) \neq 0$. Let us assume the converse. Then there exist a sequence of numbers $\{\lambda_s\}_{s=1}^{\infty} \subset R^1$ and a sequence $\{\theta_s\}_{s=1}^{\infty} \subset R^n \setminus \{0\}$ such that

$$\lambda_s > \lambda_{s+1} > \alpha_1, s = 1, 2, \ldots, \lambda_s \to \alpha_1 \text{ as } s \to \infty, \tag{5.10}$$

$$||\theta_s|| = 1, (\lambda_s \hat{X}_1 - V_1 * \hat{X}_1 - \hat{Z}_1)'\theta_s = 0, \; s = 1, 2, \ldots.$$

In view of (5.9) and property (d) from Lemma 10.5.1, we may assume that $\theta_s \to (1, 1, \ldots, 1)$ as $s \to \infty$. There exists an integer $q \geq 1$ such that $\theta_q \in K_m$.

Relations (5.9) and (5.10) imply that

$$\lambda_q((\theta_q^i X_1^i)_{i=1}^m) \in Q(a_1, V_1)((\theta_q^i X_1^i)_{i=1}^m),$$

$$\lambda_q \leq \alpha(a_1, V_1) = \alpha_1 < \lambda_q,$$

a contradiction. The contradiction we have reached proves that

$$P(\hat{X}_1, V_1 * \hat{X}_1 + \hat{Z}_1) \neq 0. \tag{5.11}$$

(5.8) implies that there is an open neighborhood U_∞ of $(\alpha_1, X_1, V_1, Z_1)$ in $\mathcal{A}$ such that for each $(\alpha, X, V, Z) \in U_\infty$,

$$Z^{si} > 0 \; (s = 1, \ldots, m, \; i \in \mathcal{I}_s), \; \alpha > ||V||. \tag{5.12}$$

Let $\epsilon \in (0, 1)$. Put

$$U(\epsilon) = \mathbf{B}((\alpha_1, X_1, V_1, Z_1), \epsilon) \cap U_\infty. \tag{5.13}$$

It follows from Lemma 10.4.6, properties (d), (e) from Lemma 10.5.1, (5.12), (5.11) and (5.8) that there exists a nonempty open set $U_1(\epsilon) \subset U(\epsilon)$ such that each $(\alpha, X, V, Z) \in U_1(\epsilon)$ has properties (a),(b),(c) from Lemma 10.4.6 with $r_0 = r_1$.

Fix $(\alpha_\epsilon, X_\epsilon, V_\epsilon, Z_\epsilon) \in U_1(\epsilon)$. By property (a) from Lemma 10.4.6 there exists $r_\epsilon \in K_n$ such that

$$||r_\epsilon|| = 1, ||r_\epsilon - r_1|| \leq \epsilon, (\alpha_\epsilon \hat{X} - V_\epsilon * \hat{X}_\epsilon - \hat{Z}_\epsilon)r_\epsilon = 0. \tag{5.14}$$

It follows from (5.12) and (5.13) that

$$Z_\epsilon^{si} > 0, \; s = 1, \ldots, m, \; j \in \mathcal{I}_s. \tag{5.15}$$

In view of (5.15) for $i \in \{1, \ldots, m\}$ there are diagonal matrices

$$D_1(i, \epsilon) \text{ and } D_2(i, \epsilon)$$

of the dimension $n \times n$ such that

$$D_1(i, \epsilon)X_\epsilon^i = X_1^i, \; D_2(i, \epsilon)Z_1^i = Z_\epsilon^i, \tag{5.16}$$

$D_2(i, \epsilon)y = y$ for each $y \in R^n$ satisfying $y^j = 0, \; j \in \mathcal{I}_i$.

For each $j \in \{1, \ldots, m\}$ there exist a function $\phi_\epsilon^j : L_j \to R^1$ of the form (5.6) and a CD-function $\psi_\epsilon^j : R_+^n \to R^1$ such that

$$(\phi_\epsilon^j)'(Z_\epsilon^j) \text{ is a projection of } r_\epsilon \text{ to } L_j, \tag{5.17}$$

$$(\psi_\epsilon^j)'(X_\epsilon^j) = < r_\epsilon, Z_\epsilon^j >^{-1} (\alpha_\epsilon I_n - V_\epsilon^j) r_\epsilon.$$

Let $\gamma, \Gamma \in (0, 1)$. For $i = 1, \ldots, m$, $x \in R_+^n$ define

$$[a^i(\epsilon)](x) = \{u \in D_2(i, \epsilon) a_1^i(D_1(i, \epsilon)x) : < (\alpha_\epsilon I_n - V_\epsilon^i) r_\epsilon, x > \geq < r_\epsilon, u >\}, \tag{5.18}$$

$$[a^i(\epsilon, \gamma)](x) = \{u \in D_2(i, \epsilon) a_1^i(D_1(i, \epsilon)x) :$$
$$< (\alpha_\epsilon I_n - V_\epsilon^i) r_\epsilon, x > \geq \gamma < r_\epsilon, u >$$
$$+ (1 - \gamma)\phi_\epsilon^i(u)\}, [a^i(\epsilon, \gamma, \Gamma)](x)$$
$$= [\Gamma a^i(\epsilon, \gamma)](x) + (1 - \Gamma)\psi_\epsilon^i(x)\{u : 0 \leq u \leq Z_\epsilon^i\}.$$

It is not difficult to see that the mappings $a^i(\epsilon), a^i(\epsilon, \gamma), a^i(\epsilon, \gamma, \Gamma)$, $i = 1, \ldots, m$ are normal convex processes and their graphs are closed. We will show that $a(\epsilon, \gamma, \Gamma) = (a^i(\epsilon, \gamma, \Gamma))_{i=1}^m \in \mathcal{M}$.

It is sufficient to show that for each $s \in \{1, \ldots, m\}$ the mapping $a^s(\epsilon, \gamma, \Gamma)$ has property (ii) (see Section 10.1).

Suppose that $x, g \in K_n$ and $i \in \{1, \ldots, m\}$. It is sufficient to show that the problem $< g, u > \to \max, u \in [a^i(\epsilon, \gamma)](x)$ has a unique solution. Let

$$\tau = \sup\{< g, u > : u \in [a^i(\epsilon, \gamma)](x)\}, \tag{5.19}$$

$$\tau_0 = \sup\{< g, u > : u \in D_2(i, \epsilon) a_1^i(D_1(i, \epsilon)x)\},$$

$$u_0 \in D_2(i, \epsilon) a_1^i(D_1(i, \epsilon)x), < g, u_0 > = \tau_0.$$

We may assume that $\tau < \tau_0$. Let $u_1, u_2 \in [a^i(\epsilon, \gamma)](x), u_1 \neq u_2$, $< g, u_s > = \tau, s = 1, 2$. Set $u_3 = 2^{-1}(u_1 + u_2)$. It is easy to see that $u_3 \in [a^i(\epsilon, \gamma)](x), < g, u_3 > = \tau$,

$$\gamma < r_\epsilon, u_3 > + (1 - \gamma)\phi_\epsilon^i(u_3) << (\alpha_\epsilon I_n - V_\epsilon^i) r_\epsilon, x > .$$

For $t \in (0, 1)$ we define $u_t = tu_3 + (1 - t)u_0$. It is easy to verify that if $|t - 1|$ is small enough then $u_t \in [a^i(\epsilon, \gamma)](x), < g, u_t >> \tau$. This contradicts (5.19). The contradiction we have reached proves that $a^i(\epsilon, \gamma, \Gamma)$ has property (ii). Therefore $a(\epsilon, \gamma, \Gamma) \in \mathcal{M}$.

For each $\epsilon, \gamma, \Gamma \in (0, 1)$ we have defined $a(\epsilon, \gamma, \Gamma) \in \mathcal{M}$, an open nonempty set $U_1(\epsilon) \subset \mathbf{B}((\alpha_1, X_1, V_1, Z_1), \epsilon) \cap U_\infty$, $(\alpha_\epsilon, X_\epsilon, V_\epsilon, Z_\epsilon) \in U_1(\epsilon)$ and $r_\epsilon \in K_n$. Let $\delta > 0$. We will show that there exist $\epsilon \in (0, \delta), \gamma, \Gamma \in (0, 1)$ such that

$$d(a^i(\epsilon, \gamma, \Gamma), a_1^i) < \delta, \; i = 1, \ldots, m.$$

Choose $\lambda > 1$ such that

$$(\lambda^2 - 1) \sup\{\|y\| : y \in a_1^i(x), \ x \in R_+^n, \quad (5.20)$$

$$\|x\| \leq 1, i = 1, \ldots, m\} < 3^{-1}\delta.$$

It follows from (5.12), (5.14), (5.16) and the definition of $(\alpha_\epsilon, X_\epsilon, V_\epsilon, Z_\epsilon)$ that there is a positive number $\epsilon < \delta$ such that

$$\lambda^{-1} r_1 \leq r_\epsilon \leq \lambda r_1, \ \lambda^{-1} I_n \leq D_1(i, \epsilon), D_2(i, \epsilon) \leq \lambda I_n, \quad (5.21)$$

$$\lambda^{-1}(\alpha_1 I_n - V_1^i) r_1 \leq (\alpha_\epsilon I_n - V_\epsilon^i) r_\epsilon \leq \lambda(\alpha_1 I_n - V_1^i) r_1, \ i = 1, \ldots, m.$$

We will show that

$$d(a^i(\epsilon), a_1^i) < 3^{-1}\delta, \ i = 1, \ldots, m.$$

Let $x \in R_+^n, i \in \{1, \ldots, m\}, \ y \in a_1^i(x)$. In view of (5.9), Lemma 10.2.3 and property (e) from Lemma 10.5.1,

$$< (\alpha_1 I_n - V_1^i) r_1, x > \geq < r_1, y > . \quad (5.22)$$

(5.21) implies that

$$D_2(i, \epsilon) a_1^i (D_1(i, \epsilon) x) \supset \lambda^{-2} a_1^i(x) \ni \lambda^{-2} y. \quad (5.23)$$

The inequalities (5.21) and (5.22) imply that

$$< (\alpha_\epsilon I_n - V_\epsilon^i) r_\epsilon, x >$$

$$\geq \lambda^{-1} < (\alpha_1 I_n - V_1^i) r_1, x >$$

$$\geq \lambda^{-1} < r_1, y > \geq \lambda^{-2} < r_\epsilon, y > .$$

Combined with (5.23) and (5.18) this implies that

$$\lambda^{-2} y \in [a^i(\epsilon)](x), \lambda^{-2} a_1^i(x) \subset [a^i(\epsilon)](x). \quad (5.24)$$

It follows from (5.18) and (5.21) that

$$[a^i(\epsilon)](x) \subset D_2(i, \epsilon) a_1^i (D_1(i, \epsilon) x) \subset \lambda^2 a_1^i(x).$$

Combined with (5.20) and (5.24) this inclusion implies that

$$d(a^i(\epsilon), a_1^i) < 3^{-1}\delta, \ i = 1, \ldots, m. \quad (5.25)$$

We will show that there exists $\gamma \in (0, 1)$ for which

$$d(a^i(\epsilon), a^i(\epsilon, \gamma)) < 3^{-1}\delta, \ i = 1, \ldots, m. \quad (5.26)$$

It follows from the definition of ϕ_ϵ^i (see (5.17)), (5.18) that for $i \in \{1, \ldots, m\}$, $x \in R^n$, numbers γ_1, γ_2 satisfying $0 < \gamma_1 < \gamma_2 < 1$ the following inclusion holds:

$$[a^i(\epsilon, \gamma_1)](x) \subset a^i(\epsilon, \gamma_2)(x) \subset a^i(\epsilon)(x).$$

On the other hand for each

$$x \in R_+^n, \ i \in \{1, \ldots, m\}, \ y \in [a^i(\epsilon)](x), \ \lambda_* \in (0, 1)$$

there exists $\gamma \in (0, 1)$ such that $\lambda_* y \in a^i(\epsilon, \gamma)(x)$. This implies the existence of $\gamma \in (0, 1)$ for which (5.26) is valid.

It is easy to see that there exists $\Gamma \in (0, 1)$ such that

$$d(a^i(\epsilon, \gamma, \Gamma), a^i(\epsilon, \gamma)) < 3^{-1}\delta, \ i = 1, \ldots, m.$$

In view of this inequality, (5.25) and (5.26),

$$d(a_1^i, a^i(\epsilon, \gamma, \Gamma)) < \delta, \ i = 1, \ldots, m. \tag{5.27}$$

We have shown that for each $\delta \in (0, 1)$ there exist $\epsilon \in (0, \delta), \gamma, \Gamma \in (0, 1)$ such that (5.27) holds. Therefore there exists $\epsilon_0, \Gamma_0, \gamma_0 \in (0, 1)$ such that

$$(a(\epsilon_0, \gamma_0, \Gamma_0), V_{\epsilon_0}) \in W_1. \tag{5.28}$$

We will show that

$$\alpha_{\epsilon_0} X_{\epsilon_0} \in Q(a(\epsilon_0, \gamma_0, \Gamma_0), V_{\epsilon_0}) X_{\epsilon_0}, \alpha_{\epsilon_0} = \alpha(a(\epsilon_0, \gamma_0, \Gamma_0), V_{\epsilon_0}). \tag{5.29}$$

It follows from (5.17), (5.18) and Lemma 10.2.1 that

$$(r_{\epsilon_0}, \ldots, r_{\epsilon_0}) \in Q(a(\epsilon_0, \gamma_0, \Gamma_0), V_{\epsilon_0})'(\alpha_{\epsilon_0}(r_{\epsilon_0}, \ldots, r_{\epsilon_0})). \tag{5.30}$$

By (5.16), property (c) from Lemma 10.5.1, (5.9), (5.17), (5.14) and (5.18) for $i \in \{1, \ldots, m\}$,

$$Z_{\epsilon_0}^i \in D_2(i, \epsilon_0) a_1^i (D_1(i, \epsilon_0) X_{\epsilon_0}^i), \quad \gamma_{\epsilon_0} < r_{\epsilon_0}, Z_{\epsilon_0}^i > \tag{5.31}$$

$$+(1 - \gamma_{\epsilon_0}) \phi_{\epsilon_0}^i (Z_{\epsilon_0}^i) = \gamma_{\epsilon_0} < r_{\epsilon_0}, Z_{\epsilon_0}^i > +(1 - \gamma_{\epsilon_0}) << (\phi_{\epsilon_0}^i)'(Z_{\epsilon_0}^i), Z_{\epsilon_0}^i >$$

$$=< r_{\epsilon_0}, Z_{\epsilon_0}^i >=< (\alpha_{\epsilon_0} I_n - V_{\epsilon_0}^i) r_{\epsilon_0}, X_{\epsilon_0}^i >, Z_{\epsilon_0}^i \in a^i(\epsilon_0, \gamma_0)(X_{\epsilon_0}^i).$$

(5.31), (4.8) and (5.30) implies (5.29). Since

$$(\alpha_{\epsilon_0}, X_{\epsilon_0}, V_{\epsilon_0}, Z_{\epsilon_0}) \in \mathcal{A}$$

it follows from (5.29), (5.28) and properties (b),(c) from Lemma 10.5.1 which holds for each $(a, V) \in W_1$ that

$$\|X_{\epsilon_0}\|^{-1} X_{\epsilon_0} = X(a(\epsilon_0, \gamma_0, \Gamma_0), V_{\epsilon_0}), \|X_{\epsilon_0}\|^{-1} Z_{\epsilon_0} = Z(a(\epsilon_0, \gamma_0, \Gamma_0), V_{\epsilon_0}). \tag{5.32}$$

Note that the functions $(a, V) \rightarrow X(a, V), Z(a, V), \alpha(a, V), (a, V) \in W_1$ are continuous and properties (b),(c) from Lemma 10.5.1 hold for each $(a, V) \in W_1$. Therefore there exists an open neighborhood W of $(a(\epsilon_0, \gamma_0, \Gamma_0), V_{\epsilon_0})$ in $\mathcal{M} \times \mathcal{L}$ such that $W \subset W_1$ and for each $(a, V) \in W$,

$$(\alpha(a, V), \|X_{\epsilon_0}\| X(a, V), V, \|X_{\epsilon_0}\| Z(a, V)) \in U_1(\epsilon_0). \tag{5.33}$$

Let $(a, V) \in W$. It follows from the definition of W, W_1 that properties (a)-(e) from Lemma 10.5.1 hold for (a, V) and (5.33) holds. By properties (a),(d) from Lemma 10.5.1,

$$\alpha(a, V) > \|V\|, \{\theta \in R^m : (\alpha(a, V)\hat{X}(a, V) - V * \hat{X}(a, V) \tag{5.34}$$

$$-\hat{Z}(a, V))'\theta = 0\} = \{\lambda(1, 1, \dots, 1) : \lambda \in R^1\}.$$

It follows from (5.33), the definition of $U_1(\epsilon_0)$ and property (c) from Lemma 10.4.6 which holds with

$$(\alpha, X, V, Z) = (\alpha(a, V), \|X_{\epsilon_0}\| X(a, V), V, \|X_{\epsilon_0}\| Z(a, V))$$

that

$$\sigma(\hat{X}'(a, V), (V * \hat{X}(a, V) + \hat{Z}(a, V))') \cap \{z \in \mathbf{C} :$$

$$Rez = \alpha(a, V)\} = \{\alpha(a, V)\}.$$

By this relation,(5.34), properties (a)-(e) from Lemma 10.5.1 which hold for (a, V), Lemma 10.3.2, the mapping $Q(a, V)$ has the turnpike property. This completes the proof of the theorem.

10.6.   Example

In this section we will construct $(a, V) \in \mathcal{M} \times \mathcal{L}$ such that there exists an equilibrium state $(\alpha(a, V), (X, \alpha(a, V)X), p)$ of $Q(a, V)$ satisfying $\alpha(a, V) > \|V\|, X, p \in (K_n)^m$ and such that $Q(a, V)$ does not have the turnpike property. We recall that $Rez = x, Imz = y$ for $z = x + iy \in \mathbf{C}$.

PROPOSITION 10.6.1 *Let*

$$V \in \mathcal{L}, \ a = (a^1, \dots, a^m) \in \mathcal{M}, \ Q = Q(a, V),$$

$$\alpha = \alpha(a, V) > \|V\|, \ X \in (K_n)^m, \ r \in K_n,$$

$$p = (r, \dots, r) \in (K_n)^m,$$

$$\alpha X \in Q(X), \ p \in Q'(\alpha p), \ Z^i \in a^i(X^i), i = 1, \dots, m, \tag{6.1}$$

$$\sum_{i=1}^{m}((\alpha I_n - V^i)X^i - Z^i) = 0.$$

Assume that there exist $\theta \in \mathbf{C}$, $\lambda \in \mathbf{C}^m \setminus \{0\}$ *such that*

$$Re\theta = \alpha, \ \theta \neq \alpha, \ \theta\hat{X}'\lambda = (V * \hat{X} + \hat{Z})'\lambda. \tag{6.2}$$

Then Q does not have the turnpike property.

Proof. Set $Re\lambda = (Re\lambda^j)_{j=1}^m$, $Im\lambda = (Im\lambda^j)_{j=1}^m$. For each $\gamma \in (0,1)$ we define $y_\gamma = (y_\gamma^j)_{j=1}^m : [0,\infty) \to (R_+^n)^m$, $u_\gamma = (u_\gamma^j)_{j=1}^m : [0,\infty) \to (R_+^n)^m$ as follows:

$$y_\gamma^j(t) = [\exp((\alpha-1)t) + \gamma Re(\exp((\theta-1)t)\lambda^j)]X^j, \tag{6.3}$$

$$u_\gamma^j(t) = [\exp((\alpha-1)t) + \gamma Im(\exp((\theta-1)t)\lambda^j)]X^j,$$

$$t \in [0,\infty), j = 1, \ldots, m.$$

It is easy to verify that for each small enough $\gamma > 0$ the functions y_γ, u_γ are trajectories of Q which have the von Neumann growth rate.

Assume that the mapping Q has the turnpike property. Then for small enough $\gamma > 0$ there exist numbers $d_\gamma^1, d_\gamma^2 > 0$ such that $\lim_{t\to\infty} \exp((1-\alpha)t)y_\gamma(t) = d_\gamma^1 X$, $\lim_{t\to\infty} \exp((1-\alpha)t)u_\gamma(t) = d_\gamma^2 X$. Together with (6.3) this implies that for $j = 1, \ldots, m$,

$\lim_{t\to\infty} Re(\exp((\theta-1-\alpha)t)\lambda^j) = \gamma^{-1}(d_\gamma^1 - 1)$,

$\lim_{t\to\infty} Im(\exp((\theta-1-\alpha)t)\lambda^j) = \gamma^{-1}(d_\gamma^2 - 1)$,

$\lim_{t\to\infty}(\exp((\theta-1-\alpha)t)\lambda^j) = \gamma^{-1}(d_\gamma^1 - 1) + i\gamma^{-1}(d^2\gamma - 1)$.

This is contradictory to (6.2). The obtained contradiction proves that Q does not have the turnpike property. The proposition is proved.

Assume that $m, n = 3$,

$$\mathcal{P} = \emptyset, \ \mathcal{I}_s = \{s\}, \ s = 1, \ldots, n,$$

$e_1 = (1,0,0), e_2 = (0,1,0), e_3 = (0,0,1)$. Assume that a number $\beta > 0$ is large enough. We set

$$c_\beta^{ij} = (3+\beta^2)/(3(1+\beta^2)),$$

$$(i,j) \in \{(1,1),(2,2),(3,3),(3,2),(1,2),(2,3),(2,1)\}.$$

It is easy to see that

$$4^{-1}[2c_\beta^{11} + ((c_\beta^{11})^2(\beta^2+1))^{-1} - 3(c_\beta^{11}(1+\beta^2))^{-1}]^2$$

$$> (c_\beta^{11})^2 - 3(1+\beta^2)^{-1}. \tag{6.4}$$

This implies that there exist $c_\beta^{13}, c_\beta^{31} > 0$ which satisfy

$$c_\beta^{31} c_\beta^{13} = (c_\beta^{11})^2 - 3(1+\beta^2)^{-1}, c_\beta^{13} + c_\beta^{31} \tag{6.5}$$

$$= 2c_\beta^{11} + [(\beta^2 + 1)(c_\beta^{11})^2]^{-1}$$

$$-3[c_\beta^{11}(1+\beta^2)]^{-1}.$$

Consider the matrix $C_\beta = (c_\beta^{ij}), i, j = 1, 2, 3$. It follows from (6.4), (6.5) that the polynomial $\det(C_\beta - \theta I_3), \theta \in \mathbf{C}$ has roots $1, (1 - i\beta)(1 + \beta^2)^{-1}, (1 + i\beta)(1 + \beta^2)^{-1}$. This implies that the polynomial $\det(\theta C_\beta' - I_3), \theta \in \mathbf{C}$ has roots $1, 1 + i\beta, 1 - i\beta$. There exist $\Lambda, r \in K_3$ such that

$$C_\beta' \Lambda = \Lambda, C_\beta r = r. \tag{6.6}$$

Set

$$X^i = \Lambda^i(c_\beta^{i1}, c_\beta^{i2}, c_\beta^{i3}), \; i = 1, 2, 3. \tag{6.7}$$

For $i = 1, 2, 3$ there exists a CD-function $\phi^i : R_+^3 \to R_+^1$ such that

$$(\phi^i)'(c_\beta^{i1}, c_\beta^{i2}, c_\beta^{i3}) = (r^i)^{-1} r. \tag{6.8}$$

Set $V = 0, a^i(x) = [0, \phi^i(x)]e_i, i = 1, 2, 3, x \in R_+^n$. It is easy to verify that

$$(a, V) \in \mathcal{M} \times \mathcal{L}, X \in Q(a, V)(X). \tag{6.9}$$

By Lemma 10.2.1 and (6.8)

$$(r, r, r) \in Q(a, V)'(r, r, r). \tag{6.10}$$

(6.10), (6.9) and (6.7) imply that $\alpha(a, V) = 1$. Since the polynomial $\det(\theta C_\beta' - I_3), \theta \in \mathbf{C}$ has a root $1 + i\beta$ there exists $\lambda = (\lambda^1, \lambda^2, \lambda^3) \in \mathbf{C}^3 \setminus \{0\}$ for which $(1 + i\beta)C_\beta' \lambda = \lambda$. Together with (6.7) this implies that

$$(1 + i\beta)\hat{X}'(\lambda^i(\Lambda^i)^{-1})_{i=1}^3 = (1 + i\beta) \sum_{j=1}^3 X^j(\lambda^j(\Lambda^j)^{-1})$$

$$= (1 + i\beta) \sum_{j=1}^3 c_\beta^j \lambda^j = (1 + i\beta)C_\beta' \lambda = \lambda = \hat{Z}'(\lambda^i(\Lambda^i)^{-1})_{i=1}^3$$

where $Z^i = \phi^i(X^i)e_i = \Lambda^i e_i, i = 1, 2, 3$. Together with (6.9), (6.10) and Proposition 10.6.1 this implies that $Q(a, V)$ does not have the turnpike property.

Chapter 11

CONVEX PROCESSES

In this chapter we study the dynamic properties of optimal trajectories of convex processes $G : K \to 2^{R^n}$ where $K \subset R^n$ is a closed convex cone. We show that optimal trajectories of a convex process G spend most of the time in a small neighborhood of a von Neumann path. The turnpike theorem that we obtain generalizes the result of Rubinov [75] which was established for a convex process $G : R_+^n \to 2^{R_+^n}$ where R_+^n is the cone of the elements of the Euclidean space R^n that have nonnegative coordinates. Also, we show that the turnpike phenomenon is stable under small perturbations of the convex process G.

11.1. Preliminaries

In this chapter we investigate the turnpike property for optimal trajectories of convex processes.

For each metric space Q denote by $\Pi(Q)$ the collection of nonempty compact subsets of Q endowed with the Hausdorff metric $d(\cdot, \cdot)$.

Let $K \subset R^n$ be a closed convex cone $(K + K \subset K, \lambda K \subset K, \lambda > 0)$ with nonempty interior in R^n which does not contain any nontrivial subspace of R^n. A set-valued mapping $G : K \to \Pi(R^n)$ is a convex process if it satisfies

$$\lambda G(x) + \mu G(y) \subset G(\lambda x + \mu y)$$

for each $x, y \in K$ and each $\lambda, \mu \geq 0$, or equivalently, if its graph is a convex cone.

Convex processes are the set-valued analogs of linear mappings and certain properties of the latter can be also established for the former [48, 71, 75, 77].

Denote by $\mathcal{M}$ the set of all convex processes $G : K \to \Pi(R^n)$ such that the set

$$\mathrm{graph}(G) = \{(x, v) \in R^n \times R^n : x \in K, v \in G(x)\}$$

is closed and $G(x) \cap K \neq \emptyset$ for each $x \in K$. Denote by $(\cdot, \cdot)$ the scalar product in R^n and by $|\cdot|$ the Euclidean norm in R^n. We say that $x \leq y$ $(x, y \in R^n)$ if and only if $y - x \in K$. Set

$$K^* = \{\eta \in R^n : (\eta, x) \geq 0 \text{ for every } x \in K\}.$$

It is a standard result (see [43]) that K^* has a nonempty interior. The interior of a set S is denoted by $\mathrm{int}S$.

Let $G \in \mathcal{M}$. A real number λ_0 is called an eigenvalue of G if there exists $x_0 \in K \setminus \{0\}$ such that $\lambda_0 x_0 \in G(x_0)$. We then call x_0 an eigenvector of G. A sequence $\{x_t\}_{t=0}^T \subset K$ where $T \in \{1, 2, \ldots\} \cup \{\infty\}$ is called a trajectory of G if $x_{t+1} \in G(x_t)$ for all nonnegative integers $t < T$.

In this chapter we study the dynamics properties of optimal trajectories of a convex process $G \in \mathcal{M}$. We show that optimal trajectories of G spend most of the time in a small neighborhood of the set $\{\lambda X : \lambda \in [0, \infty)\}$ where X is a unique (up to scalar multiplication) eigenvector corresponding to the maximal eigenvalue of G.

The turnpike theorem that we prove generalizes the result of Rubinov [75] which was established for a convex process $G : R_+^n \to \Pi(R_+^n)$ where R_+^n is the cone of the elements of the Euclidean space R^n that have nonnegative coordinates.

The results of this chapter were obtained in [101].

The chapter is organized as follows. In Section 11.2 we discuss a sufficient condition for the asymptotic turnpike property established by Leizarowitz [43]. Section 11.3 contains statements of our turnpike results which are proved in Section 11.4. In Section 11.5 we discuss the stability of the turnpike phenomenon and state our stability results which are proved in Section 11.6.

11.2. Asymptotic turnpike property

Let K be a convex closed cone with nonempty interior in R^n which does not contain any nontrivial subspace of R^n. Consider a convex process $G : K \to \Pi(R^n)$ which is a continuous mapping. It may happen that there are no eigenvalues of G in K at all. However, it is easy to see that the set of eigenvalues of G in K is compact. Denote by α_G the maximal eigenvalue of G. Set

$$\alpha_0 = \inf_{\eta \in K^*} \sup_{x \in K} \{(\eta, v)(\eta, x)^{-1} : v \in G(x)\}. \tag{2.1}$$

Leizarowitz [43] studied the existence of eigenvectors of convex processes and established the following result.

PROPOSITION 11.2.1 *Suppose that if* $x \in \partial K \setminus \{0\}$ *and* $\eta \in \partial K^* \setminus \{0\}$ *are such that* $(\eta, x) = 0$, *then* $(\eta, v) > 0$ *for some* $v \in G(x)$. *Then the infimum in (2.1) is attained inside* $\text{int} K^*$ *and there is* $x_0 \in K \setminus \{0\}$ *for which* $\alpha_0 x_0 \in G(x_0)$ *and* $\alpha_0 = \alpha_G$.

Assume that $(X, \alpha_G X) \in G$. We say that G has the asymptotic turnpike property if for any trajectory $\{x_t\}_{t=1}^{\infty}$ of G there exists a number $c_0 \geq 0$ such that

$$\lim_{t \to \infty} \alpha_G^{-t} x_t = c_0 X.$$

The asymptotic turnpike property for convex processes was studied in [48, 75, 86].

The following result which is a discrete-time analog of Theorem 6.3 in [43] gives the necessary and sufficient condition for the asymptotic turnpike property.

THEOREM 11.2.1 *Suppose that there is* $y \in K \setminus \{0\}$ *for which* $a(y) \cap (K \setminus \{0\}) \neq \emptyset$ *and if* $x \in \partial K \setminus \{0\}$, $\eta \in \partial K^* \setminus \{0\}$ *are such that* $(\eta, x) = 0$, *then* $(\eta, v) > 0$ *for some* $v \in G(x)$. *Assume that* λ_0, *the maximal eigenvalue of* G, *corresponds to a unique (up to multiplication by positive numbers) eigenvector denoted* X. *Then for every trajectory* $\{x_t\}_{t=0}^{\infty}$ *of* G *there exists a number* $c_0 \geq 0$ *such that* $\lim_{t \to \infty} \lambda_0^{-t} x_t = c_0 X$ *if and only if for each trajectory* $\{z_t\}_{t=0}^{\infty}$ *of* G *of the form*

$$z_t = \lambda_0^t (\cos(\omega t)a + \sin(\omega t)b + X), t = 0, 1, \ldots$$

for some $a, b \in R^n$, $\omega \in R^1$ *the following relation holds:* $z_t = c_0 \lambda_0 X, t = 0, 1, \ldots$ *with some constant* $c_0 \geq 0$.

The proof of Theorem 11.2.1 is analogous to the proof of Theorem 6.3 in [43]. One of the key ingredients in this proof is provided by the following result which is a discrete-time analog of Theorem 3.2 in [38].

PROPOSITION 11.2.2 *Let* $F \subset R^n \times R^n$ *be a convex compact set,*

$$domF = \{x \in R^n : \text{ there exists } y \in R^n \text{ such that } (x, y) \in F\},$$

$F(x) = \{y \in R^n : (x, y) \in F\}, x \in domF$. *Assume that the following properties hold:*

    (i) there is a unique $Y \in domF$ *such that* $Y \in F(Y)$;
    (ii) if $a, b \in R^n$, $\alpha \neq 0$ *and a sequence* $z_j = Y + \cos(\alpha j)a + \sin(\alpha j)b$, $j = 0, 1 \ldots$ *satisfies* $(z_j, z_{j+1}) \in F, j = 0, 1, \ldots$, *then* $z_j = Y, j = 0, 1, \ldots$.

Then for each $\epsilon > 0$ there exists an integer $N(\epsilon) \geq 1$ such that for each sequence $\{z_j\}_{j=0}^{\infty} \subset R^n$ which satisfies $(z_j, z_{j+1}) \in F, j = 0, 1, \ldots,$ the relation $|z_j - Y| \leq \epsilon$ holds for all $j \geq N(\epsilon)$.

11.3.   Turnpike theorems

Let $G \in \mathcal{M}$. For each $x \in K$ and each integer $T \geq 1$ denote

$$a^T(x) = \{y \in R^n : \text{ there exists a trajectory } \{x_t\}_{t=0}^{T} \text{ of } G$$

$$\text{for which } x_0 = x, x_T = y\}.$$

Evidently $a^T(x)$ is a convex compact set for each $x \in K$ and each integer $T \geq 1$.

Let T be a natural number. A trajectory $\{x_t\}_{t=0}^{T}$ of G is optimal if there exists $f \in K^* \setminus \{0\}$ such that $(f, x_T) = \sup\{(f, y) : y \in a^T(x)\}$.

For $x \in R^n, r > 0$ set $B(x, r) = \{y \in R^n : |x - y| \leq r\}$.

Suppose that

$$G \in \mathcal{M}, \alpha_G > 0, X \in intK, p \in intK^*, |X| = 1, \alpha_G X \in G(X), \quad (3.1)$$

$$\alpha_G \text{ is a maximal eigenvalue of } G \text{ and}$$

$$(\alpha_G p, x) \geq (p, y) \text{ for each } x \in K \text{ and each } y \in G(x).$$

We can easily prove the following result.

PROPOSITION 11.3.1 *Assume that $x_0 \in intK$ and $\{x_t\}_{t=0}^{T}$ is a trajectory of $G, T \geq 1$. Then this trajectory is optimal if and only if for each $\lambda > 1$ the relation $\lambda x_T \notin a^T(x_0) - K$ holds.*

Suppose that for each trajectory $\{x_t\}_{t=0}^{\infty} \subset G$ there exists a number $c_0 \geq 0$ such that

$$\alpha_G^{-t} x_t \to c_0 X \text{ as } t \to \infty. \qquad (3.2)$$

For $x \in K$ we define

$$\lambda(x) = \sup\{\beta \in [0, \infty) : \text{ there exists a trajectory } \{x_t\}_{t=0}^{\infty} \text{ of } G \quad (3.3)$$

$$\text{for which } x_0 = x, \lim_{t \to \infty} \alpha_G^{-t} x_t = \beta X\}.$$

It is easy to verify that for each $x \in K$ the number $\lambda(x)$ is finite and there exists a trajectory $\{x_t\}_{t=0}^{\infty}$ of G for which $x_0 = x$ and $\alpha_G^{-t} x_t \to \lambda(x)X$ as $t \to \infty$. It is easy to see that $\lambda(cx) = c\lambda(x)$ for each $x \in K, c > 0$, each $x \in intK$ is a continuity point of the function $\lambda : K \to [0, \infty)$ and $\lambda(x) \leq \lambda(y)$ for each $x, y \in K$ satisfying $x \leq y$.

We will establish the following results which describe the structure of optimal trajectories of G.

THEOREM 11.3.1 *Assume that a compact nonempty set* $Q \subset \text{int } K$ *and* $\epsilon > 0$. *Then there exists an integer* $L \geq 1$ *such that for each integer* $T \geq 2L$ *and each optimal trajectory* $\{x_t\}_{t=0}^T$ *of* G *satisfying* $x_0 \in Q$ *the following relation holds:*

$$|\lambda(x_0)X - \alpha_G^{-t}x_t| \leq \epsilon, \ t = L, \ldots, T - L.$$

THEOREM 11.3.2 *Let* $\epsilon > 0$. *Then there exist an integer* $L \geq 1$ *and a number* $\delta > 0$ *such that for each integer* $T \geq L$ *and each optimal trajectory* $\{x_t\}_{t=0}^T$ *of* G *satisfying* $|x_0 - X| \leq \delta$ *the following relation holds:*

$$|\alpha_G^{-t}x_t - X| \leq \epsilon, \ t = 0, \ldots, T - L.$$

11.4.   Proofs of Theorems 11.3.1 and 11.3.2

There exists $r_0 > 0$ such that

$$B(X, r_0) \subset K, B(p, r_0) \subset K^*. \tag{4.1}$$

The next auxiliary result shows that if $\{x_t\}_{t=0}^T$ is a trajectory of G and x_0, x_T are close to X, then x_t is close to X for all $t = 0, \ldots, T$.

LEMMA 11.4.1 *Let* $\alpha_G = 1$ *and* ϵ *be a positive number. Then there exists* $\delta > 0$ *such that if an integer* $T \geq 2$ *and if a trajectory* $\{x_t\}_{t=0}^T$ *of* G *satisfies* $x_0, x_T \in B(X, \delta)$, *then* $x_t \in B(X, \epsilon)$ *for* $t = 1, \ldots, T - 1$.

Proof. Let us assume the converse. Then for each $\delta \in (0, \epsilon)$ there exists an integer $T(\delta) \geq 2$ and a trajectory $\{x_t(\delta)\}_{t=0}^{T(\delta)}$ of G such that

$$x_i(\delta) \in B(X, \delta), \ i = 0, T(\delta), \ \sup\{|x_t(\delta) - X| : t = 1, \ldots, T(\delta) - 1\} > \epsilon. \tag{4.2}$$

Choose sequences $\{\gamma_i\}_{i=1}^\infty \subset (0, 1), \{\beta_i\}_{i=1}^\infty \subset (0, 1)$ such that

$$\prod_{i=1}^\infty \gamma_i > 0, \ 2r_0^{-1}(p, X)(1 - \prod_{i=1}^\infty \gamma_i^2) < 64^{-1}\epsilon, \ \sum_{i=1}^\infty \beta_i < (128(p, X))^{-1}\epsilon r_0. \tag{4.3}$$

There exists a monotone decreasing sequence of positive numbers $\{\delta_i\}_{i=1}^\infty$ such that

$$\delta_0 \leq 32^{-1}\epsilon, \ \gamma_i \leq 1 - \delta_i r_0^{-1}, \ i = 1, 2, \ldots, \tag{4.4}$$

$$(1 + \delta_i r_0^{-1})^2 (1 - \delta_i r_0^{-1})^{-2} \le 1 + \beta_i, \ i = 1, 2, \ldots.$$

(4.1) implies that for any natural number i,

$$B(X, \delta_i) \subset \{y \in R^n : (1 - \delta_i r_0^{-1})X \le y \le (1 + \delta_i r_0^{-1})X\}. \qquad (4.5)$$

Put

$$\tau_j = \sum_{i=1}^{j} T(\delta_i), \ j = 1, 2, \ldots \qquad (4.6)$$

Define a sequence $\{y_t\}_{t=0}^{\infty} \subset K$ as follows:

$$y_t = x_t(\delta_1), \ t = 0, \ldots, T(\delta_1), \qquad (4.7)$$

$$y_{\tau_j + t} = \prod_{i=1}^{j}((1 - \delta_i r_0^{-1})(1 + \delta_i r_0^{-1}))^{-1} x_t(\delta_{j+1}),$$

$$t = 1, \ldots, T(\delta_{j+1}), \ j = 1, 2, \ldots$$

Put

$$\Delta_j = \prod_{i=1}^{j}((1 - \delta_i r_0^{-1})(1 + \delta_i r_0^{-1})^{-1}), \ j = 1, 2, \ldots. \qquad (4.8)$$

It follows from (4.7), (4.5), (4.6) and (4.2) that

$$(1 - \delta_1 r_0^{-1})X \le y_0, \ y_{\tau_1} \le (1 + \delta_1 r_0^{-1})X, \qquad (4.9)$$

$$\prod_{i=1}^{j}[(1 - \delta_i r_0^{-1})(1 + \delta_i r_0^{-1})^{-1}](1 - \delta_{j+1} r_0^{-1}) \le y_{\tau_{j+1}}$$

$$\le \prod_{i=1}^{j}[(1 - \delta_i r_0^{-1})(1 + \delta_i r_0^{-1})^{-1}](1 + \delta_{j+1} r_0^{-1})X, \ j = 1, 2, \ldots,$$

$$y_{\tau_j} \ge \prod_{i=1}^{j}((1 - \delta_i r_0^{-1})(1 + \delta_i r_0^{-1})^{-1})x_0(\delta_{j+1}),$$

$$y_{\tau_j} \le (1 - \delta_j r_0^{-1})^{-2}(1 + \delta_j r_0^{-1})^2 \Delta_j x_0(\delta_{j+1}), \ j = 1, 2, \ldots.$$

For each natural number q there is a sequence $\{u_t(q)\}_{t=0}^{\infty} \subset K$ such that

$$u_t(q) = 0, \ t = 0, \ldots, \tau_q - 1, \ u_{\tau_q}(q) = y_{\tau_q} - \Delta_q x_0(\delta_{q+1}), \qquad (4.10)$$

$$u_{t+1}(q) \in G(u_t(q)), \ t = \tau_q, \tau_q + 1, \ldots.$$

Define a sequence $\{z_t\}_{t=0}^{\infty} \subset K$ by

$$z_t = y_t, \ t = 0, \ldots, \tau_1, \ z_{\tau_q+t} = y_{\tau_q+t} + \sum_{j=1}^{q} u_{\tau_q+t}(j), \qquad (4.11)$$

$$t = 1, \ldots, T(\delta_{q+1}), \ q = 1, 2, \ldots.$$

In view of (4.11), (4.7) and (4.10), $z_{t+1} \in G(z_t)$, $t = 0, 1, \ldots$. Therefore there exists a constant $c_0 \geq 0$ such that

$$\lim_{t \to \infty} z_t = c_0 X. \qquad (4.12)$$

We will show that

$$|z_{\tau(q)+t} - x_t(\delta_{q+1})| \leq 16^{-1}\epsilon, \ t = 1, \ldots, T(\delta_{q+1}), \ q = 1, 2, \ldots. \quad (4.13)$$

(4.1) implies that

$$-|x|r_0^{-1}X \leq x \leq |x|r_0^{-1}X \text{ for each } x \in R^n \setminus \{0\}, \qquad (4.14)$$

$$|x| \leq r_0^{-1}(p, x) \text{ for each } x \in K.$$

By (4.7), (4.4), (4.14), (4.2) and (4.5), for each $q \in \{1, 2, \ldots\}$ and each $t \in \{1, \ldots, T(\delta_{q+1})\}$,

$$|y_{\tau_q+t} - x_t(\delta_{q+1})| \leq |x_t(\delta_{q+1})|(1 - \prod_{i=1}^{\infty}(1 - \delta_i r_0^{-1})^2) \qquad (4.15)$$

$$\leq |x_t(\delta_{q+1})|(1 - \prod_{i=1}^{\infty}\gamma_i^2) \leq r_0^{-1}(p, x_t(\delta_{q+1}))(1 - \prod_{i=1}^{\infty}\gamma_i^2)$$

$$\leq r_0^{-1}(1 - \prod_{i=1}^{\infty}\gamma_i^2)(p, x_0(\delta_{q+1})) \leq r_0^{-1}(1 - \prod_{i=1}^{\infty}\gamma_i^2)(1 + \delta_{q+1}r_0^{-1})(p, X).$$

It follows from (4.9), (4.10), (4.14), (4.2) and (4.5) that for each natural number $q \geq 1$ and each nonnegative integer t,

$$|u_t(q)| \leq r_0^{-1}(p, u_t(q)) \leq r_0^{-1}(p, u_{\tau_q}(q))$$

$$\leq r_0^{-1}(p, x_0(\delta_{q+1}))\Delta_q[(1 + \delta_q r_0^{-1})^2(1 - \delta_q r_0^{-1})^{-2} - 1]$$

$$\leq r_0^{-1}\Delta_q[(1 + \delta_q r_0^{-1})^2(1 - \delta_q r_0^{-1})^{-2} - 1](1 + \delta_{q+1}r_0^{-1})(p, X).$$

Combined with (4.8) and (4.4) this implies that

$$|u_t(q)| \leq 2r_0^{-1}(p, X)\beta_q$$

for each natural number q and each nonnegative integer t. It follows from this inequality and (4.11) that for each natural number q and each $t \in \{1, \ldots, T(\delta_{q+1})\}$,

$$|y_{\tau_q+t} - z_{\tau_q+t}| \le 2r_0^{-1}(p, X) \sum_{i=1}^{\infty} \beta_i.$$

It follows from this relation, (4.15), (4.4) and (4.3) that (4.13) holds for each natural number q and each $t \in \{1, \ldots, T(\delta_{q+1})\}$. Combined with (4.2), (4.6) and (4.4) this implies that for each integer $q \ge 2$,

$$|z_{\tau(q)} - X| \le 3 \cdot 32^{-1}\epsilon,$$

$$\sup\{|z_{\tau(q)+t} - X| : t = 1, \ldots, T(\delta_{q+1}) - 1\} \ge \epsilon(1 - 16^{-1}).$$

This is contradictory to (4.12). The contradiction we have reached proves the lemma.

The next lemma shows that if an optimal trajectory $\{x_t\}_{t=0}^{T}$ satisfies $x_0 \in Q$ where Q is a given compact set and T is large enough, then x_τ is close to the turnpike for some natural number τ.

LEMMA 11.4.2 *Suppose that $\alpha_G = 1, \epsilon > 0$ and $Q \subset intK$ is a compact nonempty set. Then there exists a natural number $T(\epsilon)$ such that the following property holds:*

If an integer $T \ge T(\epsilon) + 1$ and if an optimal trajectory $\{x_t\}_{t=0}^{T}$ of G satisfies $x_0 \in Q$, then there is $\tau \in \{1, \ldots, T(\epsilon)\}$ for which $x_\tau \ne 0$ and $|x_\tau|^{-1}x_\tau \in B(X, \epsilon)$.

Proof. Let us assume the converse. Then for each natural number i there exist an integer $T_i \ge i + 1$ and an optimal trajectory $\{x_t(i)\}_{t=0}^{T_i}$ of G such that

$$x_0(i) \in Q \text{ and } ||x_t(i)|^{-1}x_t(i) - X| > \epsilon \qquad (4.16)$$

for each $t \in [1, \ldots, i]$ satisfying $x_t(i) \ne 0$.

It follows from Proposition 11.3.1 that for each natural number i the trajectory $\{x_t(i)\}_{t=0}^{T_i-1}$ is optimal. For each natural number i there is $p_i \in K$ such that

$$||p_i|| = 1, \ (p_i, x_{T_i-1}(i)) = \sup\{(p_i, y) : y \in a^{T_i-1}(x_0(i))\}. \qquad (4.17)$$

It is easy to see that there exists a positive number γ_0 such that

$$x \ge \gamma_0 X \text{ for all } x \in Q. \qquad (4.18)$$

By (4.17), (4.18) and (4.1) for each natural number $i \ge 1$,

$$r_0^{-1}(p, x_{T_i-1}(i)) \ge (p_i, x_{T_i-1}(i)) \ge (p_i, \gamma_0 X) \ge \gamma_0 r_0, \qquad (4.19)$$

$$(p, x_j(i)) \geq \gamma_0 r_0^2, \ j = 0, \ldots, T_i - 1.$$

We may assume without loss of generality that there exists a sequence $\{y_t\}_{t=0}^{\infty} \subset R^n$ such that

$$x_t(i) \to y_t \text{ as } i \to \infty \text{ for each } t \geq 0. \tag{4.20}$$

Clearly $\{y_t\}_{t=0}^{\infty}$ is a trajectory of G. Therefore there exists $c_0 \geq 0$ such that

$$\lim_{t \to \infty} y_t = c_0 X. \tag{4.21}$$

It follows from (4.16), (4.19) and (4.20) that

$$(p, y_t) \geq \gamma_0 r_0^2, \ t = 0, 1, \ldots, \ \||y_t|^{-1} y_t - X| \geq \epsilon, \ t = 1, 2, \ldots.$$

This is contradictory to (4.21). The contradiction we have reached proves the lemma.

The following auxiliary result shows that if x_0 is close to X, then an optimal trajectory $\{x_t\}_{t=0}^{T}$ is close to X for most $t \in [0, T]$.

LEMMA 11.4.3 *Suppose that $\alpha_G = 1, \epsilon > 0$. Then there are $\delta \in (0, r_0)$ and a natural number $T(\epsilon)$ such that the following property holds:*
If a natural number $T \geq T(\epsilon)$ and if an optimal trajectory $\{x_t\}_{t=0}^{T}$ of G satisfies $x_0 \in B(X, \delta)$, then

$$x_t \in B(X, \epsilon), \ t = 0, \ldots, T - T(\epsilon). \tag{4.22}$$

Proof. We may assume that

$$\epsilon < 4^{-1} r_0. \tag{4.23}$$

Lemma 10.4.1 implies that there exists

$$\epsilon_1 \in (0, 8^{-1}\epsilon) \tag{4.24}$$

such that the following property holds:
If an integer $T \geq 2$ and if a trajectory $\{x_t\}_{t=0}^{T}$ of G satisfies $x_0, x_T \in B(X, \epsilon_1)$, then

$$x_t \in B(X, \epsilon), \ t = 1, \ldots, T - 1. \tag{4.25}$$

Choose numbers γ_0, ϵ_0 such that

$$\gamma_0 \in (0, 4^{-1} \min\{\epsilon_1, 1\}), \ \epsilon_0 \in (0, 4^{-1}\gamma_0), \tag{4.26}$$

$$(1 - \gamma_0)(1 + \epsilon_0 r_0^{-1}) \leq (1 - 2^{-1}\gamma_0), \ (1 + \gamma_0)(1 - \epsilon_0 r_0^{-1}) > (1 + 2^{-1}\gamma_0).$$

It follows from Lemma 11.4.2 and (4.1) that there exists a natural number L_0 such that the following property holds:

If an integer $T \geq L_0 + 1$ and if an optimal trajectory $\{x_t\}_{t=0}^{T}$ of G satisfies $x_0 \in B(X, 2^{-1}r_0)$, then there exists an integer $\tau \in [1, L_0]$ such that

$$x_\tau \neq 0, \ |x_\tau|^{-1}x_\tau \in B(X, \epsilon_0). \tag{4.27}$$

Choose an integer $T(\epsilon)$ and a positive number δ such that

$$T(\epsilon) > L_0 + 1, \ \delta < 2^{-1}\epsilon_0, \ 4^{-1}r_0\gamma_0. \tag{4.28}$$

Assume that an integer $T \geq T(\epsilon)$ and an optimal trajectory $\{x_t\}_{t=0}^{T}$ of G satisfies

$$x_0 \in B(X, \delta). \tag{4.29}$$

In view of the choice of L_0 and (4.28) there exist a sequence of integers $\{T_i\}_{i=0}^{q}$ such that

$$T_0 = 0, \ T_{i+1} - T_i \in [1, L_0], \ T - T_q \in [0, L_0 + 1], \tag{4.30}$$

$$x_{T_i} \neq 0, \ |x_{T_i}|^{-1}x_{T_i} \in B(X, \epsilon_0), \ i = 1, \ldots, q.$$

We will show that

$$x_{T_i} \in B(X, \epsilon_1), \ i = 0, \ldots, q. \tag{4.31}$$

In view of (4.30), (4.29) and (4.26) in order to prove the inequality (4.31) it is sufficient to show that

$$||x_{T_i}| - 1| \leq \gamma_0, \ i = 1, \ldots, q. \tag{4.32}$$

Let us assume the converse. Then there exists $j \in \{1, \ldots, q\}$ for which

$$||x_{T_j}| - 1| > \gamma_0. \tag{4.33}$$

There are two cases: (i) $|x_{T_j}| > 1 + \gamma_0$; (ii) $|x_{T_j}| < 1 - \gamma_0$. Consider the case (i). We have

$$|x_{T_j}| > 1 + \gamma_0. \tag{4.34}$$

Put

$$u = |x_{T_j}|^{-1}x_{T_j} - X. \tag{4.35}$$

(4.30) implies that

$$|u| \leq \epsilon_0. \tag{4.36}$$

By (4.35), (4.36), (4.34) and (4.1),

$$u \geq -\epsilon_0 r_0^{-1} X, \ x_{T_j} = |x_{T_j}|X + |x_{T_j}|u \tag{4.37}$$

$$\geq (1 + \gamma_0)(X - \epsilon_0 r_0^{-1}X) = (1 + \gamma_0)(1 - \epsilon_0 r_0^{-1})X.$$

It follows from (4.1) and (4.29) that $-\delta r_0^{-1} X \le x_0 - X \le \delta r_0^{-1} X$. This inequality, (4.37) and (4.26) imply that

$$(p, (1 + \delta r_0^{-1}) X) \ge (p, x_0) \ge (p, x_{T_j}) \ge (p, X)(1 + \gamma_0)(1 - \epsilon_0 r_0^{-1}),$$

$$(1 + \delta r_0^{-1}) \ge (1 + \gamma_0)(1 - \epsilon_0 r_0^{-1}) \ge (1 + 2^{-1} \gamma_0).$$

This is contradictory to (4.28). Therefore in the case (i) we obtained a contradiction.

Consider the case (ii). We have

$$|x_{T_j}| < 1 - \gamma_0. \tag{4.38}$$

Define u by (4.35). Evidently (4.36) holds. By (4.35), (4.36), (4.38), (4.26) and (4.1),

$$u \le \epsilon_0 r_0^{-1} X, \quad x_{T_j} = |x_{T_j}| X + |x_{T_j}| u \tag{4.39}$$

$$\le |x_{T_j}|(1 + \epsilon_0 r_0^{-1}) X \le (1 - \gamma_0)(1 + \epsilon_0 r_0^{-1}) X \le (1 - 2^{-1} \gamma_0) X.$$

It follows from (4.1) and (4.29) that

$$x_0 \ge (1 - \delta r_0^{-1}) X. \tag{4.40}$$

The inequality (4.40) implies that there exists $y_0 \in a^{T_j}(x_0)$ for which $y_0 \ge (1 - \delta r_0^{-1}) X$. Combined with (4.39) and (4.28) this implies that

$$y_0 \ge (1 - 4^{-1} \gamma_0)(1 - 2^{-1} \gamma_0)^{-1} x_{T_j}.$$

In view of this inequality there exists $y_1 \in a^T(x_0)$ such that

$$y_1 \ge (1 - 4^{-1} \gamma_0)(1 - 2^{-1} \gamma_0)^{-1} x_T.$$

On the other hand $\{x_t\}_{t=0}^T$ is an optimal trajectory of G. Therefore in the case (ii) we also obtained a contradiction which proves (4.32). Hence (4.31) holds. It follows from (4.31), (4.30) and the definition of ϵ_1 (see (4.24), (4.25)) that $x_t \in B(X, \epsilon)$, $t = 0, \dots, T - L_0 - 1$. This completes the proof of the lemma.

Proof of Theorem 11.3.1. We may assume without loss of generality that $\alpha_G = 1$. It is sufficient to show that for each $y \in \text{int} K$ there exist a positive number γ and a natural number L such that $B(y, \gamma) \subset \text{int} K$ and the following property holds:

If $T \ge 2L$ and if an optimal trajectory $\{x_t\}_{t=0}^T$ of G satisfies $x_0 \in B(y, \gamma)$, then

$$x_t \in B(\lambda(y) X, 2^{-1} \epsilon), \quad t \in [L, T - L]. \tag{4.41}$$

Let $y \in \mathrm{int}K$. We may assume that

$$\lambda(y) = 1. \tag{4.42}$$

There exists $r_1 \in (0, r_0)$ for which

$$B(y, 2r_1) \subset K. \tag{4.43}$$

Lemma 11.4.3 implies that there exist a natural number L_1 and a positive number

$$\delta_1 < \min\{4^{-1}\epsilon, r_1, 4^{-1}\} \tag{4.44}$$

such that the following property holds:
If $T \geq L_1$ is an integer and if an optimal trajectory $\{x_t\}_{t=0}^T$ of G satisfies $x_0 \in B(X, \delta_1)$, then

$$x_t \in B(X, 4^{-1}\epsilon), \ t = 0, \ldots, T - L_1. \tag{4.45}$$

Fix a positive number γ which satisfies

$$\gamma < \min\{4^{-1}r_1, 32^{-1}\delta_1 r_1\}, \tag{4.46}$$

$$(1 - 32^{-1}\delta_1)(1 - 64^{-1}\delta_1)^{-1} \leq 1 - \gamma r_1^{-1}.$$

In view of Lemma 11.4.3 there exist a natural number L_2 and a positive number

$$\delta_0 < 9^{-1}\delta_1 \tag{4.47}$$

such that the following property holds:
If $T \geq L_2$ is an integer and if an optimal trajectory $\{x_t\}_{t=0}^T$ of G satisfies $x_0 \in B(X, \delta_0)$, then

$$x_t \in B(X, 64^{-1}\delta_1 r_0), \ t = 0, \ldots, T - L_2. \tag{4.48}$$

We may assume without loss of generality that

$$(1 + 8^{-1}\delta_1)(1 - \delta_0 r_0^{-1}) > (1 + 2\gamma r_1^{-1}), \tag{4.49}$$

$$(1 - 8^{-1}\delta_1)(1 + \delta_0 r_0^{-1}) < (1 - 2\gamma r_1^{-1}).$$

It follows from Lemma 11.4.2 that there exists a natural number L_3 such that the following property holds:
If $T \geq L_3 + 1$ is an integer and if an optimal trajectory $\{x_t\}_{t=0}^T$ of G satisfies $x_0 \in B(y, r_1)$, then there is $\tau \in \{1, \ldots, L_3\}$ for which

$$x_\tau \neq 0, \ |x_\tau|^{-1}x_\tau \in B(X, \delta_0). \tag{4.50}$$

By (4.42) there exists a trajectory $\{\bar{z}_t\}_{t=0}^{\infty}$ of G such that $\bar{z}_0 = y$ and $\lim_{t\to\infty} \bar{z}_t = X$. Therefore there exists a natural number L_4 such that

$$\bar{z}_t \geq (1 - 64^{-1}\delta_1)X \qquad (4.51)$$

for all integers $t \geq L_4$.

Choose an integer

$$L \geq L_1 + L_2 + L_3 + L_4 + 2. \qquad (4.52)$$

Assume that $T \geq 2L$ and $\{x_t\}_{t=0}^{T}$ is an optimal trajectory of G satisfying

$$x_0 \in B(y, \gamma). \qquad (4.53)$$

It follows from the choice of L_3 (see (4.50)), (4.46) and (4.52) that there exists a natural number $\tau \in [1, L_3]$ which satisfies (4.50). We will show that

$$||x_\tau| - 1| \leq 8^{-1}\delta_1. \qquad (4.54)$$

Let us assume the converse. There are two cases: (i) $|x_\tau| > 1 + 8^{-1}\delta_1$; (ii) $|x_\tau| < 1 - 8^{-1}\delta_1$. (4.53), (4.43) and (4.46) imply that

$$(1 - \gamma r_1^{-1})y \leq x_0 \leq (1 + \gamma r_1^{-1})y. \qquad (4.55)$$

It follows from (4.50) and (4.1) that

$$(1 - \delta_0 r_0^{-1})X \leq |x_\tau|^{-1}x_\tau \leq (1 + \delta_0 r_0^{-1})X. \qquad (4.56)$$

Consider the case (i). Then $|x_\tau| > 1 + 8^{-1}\delta_1$. Combined with (4.56) and (4.49) this inequality implies that

$$x_\tau \geq (1 + 8^{-1}\delta_1)(1 - \delta_0 r_0^{-1})X \geq (1 + 2\gamma r_1^{-1})X.$$

In view of this inequality and (4.55) there exists $u_0 \in a^\tau(y)$ such that

$$u_0 \geq (1 + 2\gamma r_1^{-1})(1 + \gamma r_1^{-1})^{-1}X.$$

This is contradictory to (4.42).

Consider the case (ii). Then

$$|x_\tau| < 1 - 8^{-1}\delta_1. \qquad (4.57)$$

By (4.50), the choice of L_2, δ_0 (see (4.47), (4.48)) and (4.52),

$$|x_\tau|^{-1}x_t \in B(X, 64^{-1}\delta_1 r_0), \ t = \tau, \ldots, T - L_2.$$

By this inclusion, (4.57) and (4.1) for each $t \in \{\tau, \ldots, T - L_2\}$,

$$|x_\tau|^{-1}x_t \leq (1 + 64^{-1}\delta_1)X, \qquad (4.58)$$

$$x_t \le (1 - 8^{-1}\delta_1)(1 + 64^{-1}\delta_1)X \le (1 - 16^{-1}\delta_1)X.$$

In view of the choice of L_4 (see (4.51)) and (4.58),

$$\tau + L_4 \in [\tau, T - L_2], \quad \bar{z}_{\tau+L_4} \in a^{\tau+L_4}(y),$$

$$\bar{z}_{\tau+L_4} \ge (1 - 64^{-1}\delta_1)(1 - 16^{-1}\delta_1)^{-1}x_{\tau+L_4}.$$

By these relations, (4.45) and (4.46), there exists $v \in a^{\tau+L_4}(x_0)$ which satisfies

$$v \ge (1 - 64^{-1}\delta_1)(1 - 16^{-1}\delta_1)^{-1}(1 - \gamma r_1^{-1})x_{\tau+L_4}$$

$$\ge (1 - 32^{-1}\delta_1)(1 - 16^{-1}\delta_1)^{-1}x_{\tau+L_4}.$$

This is contradictory to the optimality of the trajectory $\{x_t\}_{t=0}^T$. Since in both cases we obtained a contradiction we conclude that (4.54) holds. (4.50) and (4.54) imply that

$$|x_\tau - X| \le |x_\tau - |x_\tau|X| + ||x_\tau| - 1| \le |x_\tau|\delta_0 + ||x_\tau| - 1|$$

$$\le \delta_0 + (\delta_0 + 1)||x_\tau| - 1| \le \delta_0 + 4^{-1}\delta_1 \le 2^{-1}\delta_1.$$

It follows from this relation, the definition of L_1, δ_1 and (4.52) that

$$x_t \in B(X, 4^{-1}\epsilon), \quad t = \tau, \dots, T - L_1.$$

This completes the proof of the theorem.

Theorem 11.3.2 follows from Lemma 11.4.3.

11.5.   Stability of the turnpike phenomenon

A sequence $\{G_t\}_{t=0}^T \subset \mathcal{M}$ where $T \in \{0, 1, \dots\}$ is called a model.

Let $T \ge 0$ be an integer and let $\{G_t\}_{t=0}^T \subset \mathcal{M}$ be a model. A sequence $\{x_t\}_{t=0}^{T+1} \subset R^n$ is a trajectory of the model $\{G_t\}_{t=0}^T$ if $x_t \in K$ and $x_{t+1} \in G_t(x_t)$ for $t = 0, \dots, T$. A trajectory $\{x_t\}_{t=0}^{T+1}$ of the model $\{G_t\}_{t=0}^T$ is called optimal if there exists $f \in K^* \setminus \{0\}$ such that for each trajectory $\{y_t\}_{t=0}^{T+1}$ of the model satisfying $y_0 = x_0$ the relation $(f, x_{T+1}) \ge (f, y_{T+1})$ holds.

We can easily prove the following result.

PROPOSITION 11.5.1 *Assume that an integer* $T \ge 1$,

$$\{G_t\}_{t=0}^{T-1} \subset \mathcal{M}, \ Y \in intK,$$

$$\beta_0, \dots, \beta_{T-1} > 0, \ \beta_t Y \in G_t(Y) - K, \ t = 0, \dots, T-1.$$

Let $\{x_t\}_{t=0}^T$ be a trajectory of the model $\{G_t\}_{t=0}^{T-1}$ with $x_0 \in \mathrm{int}K$. Then this trajectory is optimal if and only if for each $\lambda > 1$,

$$\lambda x_T + K \cap \{y \in R^n : \text{ there exists a trajectory } \{y_t\}_{t=0}^T$$

of the model $\{G_t\}_{t=0}^{T-1}$ which satisfies $y_0 = x_0, y_T = y\} = \emptyset$.

Let $T \geq 1$ be an integer and $\{G_t\}_{t=0}^{T-1} \subset \mathcal{M}$ be a model. Assume that an integer $q \geq 1$ and a number $\gamma > 1$. A trajectory $\{x_t\}_{t=0}^T$ of the model $\{G_t\}_{t=0}^{T-1}$ is (q, γ)-optimal if for each integer τ satisfying $0 \leq \tau \leq T - q$ and each $\gamma_1 > \gamma$,

$$(\gamma_1 x_{\tau+q} + K) \cap \{y \in R^n : \text{ there exists a trajectory } \{y_t\}_{t=0}^q$$

of the model $\{G_{t+\tau}\}_{t=0}^{q-1}$ which satisfies $y_0 = x_\tau, y_q = y\} = \emptyset$.

We can easily prove the following result.

PROPOSITION 11.5.2 *Suppose that T, q are natural numbers, $\{G_t\}_{t=0}^{T-1}$ is a model, $\gamma > 1$ and $\{x_t\}_{t=0}^T$ is a (q, γ)-optimal trajectory of the model $\{G_t\}_{t=0}^{T-1}$. Then for each integer τ satisfying $0 \leq \tau \leq T - q$ there exists $f \in K^* \setminus \{0\}$ such that the following property holds:*
If $\{y_t\}_{t=0}^q$ is a trajectory of the model $\{G_{\tau+t}\}_{t=0}^{q-1}$ satisfying $y_0 = x_\tau$, then $\gamma(f, x_{\tau+q}) \geq (f, y_q)$.

For the set $\mathcal{M}$ we define a metric $\rho(\cdot, \cdot)$ as follows:

$$\rho(G_1, G_2) = d(\{(x, y) : x \in K, |x| \leq 1, y \in G_1(x)\},$$

$$\{(x, y) : x \in K, |x| \leq 1, y \in G_2(x)\}), G_1, G_2 \in \mathcal{M},$$

where $d(\cdot, \cdot)$ is the Hausdorff metric. Suppose that

$$G \in \mathcal{M}, \alpha_G > 0, X \in \mathrm{int}K, p \in \mathrm{int}K^*, |X| = 1, \qquad (5.1)$$

$$\alpha_G X \in G(X), \alpha_G \text{ is the maximal eigenvalue of } G,$$

$$(\alpha_G p, x) \geq (p, y), x \in K, y \in G(x)$$

and that for each trajectory $\{x_t\}_{t=0}^{\infty}$ of G there exists a number $c_0 \geq 0$ such that $\alpha_G^{-t} x_t \to c_0 X$ as $t \to \infty$.
For each $F \in \mathcal{M}$ we define

$$\alpha(F) = \sup\{\beta \in [0, \infty) : \text{ there exists } x \in K \setminus \{0\} \qquad (5.2)$$

$$\text{such that } \beta x \in F(x) - K\}.$$

It is easy to see that $\alpha(G) = \alpha_G$, there exists an open neighborhood V of G in $\mathcal{M}$ such that for each $F \in V$ the number $\alpha(F)$ is well defined and G is a continuity point of the mapping $F \to \alpha(F), F \in V$.

We will establish the following results which show that the turnpike phenomenon is stable under small perturbations of G.

THEOREM 11.5.1 *Suppose that* $x \in intK$ *and* ϵ *is a positive number. Then there exist numbers* $\delta > 0, \gamma > 1$, *integers* $q \geq 1, L_i \geq 0, i = 1, 2$ *and an open neighborhood* U *of* G *in* $\mathcal{M}$ *such that* $B(x, \delta) \subset intK$ *and that the following property holds:*

If $T \geq L_1 + L_2$ *is a natural number,* $\{G_t\}_{t=0}^{T-1} \subset U$ *is a model and if* $\{x_t\}_{t=0}^{T}$ *is a* (q, γ)-*optimal trajectory of the model* $\{G_t\}_{t=0}^{T-1}$ *satisfying* $x_0 \in B(x, \delta)$, *then*

$$x_t \neq 0 \text{ and } \|x_t\|^{-1}x_t - X| \leq \epsilon, \ t = L_1, \ldots, T - L_2.$$

Moreover if $x = X$, *then* $L_1 = 0$.

THEOREM 11.5.2 *Let* $x \in intK$ *and* $\epsilon > 0$. *Then there exist a number* $\delta > 0$, *integers* $L_i \geq 0, i = 1, 2$ *and a neighborhood* U *of* G *in* $\mathcal{M}$ *such that* $B(x, \delta) \subset intK$ *and for each* $F \in U$, *each natural number* $T \geq L_1 + L_2$ *and each optimal trajectory* $\{x_t\}_{t=0}^{T}$ *of* F *satisfying* $|x_0 - x| \leq \delta$ *the following relation holds:*

$$|\alpha(F)^{-t}x_t - \lambda(x)X| \leq \epsilon, \ t \in \{L_1, \ldots, T - L_2\}.$$

Moreover if $x = X$, *then* $L_1 = 0$.

Suppose that $\phi_t : K \to [0, \infty), \ t = 0, 1, \ldots$ are continuous functions,

$$\phi_t(ax) = a\phi_t(x) \text{ for each } a \in [0, \infty), x \in K \text{ and } t = 0, 1, \ldots, \quad (5.3)$$

$\phi_t(x) \leq \phi_t(y)$ for each integer $t \geq 0$ and each $x, y \in K$ satisfying $x \leq y$,

$0 < c_2 \leq \phi_t(x) \leq c_1 < \infty$ for each $t \in \{0, 1, \ldots\}$, each $x \in K$ satisfying $|x| \leq 1$ where $0 < c_2 < c_1$ are constants.

Consider an open neighborhood V of G in $\mathcal{M}$ such that for each $F \in V$ the number $\alpha(F)$ is well defined and positive.

Let $z \in intK$ and μ be a positive number. For each pair (x, F) belonging to a small neighborhood W of (z, G) in $R^n \times \mathcal{M}$ and an integer $T \geq 0$ we consider the following optimization problem:

$$\sum_{t=0}^{T} \alpha(F)^{-t}\phi_t(x_t) \to \max,$$

$\{x_t\}_{t=0}^T$ is a trajectory of F, $x_0 = x$, $\alpha(F)^{-T} x_T \geq \mu X$.

The problem (P) has a solution if the positive number μ and the neighborhood W are small enough.

THEOREM 11.5.3 *Let $z \in intK$ and let ϵ, μ be positive numbers. Then there exist a number $\delta > 0$, integers $L_0, L_1 \geq 0$ and a neighborhood U of G in $\mathcal{M}$ such that $B(z, \delta) \subset intK$, $U \subset V$ and for each $F \in U$, each natural number $T \geq L_1 + L_0$, each $x \in B(z, \delta)$ and each optimal solution $\{x_t\}_{t=0}^T$ of the problem (P) satisfying $x_0 = x$ the following relations hold:*

$$x_t \neq 0 \text{ and } ||x_t|^{-1} x_t - X| \leq \epsilon, \ t \in \{L_1, \ldots, T - L_0\}.$$

Moreover if $z = X$, then $L_1 = 0$.

11.6.   Proofs of Theorems 11.5.1, 11.5.2 and 11.5.3

Theorems 11.3.1 and 11.3.2 imply the following:

LEMMA 11.6.1 *Let $x \in intK$ and ϵ be a positive number. Then there exist $\delta > 0$ and nonnegative integers L_1, L_2 such that $B(x, \delta) \subset intK$ and the following property holds:*
    If $T \geq L_1 + L_2$ is a natural number and if $\{x_t\}_{t=0}^T$ is an optimal trajectory of G satisfying $x_0 \in B(x, \delta)$, then

$$x_t \neq 0 \text{ and } |x_t|^{-1} x_t \in B(X, \epsilon), \ t = L_1, \ldots, T - L_2.$$

Moreover if $x = X$, then $L_1 = 0$.

There exists a number $r_0 > 0$ such that

$$B(X, r_0) \subset intK, \ B(p, r_0) \subset intK^*. \tag{6.1}$$

The next auxiliary result shows that a (T, γ)-optimal trajectory of a model $\{G_t\}_{t=0}^{T-1}$ is close to an optimal trajectory of G if γ is close to 1 and $\{G_t\}_{t=0}^{T-1}$ is contained in a small neighborhood of G.

LEMMA 11.6.2 *Assume that a nonempty compact set $Q \subset intK$, T is a natural number and ϵ is a positive number. Then there exist $\gamma > 1$, and a neighborhood U of G in $\mathcal{M}$ such that the following property holds:*
    If $\{G_t\}_{t=0}^{T-1} \subset U$ and if $\{x_t\}_{t=0}^T$ is a (T, γ)-optimal trajectory of the model $\{G_t\}_{t=0}^{T-1}$ satisfying $x_0 \in Q$, then there exists an optimal trajectory $\{y_t\}_{t=0}^T$ of G for which $y_0 \in Q$ and $|y_t - x_t| \leq \epsilon$, $t = 0, \ldots, T$.

Proof. Let us assume the converse. Then for each integer $s \geq 1$ there exist

$$\{G_t^s\}_{t=0}^{T-1} \subset \{F \in \mathcal{M} : \rho(G, F) \leq s^{-1}\} \tag{6.2}$$

and a $(T, (1 + s^{-1}))$-optimal trajectory $\{x_t^s\}_{t=0}^T$ of the model $\{G_t^s\}_{t=0}^{T-1}$ such that $x_0^s \in Q$ and

$$\sup\{|y_t - x_t^s| : t = 0, \ldots, T\} > \epsilon \tag{6.3}$$

for each optimal trajectory $\{y_t\}_{t=0}^T$ of G satisfying $y_0 \in Q$.

Proposition 11.5.2 implies that for each natural number s there exists $f^s \in K^*$ such that $\|f^s\| = 1$ and the following property holds:

If a trajectory $\{y_t\}_{t=0}^T$ of the model $\{G_t^s\}_{t=0}^{T-1}$ satisfies $y_0 = x_0^s$, then

$$(1 + s^{-1})(f^s, x_T^s) \geq (f^s, y_T). \tag{6.4}$$

We may assume without loss of generality that there are a sequence $\{x_t\}_{t=0}^T \subset R^n$ and $f \in R^n$ for which

$$f^s \to f, \; x_t^s \to x_t \text{ as } s \to \infty, \; t = 0, \ldots, T. \tag{6.5}$$

It is easy to see that $x_0 \in Q$, $\{x_t\}_{t=0}^T$ is a trajectory of G. We will show that $\{x_t\}_{t=0}^T$ is an optimal trajectory of G.

Assume the contrary. Then there exists a trajectory $\{z_t\}_{t=0}^T$ of G such that

$$z_0 = x_0, \; (f, z_T) > \sup\{(f, x_T), 0\} + 4\Delta \tag{6.6}$$

where Δ is a positive number. We may assume without loss of generality that there exists $c_0 > 0$ such that

$$z_t \geq c_0 X, \; t = 0, \ldots, T. \tag{6.7}$$

There exist numbers $M, \Gamma > 0$ such that

$$|x_t^s| \leq M, \; t = 0, \ldots, T, \; s = 1, 2, \ldots, \; |z_t| \leq M, \; t = 0, \ldots, T, \tag{6.8}$$

$$\Gamma \in (0, 1), \; (1 - \Gamma^{2T+1})(f, z_T) < \Delta.$$

It is not difficult to see that there exists $\delta > 0$ such that the following property holds:

If $F \in \mathcal{M}$ satisfies $\rho(F, G) \leq \delta$ and if $u_0 \in K$, $u_1 \in G(u_0)$ satisfy

$$|u_0|, |u_1| \leq M, \; u_0, u_1 \geq c_0 X,$$

then there are $v_0 \in K, v_1 \in F(v_0)$ for which

$$v_0 \leq \Gamma^{-1} u_0, \; v_1 \geq \Gamma u_1. \tag{6.9}$$

In view of (6.5) there exists a natural number s such that

$$4s^{-1} \leq \delta, \ s^{-1} < \Delta(2M)^{-1}, |(f^s - f, z_T)| \leq \Delta, \qquad (6.10)$$

$$|(f^s, x_T^s) - (f, x_T)| \leq \Delta, \ x_0^s \geq \Gamma x_0.$$

It follows from (6.10), (6.7), (6.8), (6.2) and the choice of δ (see (6.9)) that for each $t \in \{0, \ldots, T-1\}$ there are $u_t \in K, u_{t+1} \in G_t^s(u_t)$ such that $u_t \leq \Gamma^{-1}z_t, u_{t+1} \geq \Gamma z_{t+1}$. Combined with (6.6) and (6.10) this implies that there exists a trajectory $\{v_t\}_{t=0}^T$ of the model $\{G_t^s\}_{t=0}^{T-1}$ such that

$$v_0 = x_0^s, \ v_t \geq \Gamma^{2t+1}z_t, \ t = 1, \ldots, T. \qquad (6.11)$$

By (6.11), (6.8), (6.6), (6.10) and the definition of f^s,

$$(f^s, v_T) \geq \Gamma^{2T+1}(f^s, z_T) \geq \Gamma^{2T+1}(f, z_T) - \Delta$$

$$\geq (f, z_T) - 2\Delta \geq (f, x_T) + 2\Delta \geq (f^s, x_T^s) + \Delta \geq 2^{-1}\Delta$$

$$+(1 + (2M)^{-1}\Delta)(f^s, x_T^s) \geq 2^{-1}\Delta + (1 + s^{-1})(f^s, x_T^s).$$

In view of (6.11) this is contradictory to the definition of f^s (see (6.4)). The obtained contradiction proves that $\{x_t\}_{t=0}^T$ is an optimal trajectory of G. This implies that (6.5) is contradictory to the definition of $\{x_t^s\}_{t=0}^T, \ s = 1, 2, \ldots$. The obtained contradiction proves the lemma.

COROLLARY 11.6.1 *Assume that a nonempty compact set $Q \subset \mathrm{int}K$, $T \geq 1$ is an integer, $\epsilon > 0$. Then there exist $\gamma > 1$ and a neighborhood U of G in $\mathcal{M}$ such that for each $\{G_t\}_{t=0}^{T-1} \subset U$ and each (T, γ)-optimal trajectory $\{x_t\}_{t=0}^T$ of the model $\{G_t\}_{t=0}^{T-1}$ satisfying $x_0 \in Q$, there is an optimal trajectory $\{y_t\}_{t=0}^T$ of G such that $y_0 \in Q$ and for each $t \in \{0, \ldots, T\}$,*

$$x_t, y_t \neq 0, \ ||x_t|^{-1}x_t - |y_t|^{-1}y_t| \leq \epsilon.$$

Proof of Theorem 11.5.1. We may assume that $B(X, \epsilon) \subset \mathrm{int}K$. Lemma 11.6.1 implies that there exist $\delta_1 \in (0, 2^{-1}\epsilon)$ and nonnegative integers N_1, N_2 such that the following properties hold:

(i) $B(x, \delta_1) \subset \mathrm{int}K$ and if $T \geq N_1 + N_2$ is a natural number and if $\{x_t\}_{t=0}^T$ is an optimal trajectory of G satisfying $x_0 \in B(x, \delta_1)$, then

$$x_t \neq 0, \ |x_t|^{-1}x_t \in B(X, 2^{-1}\epsilon), \ t \in \{N_1, \ldots, T - N_2\}; \qquad (6.12)$$

(ii) if $T \geq N_2$ is a natural number and if $\{x_t\}_{t=0}^T$ is an optimal trajectory of G satisfying $x_0 \in B(X, \delta_1)$, then relation (6.12) holds for each integer $t \in \{0, \ldots, T - N_2\}$. Moreover $N_1 = 0$ if $x = X$.

It follows from Lemma 11.6.1 that there exist natural numbers N_3, N_4 such that if $T \geq N_3 + N_4$ is an integer and if $\{x_t\}_{t=0}^T$ is an optimal trajectory of G satisfying $x_0 \in B(X\epsilon)$, then

$$x_t \neq 0, \; |x_t|^{-1}x_t \in B(X, 2^{-1}\delta_1), \; t \in \{N_3, \ldots, T - N_4\}. \tag{6.13}$$

Put

$$q = 2(N_1 + N_2 + N_3 + N_4). \tag{6.14}$$

In view of Corollary 11.6.1 there exist a number $\gamma > 1$ and a neighborhood U of G in $\mathcal{M}$ such that the following property holds:

(iii) If $\{G_t\}_{t=0}^{q-1} \subset U$ is a model and if $\{x_t\}_{t=0}^q$ is a (q, γ)-optimal trajectory of the model $\{G_t\}_{t=0}^{q-1}$ satisfying $x_0 \in B(X, \epsilon)$ (respectively $x_0 \in B(x, \delta_1)$, $x_0 \in B(X, \delta_1)$), then there exists an optimal trajectory $\{y_t\}_{t=0}^q$ of G satisfying $y_0 \in B(X, \epsilon)$ (respectively $y_0 \in B(x, \delta_1)$, $y_0 \in B(X, \delta_1)$) such that

$$x_t, y_t \neq 0, \; \||x_t|^{-1}x_t - |y_t|^{-1}y_t| \leq 8^{-1}\delta_1, \; t = 0, \ldots, q. \tag{6.15}$$

Choose $\delta \in (0, 2^{-1}\delta_1)$ and put

$$L_1 = N_1, \; L_2 = q. \tag{6.16}$$

Assume that an integer $T \geq L_1 + L_2$, $\{G_t\}_{t=0}^{T-1} \subset U$ and $\{x_t\}_{t=0}^T$ is a (q, γ)-optimal trajectory of the model $\{G_t\}_{t=0}^{T-1}$ satisfying $x_0 \in B(x, \delta)$. We will show that for each $t \in \{L_1, \ldots, T - L_2\}$,

$$x_t \neq 0, \; |x_t|^{-1}x_t \in B(X, \epsilon). \tag{6.17}$$

Assume the contrary. Then there exists an integer $t_1 \in \{N_1, \ldots, T - q\}$ such that (6.17) does not hold with $t = t_1$.

Consider a (q, γ)-optimal trajectory $\{x_t\}_{t=0}^q$ of the model $\{G_t\}_{t=0}^{T-1}$. By property (iii) there exists an optimal trajectory $\{y_t\}_{t=0}^q$ of G such that $y_0 \in B(x, \delta_1)$ and (6.15) holds for $t = 0, \ldots, q$. Combined with property (i) this implies that

$$|y_t|^{-1}y_t \in B(X, 2^{-1}\epsilon), \; t \in \{N_1, \ldots, q - N_2\}, \tag{6.18}$$

$$|x_t|^{-1}x_t \in B(X, \epsilon), \; t \in \{N_1, \ldots, q - N_2\}.$$

Therefore $t_1 > q - N_2$. Clearly a sequence $\{|x_{N_1}|^{-1}x_{N_1+t}\}_{t=0}^q$ is a (q, γ)-optimal trajectory of the model $\{G_{t+N_1}\}_{t=0}^{q-1}$. By (6.18) and property (iii) there exists an optimal trajectory $\{z_t\}_{t=0}^q$ of G such that $z_0 \in B(X, \epsilon)$ and for each $t \in \{0, \ldots, q\}$,

$$z_t \neq 0, \; x_{t+N_1} \neq 0, \; \||z_t|^{-1}z_t - |x_{t+N_1}|^{-1}x_{t+N_1}| \leq 8^{-1}\delta_1.$$

In view of these inequalities and the choice of N_3, N_4 (see (6.13)),

$$|z_t|^{-1}z_t \in B(X, 2^{-1}\delta_1), \ t \in \{N_3, \ldots, q - N_4\}, \tag{6.19}$$

$$|x_t|^{-1}x_t \in B(X, \delta_1), \ t \in \{N_1 + N_3, \ldots, q + N_1 - N_4\}.$$

Define

$$E = \{t \in \{N_1, \ldots, t_1\} : \ x_\tau \neq 0, \ |x_\tau|^{-1}x_\tau \in B(X, \epsilon) \text{ for}$$

$$\tau = N_1, \ldots, t \text{ and } |x_t|^{-1}x_t \in B(X, \delta_1)\}.$$

(6.15), (6.18), (6.14) and (6.19) imply that $N_1 + N_3 \in E$. Denote by t_2 the maximal element of E. Consider a (q, γ)-optimal trajectory $\{|x_{t_2}|^{-1}x_{t+t_2}\}_{t=0}^q$ of the model $\{G_{t+t_2}\}_{t=0}^{q-1}$. By the definition of E, t_2 and property (iii) there exists an optimal trajectory $\{h_t\}_{t=0}^q$ of G satisfying $h_0 \in B(X, \delta_1)$ such that

$$x_{t+t_2} \neq 0, \ h_t \neq 0, \ \||x_{t+t_2}|^{-1}x_{t+t_2} - |h_t|^{-1}h_t| \le 8^{-1}\delta_1, \ t = 0, \ldots, q. \tag{6.20}$$

It follows from (6.20) and the property (ii) that for each $t \in \{0, \ldots, q - N_2\}$

$$|h_t|^{-1}h_t \in B(X, 2^{-1}\epsilon), \ |x_{t+t_2}|^{-1}x_{t+t_2} \in B(X, \epsilon). \tag{6.21}$$

In view if (6.14), (6.20) and the choice of N_3, N_4 (see (6.13)) for each $t \in \{N_3, \ldots, q - N_4\}$,

$$|h_t|^{-1}h_t \in B(X, 2^{-1}\delta_1), \ |x_{t+t_2}|^{-1}x_{t+t_2} \in B(X, \delta_1).$$

Combined with (6.21) and the definition of t_1 this implies that $t_2 + N_3 \in E$. This is contradictory to the definition of t_2. The obtained contradiction proves the theorem.

For each $u, v \in R^n$ satisfying $u \le v$ we put

$$< u, v >= \{z \in R^n : u \le z \le v\}.$$

Proof of Theorem 11.5.2. There exists a neighborhood U_0 of G in $\mathcal{M}$ such that for each $F \in U_0$ the number $\alpha(F)$ is well defined and positive. We may assume that $B(\lambda(x)X, \epsilon) \subset \text{int}K$. Put

$$\Lambda = \lambda(x). \tag{6.22}$$

There exists a number $\theta > 1$ for which

$$\Lambda < \theta^{-10}X, \theta^{10}X > \subset B(\Lambda X, \epsilon). \tag{6.23}$$

It follows from Theorem 11.5.1 and Proposition 11.3.1 that there exist a positive number $\delta_1 < \epsilon$, a nonnegative integer N_1 and a neighborhood U_1 of G in $\mathcal{M}$ such that

$$U_1 \subset U_0, \ B(x, \delta_1) \subset \text{int} K, \ B(\Lambda X, \delta_1) \subset \Lambda < \theta^{-1} X, \theta X > \qquad (6.24)$$

and the following property holds:

(i) If $F \in U_1$, $T \geq N_1$ is a natural number and if an optimal trajectory $\{x_t\}_{t=0}^T$ of F satisfies $x_0 \in B(\Lambda X, \delta_1)$, then

$$x_t \in |x_t| < \theta^{-1} X, \theta X >$$

holds for $t \in \{0, \ldots, T - N_1\}$.

Since the function $F \to \alpha(F), F \in U_0$ is continuous at the point G there exists a neighborhood $U_2 \subset U_1$ of G in $\mathcal{M}$ such that for each $F \in U_2$,

$$\{x \in K : |x| = 1, \ \alpha(F)x \in F(x) - K\} \subset < \theta^{-1} X, \theta X > . \qquad (6.25)$$

It follows from Theorems 11.3.1 and 11.3.2 that there exist a positive number $\delta_2 < 2^{-1}\delta_1$, a nonnegative integer N_2 and a natural number N_3 such that the following property holds:

(ii) If $T \geq N_2 + N_3$ is a natural number and if an optimal trajectory $\{x_t\}_{t=0}^T$ of G satisfies $x_0 \in B(x, \delta_2)$, then

$$\alpha(G)^{-t} x_t \in B(\Lambda X, 2^{-1}\delta_1), \ t \in \{N_2, \ldots, T - N_3\};$$

moreover if $x = X$, then $N_2 = 0$.

In view of Lemma 11.6.2, the choice of N_2, N_3 and the continuity of the function $F \to \alpha(F), F \in U_0$ at the point G there exist a positive number $\delta < \delta_2$ and a neighborhood $U \subset U_2$ of G in $\mathcal{M}$ such that the following property holds:

(iii) If $F \in U$ and if an optimal trajectory $\{x_t\}_{t=0}^{N_2+N_3}$ of F satisfies $x_0 \in B(x, \delta)$, then

$$\alpha(F)^{-N_2} x_{N_2} \in B(\Lambda X, \delta_1).$$

Put

$$L_1 = N_2, L_2 = 2(N_1 + N_2 + N_3). \qquad (6.26)$$

Assume that $F \in U$, an integer $T \geq L_1 + L_2$ and $\{x_t\}_{t=0}^T$ is an optimal trajectory of F such that $x_0 \in B(x, \delta)$. There exists $Y \in K$ such that

$$|Y| = 1, \ \alpha(F)Y \in F(Y) - K. \qquad (6.27)$$

In view of the choice of U_2 (see (6.25))

$$Y \in < \theta^{-1} X, \theta X > . \qquad (6.28)$$

Let

$$t_0 \in \{L_1, \ldots, T - L_2\}. \tag{6.29}$$

We will show that $\alpha(F)^{-t_0} x_{t_0} \in B(\Lambda X, \epsilon)$. Property (iii), (6.24) and (6.28) imply that

$$\alpha(F)^{-N_2} x_{N_2} \in B(\Lambda X, \delta_1) \subset \Lambda < \theta^{-2} Y, \theta^2 Y > . \tag{6.30}$$

We may assume that $t_0 > N_2$.

Consider an optimal trajectory $\{\alpha(F)^{-N_2} x_{N_2 + t}\}_{t=0}^{T-N_2}$ of F. By property (i) and (6.30),

$$x_t \in |x_t| < \theta^{-1} X, \theta X >, \ t \in \{N_2, \ldots, T - N_1\}. \tag{6.31}$$

Combined with (6.30), (6.29) and (6.28) this implies that there exists $h \in R^n$ such that

$$h \in F^{t_0 - N_2}(\theta^2 \Lambda Y), \ h \ge \alpha(F)^{-N_2} x_{t_0} \ge \alpha(F)^{-N_2} |x_{t_0}| \theta^{-2} Y, \tag{6.32}$$

$$F^{t_0 - N_2}(Y) - K \ni (\Lambda \theta^4 \alpha(F)^{N_2})^{-1} |x_{t_0}| Y.$$

There exists $p_F \in K^* \setminus \{0\}$ such that

$$(\alpha(F) p_F, u) \ge (p_F, v) \text{ for each } u \in K, v \in F(u).$$

Combined with (6.32) this inequality implies that

$$(\Lambda \theta^4 \alpha(F)^{N_2})^{-1} |x_{t_0}| \le \alpha(F)^{t_0 - N_2}, \ |x_{t_0}| \le \alpha(F)^{t_0} \Lambda \theta^4. \tag{6.33}$$

We will show that $|x_{t_0}| > \alpha(F)^{t_0} \Lambda \theta^{-8}$. Assume the contrary. Then by (6.31), (6.28), (6.29) and (6.26),

$$x_{t_0} \le |x_{t_0}| \theta^2 Y \le \theta^{-6} \alpha(F)^{t_0} \Lambda Y. \tag{6.34}$$

The inclusion (6.30) implies that

$$\alpha(F)^{t_0 - N_2} \Lambda Y \in F^{t_0 - N_2}(\Lambda Y) - K \subset F^{t_0 - N_2}(\theta^2 \alpha(F)^{-N_2} x_{N_2}) - K,$$

$$\theta^{-2} \alpha(F)^{t_0} \Lambda Y \in F^{t_0 - N_2}(x_{N_2}) - K. \tag{6.35}$$

Since $\{x_t\}_{t=0}^T$ is an optimal trajectory of F relation (6.35) is contradictory to (6.34). The obtained contradiction proves that

$$|x_{t_0}| > \alpha(F)^{t_0} \Lambda \theta^{-8}.$$

It follows from this relation, (6.31), (6.29), (6.33) and (6.23) that

$$x_{t_0} \in |x_{t_0}| < \theta^{-1} X, \theta X > \subset \Lambda \alpha(F)^{t_0} < \theta^{-9} X, \theta^9 X >,$$

$$\alpha(F)^{-t_0} x_{t_0} \in \Lambda < \theta^{-9} X, \theta^9 X > \subset B(\Lambda X, \epsilon).$$

This completes the proof of the theorem.

Proof of Theorem 11.5.3.
We preface the proof of Theorem 11.5.3 with two auxiliary results.

LEMMA 11.6.3 *Suppose that $z \in \text{int} K$ and $\mu > 0$. Then there exist a pair of positive numbers δ, c_3 and a neighborhood U of G in $\mathcal{M}$ such that $B(z, \delta) \subset \text{int} K$, $U \subset V$, and the following property holds:*
If $T \geq 1$ is an integer, $F \in U$, $x \in B(z, \delta)$ and if an optimal solution $\{x_t\}_{t=0}^{T}$ of the problem (P) satisfies $x_0 = x$, then

$$|\alpha(F)^{-t} x_t| \geq c_3, \ t = 0, \dots, T. \tag{6.36}$$

Proof. There exists a neighborhood U of G in $\mathcal{M}$ such that $U \subset V$ and for each $F \in U$,

$$\{x \in K : |x| = 1, \alpha(F) x \in F(x) - K\} \subset < 2^{-1} X, 2X > . \tag{6.37}$$

Choose $\delta > 0$ such that $B(z, \delta) \subset \text{int} K$ and choose

$$c_3 \in (0, 16^{-1} \mu r_0) \tag{6.38}$$

(recall r_0 in (6.1)).
   Assume that $F \in U, x \in B(z, \delta)$, an integer $T \geq 1$ and $\{x_t\}_{t=0}^{T}$ is an optimal solution of the problem (P) satisfying $x_0 = x$. We will show that (6.36) holds. Let us assume the converse. Then there exists $t \in \{0, \dots, T\}$ for which

$$|\alpha(F)^{-t} x_t| \leq c_3. \tag{6.39}$$

There exists $Y \in K$ such that

$$|Y| = 1, \ \alpha(F) Y \in F(Y) - K. \tag{6.40}$$

In view of the choice of U (see (6.37)),

$$Y \in < 2^{-1} X, 2X > . \tag{6.41}$$

There exists $p_F \in K^* \setminus \{0\}$ such that

$$(\alpha(F) p_F, u) \geq (p_F, v) \text{ for each } u \in K, v \in F(u). \tag{6.42}$$

(6.39), (6.1), (6.41), (6.38) imply that

$$\alpha(F)^{-t} x_t \leq c_3 r_0^{-1} X \leq 2 c_3 r_0^{-1} Y \leq 8^{-1} \mu Y,$$

$$2^{-1}\alpha(F)^T\mu Y \le \alpha(F)^T\mu X \in F^{T-t}(x_t) - K$$
$$\subset F^{T-t}(8^{-1}\mu\alpha(F)^t Y) - K \subset 8^{-1}\mu\alpha(F)^t F^{T-t}(Y) - K,$$
$$\alpha(F)^{T-t}Y \in 4^{-1}F^{T-t}(Y) - K.$$

This is contradictory to (6.42). The contradiction we have reached proves the lemma.

LEMMA 11.6.4 *Suppose that $z \in \mathrm{int}K$. Then there exist a pair of positive numbers δ, c_4 and a neighborhood U of G in $\mathcal{M}$ such that $B(z,\delta) \subset \mathrm{int}K$, $U \subset V$ and the following property holds:*
If $F \in U$, $T \ge 1$ is an integer and if a trajectory $\{x_t\}_{t=0}^T$ of F satisfies $x_0 \in B(z,\delta)$, then

$$(\alpha(F)^{-t}x_t, p) \le c_4, \ t = 0, \dots, T.$$

Proof. Choose a positive number $\epsilon < \lambda(z)$. By Theorem 11.5.2 there exist nonnegative integers L_1, L_2, a neighborhood U of G in $\mathcal{M}$ and a positive number δ such that

$$U \subset V \cap \{F \in \mathcal{M} : \rho(F,G) \le 4^{-1}\min\{1,\delta\}\}, \tag{6.43}$$

$$\delta < |z|, \ \sup\{|u| : u \in G(v) \text{ for some } v \in K \text{ satisfying } |v| \le 1\},$$

$$B(z,\delta) \subset \mathrm{int}K, \ \{\alpha(F) : F \in U\} \subset (2^{-1}\alpha(G), 2\alpha(G))$$

and that the following property holds:
If $F \in U$, $T \ge L_1 + L_2$ is a natural number and if an optimal trajectory $\{x_t\}_{t=0}^T$ of F satisfies $x_0 \in B(z,\delta)$, then

$$\alpha(F)^{-t}x_t \in B(\lambda(z)X, \epsilon), \ |\alpha(F)^{-t}x_t| \le 2\lambda(z) \tag{6.44}$$

for all $t \in \{L_1, \dots, T - L_2\}$.
Choose a number

$$c_5 > \sup\{\alpha(F)^{-t}|y| : y \in F^t(x), F \in U, \tag{6.45}$$

$$x \in B(z,\delta), \ t \in \{0, \dots, L_1 + L_2\}\}(1 + |p|)(1 + 2\lambda(z))$$

and choose a number

$$c_4 > \sup\{c_5, 2\lambda(z), 2\lambda(z)(4\alpha(G)^{-1})\sup\{|v| : \tag{6.46}$$

$$v \in G(u), \ u \in K, |u| \le 1\}\}(1 + |p|).$$

Assume that $F \in U$, an integer $T \ge 1$ and $\{x_t\}_{t=0}^T$ is a trajectory of F satisfying $x_0 \in B(z,\delta)$. To prove the lemma it is sufficient to show that

$$(\alpha(F)^{-T}x_T, p) \le c_4. \tag{6.47}$$

We may assume that $(x_T, p) = \sup\{(p, y) : y \in F^T(x_0)\}$. This implies
that $\{x_t\}_{t=0}^T$ is an optimal trajectory of F. If $T < L_1 + L_2$, then (6.47)
follows from (6.45) and (6.46). Assume that $T \geq L_1 + L_2$. In view of
the choice of U, δ, L_1, L_2 (see (6.44))

$$|\alpha(F)^{L_2-T} x_{T-L_2}| \leq 2\lambda(z).$$

Combined with (6.46), (6.43), (6.45) this implies (6.47). This completes
the proof of the lemma.

We will complete the proof of Theorem 11.5.3. Lemmas 11.6.3 and
11.6.4 imply that there exist positive numbers δ_1, c_3, c_4 and a neighbor-
hood U_0 of G in $\mathcal{M}$ such that $U_0 \subset V, B(z, \delta_1) \subset \text{int} K$ and the following
property holds:
  (a) If $F \in U_0$, $T \geq 1$ is a natural number, $x \in B(z, \delta_1)$, and if an
optimal solution $\{x_t\}_{t=0}^T$ of the problem (P) satisfies $x_0 = x$, then

$$c_3 \leq |\alpha(F)^{-t} x_t| \leq c_4, \quad t = 0, \ldots, T.$$

In view of Theorem 11.5.1 there exist $\delta \in (0, \delta_1), \gamma > 1$, a natural number
$q \geq 1$, a pair of nonnegative integers L_i, $i = 1, 2$ and a neighborhood
$U \subset U_0$ of G in $\mathcal{M}$ such that the following property holds:
  (b) If $T \geq L_1 + L_2$ is a natural number, $F \in U$ and if a (q, γ)-optimal
trajectory $\{x_t\}_{t=0}^T$ of F satisfies $x_0 \in B(z, \delta)$, then

$$x_t \neq 0, |x_t|^{-1} x_t \in B(X, \epsilon), \quad t \in \{L_1, \ldots, T - L_2\};$$

moreover if $z = X$, then $L_1 = 0$.
  Choose a natural number L_3 such that

$$(\gamma - 1) L_3 c_3 c_2 > q c_1 c_4$$

and put

$$L_0 = L_2 + 2L_3 + 2q.$$

Assume that $F \in U$, a natural number $T \geq L_1 + L_0, x \in B(z, \delta)$ and
$\{x_t\}_{t=0}^T$ is an optimal solution of the problem (P) satisfying $x_0 = x$. We
will show that $\{x_t\}_{t=0}^{T-L_3}$ is a (q, γ)-optimal trajectory of F.
  Let us assume the converse. Then there exists an integer $\tau \in [0, T - L_3 - q]$ and a number $\gamma_1 > \gamma$ such that $\gamma_1 x_{\tau+q} \in F^q(x_\tau) - K$. This
implies that there exists a trajectory $\{y_t\}_{t=0}^T$ of F for which $y_t = x_t$, $t = 0, \ldots, \tau$, $y_t \geq \gamma_1 x_t$, $t = \tau + q, \ldots, T$. It follows from the optimality of
$\{x_t\}_{t=0}^T$, the definition of $\{y_t\}_{t=0}^T$, L_3, (5.3) and property (a) that

$$0 \geq \sum_{t=0}^T \alpha(F)^{-t} \phi_t(y_t) - \sum_{t=0}^T \alpha(F)^{-t} \phi_t(x_t) \geq (\gamma - 1) \sum_{t=\tau+q}^T \phi_t(x_t) \alpha(F)^{-t}$$

$$-\sum_{t=\tau}^{\tau+q-1} \alpha(F)^{-t}\phi_t(x_t) \geq (\gamma-1)L_3 c_3 c_2 - q c_4 c_1 > 0.$$

The obtained contradiction proves that $\{x_t\}_{t=0}^{T-L_3}$ is a (q, γ)-optimal trajectory of F. By property (b),

$$x_t \neq 0, \|x_t\|^{-1}x_t - X| \leq \epsilon, \ t = L_1, \ldots, T - L_2 - L_3.$$

This completes the proof of the theorem.

Chapter 12

A DYNAMIC ZERO-SUM GAME

In this chapter we consider a class of dynamic discrete-time two-player zero-sum games. We show that for a generic cost function and each initial state there exists a pair of overtaking equilibria strategies over an infinite horizon. We also establish that for a generic cost function f there exists a pair of stationary equilibria strategies (x_f, y_f) such that each pair of "approximate" equilibria strategies spends almost all of its time in a small neighborhood of (x_f, y_f).

12.1. Preliminaries

Denote by $|\cdot|$ the Euclidean norm in R^m. Let $X \subset R^{m_1}$ and $Y \subset R^{m_2}$ be nonempty convex compact sets. Denote by $\mathcal{M}$ the set of all continuous functions $f : X \times X \times Y \times Y \to R^1$ such that:

for each $(y_1, y_2) \in Y \times Y$ the function $(x_1, x_2) \to f(x_1, x_2, y_1, y_2)$, $(x_1, x_2) \in X \times X$ is convex;

for each $(x_1, x_2) \in X \times X$ the function $(y_1, y_2) \to f(x_1, x_2, y_1, y_2)$, $(y_1, y_2) \in Y \times Y$ is concave.

For the set $\mathcal{M}$ we define a metric $\rho : \mathcal{M} \times \mathcal{M} \to R^1$ by

$$\rho(f, g) = \sup\{|f(x_1, x_2, y_1, y_2) - g(x_1, x_2, y_1, y_2)| : \qquad (1.1)$$

$$x_1, x_2 \in X, \quad y_1, y_2 \in Y\}, \quad f, g \in \mathcal{M}.$$

Clearly $\mathcal{M}$ is a complete metric space.

Given $f \in \mathcal{M}$ and an integer $n \geq 1$ we consider a discrete-time two-player zero-sum game over the interval $[0, n]$. For this game $\{\{x_i\}_{i=0}^n : x_i \in X, i = 0, \ldots, n\}$ is the set of strategies for the first player, $\{\{y_i\}_{i=0}^n :$

$y_i \in Y$, $i = 0, \ldots, n\}$ is the set of strategies for the second player, and the cost for the first player associated with the strategies $\{x_i\}_{i=0}^n$, $\{y_i\}_{i=0}^n$ is given by $\sum_{i=0}^{n-1} f(x_i, x_{i+1}, y_i, y_{i+1})$.

Definition 1.1 Let $f \in \mathcal{M}$, $n \geq 1$ be an integer and let $M \in [0, \infty)$. A pair of sequences $\{\bar{x}_i\}_{i=0}^n \subset X$, $\{\bar{y}_i\}_{i=0}^n \subset Y$ is called (f, M)-good if the following properties hold:

for each sequence $\{x_i\}_{i=0}^n \subset X$ satisfying $x_0 = \bar{x}_0$, $x_n = \bar{x}_n$,

$$M + \sum_{i=0}^{n-1} f(x_i, x_{i+1}, \bar{y}_i, \bar{y}_{i+1}) \geq \sum_{i=0}^{n-1} f(\bar{x}_i, \bar{x}_{i+1}, \bar{y}_i, \bar{y}_{i+1}); \qquad (1.2)$$

for each sequence $\{y_i\}_{i=0}^n \subset Y$ satisfying $y_0 = \bar{y}_0$, $y_n = \bar{y}_n$,

$$M + \sum_{i=0}^{n-1} f(\bar{x}_i, \bar{x}_{i+1}, \bar{y}_i, \bar{y}_{i+1}) \geq \sum_{i=0}^{n-1} f(\bar{x}_i, \bar{x}_{i+1}, y_i, y_{i+1}). \qquad (1.3)$$

If a pair of sequences $\{x_i\}_{i=0}^n \subset X$, $\{y_i\}_{i=0}^n \subset Y$ is $(f, 0)$-good, it is called (f)-optimal.

Our first main result in this chapter deals with the so-called "turnpike property" of "good" pairs of sequences.

Consider any $f \in \mathcal{M}$. We say that the function f has the *turnpike property* if there exists a unique pair $(x_f, y_f) \in X \times Y$ for which the following assertion holds:

For each $\epsilon > 0$ there exist an integer $n_0 \geq 2$ and a number $\delta > 0$ such that for each integer $n \geq 2n_0$ and each (f, δ)-good pair of sequences $\{x_i\}_{i=0}^n \subset X$, $\{y_i\}_{i=0}^n \subset Y$ the relations $|x_i - x_f|$, $|y_i - y_f| \leq \epsilon$ hold for all integers $i \in [n_0, n - n_0]$.

In this chapter our goal is to show that the turnpike property holds for a generic $f \in \mathcal{M}$. We will prove the existence of a set $\mathcal{F} \subset \mathcal{M}$ which is a countable intersection of open everywhere dense sets in $\mathcal{M}$ such that each $f \in \mathcal{F}$ has the turnpike property (see Theorem 12.2.1).

We also study the existence of equilibria over an infinite horizon for the class of zero-sum games considered in the chapter.

Definition 1.2 Let $f \in \mathcal{M}$. A pair of sequences

$$\{\bar{x}_i\}_{i=0}^\infty \subset X, \ \{\bar{y}_i\}_{i=0}^\infty \subset Y$$

is called (f)-*overtaking optimal* if the following properties hold:

for each sequence $\{x_i\}_{i=0}^\infty \subset X$ satisfying $x_0 = \bar{x}_0$,

$$\limsup_{T \to \infty} [\sum_{i=0}^{T-1} f(\bar{x}_i, \bar{x}_{i+1}, \bar{y}_i, \bar{y}_{i+1}) - \sum_{i=0}^{T-1} f(x_i, x_{i+1}, \bar{y}_i, \bar{y}_{i+1})] \leq 0;$$

for each sequence $\{y_i\}_{i=0}^{\infty} \subset Y$ satisfying $y_0 = \bar{y}_0$,

$$\limsup_{T \to \infty} [\sum_{i=0}^{T-1} f(\bar{x}_i, \bar{x}_{i+1}, y_i, y_{i+1}) - \sum_{i=0}^{T-1} f(\bar{x}_i, \bar{x}_{i+1}, \bar{y}_i, \bar{y}_{i+1})] \leq 0.$$

Our second main result (see Theorem 12.2.2) shows that for a generic $f \in \mathcal{M}$ and each $(x, y) \in X \times Y$ there exists an (f)-overtaking optimal pair of sequences $\{x_i\}_{i=0}^{\infty} \subset X$, $\{y_i\}_{i=0}^{\infty} \subset Y$ such that $x_0 = x$, $y_0 = y$. The results of this chapter were obtained in [102].

12.2. Main results

In this section we present the main results of the chapter.

THEOREM 12.2.1 *There exists a set $\mathcal{F} \subset \mathcal{M}$ which is a countable intersection of open everywhere dense sets in $\mathcal{M}$ such that for each $f \in \mathcal{F}$ the following assertions hold:*

1. There exists a unique pair $(x_f, y_f) \in X \times Y$ for which

$$\sup_{y \in Y} f(x_f, x_f, y, y) = f(x_f, x_f, y_f, y_f) = \inf_{x \in X} f(x, x, y_f, y_f).$$

2. For each $\epsilon > 0$ there exist a neighborhood U of f in $\mathcal{M}$, an integer $n_0 \geq 2$ and a number $\delta > 0$ such that for each $g \in U$, each integer $n \geq 2n_0$ and each (g, δ)-good pair of sequences $\{x_i\}_{i=0}^n \subset X$, $\{y_i\}_{i=0}^n \subset Y$ the relation

$$|x_i - x_f|, \; |y_i - y_f| \leq \epsilon \tag{2.1}$$

holds for all integers $i \in [n_0, n-n_0]$. Moreover, if $|x_0 - x_f|, |y_0 - y_f| \leq \delta$, then (2.1) holds for all integers $i \in [0, n-n_0]$, and if $|x_n - x_f|, |y_n - y_f| \leq \delta$, then (2.1) is valid for all integers $i \in [n_0, n]$.

THEOREM 12.2.2 *There exists a set $\mathcal{F} \subset \mathcal{M}$ which is a countable intersection of open everywhere dense sets in $\mathcal{M}$ such that for each $f \in \mathcal{F}$ the following assertion holds:*

For each $x \in X$ and each $y \in Y$ there exists an (f)-overtaking optimal pair of sequences $\{x_i\}_{i=0}^{\infty} \subset X$, $\{y_i\}_{i=0}^{\infty} \subset Y$ such that $x_0 = x$, $y_0 = y$.

The chapter is organized as follows. Section 12.3 contains definitions and notation while Section 12.4 contains preliminary results. In Section 12.5 we prove the existence of a minimal pair of sequences. Section 12.6 contains auxiliary results for Theorem 12.2.1 while Section 12.7 contains auxiliary results for Theorem 12.2.2. Theorems 12.2.1 and 12.2.2 are proved in Section 12.8.

12.3.   Definitions and notation

Let $f \in \mathcal{M}$. Define a function $\bar{f} : X \times Y \to R^1$ by

$$\bar{f}(x, y) = f(x, x, y, y), \quad x \in X, \ y \in Y. \tag{3.1}$$

Then there exists a saddle point $(x_f, y_f) \in X \times Y$ for $\bar{f}$. We have

$$\sup_{y \in Y} \bar{f}(x_f, y) = \bar{f}(x_f, y_f) = \inf_{x \in X} \bar{f}(x, y_f). \tag{3.2}$$

Set

$$\mu(f) = \bar{f}(x_f, y_f). \tag{3.3}$$

Definition 3.1. Let $f \in \mathcal{M}$. A pair of sequences

$$\{x_i\}_{i=0}^{\infty} \subset X, \ \{y_i\}_{i=0}^{\infty} \subset Y$$

is called (f)-minimal if for each integer $n \geq 2$ the pair of sequences $\{x_i\}_{i=0}^{n}$, $\{y_i\}_{i=0}^{n}$ is (f)-optimal.

We will show in Section 12.5 (see Proposition 12.5.3) that for each $f \in \mathcal{M}$, each $x \in X$ and each $y \in Y$ there exists an (f)-minimal pair of sequences $\{x_i\}_{i=0}^{\infty} \subset X$, $\{y_i\}_{i=0}^{\infty} \subset Y$ such that $x_0 = x$, $y_0 = y$.

Let $f \in \mathcal{M}$, $n \geq 1$ be an integer, and let $\xi = (\xi_1, \xi_2, \xi_3, \xi_4) \in X \times X \times Y \times Y$. Define

$$\Lambda_X(\xi, n) = \{\{x_i\}_{i=0}^{n} \subset X : \quad x_0 = \xi_1, \ x_n = \xi_2\}, \tag{3.4}$$

$$\Lambda_Y(\xi, n) = \{\{y_i\}_{i=0}^{n} \subset Y : \quad y_0 = \xi_3, \ y_n = \xi_4\}, \tag{3.5}$$

$$f^{(\xi, n)}((x_0, \ldots, x_i, \ldots, x_n), (y_0, \ldots, y_i, \ldots y_n)) = \sum_{i=0}^{n-1} f(x_i, x_{i+1}, y_i, y_{i+1}),$$

$$\tag{3.6}$$

$$\{x_i\}_{i=0}^{n} \in \Lambda_X(\xi, n), \quad \{y_i\}_{i=0}^{n} \in \Lambda_Y(\xi, n).$$

For $x \in X$, $y \in Y$, $r > 0$ put

$$B_X(x, r) = \{z \in X : |x - z| \leq r\},$$

$$B_Y(y, r) = \{z \in Y : |y - z| \leq r\}.$$

12.4.   Preliminary results

Let M, N be nonempty sets and let $f : M \times N \to R^1$. Set

$$f^a(x) = \sup_{y \in N} f(x, y), \ x \in M, \quad f^b(y) = \inf_{x \in M} f(x, y), \mathrm{y} \in N, \quad (4.1)$$

$$v_f^a = \inf_{x \in M} \sup_{y \in N} f(x, y), \quad v_f^b = \sup_{y \in N} \inf_{x \in M} f(x, y). \quad (4.2)$$

Clearly

$$v_f^b \leq v_f^a. \quad (4.3)$$

We have the following result (see Chapter 6, Section 2, Proposition 1 of [5]).

PROPOSITION 12.4.1 *Let* $f : M \times N \to R^1$, $\bar{x} \in M$, $\bar{y} \in N$. *Then*

$$\sup_{y \in N} f(\bar{x}, y) = f(\bar{x}, \bar{y}) = \inf_{x \in M} f(x, \bar{y}) \quad (4.4)$$

if and only if

$$v_f^a = v_f^b, \ \sup_{y \in N} f(\bar{x}, y) = v_f^a, \ \inf_{x \in M} f(x, \bar{y}) = v_f^b.$$

Let $f : M \times N \to R^1$. If $(\bar{x}, \bar{y}) \in M \times N$ satisfies (4.4), it is called a saddle point (for f). We have the following result (see Chapter 6, Section 2, Theorem 8 of [5]).

PROPOSITION 12.4.2 *Let* $M \subset R^m$, $N \subset R^n$ *be convex compact sets and let* $f : M \times N \to R^1$ *be a continuous function. Assume that for each* $y \in N$ *the function* $x \to f(x, y)$, $x \in M$ *is convex and for each* $x \in M$ *the function* $y \to f(x, y)$, $y \in N$ *is concave. Then there exists a saddle point for* f.

PROPOSITION 12.4.3 *Let* M, N *be nonempty sets,* $f : M \times N \to R^1$ *and*

$$-\infty < v_f^a = v_f^b < +\infty, \ x_0 \in M, \ y_0 \in N, \ \Delta_1, \Delta_2 \in [0, \infty), \quad (4.5)$$

$$\sup_{y \in N} f(x_0, y) \leq v_f^a + \Delta_1, \quad \inf_{x \in M} f(x, y_0) \geq v_f^b - \Delta_2. \quad (4.6)$$

Then

$$\sup_{y \in N} f(x_0, y) - \Delta_1 - \Delta_2 \leq f(x_0, y_0) \leq \inf_{x \in M} f(x, y_0) + \Delta_1 + \Delta_2. \quad (4.7)$$

Proof. By (4.6) and (4.5),

$$\sup_{y \in N} f(x_0, y) - \Delta_1 - \Delta_2 \leq v_f^a - \Delta_2 = v_f^b - \Delta_2 \leq \inf_{x \in M} f(x, y_0) \leq f(x_0, y_0)$$

$$\leq \sup_{y \in N} f(x_0, y) \leq v_f^a + \Delta_1 = v_f^b + \Delta_1 \leq \inf_{x \in M} f(x, y_0) + \Delta_1 + \Delta_2.$$

The proposition is proved.

PROPOSITION 12.4.4 *Let M, N be nonempty sets and let $f : M \times N \to R^1$. Assume that (4.5) is valid, $x_0 \in M$, $y_0 \in N$, $\Delta_1, \Delta_2 \in [0, \infty)$ and*

$$\sup_{y \in N} f(x_0, y) - \Delta_2 \leq f(x_0, y_0) \leq \inf_{x \in M} f(x, y_0) + \Delta_1. \tag{4.8}$$

Then

$$\sup_{y \in N} f(x_0, y) \leq v_f^a + \Delta_1 + \Delta_2, \quad \inf_{x \in M} f(x, y_0) \geq v_f^b - \Delta_1 - \Delta_2. \tag{4.9}$$

Proof. It follows from (4.8), (4.2), (4.5) and (4.3) that

$$v_f^b - \Delta_2 = v_f^a - \Delta_2 \leq \sup_{y \in N} f(x_0, y) - \Delta_2$$

$$\leq \inf_{x \in M} f(x, y_0) + \Delta_1 \leq v_f^b + \Delta_1.$$

This implies (4.9). The proposition is proved.

12.5.   The existence of a minimal pair of sequences

Let $f \in \mathcal{M}$, $x_f \in X$, $y_f \in Y$ and

$$\sup_{y \in Y} \bar{f}(x_f, y) = \bar{f}(x_f, y_f) = \inf_{x \in X} \bar{f}(x, y_f). \tag{5.1}$$

PROPOSITION 12.5.1 *Suppose that $n \geq 2$ is a natural number and*

$$\bar{x}_i = x_f, \quad \bar{y}_i = y_f, \quad i = 0, \dots, n. \tag{5.2}$$

Then the pair of sequences $\{\bar{x}_i\}_{i=0}^n$, $\{\bar{y}_i\}_{i=0}^n$ is (f)-optimal.

Proof. Assume that $\{x_i\}_{i=0}^n \subset X$, $\{y_i\}_{i=0}^n \subset Y$ and

$$x_0, x_n = x_f, \quad y_0, y_n = y_f. \tag{5.3}$$

The equalities (5.3), (5.2) and (5.1) imply that

$$\sum_{i=0}^{n-1} f(x_i, x_{i+1}, \bar{y}_i, \bar{y}_{i+1}) = \sum_{i=0}^{n-1} f(x_i, x_{i+1}, y_f, y_f)$$

$$\geq nf(n^{-1} \sum_{i=0}^{n-1} x_i, n^{-1} \sum_{i=0}^{n-1} x_{i+1}, y_f, y_f)$$

$$= nf(n^{-1} \sum_{i=0}^{n-1} x_i, n^{-1} \sum_{i=0}^{n-1} x_i, y_f, y_f) \geq nf(x_f, x_f, y_f, y_f),$$

$$\sum_{i=0}^{n-1} f(\bar{x}_i, \bar{x}_{i+1}, y_i, y_{i+1}) = \sum_{i=0}^{n-1} f(x_f, x_f, y_i, y_{i+1})$$

$$\leq nf(x_f, x_f, n^{-1} \sum_{i=0}^{n-1} y_i, n^{-1} \sum_{i=0}^{n-1} y_{i+1})$$

$$= nf(x_f, x_f, n^{-1} \sum_{i=0}^{n-1} y_i, n^{-1} \sum_{i=0}^{n-1} y_i) \leq nf(x_f, x_f, y_f, y_f).$$

This completes the proof of the proposition.

PROPOSITION 12.5.2 *Let $n \geq 2$ be an integer and let*

$$(\{x_i^{(k)}\}_{i=0}^n, \{y_i^{(k)}\}_{i=0}^n) \subset X \times Y, \quad k = 1, 2, \ldots$$

be a sequence of (f)-optimal pairs. Assume that

$$\lim_{k \to \infty} x_i^{(k)} = x_i, \quad \lim_{k \to \infty} y_i^{(k)} = y_i, \quad i = 0, 1, 2, \ldots, n. \quad (5.4)$$

Then the pair of sequences $(\{x_i\}_{i=0}^n, \{y_i\}_{i=0}^n)$ is (f)-optimal.

Proof. Let

$$\{u_i\}_{i=0}^n \subset X, \quad u_0 = x_0, u_n = x_n. \quad (5.5)$$

We will show that

$$\sum_{i=0}^{n-1} f(x_i, x_{i+1}, y_i, y_{i+1}) \leq \sum_{i=0}^{n-1} f(u_i, u_{i+1}, y_i, y_{i+1}). \quad (5.6)$$

Assume the contrary. Then there exists a positive number ϵ such that

$$\sum_{i=0}^{n-1} f(x_i, x_{i+1}, y_i, y_{i+1}) > \sum_{i=0}^{n-1} f(u_i, u_{i+1}, y_i, y_{i+1}) + 8\epsilon. \quad (5.7)$$

There is a positive number $\delta < \epsilon$ such that the following property holds:
If $z_1, z_2, \bar{z}_1, \bar{z}_2 \in X$, $\xi_1, \xi_2, \bar{\xi}_1, \bar{\xi}_2 \in Y$ satisfy

$$|z_i - \bar{z}_i|, \ |\xi_i - \bar{\xi}_i| \leq \delta, \quad i = 1, 2,$$

then

$$|f(z_1, z_2, \xi_1, \xi_2) - f(\bar{z}_1, \bar{z}_2, \bar{\xi}_1, \bar{\xi}_2)| \leq \epsilon(8n)^{-1}. \tag{5.8}$$

There is a natural number q such that

$$|x_i - x_i^{(q)}|, \ |y_i - y_i^{(q)}| \leq \delta, \quad i = 0, \ldots, n. \tag{5.9}$$

Set

$$u_0^{(q)} = x_0^{(q)}, \ u_n^{(q)} = x_n^{(q)}, \quad u_i^{(q)} = u_i, \ i = 1, \ldots, n - 1. \tag{5.10}$$

Since the pair of sequences $(\{x_i^{(q)}\}_{i=0}^n, \{y_i^{(q)}\}_{i=0}^n)$ is (f)-optimal it follows from (5.10) that

$$\sum_{i=0}^{n-1} f(x_i^{(q)}, x_{i+1}^{(q)}, y_i^{(q)}, y_{i+1}^{(q)}) \leq \sum_{i=0}^{n-1} f(u_i^{(q)}, u_{i+1}^{(q)}, y_i^{(q)}, y_{i+1}^{(q)}). \tag{5.11}$$

In view of the choice of δ (see (5.8)), (5.9), (5.10) and (5.5) for $i = 0, \ldots, n - 1$,

$$|f(x_i^{(q)}, x_{i+1}^{(q)}, y_i^{(q)}, y_{i+1}^{(q)}) - f(x_i, x_{i+1}, y_i, y_{i+1})| \leq (8n)^{-1}\epsilon,$$

$$|f(u_i^{(q)}, u_{i+1}^{(q)}, y_i^{(q)}, y_{i+1}^{(q)}) - f(u_i, u_{i+1}, y_i, y_{i+1})| \leq (8n)^{-1}\epsilon.$$

Combined with (5.7) these inequalities imply that

$$\sum_{i=0}^{n-1} f(x_i^{(q)}, x_{i+1}^{(q)}, y_i^{(q)}, y_{i+1}^{(q)}) - \sum_{i=0}^{n-1} f(u_i^{(q)}, u_{i+1}^{(q)}, y_i^{(q)}, y_{i+1}^{(q)}) > \epsilon.$$

This is contradictory to (5.11). The obtained contradiction proves that (5.6) is valid. Analogously we can show that for each $\{u_i\}_{i=0}^n \subset Y$ satisfying $u_0 = y_0$, $u_n = y_n$, the following relation holds:

$$\sum_{i=0}^{n-1} f(x_i, x_{i+1}, y_i, y_{i+1}) \geq \sum_{i=0}^{n-1} f(x_i, x_{i+1}, u_i, u_{i+1}).$$

This completes the proof of the proposition.

The following proposition is the main result of this section.

PROPOSITION 12.5.3 *Let $f \in \mathcal{M}$ and let $x \in X$, $y \in Y$. Then there exists an (f)-minimal pair of sequences $\{x_i\}_{i=0}^\infty \subset X$, $\{y_i\}_{i=0}^\infty \subset Y$ such that $x_0 = x$, $y_0 = y$.*

Proof. By Proposition 12.4.2 for each integer $n \geq 2$ there exists an (f)-optimal pair of sequences $\{x_i^{(n)}\}_{i=0}^n \subset X$, $\{y_i^{(n)}\}_{i=0}^n \subset Y$ such that $x_0^{(n)} =$

x, $y_0^{(n)} = y$. There exist a pair of sequences $\{x_i\}_{i=0}^{\infty} \subset X$, $\{y_i\}_{i=0}^{\infty} \subset Y$ and a strictly increasing sequence of natural numbers $\{n_k\}_{k=1}^{\infty}$ such that for each integer $i \geq 0$,

$$x_i^{(n_k)} \to x_i, \quad y_i^{(n_k)} \to y_i \text{ as } k \to \infty.$$

It follows from Proposition 12.5.2 that the pair of sequences $\{x_i\}_{i=0}^{\infty}$, $\{y_i\}_{i=0}^{\infty}$ is (f)-minimal. The proposition is proved.

12.6. Preliminary lemmas for Theorem 12.2.1

Let $f \in \mathcal{M}$. There exist $x_f \in X$, $y_f \in Y$ such that

$$\sup_{y \in Y} f(x_f, x_f, y, y) = f(x_f, x_f, y_f, y_f) = \inf_{x \in X} f(x, x, y_f, y_f). \qquad (6.1)$$

For each $r \in (0, 1)$ we define $f_r \in \mathcal{M}$ which has the turnpike property such that $f_r \to f$ as $r \to 0^+$ in $\mathcal{M}$.

Let $r \in (0, 1)$. Define $f_r : X^2 \times Y^2 \to R^1$ by

$$f_r(x_1, x_2, y_1, y_2) = f(x_1, x_2, y_1, y_2) + r|x_1 - x_f| - r|y_1 - y_f|, \qquad (6.2)$$

$$x_1, x_2 \in X, \quad y_1, y_2 \in Y.$$

Clearly $f_r \in \mathcal{M}$,

$$\sup_{y \in Y} f_r(x_f, x_f, y, y) = f_r(x_f, x_f, y_f, y_f) = \inf_{x \in X} f_r(x, x, y_f, y_f). \qquad (6.3)$$

We show that f_r has the turnpike property. We begin with the following lemma which establishes that if a good pair of sequences

$$\{x_i\}_{i=0}^{n}, \quad \{y_i\}_{i=0}^{n}$$

satisfies $x_n, x_0 = x_f$ and $y_n, y_0 = y_f$, then $\{x_i\}_{i=0}^{n}$ is contained in a small neighborhood of x_f and $\{y_i\}_{i=0}^{n}$ is contained in a small neighborhood of y_f.

LEMMA 12.6.1 *For each $\epsilon \in (0, 1)$ there exists a positive number $\delta < \epsilon$ such that the following property holds:*

If $n \geq 2$ is a natural number and if an (f_r, δ)-good pair of sequences $\{x_i\}_{i=0}^{n} \subset X$, $\{y_i\}_{i=0}^{n} \subset Y$ satisfies

$$x_n, x_0 = x_f, \quad y_n, y_0 = y_f, \qquad (6.4)$$

then

$$x_i \in B_X(x_f, \epsilon), \quad y_i \in B_Y(y_f, \epsilon), \quad i = 0, \ldots, n. \qquad (6.5)$$

Proof. Let $\epsilon \in (0,1)$. Choose a positive number

$$\delta < 8^{-1} r\epsilon. \tag{6.6}$$

Assume that an integer $n \geq 2$, $\{x_i\}_{i=0}^{n} \subset X$, $\{y_i\}_{i=0}^{n} \subset Y$ is an (f_r, δ)-good pair of sequences and (6.4) is valid. Set

$$\xi_1, \xi_2 = x_f, \quad \xi_3, \xi_4 = y_f, \quad \xi = (\xi_1, \xi_2, \xi_3, \xi_4). \tag{6.7}$$

Consider the sets $\Lambda_X(\xi, n)$, $\Lambda_Y(\xi, n)$ and the functions $(f_r)^{(\xi,n)}$, $f^{(\xi,n)}$. (see (3.4)-(3.6)). Proposition 12.5.1 and (6.1) imply that

$$\sup\{\sum_{i=0}^{n-1} f(x_f, x_f, u_i, u_{i+1}) : \{u_i\}_{i=0}^{n} \in \Lambda_Y(\xi, n)\} = nf(x_f, x_f, y_f, y_f)$$

$$\tag{6.8}$$

$$= \inf\{\sum_{i=0}^{n-1} f(p_i, p_{i+1}, y_f, y_f) : \{p_i\}_{i=0}^{n} \in \Lambda_X(\xi, n)\}.$$

In view of (6.8) and Proposition 12.4.1,

$$\sup\{\sum_{i=0}^{n-1} f(x_f, x_f, u_i, u_{i+1}) : \{u_i\}_{i=0}^{n} \in \Lambda_Y(\xi, n)\} \tag{6.9}$$

$$= \inf\{\sup\{\sum_{i=0}^{n-1} f(p_i, p_{i+1}, u_i, u_{i+1}) : \{u_i\}_{i=0}^{n} \in \Lambda_Y(\xi, n)\} :$$

$$\{p_i\}_{i=0}^{n} \in \Lambda_X(\xi, n)\},$$

$$\inf\{\sum_{i=0}^{n-1} f(p_i, p_{i+1}, y_f, y_f) : \{p_i\}_{i=0}^{n} \in \Lambda_X(\xi, n)\} \tag{6.10}$$

$$= \sup\{\inf\{\sum_{i=0}^{n-1} f(p_i, p_{i+1}, u_i, u_{i+1}) : \{p_i\}_{i=0}^{n} \in \Lambda_X(\xi, n)\} :$$

$$\{u_i\}_{i=0}^{n} \in \Lambda_Y(\xi, n)\}.$$

Proposition 12.5.1 and (6.3) imply that

$$\sup\{\sum_{i=0}^{n-1} f_r(x_f, x_f, u_i, u_{i+1}) : \{u_i\}_{i=0}^{n} \in \Lambda_Y(\xi, n)\} = nf_r(x_f, x_f, y_f, y_f)$$

$$\tag{6.11}$$

$$= \inf\{\sum_{i=0}^{n-1} f_r(p_i, p_{i+1}, y_f, y_f) : \{p_i\}_{i=0}^{n} \in \Lambda_X(\xi, n)\}.$$

It follows from Proposition 12.4.1 and (6.11) that

$$\sup\{\sum_{i=0}^{n-1} f_r(x_f, x_f, u_i, u_{i+1}) : \{u_i\}_{i=0}^n \in \Lambda_Y(\xi, n)\} \qquad (6.12)$$

$$= \inf\{\sup\{\sum_{i=0}^{n-1} f_r(p_i, p_{i+1}, u_i, u_{i+1}) : \{u_i\}_{i=0}^n \in \Lambda_Y(\xi, n)\} :$$

$$\{p_i\}_{i=0}^n \in \Lambda_X(\xi, n)\},$$

$$\inf\{\sum_{i=0}^{n-1} f_r(p_i, p_{i+1}, y_f, y_f) : \{p_i\}_{i=0}^n \in \Lambda_X(\xi, n)\} \qquad (6.13)$$

$$= \sup\{\inf\{\sum_{i=0}^{n-1} f_r(p_i, p_{i+1}, u_i, u_{i+1}) : \{p_i\}_{i=0}^n \in \Lambda_X(\xi, n)\} :$$

$$\{u_i\}_{i=0}^n \in \Lambda_Y(\xi, n)\}.$$

(6.7) and (6.4) imply that

$$\{x_i\}_{i=0}^n \in \Lambda_X(\xi, n), \quad \{y_i\}_{i=0}^n \in \Lambda_Y(\xi, n). \qquad (6.14)$$

Since $(\{x_i\}_{i=0}^n, \{y_i\}_{i=0}^n)$ is an (f_r, δ)-good pair of sequences we conclude that

$$\sup\{\sum_{i=0}^{n-1} f_r(x_i, x_{i+1}, u_i, u_{i+1}) : \{u_i\}_{i=0}^n \in \Lambda_Y(\xi, n)\} - \delta \qquad (6.15)$$

$$\leq \sum_{i=0}^{n-1} f_r(x_i, x_{i+1}, y_i, y_{i+1})$$

$$\leq \inf\{\sum_{i=0}^{n-1} f_r(p_i, p_{i+1}, y_i, y_{i+1}) : \{p_i\}_{i=0}^n \in \Lambda_X(\xi, n)\} + \delta.$$

In view of Proposition 12.4.4, (6.15), (6.12) and (6.13),

$$\sup\{\sum_{i=0}^{n-1} f_r(x_i, x_{i+1}, u_i, u_{i+1}) : \{u_i\}_{i=0}^n \in \Lambda_Y(\xi, n)\} \qquad (6.16)$$

$$\leq \sup\{\sum_{i=0}^{n-1} f_r(x_f, x_f, u_i, u_{i+1}) : \{u_i\}_{i=0}^n \in \Lambda_Y(\xi, n)\} + 2\delta,$$

$$\inf\{\sum_{i=0}^{n-1} f_r(p_i, p_{i+1}, y_i, y_{i+1}) : \{p_i\}_{i=0}^n \in \Lambda_X(\xi, n)\} \qquad (6.17)$$

$$\geq \inf\{\sum_{i=0}^{n-1} f_r(p_i, p_{i+1}, y_f, y_f) : \{p_i\}_{i=0}^{n} \in \Lambda_X(\xi, n)\} - 2\delta.$$

It follows from (6.2), (6.11), (6.16) and (6.8) that

$$n f(x_f, x_f, y_f, y_f) = n f_r(x_f, x_f, y_f, y_f) \tag{6.18}$$

$$\geq \sup\{\sum_{i=0}^{n-1} f_r(x_i, x_{i+1}, u_i, u_{i+1}) : \{u_i\}_{i=0}^{n} \in \Lambda_Y(\xi, n)\} - 2\delta$$

$$\geq -2\delta + \sum_{i=0}^{n-1} f_r(x_i, x_{i+1}, y_f, y_f) = -2\delta$$

$$+ r \sum_{i=0}^{n-1} |x_i - x_f| + \sum_{i=0}^{n-1} f(x_i, x_{i+1}, y_f, y_f)$$

$$\geq -2\delta + r \sum_{i=0}^{n-1} |x_i - x_f| + n f(x_f, x_f, y_f, y_f).$$

(6.2), (6.11), (6.17) and (6.8) imply that

$$n f(x_f, x_f, y_f, y_f) = n f_r(x_f, x_f, y_f, y_f) \tag{6.19}$$

$$\leq \inf\{\sum_{i=0}^{n-1} f_r(p_i, p_{i+1}, y_i, y_{i+1}) : \{p_i\}_{i=0}^{n} \in \Lambda_X(\xi, n)\}$$

$$+ 2\delta \leq 2\delta + \sum_{i=0}^{n-1} f_r(x_f, x_f, y_i, y_{i+1})$$

$$= 2\delta - r \sum_{i=0}^{n-1} |y_i - y_f| + \sum_{i=0}^{n-1} f(x_f, x_f, y_i, y_{i+1})$$

$$\leq 2\delta - r \sum_{i=0}^{n-1} |y_i - y_f| + n f(x_f, x_f, y_f, y_f).$$

(6.18), (6.19) and (6.6) imply that for $i = 1, \ldots, n-1$,

$$|x_i - x_f| \leq r^{-1}(2\delta) < \epsilon, \quad |y_i - y_f| \leq 2\delta r^{-1} < \epsilon.$$

This completes the proof of the lemma.

Choose a number

$$D_0 \geq \sup\{|f_r(x_1, x_2, y_1, y_2)| : x_1, x_2 \in X, \ y_1, y_2 \in Y\}. \tag{6.20}$$

We can easily prove the following:

LEMMA 12.6.2 *Suppose that $n \geq 2$ is a natural number, $M > 0$ and $\{x_i\}_{i=0}^n \subset X$, $\{y_i\}_{i=0}^n \subset Y$ is an (f_r, M)-good pair of sequences. Then the pair of sequences $\{\bar{x}_i\}_{i=0}^n \subset X$, $\{\bar{y}_i\}_{i=0}^n \subset Y$ defined by*

$$\bar{x}_i = x_i, \ \bar{y}_i = y_i, \ i = 1, \ldots, n-1, \quad \bar{x}_0, \bar{x}_n = x_f, \quad \bar{y}_0, \bar{y}_n = y_f$$

is $(f_r, M + 8D_0)$-good.

By using the uniform continuity of the function $f_r : X \times X \times Y \times Y$ we can easily prove

LEMMA 12.6.3 *For each $\epsilon > 0$ there exists a positive number δ such that the following property holds:*
If $n \geq 2$ is a natural number and if sequences $\{x_i\}_{i=0}^n$, $\{\bar{x}_i\}_{i=0}^n \subset X$, $\{y_i\}_{i=0}^n$, $\{\bar{y}_i\}_{i=0}^n \subset Y$ satisfy

$$|\bar{x}_j - x_j|, \ |\bar{y}_j - y_j| \leq \delta, \ j = 0, n, \quad x_j = \bar{x}_j, \ y_j = \bar{y}_j, \ j = 1, \ldots, n-1, \tag{6.21}$$

then

$$\left| \sum_{i=0}^{n-1} [f_r(x_i, x_{i+1}, y_i, y_{i+1}) - f_r(\bar{x}_i, \bar{x}_{i+1}, \bar{y}_i, \bar{y}_{i+1})] \right| \leq \epsilon. \tag{6.22}$$

Lemma 12.6.3 implies the following result.

LEMMA 12.6.4 *For each positive number ϵ there exists a positive number δ such that the following property holds:*
If $n \geq 2$ is a natural number, $\{x_i\}_{i=0}^n \subset X$, $\{y_i\}_{i=0}^n \subset Y$ is an (f_r, ϵ)-good pair of sequences, then each pair of sequences

$$\{\bar{x}_i\}_{i=0}^n \subset X, \ \{\bar{y}_i\}_{i=0}^n \subset Y$$

satisfying (6.21) is $(f_r, 2\epsilon)$-good.

Lemmas 12.6.4 and 12.6.1 imply the following auxiliary result which shows that the property established in Lemma 12.6.1 also holds if x_0, x_n belong to a small neighborhood of x_f and y_0, y_n belong to a small neighborhood of y_f.

LEMMA 12.6.5 *For each $\epsilon \in (0, 1)$ there exists a positive number $\delta < \epsilon$ such that the following property holds:*
If $n \geq 2$ is a natural number and if a (f_r, δ)-good pair of sequences $\{x_i\}_{i=0}^n \subset X$, $\{y_i\}_{i=0}^n \subset Y$ satisfies $|x_j - x_f|, \ |y_j - y_f| \leq \delta, \ j = 0, n$, then $|x_i - x_f|, \ |y_i - y_f| \leq \epsilon, \ i = 0, \ldots, n$.

Denote by $\mathrm{Card}(E)$ the cardinality of a set E.

The next auxiliary result shows that if an integer n is large enough and if a good pair of sequences $\{x_i\}_{i=0}^n, \{y_i\}_{i=0}^n$ satisfies $x_0, x_n = x_f$ and $y_0, y_n = y_f$, then (x_j, y_j) belongs to a small neighborhood of (x_f, y_f) for some $j \in \{1, \ldots, n-1\}$.

LEMMA 12.6.6 *Suppose that $M > 0$ and let $\epsilon \in (0,1)$. Then there exists a natural number $n_0 \geq 4$ such that the following property holds:*

If an (f_r, M)-good pair of sequences $\{x_i\}_{i=0}^{n_0} \subset X$, $\{y_i\}_{i=0}^{n_0} \subset Y$ satisfies

$$x_0, x_{n_0} = x_f, \quad y_0, y_{n_0} = y_f, \tag{6.23}$$

then there exists $j \in \{1, \ldots, n_0 - 1\}$ such that

$$x_j \in B_X(x_f, \epsilon), \quad y_j \in B_Y(y_f, \epsilon). \tag{6.24}$$

Proof. Choose an integer

$$n_0 > 8 + 8(r\epsilon)^{-1}M \tag{6.25}$$

and put

$$\xi_1, \xi_2 = x_f, \quad \xi_3, \xi_4 = y_f, \quad \xi = \{\xi_i\}_{i=1}^4. \tag{6.26}$$

Assume that $\{x_i\}_{i=0}^{n_0} \subset X$, $\{y_i\}_{i=0}^{n_0} \subset Y$ is an (f_r, M)-good pair of sequences and (6.23) holds. Proposition 12.4.4 implies that

$$\sup\{\sum_{i=0}^{n_0-1} f_r(x_i, x_{i+1}, u_i, u_{i+1}) : \{u_i\}_{i=0}^{n_0} \in \Lambda_Y(\xi, n_0)\} \tag{6.27}$$

$$\leq \inf\{\sup\{\sum_{i=0}^{n_0-1} f_r(p_i, p_{i+1}, u_i, u_{i+1}) : \{u_i\}_{i=0}^{n_0} \in \Lambda_Y(\xi, n_0)\} :$$

$$\{p_i\}_{i=0}^{n_0} \in \Lambda_X(\xi, n_0)\} + 2M,$$

$$\inf\{\sum_{i=0}^{n_0-1} f_r(p_i, p_{i+1}, y_i, y_{i+1}) : \{p_i\}_{i=0}^{n_0} \in \Lambda_X(\xi, n_0)\} \tag{6.28}$$

$$\geq \sup\{\inf\{\sum_{i=0}^{n_0-1} f_r(p_i, p_{i+1}, u_i, u_{i+1}) : \{p_i\}_{i=0}^{n_0} \in \Lambda_X(\xi, n_0)\} :$$

$$\{u_i\}_{i=0}^{n_0} \in \Lambda_Y(\xi, n_0)\} - 2M.$$

It follows from Proposition 12.5.1, (6.3), Propositions 12.4.1 and 12.4.2 that

$$\inf\{\sup\{\sum_{i=0}^{n_0-1} f_r(p_i, p_{i+1}, u_i, u_{i+1}) : \{u_i\}_{i=0}^{n_0} \in \Lambda_Y(\xi, n_0)\} : \tag{6.29}$$

$$\{p_i\}_{i=0}^{no} \in \Lambda_X(\xi, n_0)\}$$

$$= \sup\{\inf\{\sum_{i=0}^{n_0-1} f_r(p_i, p_{i+1}, u_i, u_{i+1}) : \{p_i\}_{i=0}^{no} \in \Lambda_X(\xi, n_0)\} :$$

$$\{u_i\}_{i=0}^{no} \in \Lambda_Y(\xi, n_0)\} = n_0 f_r(x_f, x_f, y_f, y_f).$$

In view of (6.29), (6.27), (6.2) and (6.28),

$$n_0 f(x_f, x_f, y_f, y_f) = n_0 f_r(x_f, x_f, y_f, y_f) \geq -2M \qquad (6.30)$$

$$+ \sup\{\sum_{i=0}^{n_0-1} f_r(x_i, x_{i+1}, u_i, u_{i+1}) : \{u_i\}_{i=0}^{no} \in \Lambda_Y(\xi, n_0)\}$$

$$\geq -2M + \sum_{i=0}^{n_0-1} f_r(x_i, x_{i+1}, y_f, y_f) = -2M$$

$$+ \sum_{i=0}^{n_0-1} f(x_i, x_{i+1}, y_f, y_f) + r \sum_{i=0}^{n_0-1} |x_i - x_f|,$$

$$n_0 f(x_f, x_f, y_f, y_f) = n_0 f_r(x_f, x_f, y_f, y_f) \leq 2M \qquad (6.31)$$

$$+ \inf\{\sum_{i=0}^{n_0-1} f_r(p_i, p_{i+1}, y_i, y_{i+1}) : \{p_i\}_{i=0}^{no} \in \Lambda_X(\xi, n_0)\}$$

$$\leq 2M + \sum_{i=0}^{n_0-1} f_r(x_f, x_f, y_i, y_{i+1}) = 2M$$

$$+ \sum_{i=0}^{n_0-1} f(x_f, x_f, y_i, y_{i+1}) - r \sum_{i=0}^{n_0-1} |y_i - y_f|.$$

Proposition 12.5.1 and (6.1) imply that

$$\sum_{i=0}^{n_0-1} f(x_i, x_{i+1}, y_f, y_f) \geq n_0 f(x_f, x_f, y_f, y_f) \geq \sum_{i=0}^{n_0-1} f(x_f, x_f, y_i, y_{i+1}).$$

Combined with (6.30) and (6.31) this implies that

$$n_0 f(x_f, x_f, y_f, y_f) \geq -2M + n_0 f(x_f, x_f, y_f, y_f) + r \sum_{i=0}^{n_0-1} |x_i - x_f|,$$

$$n_0 f(x_f, x_f, y_f, y_f) \leq 2M + n_0 f(x_f, x_f, y_f, y_f) - r \sum_{i=0}^{n_0-1} |y_i - y_f|,$$

$$r \sum_{i=0}^{n_0-1} |x_i - x_f| \leq 2M, \quad r \sum_{i=0}^{n_0-1} |y_i - y_f| \leq 2M. \tag{6.32}$$

It follows from (6.23), (6.32) and (6.25) that

$$\epsilon \mathrm{Card}\{i \in \{1, \ldots, n_0 - 1\} : |x_i - x_f| \geq \epsilon\} \leq 2Mr^{-1},$$

$$\epsilon \mathrm{Card}\{i \in \{1, \ldots, n_0 - 1\} : |y_i - y_f| \geq \epsilon\} \leq 2Mr^{-1},$$

$$\mathrm{Card}\{i \in \{1, \ldots, n_0 - 1\} : |x_i - x_f| < \epsilon,$$

$$|y_i - y_f| < \epsilon\} \geq n_0 - 1 - 4M(\epsilon r)^{-1} > 6.$$

This completes the proof of the lemma.

Lemmas 12.6.6 and 12.6.2 imply the following.

LEMMA 12.6.7 *For each $\epsilon \in (0,1)$ and each $M \in (0,\infty)$ there exists a natural number $n_0 \geq 4$ such that the following property holds:*
  If $\{x_i\}_{i=0}^{n_0} \subset X$, $\{y_i\}_{i=0}^{n_0} \subset Y$ is an (f_r, M)-good pair of sequences, then there is $j \in \{1, \ldots, n_0 - 1\}$ for which $x_j \in B_X(x_f, \epsilon)$, $y_j \in B(y_f, \epsilon)$.

The next lemma shows that if an integer n is large enough,

$$\{x_i\}_{i=0}^{n}, \quad \{y_i\}_{i=0}^{n}$$

is a good pair of sequences with respect to g belonging to a small neighborhood of f_r, then (x_j, y_j) belongs to a small neighborhood of (x_f, y_f) for some $j \in \{1, \ldots, n - 1\}$.

LEMMA 12.6.8 *For each $\epsilon \in (0,1)$ and each $M \in (0,\infty)$ there exists a natural number $n_0 \geq 4$ and a neighborhood U of f_r in $\mathcal{M}$ such that the following property holds:*
  If $g \in U$ and if $\{x_i\}_{i=0}^{n_0} \subset X$, $\{y_i\}_{i=0}^{n_0} \subset Y$ is a (g, M)-good pair of sequences, then there is $j \in \{1, \ldots, n_0 - 1\}$ for which

$$x_j \in B_X(x_f, \epsilon), \quad y_j \in B_Y(y_f, \epsilon). \tag{6.33}$$

Proof. Let $\epsilon \in (0,1)$ and $M \in (0,\infty)$. It follows from Lemma 12.6.7 that there exists a natural number $n_0 \geq 4$ such that for each $(f_r, M + 8)$-good pair of sequences $\{x_i\}_{i=0}^{n_0} \subset X$, $\{y_i\}_{i=0}^{n_0} \subset Y$ there is $j \in \{1, \ldots, n_0 - 1\}$ for which (6.33) is valid. Put

$$U = \{g \in \mathcal{M} : \rho(f_r, g) \leq (16 n_0)^{-1}\}. \tag{6.34}$$

Assume that $g \in U$ and $\{x_i\}_{i=0}^{n_0} \subset X$, $\{y_i\}_{i=0}^{n_0} \subset Y$ is a (g, M)-good pair of sequences. By (6.34) the pair of sequences $\{x_i\}_{i=0}^{n_0}$, $\{y_i\}_{i=0}^{n_0}$ is

$(f_r, M + 8)$-good. It follows from the definition of n_0 that there exists $j \in \{1, \ldots, n_0 - 1\}$ for which (6.33) is valid. The lemma is proved.

The following auxiliary result is the main ingredient for the proof of Theorem 12.2.1.

LEMMA 12.6.9 *For each $\epsilon \in (0, 1)$ there exist a neighborhood U of f_r in $\mathcal{M}$, a number $\delta \in (0, \epsilon)$ and a natural number $n_1 \geq 4$ such that the following property holds:*

If $g \in U$, $n \geq 2n_1$ is a natural number and if $\{x_i\}_{i=0}^n \subset X$, $\{y_i\}_{i=0}^n \subset Y$ is a (g, δ)-good pair of sequences, then

$$x_i \in B_X(x_f, \epsilon), \ y_i \in B_Y(y_f, \epsilon), \qquad (6.35)$$

for all $i \in [n_1, n - n_1]$. Moreover if

$$x_0 \in B_X(x_f, \delta), \ y_0 \in B_Y(y_f, \delta),$$

then (6.35) holds for all $i \in [0, n - n_1]$, and if

$$x_n \in B_X(x_f, \delta), \ y_n \in B_Y(y_f, \delta),$$

then (6.35) is valid for all $i \in [n_1, n]$.

Proof. Let $\epsilon \in (0, 1)$. Lemma 12.6.5 implies that there exists $\delta_0 \in (0, \epsilon)$ such that for each natural number $n \geq 2$ and each (f_r, δ_0)-good pair of sequences $\{x_i\}_{i=0}^n \subset X$, $\{y_i\}_{i=0}^n \subset Y$ satisfying

$$x_j \in B_X(x_f, \delta_0), \ y_j \in B(y_f, \delta_0), \ j = 0, n, \qquad (6.36)$$

relation (6.35) is valid for $i = 0, \ldots, n$. By Lemma 12.6.8 there exists an integer $n_0 \geq 4$ and a neighborhood U_0 of f_r in $\mathcal{M}$ such that for each $g \in U_0$ and each $(g, 8)$-good pair of sequences $\{x_i\}_{i=0}^{n_0} \subset X$, $\{y_i\}_{i=0}^{n_0} \subset Y$ there is $j \in \{1, \ldots, n_0 - 1\}$ for which

$$x_j \in B_X(x_f, \delta_0), \ y_j \in B_Y(y_f, \delta_0). \qquad (6.37)$$

Fix an integer

$$n_1 \geq 4n_0 \qquad (6.38)$$

and a number

$$\delta \in (0, 4^{-1}\delta_0). \qquad (6.39)$$

Define

$$U = U_0 \cap \{g \in \mathcal{M} : \ \rho(g, f_r) \leq 16^{-1}\delta n_1^{-1}\}. \qquad (6.40)$$

Assume that $g \in U$, an integer $n \geq 2n_1$ and $\{x_i\}_{i=0}^n \subset X$, $\{y_i\}_{i=0}^n \subset Y$ is a (g, δ)-good pair of sequences. It follows from (6.39), (6.38) and the definition of n_0, U_0 that there exists a sequence of integers $\{t_i\}_{i=1}^k \subset [0, n]$ such that

$$t_1 \leq n_0, \ t_{i+1} - t_i \in [n_0, 3n_0], \ i = 1, \ldots, k-1, \tag{6.41}$$

$$n - t_k \leq n_0, \quad |x_{t_i} - x_f|, \ |y_{t_i} - y_f| \leq \delta_0, \ i = 1, \ldots, k$$

and, moreover, if $|x_0 - x_f|$, $|y_0 - y_f| \leq \delta$, then $t_1 = 0$, and if $|x_n - x_f|$, $|y_n - y_f| \leq \delta$, then $t_k = n$. Clearly $k \geq 2$. Fix $q \in \{1, \ldots, k-1\}$. To complete the proof of the lemma it is sufficient to show that for each integer $i \in [t_q, t_{q+1}]$ the relation (6.35) holds.

Define sequences $\{x_i^{(q)}\}_{i=0}^{t_{q+1}-t_q} \subset X$, $\{y_i^{(q)}\}_{i=0}^{t_{q+1}-t_q} \subset Y$ by

$$x_i^{(q)} = x_{i+t_q}, \ y_i^{(q)} = y_{i+t_q}, \quad i \in [0, t_{q+1} - t_q]. \tag{6.42}$$

It is easy to see that $\{x_i^{(q)}\}_{i=0}^{t_{q+1}-t_q}$, $\{y_i^{(q)}\}_{i=0}^{t_{q+1}-t_q}$ is a (g, δ)-good pair of sequences. Combined with (6.40), (6.39) and (6.41) this implies that the pair of sequences $\{x_i^{(q)}\}_{i=0}^{t_{q+1}-t_q}$, $\{y_i^{(q)}\}_{i=0}^{t_{q+1}-t_q}$ is (f_r, δ_0)-good.

It follows from (6.42), (6.41), (6.39) and the definition of δ_0 (see (6.36)) that

$$x_i^{(q)} \in B_X(x_f, \epsilon), \ y_i^{(q)} \in B_Y(y_f, \epsilon), \quad i = 0, \ldots, t_{q+1} - t_q.$$

Together with (6.42) this implies that

$$x_i \in B_X(x_f, \epsilon), \ y_i \in B_Y(y_f, \epsilon), \ i = t_q, \ldots, t_{q+1}.$$

This completes the proof of the lemma.

12.7.   Preliminary lemmas for Theorem 12.2.2

For each metric space K denote by $C(K)$ the space of all continuous functions on K with the topology of uniform convergence ($\|\phi\| = \sup\{|\phi(z)| : z \in K\}$, $\phi \in C(K)$).

Let $f \in \mathcal{M}$. There exist $x_f \in X$, $y_f \in Y$ such that

$$\sup_{y \in Y} f(x_f, x_f, y, y) = f(x_f, x_f, y_f, y_f) \tag{7.1}$$

$$= \inf_{x \in X} f(x, x, y_f, y_f)$$

(see (6.1)).

Let $r \in (0, 1)$. Define $f_r : X \times X \times Y \times Y \to R^1$ by

$$f_r(x_1, x_2, y_1, y_2) = f(x_1, x_2, y_1, y_2) + r|x_1 - x_f| \tag{7.2}$$

$$-r|y_1 - y_f|, \quad x_1, x_2 \in X, \ y_1, y_2 \in Y$$

(see (6.2)). Clearly $f_r \in \mathcal{M}$. Define functions $f_r^{(X)} : X \times X \to R^1$, $f_r^{(Y)} : Y \times Y \to R^1$ by

$$f_r^{(X)}(x_1, x_2) = f_r(x_1, x_2, y_f, y_f), \quad x_1, x_2 \in X, \tag{7.3}$$

$$f_r^{(Y)}(y_1, y_2) = f_r(x_f, x_f, y_1, y_2), \quad y_1, y_2 \in Y. \tag{7.4}$$

LEMMA 12.7.1 *For each $\epsilon \in (0,1)$ there exists a positive number $\delta < \epsilon$ for which the following condition holds:*

If $n \geq 2$ is a natural number,

$$\{x_i\}_{i=0}^n \subset X, \quad x_0, x_n = x_f \tag{7.5}$$

and if for each $\{z_i\}_{i=0}^n \subset X$ satisfying

$$z_0 = x_0, \quad z_n = x_n \tag{7.6}$$

the inequality

$$\sum_{i=0}^{n-1} f_r^{(X)}(x_i, x_{i+1}) \leq \sum_{i=0}^{n-1} f_r^{(X)}(z_i, z_{i+1}) + \delta \tag{7.7}$$

holds, then

$$|x_i - x_f| \leq \epsilon, \quad i = 0, \dots, n. \tag{7.8}$$

Proof. Let $\epsilon \in (0,1)$. Choose a number

$$\delta \in (0, 8^{-1} r \epsilon). \tag{7.9}$$

Assume that an integer $n \geq 2$, $\{x_i\}_{i=0}^n \subset X$, (7.5) is valid and for each sequence $\{z_i\}_{i=0}^n \subset X$ satisfying (7.6), relation (7.7) holds. This implies that

$$\sum_{i=0}^{n-1} f_r(x_i, x_{i+1}, y_f, y_f) \leq n f_r(x_f, x_f, y_f, y_f) + \delta \tag{7.10}$$

$$= n f(x_f, x_f, y_f, y_f) + \delta.$$

By (7.2), (7.5) and (7.1),

$$\sum_{i=0}^{n-1} f_r(x_i, x_{i+1}, y_f, y_f) = r \sum_{i=0}^{n-1} |x_i - x_f|$$

$$+ \sum_{i=0}^{n-1} f(x_i, x_{i+1}, y_f, y_f)$$

$$\geq r \sum_{i=0}^{n-1} |x_i - x_f| + nf(n^{-1} \sum_{i=0}^{n-1} x_i, n^{-1} \sum_{i=0}^{n-1} x_i, y_f, y_f)$$

$$\geq r \sum_{i=0}^{n-1} |x_i - x_f| + nf(x_f, x_f, y_f, y_f).$$

It follows from this relation, (7.10) and (7.9) that for each $i \in \{0, \ldots, n-1\}$ the inequality $|x_i - x_f| \leq r^{-1}\delta < \epsilon$ is true. This completes the proof of the lemma.

Definition 7.1. Let $g \in C(X \times X)$, n be a natural number, and let M be a nonnegative number. A sequence $\{\bar{x}_i\}_{i=0}^n \subset X$ is called (g, X, M)-good if

$$M + \sum_{i=0}^{n-1} g(x_i, x_{i+1}) \geq \sum_{i=0}^{n-1} g(\bar{x}_i, \bar{x}_{i+1})$$

for each sequence $\{x_i\}_{i=0}^n \subset X$ satisfying $x_0 = \bar{x}_0$, $x_n = \bar{x}_n$.

Definition 7.2. Let $g \in C(Y \times Y)$, n be a natural number and let M be a nonnegative number. A sequence $\{\bar{y}_i\}_{i=0}^n \subset Y$ is called (g, Y, M)-good if

$$\sum_{i=0}^{n-1} g(y_i, y_{i+1}) \leq M + \sum_{i=0}^{n-1} g(\bar{y}_i, \bar{y}_{i+1})$$

for each sequence $\{y_i\}_{i=0}^n \subset Y$ satisfying $y_0 = \bar{y}_0$, $y_n = \bar{y}_n$.

Definition 7.3. Let $n_1 \geq 0$, $n_2 > n_1$ be a pair of integers, M be a nonnegative number, and let

$$\{g_i\}_{i=n_1}^{n_2-1} \subset C(X \times X).$$

A sequence $\{\bar{x}_i\}_{i=n_1}^{n_2} \subset X$ is called $(\{g_i\}_{i=n_1}^{n_2-1}, X, M)$-good if

$$M + \sum_{i=n_1}^{n_2-1} g_i(x_i, x_{i+1}) \geq \sum_{i=n_1}^{n_2-1} g_i(\bar{x}_i, \bar{x}_{i+1})$$

for each sequence $\{x_i\}_{i=n_1}^{n_2} \subset X$ satisfying $x_{n_1} = \bar{x}_{n_1}$, $x_{n_2} = \bar{x}_{n_2}$.

Definition 7.4. Let $n_1 \geq 0$, $n_2 > n_1$ be integers, and let

$$\{g_i\}_{i=n_1}^{n_2-1} \subset C(Y \times Y), \ M \in [0, \infty).$$

A sequence $\{\bar{y}_i\}_{i=n_1}^{n_2} \subset Y$ is called $(\{g_i\}_{i=n_1}^{n_2-1}, Y, M)$-good if for each sequence $\{y_i\}_{i=n_1}^{n_2} \subset Y$ satisfying

$$y_{n_1} = \bar{y}_{n_1}, \ y_{n_2} = \bar{y}_{n_2}$$

the following inequality holds:

$$\sum_{i=n_1}^{n_2-1} g_i(y_i, y_{i+1}) \le \sum_{i=n_1}^{n_2-1} g_i(\bar{y}_i, \bar{y}_{i+1}) + M.$$

Analogously to Lemma 12.7.1 we can establish the following

LEMMA 12.7.2 *For each $\epsilon \in (0,1)$ there exists a positive number $\delta < \epsilon$ such that the following property holds:*

If $n \ge 2$ is a natural number and if an $(f_r^{(Y)}, Y, \delta)$-good sequence $\{y_i\}_{i=0}^n \subset Y$ satisfies $y_0, y_n = y_f$, then $y_i \in B_Y(y_f, \epsilon)$, $i = 0, \ldots, n$.

By using Lemmas 12.7.1 and 12.6.3 we can easily deduce

LEMMA 12.7.3 *For each $\epsilon \in (0,1)$ there exists a positive number δ such that the following property holds:*

If $n \ge 2$ is a natural number and if an $(f_r^{(X)}, X, \delta)$-good sequence $\{x_i\}_{i=0}^n \subset X$ satisfies $x_0, x_n \in B_X(x_f, \delta)$, then

$$x_i \in B_X(x_f, \epsilon), \ i = 0, \ldots, n.$$

By using Lemmas 12.7.2 and 12.6.3 we can easily deduce

LEMMA 12.7.4 *For each $\epsilon \in (0,1)$ there exists a positive number δ such that the following property holds:*

If $n \ge 2$ is a natural number and if an $(f_r^{(Y)}, Y, \delta)$-good sequence $\{y_i\}_{i=0}^n \subset Y$ satisfies $y_0, y_n \in B_Y(y_f, \delta)$, then

$$y_i \in B_Y(y_f, \epsilon), \ i = 0, \ldots, n.$$

LEMMA 12.7.5 *For each $\epsilon \in (0,1)$ and each $M > 0$ there exists a natural number $n_0 \ge 4$ such that the following property holds:*

If an $(f_r^{(X)}, X, M)$-good sequence $\{x_i\}_{i=0}^{n_0} \subset X$ satisfies

$$x_0 = x_f, \ x_{n_0} = x_f, \tag{7.11}$$

then there is $j \in \{1, \ldots, n_0 - 1\}$ such that

$$x_j \in B_X(x_f, \epsilon). \tag{7.12}$$

Proof. Let $\epsilon \in (0,1)$ and $M > 0$. Choose an integer

$$n_0 > 8 + 8M(r\epsilon)^{-1}. \tag{7.13}$$

Assume that $\{x_i\}_{i=0}^{n_0} \subset X$ is an $(f_r^{(X)}, X, M)$-good sequence and (7.11) is valid. It is not difficult to see that

$$M + n_0 f(x_f, x_f, y_f, y_f) = n_0 f_r(x_f, x_f, y_f, y_f) + M$$

$$\geq \sum_{i=0}^{n_0-1} f_r(x_i, x_{i+1}, y_f, y_f) = r \sum_{i=0}^{n_0-1} |x_i - x_f|$$

$$+ \sum_{i=0}^{n_0-1} f(x_i, x_{i+1}, y_f, y_f) \geq r \sum_{i=0}^{n_0-1} |x_i - x_f|$$

$$+ n_0 f(n_0^{-1} \sum_{i=0}^{n_0-1} x_i, n_0^{-1} \sum_{i=0}^{n_0-1} x_i, y_f, y_f)$$

$$\geq r \sum_{i=0}^{n_0-1} |x_i - x_f| + n_0 f(x_f, x_f, y_f, y_f).$$

Combined with (7.13) this implies that there is $j \in \{1, \ldots, n_0 - 1\}$ for which (7.12) is valid. This completes the proof of the lemma.

Analogously to Lemma 12.7.5 we can establish the following

LEMMA 12.7.6 *For each $\epsilon \in (0,1)$ and each $M > 0$ there exists a natural number $n_0 \geq 4$ such that the following property holds:*
If an $(f_r^{(Y)}, Y, M)$-good sequence $\{y_i\}_{i=0}^{n_0} \subset Y$ satisfies $y_0 = y_f$, $y_{n_0} = y_f$, then there exists $j \in \{1, \ldots, n_0 - 1\}$ such that $y_j \in B_Y(y_f, \epsilon)$.

Choose a number

$$D_0 \geq \sup\{|f_r(x_1, x_2, y_1, y_2)| : x_1, x_2 \in X, \ y_1, y_2 \in Y\}.$$

We can easily prove the following lemma.

LEMMA 12.7.7 *1. Assume that $n \geq 2$ is an integer, M is a positive number, a sequence $\{x_i\}_{i=0}^{n} \subset X$ is $(f_r^{(X)}, X, M)$-good and $\bar{x}_0 = x_f$, $\bar{x}_n = x_f$, $\bar{x}_i = x_i$, $i = 1, \ldots, n-1$. Then the sequence $\{\bar{x}_i\}_{i=0}^{n}$ is $(f_r^{(X)}, X, M + 8D_0)$-good.*
2. Assume that $n \geq 2$ is an integer, M is a positive number, a sequence $\{y_i\}_{i=0}^{n} \subset Y$ is $(f_r^{(Y)}, Y, M)$-good and $\bar{y}_0 = y_f$, $\bar{y}_n = y_f$, $\bar{y}_i = y_i$, $i = 1, \ldots, n-1$. Then the sequence $\{\bar{y}_i\}_{i=0}^{n}$ is $(f_r^{(Y)}, Y, M + 8D_0)$-good.

Lemmas 12.7.5, 12.7.6 and 12.7.7 imply the following two results.

LEMMA 12.7.8 *For each $\epsilon \in (0,1)$ and each $M > 0$ there exists a natural number $n_0 \geq 4$ such that the following condition holds:*
For each $(f_r^{(X)}, X, M)$-good sequence $\{x_i\}_{i=0}^{n_0} \subset X$,

$$\min\{|x_j - x_f| : j \in \{1, \ldots, n_0 - 1\}\} \leq \epsilon.$$

LEMMA 12.7.9 *For each $\epsilon \in (0,1)$ and each $M > 0$ there exists a natural number $n_0 \geq 4$ such that the following condition holds:*
For each $(f_r^{(Y)}, Y, M)$-good sequence $\{y_i\}_{i=0}^{n_0} \subset Y$,

$$\min\{|y_j - y_f| : j \in \{1, \ldots, n_0 - 1\}\} \leq \epsilon.$$

By using Lemmas 12.7.8 and 12.7.9 analogously to the proof of Lemma 12.6.8, we can establish the following two results.

LEMMA 12.7.10 *For each $\epsilon \in (0,1)$ and each $M > 0$ there exists a natural number $n_0 \geq 4$ and a neighborhood U of $f_r^{(X)}$ in $C(X \times X)$ such that the following property holds:*
If $\{g_i\}_{i=0}^{n_0-1} \subset U$ and if $\{x_i\}_{i=0}^{n_0} \subset X$ is a $(\{g_i\}_{i=0}^{n_0-1}, X, M)$-good sequence, then there is $j \in \{1, \ldots, n_0 - 1\}$ for which $x_j \in B_X(x_f, \epsilon)$.

LEMMA 12.7.11 *For each $\epsilon \in (0,1)$ and each $M > 0$ there exists a natural number $n_0 \geq 4$ and a neighborhood U of $f_r^{(Y)}$ in $C(Y \times Y)$ such that the following property holds:*
If $\{g_i\}_{i=0}^{n_0-1} \subset U$ and if $\{y_i\}_{i=0}^{n_0} \subset Y$ is a $(\{g_i\}_{i=0}^{n_0-1}, Y, M)$-good sequence, then there is $j \in \{1, \ldots, n_0 - 1\}$ for which $y_j \in B_Y(y_f, \epsilon)$.

LEMMA 12.7.12 *For each $\epsilon \in (0,1)$ there exist a neighborhood U of $f_r^{(X)}$ in $C(X \times X)$, a positive number $\delta < \epsilon$ and a natural number $n_1 \geq 4$ such that the following property holds:*
If $n \geq 2n_1$ is a natural number, $\{g_i\}_{i=0}^{n-1} \subset U$ and if $\{x_i\}_{i=0}^{n} \subset X$ is a $(\{g_i\}_{i=0}^{n-1}, X, \delta)$-good sequence, then

$$x_i \in B_X(x_f, \epsilon) \tag{7.14}$$

for all integers $i \in [n_1, n - n_1]$. Moreover if $x_0 \in B_X(x_f, \delta)$, then (7.14) holds for all integers $i \in [0, n - n_1]$, and if $x_n \in B_X(x_f, \delta)$, then (7.14) is valid for all integers $i \in [n_1, n]$.

Proof. Let $\epsilon \in (0,1)$. Lemma 12.7.3 implies that there exists a positive number $\delta_0 < \epsilon$ such that the following property holds:
If $n \geq 2$ is a natural number and if an $(f_r^{(X)}, X, \delta_0)$-good sequence $\{x_i\}_{i=0}^{n} \subset X$ satisfies $x_0, x_n \in B_X(x_f, \delta_0)$, then the relation (7.14) is valid for $i = 0, \ldots, n$.

In view of Lemma 12.7.10 there exist a natural number $n_0 \geq 4$ and a neighborhood U_0 of $f_r^{(X)}$ in $C(X \times X)$ such that the following property holds:
If $\{g_i\}_{i=0}^{n_0-1} \subset U_0$ and if $\{x_i\}_{i=0}^{n} \subset X$ is a $(\{g_i\}_{i=0}^{n_0-1}, X, 8)$-good sequence, then there is $j \in \{1, \ldots, n_0 - 1\}$ for which $x_j \in B_X(x_f, \delta_0)$.

Choose a natural number $n_1 \geq 4n_0$ and a positive number $\delta < 4^{-1}\delta_0$ and set

$$U = U_0 \cap \{g \in C(X \times X) : \|g - f_r^{(X)}\| \leq (16n_1)^{-1}\delta\}.$$

Assume that an integer $n \geq 2n_1$, $\{g_i\}_{i=0}^{n-1} \subset U$ and a sequence $\{x_i\}_{i=0}^{n} \subset X$ is $(\{g_i\}_{i=0}^{n-1}, X, \delta)$-good. Arguing as in the proof of Lemma 12.6.9 we can show that (7.14) is valid for all integers $i \in [n_1, n-n_1]$ and, moreover, if $x_0 \in B_X(x_f, \delta)$, then (7.14) holds for all integers $i \in [0, n - n_1]$, and if $x_n \in B_X(x_f, \delta)$, then (7.14) is valid for all integers $i \in [n_1, n]$. The lemma is proved.

Analogously to Lemma 12.7.12 we can prove the following

LEMMA 12.7.13 *For each $\epsilon \in (0,1)$ there exist a neighborhood U of $f_r^{(Y)}$ in $C(Y \times Y)$, a positive number $\delta < \epsilon$ and a natural number $n_1 \geq 4$ such that the following property holds:*

If $n \geq 2n_1$ is a natural number, $\{g_i\}_{i=0}^{n-1} \subset U$ and if $\{y_i\}_{i=0}^{n} \subset Y$ is a $(\{g_i\}_{i=0}^{n-1}, Y, \delta)$-good sequence, then

$$y_i \in B_Y(y_f, \epsilon) \tag{7.15}$$

for all integers $i \in [n_1, n - n_1]$. Moreover if $y_0 \in B_Y(y_f, \delta)$, then (7.15) holds for all integers $i \in [0, n - n_1]$, and if $y_n \in B_Y(y_f, \delta)$, then (7.15) is valid for all integers $i \in [n_1, n]$.

12.8.   Proofs of Theorems 12.2.1 and 12.2.2

We will use the notation from sections 12.1-12.7.

Let $f \in \mathcal{M}$. There exists a pair $(x_f, y_f) \in X \times Y$ such that (6.1) holds. Let $r \in (0,1)$ and let i be a natural number. Consider the function $f_r : X \times X \times Y \times Y$ defined by (6.2). Clearly all lemmas from sections 12.6 and 12.7 are valid for f_r.

By Lemma 12.7.12 there exist a number

$$\gamma_1(f, r, i) \in (0, 2^{-i}), \tag{8.1}$$

a number

$$\delta_1(f, r, i) \in (0, 2^{-i}) \tag{8.2}$$

and an integer $n_1(f, r, i) \geq 4$ such that the following property holds:
    (a) If $n \geq 2n_1(f, r, i)$ is a natural number, $\{g_j\}_{j=0}^{n-1} \subset C(X \times X)$ satisfies

$$\|g_j - f_r^{(X)}\| \leq \gamma_1(f, r, i), \quad j = 0, \ldots, n - 1,$$

and if $\{x_j\}_{j=0}^n \subset X$ is a $(\{g_j\}_{j=0}^{n-1}, X, \delta_1(f, r, i))$-good sequence, then

$$x_j \in B_X(x_f, 2^{-i}), \quad j \in [n_1(f, r, i), n - n_1(f, r, i)]. \tag{8.3}$$

Lemma 12.7.13 implies that there exist numbers

$$\delta_2(f, r, i), \ \gamma_2(f, r, i) \in (0, 2^{-i}) \tag{8.4}$$

and a natural number $n_2(f, r, i) \geq 4$ such that the following property holds:

(b) If $n \geq 2n_2(f, r, i)$ is a natural number, $\{g_j\}_{j=0}^{n-1} \subset C(Y \times Y)$ satisfies

$$\|g_j - f_r^{(Y)}\| \leq \gamma_2(f, r, i), \quad j = 0, \dots, n-1$$

and if $\{y_j\}_{j=0}^n \subset Y$ is a $(\{g_j\}_{j=0}^{n-1}, Y, \delta_2(f, r, i))$-good sequence, then

$$y_j \in B_Y(y_f, 2^{-i}), \quad j \in [n_2(f, r, i), n - n_2(f, r, i)]. \tag{8.5}$$

Put

$$n_3(f, r, i) = n_1(f, r, i) + n_2(f, r, i), \tag{8.6}$$
$$\delta_3(f, r, i) = \min\{\delta_1(f, r, i), \delta_2(f, r, i)\},$$
$$\gamma_3(f, r, i) = \min\{\gamma_1(f, r, i), \gamma_2(f, r, i)\}.$$

In view of the uniform continuity of the function f_r there exists a number

$$\delta_4(f, r, i) \in (0, \delta_3(f, r, i)) \tag{8.7}$$

such that if $x_1, x_2, \bar{x}_1, \bar{x}_2 \in X$, $y_1, y_2, \bar{y}_1, \bar{y}_2 \in Y$ satisfy

$$|x_j - \bar{x}_j|, \ |y_j - \bar{y}_j| \leq \delta_4(f, r, i), \quad j = 1, 2,$$

then

$$|f_r(x_1, x_2, y_1, y_2) - f_r(\bar{x}_1, \bar{x}_2, \bar{y}_1, \bar{y}_2)| \leq 16^{-1}\gamma_3(f, r, i). \tag{8.8}$$

It follows from Lemma 12.6.9 that there exist numbers

$$\gamma_4(f, r, i) \in (0, 16^{-1}\gamma_3(f, r, i)), \ \delta_5(f, r, i) \in (0, 8^{-1}\delta_4(f, r, i)) \tag{8.9}$$

and a natural number $n_4(f, r, i) \geq 4$ such that the following property holds:

(c) If $g \in \mathcal{M}$ satisfies $\rho(g, f_r) \leq \gamma_4(f, r, i)$, $n \geq 2n_4(f, r, i)$ is a natural number and if

$$\{x_j\}_{j=0}^n \subset X, \ \{y_j\}_{j=0}^n \subset Y$$

is a $(g, \delta_5(f, r, i))$-good pair of sequences, then

$$x_j \in B_X(x_f, 8^{-1}\delta_4(f, r, i)), \ y_j \in B_Y(y_f, 8^{-1}\delta_4(f, r, i)) \tag{8.10}$$

for all $j \in [n_4(f, r, i), n - n_4(f, r, i)]$; moreover if

$$x_0 \in B_X(x_f, \delta_5(f, r, i)), \quad y_0 \in B_Y(y_f, \delta_5(f, r, i)),$$

then (8.10) holds for all integers $j \in [0, n - n_4(f, r, i)]$, and if

$$x_n \in B_X(x_f, \delta_5(f, r, i)), \quad y_n \in B_Y(y_f, \delta_5(f, r, i)),$$

then (8.10) is valid for all integers $j \in [n_4(f, r, i), n]$.

Lemma 12.6.9 implies that there exist numbers

$$\gamma(f, r, i) \in (0, 8^{-1}\gamma_4(f, r, i)), \quad \delta(f, r, i) \in (0, 8^{-1}\delta_5(f, r, i)) \qquad (8.11)$$

and a natural number $n_5(f, r, i) \geq 4$ such that the following property holds:

(d) If $g \in \mathcal{M}$ satisfies $\rho(g, f_r) \leq \gamma(f, r, i)$, $n \geq 2n_5(f, r, i)$ is a natural number and if

$$\{x_j\}_{j=0}^n \subset X, \quad \{y_j\}_{j=0}^n \subset Y$$

is a $(g, \delta(f, r, i))$-good pair of sequences, then

$$x_j \in B_X(x_f, 8^{-1}\delta_5(f, r, i)), \quad y_j \in B_Y(y_f, 8^{-1}\delta_5(f, r, i)) \qquad (8.12)$$

for all $j \in [n_5(f, r, i), n - n_5(f, r, i)]$. Set

$$U(f, r, i) = \{g \in \mathcal{M} : \rho(g, f_r) < \gamma(f, r, i)\}. \qquad (8.13)$$

Define

$$\mathcal{F} = \cap_{k=1}^{\infty} \cup \{U(f, r, i) : f \in \mathcal{M}, \ r \in (0, 1), \ i = k, k+1, \ldots\}. \qquad (8.14)$$

It is easy to see that $\mathcal{F}$ is a countable intersection of open everywhere dense sets in $\mathcal{M}$.

Proof of Theorem 12.2.1. Let $h \in \mathcal{F}$. There exists a pair $(x_1, y_1) \in X \times Y$ such that

$$\sup_{y \in Y} h(x_1, x_1, y, y) = h(x_1, x_1, y_1, y_1) = \inf_{x \in X} h(x, x, y_1, y_1) \qquad (8.15)$$

(see (3.1) and (3.2)).

Assume that $(x_2, y_2) \in X \times Y$ and

$$\sup_{y \in Y} h(x_2, x_2, y, y) = h(x_2, x_2, y_2, y_2) = \inf_{x \in X} h(x, x, y_2, y_2). \qquad (8.16)$$

We will show that

$$x_2 = x_1, \quad y_2 = y_1. \qquad (8.17)$$

Define sequences $\{x_j^{(1)}\}_{j=0}^{\infty}, \{x_j^{(2)}\}_{j=0}^{\infty} \subset X, \{y_j^{(1)}\}_{j=0}^{\infty}, \{y_j^{(2)}\}_{j=0}^{\infty} \subset Y$ by

$$x_j^{(1)} = x_1, \ x_j^{(2)} = x_2, \ y_j^{(1)} = y_1, \ y_j^{(2)} = y_2, \quad j = 0, 1, \dots. \quad (8.18)$$

It follows from (8.15), (8.18) and Proposition 12.5.1 that the pairs of sequences

$$(\{x_j^{(1)}\}_{j=0}^{\infty}, \{y_j^{(1)}\}_{j=0}^{\infty}), \quad (\{x_j^{(2)}\}_{j=0}^{\infty}, \{y_j^{(2)}\}_{j=0}^{\infty})$$

are (h)-minimal. Let $\epsilon \in (0, 1)$. Choose a natural number k such that

$$2^{-k} < 64^{-1}\epsilon. \quad (8.19)$$

There exist $f \in \mathcal{M}$, $r \in (0, 1)$ and an integer $i \geq k$ such that

$$h \in U(f, r, i). \quad (8.20)$$

Since the pairs of sequences $(\{x_j^{(1)}\}_{j=0}^{\infty}, \{y_j^{(1)}\}_{j=0}^{\infty}), (\{x_j^{(2)}\}_{j=0}^{\infty}, \{y_j^{(2)}\}_{j=0}^{\infty})$ are (h)-minimal it follows from (8.19), (8.20), (8.18), property (d) and (8.13) that

$$|x_1 - x_f|, \ |x_2 - x_f|, \ |y_1 - y_f|, \ |y_2 - y_f| \leq 8^{-1}\delta_5(f, r, i) < 2^{-i} < \epsilon,$$

$$|x_1 - x_2|, \ |y_1 - y_2| \leq 2\epsilon.$$

Since ϵ is an arbitrary number in the interval $(0, 1)$ we conclude that (8.17) is valid. Therefore we have shown that there exists a unique pair $(x_h, y_h) \in X \times Y$ such that

$$\sup_{y \in Y} h(x_h, x_h, y, y) = h(x_h, x_h, y_h, y_h) = \inf_{x \in X} h(x, x, y_h, y_h). \quad (8.21)$$

Let $\epsilon > 0$. Choose a natural number k for which (8.19) holds. There exist $f \in \mathcal{M}$, $r \in (0, 1)$ and an integer $i \geq k$ for which (8.20) is valid. Consider the sequences $\{x_j^{(h)}\}_{j=0}^{\infty} \subset X, \{y_j^{(h)}\}_{j=0}^{\infty} \subset Y$ defined by

$$x_j^{(h)} = x_h, \ y_j^{(h)} = y_h, \quad j = 0, 1, \dots. \quad (8.22)$$

It was shown above that the pair of sequences $\{x_j^{(h)}\}_{j=0}^{\infty}, \{y_j^{(h)}\}_{j=0}^{\infty}$ is (h)-minimal. It follows from (8.19), (8.22), (8.13) and property (d) that

$$x_h \in B_X(x_f, 8^{-1}\delta_5(f, r, i)), \ y_h \in B_Y(y_f, 8^{-1}\delta_5(f, r, i)). \quad (8.23)$$

Assume that $g \in U(f, r, i)$, an integer $n \geq 2n_4(f, r, i)$ and $\{x_j\}_{j=0}^{n} \subset X$, $\{y_j\}_{j=0}^{n} \subset Y$ is a $(g, \delta_5(f, r, i))$-good pair of sequences. It follows from property (c), (8.11), (8.13) and (8.23) that the following properties hold:

$$x_j \in B_X(x_f, 8^{-1}\delta_4(f, r, i)), \ y_j \in B_Y(y_f, 8^{-1}\delta_4(f, r, i)), \qquad (8.24)$$

and $|x_j - x_h|, \ |y_j - y_h| \leq \epsilon$ for all integers $j \in [n_4(f, r, i), n - n_4(f, r, i)]$;

if $|x_0 - x_f|, \ |y_0 - y_f| \leq \delta_5(f, r, i)$, then (8.24) holds for all integers $j \in [0, n - n_4(f, r, i)]$;

if $|x_n - x_f|, \ |y_n - y_f| \leq \delta_5(f, r, i)$, then (8.24) holds for all integers $j \in [n_4(f, r, i), n]$.

Together with (8.23) this implies that the following properties hold:

if

$$x_0 \in B_X(x_h, 2^{-1}\delta_5(f, r, i)), \ y_0 \in B_Y(y_f, 2^{-1}\delta_5(f, r, i)),$$

then (8.24) hold for all integers $j \in [0, n - n_4(f, r, i)]$;

if

$$x_n \in B_X(x_h, 2^{-1}\delta_5(f, r, i)), \ y_n \in B_Y(y_f, 2^{-1}\delta_5(f, r, i)),$$

then (8.24) is valid for all integers $j \in [n_4(f, r, i), n]$.

This completes the proof of the theorem.

Proof of Theorem 12.2.2. Let $h \in \mathcal{F}$, $z \in X$, $\xi \in Y$. By Theorem 12.2.1 there exists a unique pair $(x_h, y_h) \in X \times Y$ such that

$$\sup_{y \in Y} h(x_h, x_h, y, y) = h(x_h, x_h, y_h, y_h)$$

$$= \inf_{x \in X} h(x, x, y_h, y_h). \qquad (8.25)$$

By Proposition 12.5.3 there is an (h)-minimal pair of sequences

$$\{\bar{x}_j\}_{j=0}^{\infty} \subset X, \ \{\bar{y}_j\}_{j=0}^{\infty} \subset Y$$

for which

$$\bar{x}_0 = z, \quad \bar{y}_0 = \xi. \qquad (8.26)$$

We show that the pair of sequences $(\{\bar{x}_j\}_{j=0}^{\infty}, \{\bar{y}_j\}_{j=0}^{\infty})$ is (h)-overtaking optimal. Theorem 12.2.1 implies that

$$\bar{x}_j \to x_h, \quad \bar{y}_j \to y_h \text{ as } j \to \infty. \qquad (8.27)$$

Let $\{x_i\}_{i=0}^{\infty} \subset X$ and $x_0 = z$. We will show that

$$\limsup_{T \to \infty} \left[\sum_{j=0}^{T-1} h(\bar{x}_j, \bar{x}_{j+1}, \bar{y}_j, \bar{y}_{j+1}) - \sum_{j=0}^{T-1} h(x_j, x_{j+1}, \bar{y}_j, \bar{y}_{j+1}) \right] \leq 0. \quad (8.28)$$

Assume the contrary. Then there exists a number $\Gamma_0 > 0$ and a strictly increasing sequence of natural numbers $\{T_k\}_{k=1}^{\infty}$ such that for all integers $k \geq 1$,

$$\sum_{j=0}^{T_k-1} h(\bar{x}_j, \bar{x}_{j+1}, \bar{y}_j, \bar{y}_{j+1}) - \sum_{j=0}^{T_k-1} h(x_j, x_{j+1}, \bar{y}_j, \bar{y}_{j+1}) \geq \Gamma_0. \qquad (8.29)$$

We will show that

$$x_j \to x_h \text{ as } j \to \infty. \qquad (8.30)$$

For $j = 0, 1, \ldots$ define a function $g_j : X \times X \to R^1$ by

$$g_j(u_1, u_2) = h(u_1, u_2, \bar{y}_j, \bar{y}_{j+1}), \quad u_1, u_1 \in X. \qquad (8.31)$$

Clearly $g_j \in C(X \times X)$, $j = 0, 1, \ldots$. Let $\epsilon > 0$. Choose a natural number q such that

$$2^{-q} < 64^{-1}\epsilon. \qquad (8.32)$$

There exist $f \in \mathcal{M}$, $r \in (0, 1)$ and an integer $p \geq q$ such that

$$h \in U(f, r, p). \qquad (8.33)$$

Since the pair of sequences $(\{\bar{x}_j\}_{j=0}^{\infty}, \{\bar{y}_j\}_{j=0}^{\infty})$ is (h)-minimal it follows from the definition of $U(f, r, p)$ (see (8.13)), (8.33) and property (d) that for all integers $j \geq n_5(f, r, p)$,

$$\bar{x}_j \in B_X(x_f, 8^{-1}\delta_5(f, r, p)), \quad \bar{y}_j \in B_Y(y_f, 8^{-1}\delta_5(f, r, p)). \qquad (8.34)$$

By (8.25), Proposition 12.5.1, (8.33) and property (d),

$$x_h \in B_X(x_f, 8^{-1}\delta_5(f, r, p)), \quad y_h \in B_Y(y_f, 8^{-1}\delta_5(f, r, p)). \qquad (8.35)$$

Since the pair of sequences $(\{\bar{x}_j\}_{j=0}^{\infty}, \{\bar{y}_j\}_{j=0}^{\infty})$ is (h)-minimal there exists a constant $c_0 > 0$ such that for each integer $T \geq 1$,

$$\sum_{j=0}^{T-1} h(\bar{x}_j, \bar{x}_{j+1}, \bar{y}_j, \bar{y}_{j+1}) \leq \inf\{\sum_{j=0}^{T-1} h(u_j, u_{j+1}, \bar{y}_j, \bar{y}_{j+1}) : \qquad (8.36)$$

$$\{u_j\}_{j=0}^{T} \subset X, \quad u_0 = z\} + c_0.$$

(8.36), (8.31) and (8.29) imply that the following property holds:

(e) For each $\Delta > 0$ there exists an integer $j(\Delta) \geq 1$ such that for each pair of integers $n_1 \geq j(\Delta)$, $n_2 > n_1$ the sequence $\{x_j\}_{j=n_1}^{n_2}$ is $(\{g_j\}_{j=n_1}^{n_2-1}, X, \Delta)$-good.

Consider the function $f_r^{(X)} : X \times X \to R^1$ defined by (7.3). For $j = 0, 1, \ldots$ define a function $\bar{g}_j : X \times X \to R^1$ by

$$\bar{g}_j(u_1, u_2) = f_r(u_1, u_2, \bar{y}_j, \bar{y}_{j+1}), \quad u_1, u_2 \in X. \tag{8.37}$$

It follows from (7.3), (8.37), (8.34), (8.9) and the definition of $\delta_4(f, r, p)$ (see (8.7), (8.8)) that for all integers $j \geq n_5(f, r, p)$,

$$\|\bar{g}_j - f_r^{(X)}\| \leq 16^{-1}\gamma_3(f, r, p). \tag{8.38}$$

By (8.38), (8.37), (8.31), (8.33), (8.13), (8.11), (8.9) for all integers $j \geq n_5(f, r, p)$,

$$\|g_j - f_r^{(X)}\| \leq 16^{-1}\gamma_3(f, r, p) + \gamma(f, r, p) < \gamma_3(f, r, p). \tag{8.39}$$

It follows from (8.39), properties (e) and (a) and (8.6) that there exists an integer $m_0 \geq 1$ such that $|x_j - x_f| \leq 2^{-p}$ for all integers $j \geq m_0$. Together with (8.32) and (8.35) this implies that for all integers $j \geq m_0$ the relation $|x_j - x_h| \leq 2^{-p} + 2^{-p} < \epsilon$ is true. Since ϵ is an arbitrary positive number we conclude that

$$\lim_{j \to \infty} x_j = x_h. \tag{8.40}$$

There exists a number $\epsilon_0 > 0$ such that for each $z_1, z_2, \bar{z}_1, \bar{z}_2 \in X$ and each $\xi_1, \xi_2, \bar{\xi}_1, \bar{\xi}_2 \in Y$ which satisfy

$$|z_j - \bar{z}_j|, \ |\xi_j - \bar{\xi}_j| \leq 2\epsilon_0, \quad j = 1, 2 \tag{8.41}$$

the following relation holds:

$$|h(z_1, z_2, \xi_1, \xi_2) - h(\bar{z}_1, \bar{z}_2, \bar{\xi}_1, \bar{\xi}_2)| \leq 8^{-1}\Gamma_0. \tag{8.42}$$

By (8.40) and (8.27) there exists an integer $j_0 \geq 8$ such that for all integers $j \geq j_0$,

$$|x_j - x_h| \leq 2^{-1}\epsilon_0, \quad |\bar{x}_j - x_h| \leq 2^{-1}\epsilon_0. \tag{8.43}$$

There exists an integer $s \geq 1$ such that

$$T_s > j_0. \tag{8.44}$$

Define a sequence $\{x_j^*\}_{j=0}^s \subset X$ by

$$x_j^* = x_j, \ j = 0, \ldots, T_s - 1, \quad x_{T_s}^* = \bar{x}_{T_s}. \tag{8.45}$$

Since the pair of sequences $(\{\bar{x}_j\}_{j=0}^{\infty}, \{\bar{y}_j\}_{j=0}^{\infty})$ is (h)-minimal we conclude that, by (8.45),

$$\sum_{j=0}^{T_s-1} h(\bar{x}_j, \bar{x}_{j+1}, \bar{y}_j, \bar{y}_{j+1}) - \sum_{j=0}^{T_s-1} h(x_j^*, x_{j+1}^*, \bar{y}_j, \bar{y}_{j+1}) \le 0. \qquad (8.46)$$

On the other hand it follows from (8.45), (8.29), (8.43), (8.44) and the definition of ϵ_0 (see (8.41),(8.42)) that

$$\sum_{j=0}^{T_s-1} h(\bar{x}_j, \bar{x}_{j+1}, \bar{y}_j, \bar{y}_{j+1}) - \sum_{j=0}^{T_s-1} h(x_j^*, x_{j+1}^*, \bar{y}_j, \bar{y}_{j+1})$$

$$= \sum_{j=0}^{T_s-1} h(\bar{x}_j, \bar{x}_{j+1}, \bar{y}_j, \bar{y}_{j+1}) - \sum_{j=0}^{T_s-1} h(x_j, x_{j+1}, \bar{y}_j, \bar{y}_{j+1})$$

$$+ h(x_{T_s-1}, x_{T_s}, \bar{y}_{T_s-1}, \bar{y}_{T_s}) - h(x_{T_s-1}^*, x_{T_s}^*, \bar{y}_{T_s-1}, \bar{y}_{T_s})$$

$$\ge \Gamma_0 + h(x_{T_s-1}, x_{T_s}, \bar{y}_{T_s-1}, \bar{y}_{T_s}) - h(x_{T_s-1}, \bar{x}_{T_s}, \bar{y}_{T_s-1}, \bar{y}_{T_s})$$

$$\ge \Gamma_0 - 8^{-1}\Gamma_0.$$

This is contradictory to (8.46). The obtained contradiction proves that (8.28) holds.

Analogously we can show that for each sequence $\{y_j\}_{j=0}^{\infty} \subset Y$ satisfying $y_0 = \xi$,

$$\limsup_{T \to \infty} \left[\sum_{j=0}^{T-1} h(\bar{x}_j, \bar{x}_{j+1}, y_j, y_{j+1}) \right.$$

$$\left. - \sum_{j=0}^{T-1} h(\bar{x}_j, \bar{x}_{j+1}, \bar{y}_j, \bar{y}_{j+1}) \right] \le 0.$$

This implies that the pair of sequences

$$(\{\bar{x}_j\}_{j=0}^{\infty}, \{\bar{y}_j\}_{j=0}^{\infty})$$

is (h)-overtaking optimal. This completes the proof of the theorem.

Comments

Chapter 1.

In this chapter we discuss three notions of optimality for infinite horizon problems. The notion of (f)-minimal solutions was introduced by Aubry and Le Daeron [6] in their study of the discrete Frenkel–Kontorova model related to dislocations in one-dimensional crystals. In [6] Aubry and Le Daeron established the existence of (f)-minimal solutions and obtained a full description of their structure. A minimal solution was called in [6] a minimal energy configuration. The theory developed in [6] is of great interest from the point of view of infinite horizon optimal control as well as from the point of view of the theory of dynamical systems. Leizarowitz and Mizel [44] used the notion of (f)-minimal solutions in their study of a class of variational problems arising in continuum mechanics. They established the existence of periodic (f)-minimal solutions under a certain technical assumption which was removed in [90].

The notions of overtaking optimal solutions and good solutions were introduced in the economics literature [4, 33, 81]. Usually the existence of overtaking optimal solutions is a difficult problem and we solve it for nonconvex variational problems only in a generic setting. But good solutions exist for all infinite horizon variational problems considered in the book. Note that for practical needs it is enough to obtain approximate solutions which are good functions. The results which we obtained for good solutions are an important tool in our existence theory of overtaking optimal solutions. In order to establish the existence of overtaking optimal solutions we usualy verify that all good functions have the same asymptotic behavior and then show that a good minimal function is overtaking optimal. Note that many results on infinite horizon optimal control problems are collected in [16, 26, 48, 60, 61, 67].

Chapter 2.

This chapter contains the turnpike results for nonautonomous non-convex variational problems, the strongest and the most general results of this book. These results allow us to consider the turnpike property as a general phenomenon which holds for large classes of problems. We consider the complete metric space of integrands $\mathcal{M}$ and show that the turnpike property holds for most integrands of $\mathcal{M}$ in the sense of the Baire category. Such an approach is common in many areas of Mathematical Analysis [19, 21-23, 25, 35, 70, 106, 107].

The example of an integrand belonging to the space $\mathcal{M}$ which does not have the turnpike property given in Section 2.6 shows that the main results of the chapter cannot be improved. In [109] we obtained results which may help us to verify if a given integrand has the turnpike property.

The turnpike property is very important for applications. Suppose that our integrand has the turnpike property and we know a finite number of "approximate" solutions of the variational problems with this integrand. Then we know the turnpike X, or at least its approximation, and the constant τ which is an estimate for the time period required to reach the turnpike. This information can be useful if we need to find an "approximate" solution of a new variational problem with a new time interval $[T_1, T_2]$ and the new values y, z at the end points T_1 and T_2. Namely, instead of solving this new problem on the "large" interval $[T_1, T_2]$ we can find an "approximate" solution of the variational problem on the "small" interval $[T_1, T_1 + \tau]$ with the values $y, X(T_1 + \tau)$ at the end points and an "approximate" solution of the variational problem on the "small" interval $[T_2 - \tau, T_2]$ with the values $X(T_2 - \tau), z$ at the end points. Then the concatenation of the first solution, the function $X(t)$, $t \in [T_1 + \tau, T_2 - \tau]$ and the second solution is an "approximate" solution of the variational problem on the interval $[T_1, T_2]$ with the values y, z at the end points. Numerical applications of the turnpike theory are discussed in [62, 63, 67].

Chapter 3.

The notion of a weakly optimal function was introduced in the economic literature by Brock [12]. The results of this chapter are obtained for the subspace $\mathcal{A}$ of the space of integrands $\mathcal{M}$ considered in Chapters 1 and 2. This subspace $\mathcal{A}$ consists of all time-independent integrands belonging to $\mathcal{M}$. Since the subspace $\mathcal{A}$ is a small set in $\mathcal{M}$ and the results of Chapter 2 are of generic nature, they cannot be applied for the subspace $\mathcal{A}$. The results which we obtain in this chapter are weaker than their analogs in Chapter 2. In Chapter 2 we establish a generic existence of an overtaking optimal function, while in this chapter we obtain a generic existence of a weakly optimal function. Also the convergence in the turnpike results for the autonomous case are weaker than the convergence in their analogs for the nonautonomous case. This fact is natural since in a large space we have more possibilities for perturbations. In Chapter 2 for a given integrand we use a perturbation which depends on t and obtain a new integrand which has the strong turnpike property. In Chapter 3 for a given integrand we may only use a perturbation which does not depend on t. As a result, for the space $\mathcal{A}$ we obtain a weaker turnpike property than for the space $\mathcal{M}$. Some nongeneric turnpike results for the space $\mathcal{A}$ were obtained in [110].

In this chapter we show that the turnpike property holds for approximate solutions on finite intervals if the integrand has the so-called asymptotic turnpike property, which means that all good functions on the infinite interval $[0, \infty)$ have the same asymptotic behavior. An analogous result for one-dimensional second order variational problems arising in continuum mechanics was obtained in [55].

Chapter 4.

Chapter 4 is a continuation of Chapter 3. It contains a detailed analysis of the structure of optimal solutions on infinite horizon problems for an integrand which has the asymptotic turnpike property. In Chapter 3 we associate with any integrand the so-called long run average cost growth rate and a certain continuous function. In this chapter we show that this long run average cost growth rate and the continuous function depend on the integrand continuously if the integrand has the asymptotic turnpike property. This result may be useful if we try to apply some numerical procedures in order to calculate weakly optimal solutions. It implies stability of such procedures.

In this chapter we improve some turnpike results for optimal solutions of infinite horizon problems. For example, we show that all optimal solutions with initial points belonging to a given bounded set converge to the turnpike uniformly.

Chapter 5.

In Chapter 5 our goal is to improve the turnpike results obtained in
Chapter 2 for approximate solutions of autonomous variational problems
on finite intervals. In order to obtain this improvement we need to suppose additional assumptions on integrands. Namely, we assume that the
integrands are smooth and their partial derivatives satisfy certain growth
conditions. These assumptions imply, in particular, that minimizers of
variational problems are solutions of the corresponding Euler–Lagrange
equations. We establish the turnpike property for integrands which have
the asymptotic turnpike property.

Chapter 6.

In this chapter we study the turnpike properties of a class of linear
control problems arising in engineering. This class includes an infinite
horizon problem of tracking of the periodic trajectory studied in [3]. An
integrand is assumed to be periodic with respect to the time variable
and strictly convex as the function of the state variable and the control variable. This assumption implies that the turnpike is a periodic
trajectory. The strict convexity assumption implies the uniqueness of
overtaking optimal solutions. As in the previous chapters we associate
with our linear control problem a related discrete time optimal control
problem which is in this case autonomous and strictly convex.

Chapter 7.

This chapter is devoted to the study of the turnpike properties for
a class of linear control problems arising in engineering. This class includes the class of linear control problems discussed in Chapter 6 and,
in particular, the infinite horizon problem of tracking of the periodic
trajectory studied in [3]. An integrand is assumed to be strictly convex
as the function of the state variable and the control variable, but we
do not assume that it is periodic with respect to the time variable. The
strict convexity assumption implies the uniqueness of overtaking optimal
solutions which are not periodic in general.

Chapter 8.

Discrete-time control systems considered in this chapter appear in
many areas of applied mathematics: in mathematical economics [47, 48,
56-61], continuum mechanics [20, 44] and in the theory of dynamical
systems [6]. As we have already seen in the previous chapters these

systems are also useful tools in the study of continuous-time optimal control problems. With any continuous-time control problem we associate a related discrete-time control problem. It turns out that there is a simple correspondence between solutions of the continuous-time problem and solutions of the related discrete-time problem. In Chapter 8 we study an autonomous problem with convex cost function on a bounded closed convex subset of a Banach space and a nonautonomous nonconvex problem on a complete metric space. Some results on infinity horizon autonomous nonconvex problems on a complete metric space were established in [108].

Chapter 9.

The primary area of appications of infinite-dimensional optimal continuous-time control problems concerns models of regional economic growth discussed in [36], cattle ranching models proposed in [24], and systems with distributed parameters and boundary controls related to engineering [8, 31] and to water resources problems [62, 63]. In this chapter we obtain the convergence to the turnpike in the weak topology. In order to establish the convergence to the turnpike in the strong topology we need to assume that the integrand is strictly convex.

Chapter 10.

This chapter is devoted to applications of the turnpike theory to mathematical economics. We consider a large class of nonlinear Leontjev type models of multisector economics. It should be mentioned that we obtain a sufficient condition for the turnpike property which can be verified if we know a generalized equilibrium state of the model. A version of this model which takes into account lag was considered in [87].

Chapter 11.

The general Neumann–Gale model has been studied in many publications. Many results on this model are collected in [48, 75, 77]. A stochastic version of the Neumann–Gale model was studied in [2].

Chapter 12.

This chapter is devoted to applications of the turnpike theory to game theory. It is based on the paper [102]. For other applications see [15, 17]. In [17] overtaking equilibria was studied for switching regulator and tracking games. Turnpike properties for infinite horizon open-loop competitive processes were discussed in [15]. It is not clear if it is possible to obtain turnpike results in game theory without convexity assumptions.

References

[1] Anderson, B.D.O. and Moore, J.B. (1971). *Linear Optimal Control.* Englewood Cliffs, NJ: Prentice-Hall.

[2] Arkin, V. and Evstigneev, I. (1979). *Probabilistic Models of Control and Economic Dynamics.* Nauka, Moscow.

[3] Artstein, Z. and Leizarowitz, A. (1985). Tracking periodic signals with overtaking criterion, *IEEE Trans. on Autom. Control AC*, Vol. 30, pp. 1122-1126.

[4] Atsumi, H. (1965). Neoclassical growth and the efficient program of capital accumulation, *Review of Economic Studies*, Vol. 32, pp. 127-136.

[5] Aubin, J.P. and Ekeland, I. (1984). *Applied Nonlinear Analysis.* Wiley Interscience, New York.

[6] Aubry, S. and Le Daeron, P.Y. (1983). The discrete Frenkel-Kontorova model and its extensions I, *Physica D*, Vol. 8, pp. 381-422.

[7] Balakrishnan, A.V. (1976). *Applied Functional Analysis.* Springer-Verlag, Berlin.

[8] Barbu, V. (1980). Boundary control problems with convex cost criterion, *SIAM Journal on Control and Optimization*, Vol. 18, pp. 227-243.

[9] Berkovitz, L.D. (1974). Lower semicontinuity of integral functionals, *Trans. Amer. Math. Soc.*, Vol. 192, pp. 51-57.

[10] Blot, J. and Cartigny, P. (2000). Optimality in infinite-horizon variational problems under sign conditions, *J. Optim. Theory Appl.*, Vol. 106, pp. 411-419.

[11] Blot, J. and Michel, P. (2003). The value-function of an infinite-horizon linear quadratic problem, *Appl. Math. Lett.*, Vol. 16, pp. 71-78.

[12] Brock, W.A. (1970). On existence of weakly maximal programmes in a multi-sector economy, *Review of Economic Studies*, Vol. 37, pp. 275-280.

[13] Brock, W.A. and Haurie, A. (1976). On existence of overtaking optimal trajectories over an infinite horizon, *Math. Op. Res.*, Vol. 1, pp. 337-346.

[14] Carlson, D.A. (1990). The existence of catching-up optimal solutions for a class of infinite horizon optimal control problems with time delay, *SIAM Journal on Control and Optimization*, Vol. 28, pp. 402-422.

[15] Carlson, D.A. and Haurie, A. (1996). A turnpike theory for infinite-horizon open-loop competitive processes, *SIAM Journal on Control and Optimization*, Vol. 34, pp. 1405-1419.

[16] Carlson, D.A., Haurie, A. and Leizarowitz, A. (1991). *Infinite Horizon Optimal Control*. Springer-Verlag, Berlin.

[17] Carlson, D.A., Haurie, A. and Leizarowitz, A. (1994). Overtaking equilibria for switching regulator and tracking games, *Advances in Dynamic Games and Applications*, Vol. 1, pp. 247-268.

[18] Carlson, D.A., Jabrane, A and Haurie, A. (1987). Existence of overtaking solutions to infinite dimensional control problems on unbounded time intervals, *SIAM Journal on Control and Optimization*, Vol. 25, pp. 1517-1541.

[19] Cobzas, S. (2000). Generic existence of solutions for some perturbed optimization problems, *J. Math. Anal. Appl.*, Vol. 243, pp. 344-356.

[20] Coleman, B.D., Marcus, M. and Mizel, V.J. (1992). On the thermodynamics of periodic phases, *Arch. Rat. Mech. Anal.*, Vol. 117, pp. 321-347.

[21] De Blasi, F.S. and Myjak, J. (1976). Sur la convergence des approximations successives pour les contractions non linéaires dans un espace de Banach, *C.R. Acad. Sc. Paris*, Vol. 283, pp. 185-187.

[22] De Blasi, F.S. and Myjak, J. (1989). Sur la porosité des contractions sans point fixe, *C. R. Acad. Sci. Paris*, Vol. 308, pp. 51-54.

[23] De Blasi, F.S., Myjak, J. and Papini, P.L. (1991). Porous sets in best approximation theory, *J. London Math. Soc.*, Vol. 44, pp. 135-142.

[24] Derzko, N. and Sethi, S.P. (1980). Distributed parameter systems approach to the optimal cattle ranching problem, *Optimal Control Appl. Methods*, Vol. 1, pp. 3-10.

[25] Deville, R. and Revalski, J. (2000). Porosity of ill-posed problems, *Proc. Amer. Math. Soc.*, Vol. 128, pp. 1117-1124.

[26] Dyukalov, A. (1983). *Problems of Applied Mathematical Economics.* Nauka, Moscow, 1983.

[27] Dyukalov, A. and Ilyutovich, A. (1974). Mail-line properties of optimal trajectories of a dynamic model of interindustry balance in continuous time, *Automat. Remote Control,* Vol. 35, pp. 1973-1981.

[28] Dzalilov, Z., Ivanov, A.F. and Rubinov, A.M. (2001). Difference inclusions with delay of economic growth, *Dynam. Systems Appl.,* Vol. 10, pp. 283-293.

[29] Dzalilov, Z., Rubinov, A.M. and Kloeden, P.E. (1998). Lyapunov sequences and a turnpike theorem without convexity, *Set-Valued Analysis,* Vol. 6, pp. 277-302.

[30] Ekeland, I. and Temam, R. (1976). *Convex Analysis and Variational Problems.* North-Holland, Amsterdam.

[31] Fattorini, H.O. (1968). Boundary control systems, *SIAM Journal on Control and Optimization,* Vol. 6, pp. 349-385.

[32] Feinstein, C.D. and Luenberger, D.G. (1981). Analysis of the asymptotic behavior of optimal control trajectories: the implicit programming problem, *SIAM Journal on Control and Optimization,* Vol. 19, pp. 561-585.

[33] Gale, D. (1967). On optimal development in a multisector economy, *Rev. of Econ. Studies,* Vol. 34, pp. 1-19.

[34] Halkin, H. (1974). Necessary conditions for optimal control problems with infinite horizon, *Econometrica,* Vol. 42, pp. 267-273.

[35] Ioffe, A.D. and Zaslavski, A.J. (2000). Variational principles and well-posedness in optimization and calculus of variations, *SIAM J. Control Optim.,* Vol. 38, pp. 566-581.

[36] Isard, W. and Liossatos, P. (1979). *Spatial Dynamics and Space-Time Development.* North-Holland, Amsterdam.

[37] Kelley, J.L. (1955). *General Topology.* Van Nostrand, Princeton NJ.

[38] Leizarowitz, A. (1985). Convergence of viable solutions of differential inclusions with convex compact graphs, *SIAM Journal on Control and Optimization,* Vol. 22, pp. 514-522.

[39] Leizarowitz, A. (1985). Infinite horizon autonomous systems with unbounded cost, *Appl. Math. and Opt.,* Vol, 13, pp. 19-43.

[40] Leizarowitz, A. (1985). Existence of overtaking optimal trajectories for problems with convex integrands, *Math. Op. Res.,* Vol. 10, pp. 450-461.

[41] Leizarowitz, A. (1986). Tracking nonperiodic trajectories with the overtaking criterion, *Appl. Math. and Opt.,* Vol. 14, pp. 155-171.

[42] Leizarowitz, A. (1989). Optimal trajectories of infinite horizon deterministic control systems, *Appl. Math. and Opt.*, Vol, 19, pp. 11-32.

[43] Leizarowitz, A. (1994). Eigenvalues of convex processes and convergence properties of differential inclusions, *Set-Valued Analysis*, Vol. 2, pp. 505-527.

[44] Leizarowitz, A. and Mizel, V.J. (1989). One-dimensional infinite horizon variational problems arising in continuum mechanics, *Arch. Rational Mech. Anal.*, Vol. 106, pp. 161-194.

[45] Lyapunov, A.N. (1989). Asymptotic criteria in models of economic dynamics, *Mat. Metody v Sotsial. Nauk*, No. 22, pp. 33-37.

[46] Magill, M.J.P. and Scheinkman, J.A. (1979). Stability of regular equilibria and the correspondence principle for symmetric variational problems, *International Econom. Rev.*, Vol. 20, pp. 279-315.

[47] Makarov, V.L., Levin, M.J. and Rubinov, A.M. (1995). *Mathematical economic theory: pure and mixed types of economic mechanisms*, North-Holland, Amsterdam.

[48] Makarov, V.L. and Rubinov, A.M. (1977). *Mathematical Theory of Economic Dynamics and Equilibria.* Springer-Verlag, New York.

[49] Mamedov, M.A. (1992). Turnpike theorems in continuous systems with integral functionals, *Russian Acad. Sci. Dokl. Math.*, Vol. 45, pp. 432-435.

[50] Mamedov, M.A. (1993). Turnpike theorems for integral functionals, *Russian Acad. Sci. Dokl. Math.*, Vol. 46, pp. 174-177.

[51] Mamedov, M.A. and Pehlivan, S. (2000). Statistical convergence of optimal paths, *Math. Japon.*, Vol. 52, pp. 51-55.

[52] Mamedov, M.A. and Pehlivan, S. (2001). Statistical cluster points and turnpike theorem in nonconvex problems, *J. Math. Anal. Appl.*, Vol. 256, pp. 686-693.

[53] Marcus, M. (1993). Uniform estimates for variational problems with small parameters, *Arch. Rational Mech. Anal.*, Vol. 124, pp. 67-98.

[54] Marcus, M. (1998). Universal properties of stable states of a free energy model with small parameters, *Calc. Var.*, Vol. 6, pp. 123-142.

[55] Marcus, M. and Zaslavski, A.J. (1999). The structure of extremals of a class of second order variational problems, *Ann Inst H. Poincare, Anal Non Lineare*, Vol. 16, pp. 593-629.

[56] McKenzie, L.W. (1963). Turnpike theorems for a generalized Leontief model, *Econometrica*, Vol. 31, pp. 165-180.

[57] McKenzie, L.W. (1968). Accumulation programs of maximal utility and the von Neumann facet, *Value, Capital and Growth*, Edinburgh, pp. 353-383.

[58] McKenzie, L.W. (1974). Turnpike theorems with technology and welfare function variable, *Mathematical Models in Economics*, North-Holland, Amsterdam, pp. 271-287.

[59] McKenzie, L.W. (1976). Turnpike theory, *Econometrica*, Vol. 44, pp. 841-866.

[60] McKenzie, L.W. (1986). Optimal economic growth, turnpike theorems and comparative dynamics, *Handbook of Mathematical Economics*, Vol. 3, North-Holland, New York, pp. 1281-1355.

[61] McKenzie, L.W. (2002). *Classical General Equilibrium Theory*, The MIT press, Cambridge, Massachusetts.

[62] Mordukhovich, B. (1990). Minimax design for a class of distributed parameter systems, *Automat. Remote Control*, Vol. 50, pp. 1333-1340.

[63] Mordukhovich, B. and Shvartsman, I. (2004). Optimization and feedback control of constrained parabolic systems under uncertain perturbations, *Optimal Control, Stabilization and Nonsmooth Analysis*, Lecture Notes Control Inform. Sci., Springer, pp. 121-132.

[64] Morrey, C.B. (1966). *Multiple Integrals in the Calculus of Variations*. Springer, Berlin.

[65] Moser, J. (1986). Minimal solutions of variational problems on a torus, *Ann. Inst. H. Poincare, Anal. non lineare*, Vol. 3, pp. 229-272.

[66] Moser, J. (1986). Recent developments in the theory of Hamiltonian systems, *SIAM Review*, Vol. 28, pp. 459-485.

[67] Panasyuk, A. and Panasyuk, V. (1986). *Asymptotic Turnpike Optimization of Control Systems*. Nauka i Technika, Minsk.

[68] Pehlivan, S. and Mamedov, M.A. (2000). Statistical cluster points and turnpike, *Optimization*, Vol. 48, pp. 93-106.

[69] Radner, R. (1961). Path of economic growth that are optimal with regard only to final states; a turnpike theorem, *Rev. Econom. Stud.*, Vol. 28, pp. 98-104.

[70] Reich, S. and Zaslavski, A.J. (2000). Generic convergence of descent methods in Banach spaces, *Math. Oper. Research*, Vol. 25, pp. 231-242.

[71] Rockafellar, R.T. (1969). *Convex Analysis*. Princeton University Press, Princeton, NJ.

[72] Rockafellar, R.T. (1973). Saddle points of Hamiltonian systems in convex problem of Lagrange, *Journal of Optim. Theory and Appl.*, Vol. 12, pp. 367-389.

[73] Rockafellar, R.T. (1974). Convex algebra and duality in dynamic models of production, *Mathematical Models in Economics*, North-Holland, Amsterdam, pp. 351-378.

[74] Rockafellar, R.T. (1976). Saddle points of Hamiltonian systems in convex Lagrange problems having a nonzero discount rate. Hamiltonian dynamics in economics, *J. Econom. Theory*, Vol. 12, pp. 71-113.

[75] Rubinov, A.M. (1980). *Superlinear Multivalued Mappings and Their Applications in Economic Mathematical Problems.* Nauka, Leningrad.

[76] Rubinov, A.M. (1982). On a nonlinear Leontjev type model, *Optimization, Institute of Mathematics, Siberian Branch Acad. Nauk SSSR*, Vol. 32 (49), pp. 109-127.

[77] Rubinov, A.M. (1984). Economic dynamics, *J. Soviet Math.*, Vol. 26, pp. 1975-2012.

[78] Samuelson P.A. (1965). A catenary turnpike theorem involving consumption and the golden rule, *American Economic Review*, Vol. 55, pp. 486-496.

[79] Samuelson P.A. (1966). *The Collected Scientific Papers of Paul A. Samuelson.* MIT Press.

[80] Scheinkman, J.A. (1976). On optimal steady states of n-sector growth models when utility is discounted. Hamiltonian dynamics in economics, *J. Econom. Theory*, Vol. 12, pp. 11-30.

[81] von Weizsacker, C.C. (1965). Existence of optimal programs of accumulation for an infinite horizon, *Rev. Econ. Studies*, Vol. 32, pp. 85-104.

[82] Zalinescu, C. (1983). On uniformly convex functions, *J. Math. Anal. Appl.*, Vol. 95, pp. 344-374.

[83] Zaslavski, A.J. (1985). Asymptotics of optimal trajectories of a Leontjev type model 1, *Optimization, Institute of Mathematics, Siberian Branch Acad. Nauk SSSR*, Vol. 36 (53), pp. 87-100.

[84] Zaslavski, A.J. (1986). Asymptotic of optimal trajectories of a Leontjev type model 2, *Optimization, Institute of Mathematics, Siberian Branch Acad. Nauk SSSR*, Vol. 37 (54), pp. 108-120.

[85] Zaslavski, A.J. (1987). Ground states in Frenkel-Kontorova model, *Math. USSR Izvestiya*, Vol. 29, pp. 323-354.

[86] Zaslavski, A.J. (1988). Turnpike sets in models of economic dynamics, *Optimization*, Vol. 19, pp. 427-441.

[87] Zaslavski, A.J. (1988). Production model taking into account lag. *Mathematical Models of Economic Dynamics*, Institute of Economics of Academy of Sciences of Lithuanian SSR, pp. 70-96.

[88] Zaslavski, A.J. (1995) Optimal programs on infinite horizon 1, *SIAM Journal on Control and Optimization*, Vol. 33, pp. 1643-1660.

[89] A.J. Zaslavski, A.J. (1995) Optimal programs on infinite horizon 2, *SIAM Journal on Control and Optimization*, Vol. 33, pp. 1661-1686.

[90] Zaslavski, A.J. (1995) The existence of periodic minimal energy configurations for one dimensional infinite horizon variational problems arising in continuum mechanics, *Journal of Mathematical Analysis and Applications*, Vol. 194, pp. 459-476.

[91] Zaslavski, A.J. (1995). The existence and structure of extremals for a class of second order infinite horizon variational problems, *Journal of Mathematical Analysis and Applications*, Vol. 194, pp. 660-696.

[92] Zaslavski, A.J. (1996). Structure of extremals for one-dimensional variational problems arising in continuum mechanics, *Journal of Mathematical Analysis and Applications*, Vol. 198, pp. 893-921.

[93] Zaslavski, A.J. (1996). Dynamic properties of optimal solutions of variational problems, *Nonlinear Analysis: Theory, Methods and Applications*, Vol. 27, pp. 895-931.

[94] Zaslavski, A.J. (1996). Turnpike theorem for a class of set-valued mappings, *Numerical Functional Analysis and Optimization*, Vol. 17, pp. 215-240.

[95] Zaslavski, A.J. (1997). Existence and structure of optimal solutions of variational problems, *Proceedings of the Special Session on Optimization and Nonlinear Analysis, Joint AMS-IMU Conference, Jerusalem, May 1995, Contemporary Mathematics*, Vol. 204, pp. 247-278.

[96] Zaslavski, A.J. (1997). Turnpike theorem for a class of differential inclusions arising in economic dynamics, *Optimization*, Vol. 42, pp. 139-168.

[97] Zaslavski, A.J. (1997). Turnpike property of optimal solutions of infinite-horizon variational problems, *SIAM Journal on Control and Optimization*, Vol. 35, pp. 1169-1203.

[98] Zaslavski, A.J. (1998). Existence and uniform boundedness of optimal solutions of variational problems, *Abstract and Applied Analysis*, Vol. 3, pp. 265-292.

[99] Zaslavski, A.J. (1998). Turnpike theorem for convex infinite dimensional discrete-time control systems, *Convex Analysis*, Vol. 5, pp. 237-248.

[100] Zaslavski, A.J. (1999). Turnpike property for extremals of variational problems with vector-valued functions. *Trans. Amer. Math. Soc.*, Vol. 351, pp. 211-231.

[101] Zaslavski, A.J. (1999). Structure of optimal trajectories of convex processes, *Numerical Functional Analysis and Optimization*, Vol. 20, pp. 175-200.

[102] Zaslavski, A.J. (1999). The Turnpike Property for Dynamic Discrete-Time Zero-Sum Game, *Abstract and Applied Analysis*, Vol. 4, pp. 21-48.

[103] Zaslavski, A.J. (2000). The turnpike property for extremals of nonautonomous variational problems with vector-valued functions. *Nonlinear Analysis: Theory, Methods and Applications*, Vol. 42, pp. 1465-1498.

[104] Zaslavski, A.J. (2000). Turnpike theorem for nonautonomous infinite dimensional discrete-time control systems, *Optimization*, Vol. 48, pp. 69-92.

[105] Zaslavski, A.J. (2000). Existence and structure of optimal solutions of infinite dimensional control problems, *Appl. Math. Opt.*, Vol. 42, pp. 291-313.

[106] Zaslavski, A.J. (2000). Generic well-posedness of optimal control problems without convexity assumptions, *SIAM J. Control Optim.*, Vol. 39, pp. 250-280.

[107] Zaslavski, A.J. (2001). Well-posedness and porosity in optimal control without convexity assumptions, *Calc. Var.*, Vol. 13, pp. 265-293.

[108] Zaslavski, A.J. (2003). Minimal solutions for discrete-time control systems in metric spaces, *Numerical Func. Anal. Optim.*, Vol. 24, pp. 637-651.

[109] Zaslavski, A.J. (2004). The structure of approximate solutions of variational problems without convexity, *J. Math. Anal. Appl.*, Vol. 296, pp. 578-593.

[110] Zaslavski, A.J. (2004). A turnpike result for autonomous variational problems, *Optimization,* Vol. 53, pp. 377-391.

[111] Zaslavski, A.J. and Leizarowitz, A. (1997). Optimal solutions of linear periodic control systems with convex integrands, *Appl. Math. Opt.*, Vol. 10, pp. 450-461.

[112] Zaslavski, A.J. and Leizarowitz, A. (1997). Optimal solutions of linear control systems with nonperiodic integrands, *Math. Op. Res.*, Vol. 22, pp. 726-746.

Index